T0324135

Handbook of Computer Programming with Python

This handbook provides a hands-on experience based on the underlying topics, and assists students and faculty members in developing their algorithmic thought process and programs for given computational problems. It can also be used by professionals who possess the necessary theoretical and computational thinking background but are presently making their transition to Python.

Key Features:

- Discusses concepts such as *basic programming principles, OOP principles, database programming, GUI programming, application development, data analytics* and *visualization, statistical analysis, virtual reality, data structures* and *algorithms, machine learning,* and *deep learning.*
- Provides the code and the output for all the concepts discussed.
- Includes a case study at the end of each chapter.

This handbook will benefit students of computer science, information systems, and information technology, or anyone who is involved in computer programming (entry-to-intermediate level), data analytics, HCI-GUI, and related disciplines.

Handbook of Computer Programming with Python

Edited by
Dimitrios Xanthidis
Christos Manolas
Ourania K. Xanthidou
Han-I Wang

CRC Press
Taylor & Francis Group
Boca Raton London New York

CRC Press is an imprint of the
Taylor & Francis Group, an **informa** business

A CHAPMAN & HALL BOOK

First edition published 2023
by CRC Press
6000 Broken Sound Parkway NW, Suite 300, Boca Raton, FL 33487-2742

and by CRC Press
4 Park Square, Milton Park, Abingdon, Oxon, OX14 4RN

CRC Press is an imprint of Taylor & Francis Group, LLC

ISBN: 978-0-367-68777-9 (hbk)
ISBN: 978-0-367-68778-6 (pbk)
ISBN: 978-1-003-13901-0 (ebk)

DOI: 10.1201/9781003139010

Typeset in Times
by codeMantra

Access the Support Material: https://www.routledge.com/9780367687779

Contents

Editors...vii

Contributors ix

Chapter 1   Introduction 1

Dimitrios Xanthidis, Christos Manolas, Ourania K. Xanthidou,
and Han-I Wang

Chapter 2   Introduction to Programming with Python...9

Ameur Bensefia, Muath Alrammal, and Ourania K. Xanthidou

Chapter 3   Object-Oriented Programming in Python...59

Ghazala Bilquise, Thaeer Kobbaey, and Ourania K. Xanthidou

Chapter 4   Graphical User Interface Programming with Python 107

Ourania K. Xanthidou, Dimitrios Xanthidis, and Sujni Paul

Chapter 5   Application Development with Python 161

Dimitrios Xanthidis, Christos Manolas, and Hanêne Ben-Abdallah

Chapter 6   Data Structures and Algorithms with Python..207

Thaeer Kobbaey, Dimitrios Xanthidis, and Ghazala Bilquise

Chapter 7   Database Programming with Python 273

Dimitrios Xanthidis, Christos Manolas, and Tareq Alhousary

Chapter 8   Data Analytics and Data Visualization with Python 319

Dimitrios Xanthidis, Han-I Wang, and Christos Manolas

Chapter 9   Statistical Analysis with Python 373

Han-I Wang, Christos Manolas, and Dimitrios Xanthidis

Chapter 10 Machine Learning with Python 409

Muath Alrammal, Dimitrios Xanthidis, and Munir Naveed

Chapter 11 Introduction to Neural Networks and Deep Learning 449

Dimitrios Xanthidis, Muhammad Fahim, and Han-I Wang

Chapter 12 Virtual Reality Application Development with Python 485

    Christos Manolas, Ourania K. Xanthidou, and Dimitrios Xanthidis

Appendix: Case Studies Solutions 527

Index 617

Editors

Dimitrios Xanthidis holds a PhD in Information Systems from University College London. For the past 25 years, he has been teaching computer science subjects with a focus on programming and software development, and data structures and databases in various tertiary education institutions. Currently, he is working in Higher Colleges of Technology in Dubai, U.A.E. Dimitrios' research interests and work revolve around the topics of data science, machine learning/deep learning, virtual/augmented reality, and emerging technologies.

Christos Manolas holds a PhD in Stereoscopic 3D Media (University of York, UK), and degrees and qualifications in Postproduction (MA), Music Technology (MSc), Music Performance, Software Development, and Media Production. Christos' career includes work as a software developer, musician, audio producer, and educator for over 20 years. His research interests include multimodal (audiovisual) perception, spatial audio, interactive and immersive media (VR/AR/XR), and generally the impact and role of digital technologies on media production.

Ourania K. Xanthidou is a PhD researcher at Brunel University, London. She holds an MSc in Computer Science from the University of Malaya, Kuala Lumpur, Malaysia. She has more than 15 years of involvement with the IT industry in the form of supporting IT departments of SMEs and more than 5 years of teaching experience in tertiary education. Ourania's research interests are in the areas of eHealth, smart health, databases, web application development, and object-oriented programming with a focus on application development for VR/AR/XR.

Han-I Wang holds a PhD in Health Economics from the University of York, UK. Han-I has been working as a research fellow for over 10 years, starting at the Epidemiology & Cancer Statistics Group (ECSG) before joining the Mental Health and Addiction Research Group (MHARG) at the University of York, UK. Her area of expertise spans across cost analysis, health outcome research, and decision modeling using complex patient-level data, and her main research interests are related with the exploration of different decision-modeling techniques and their application to predict healthcare expenditure, patients' quality of life, and life expectancy.

Contributors

Tareq Alhousary
Business Information Systems
University of Salford
Manchester, United Kingdom
and
Department of Management Information
    Systems
Dhofar University, College of Commerce and
    Business Administration
Salalah, Oman

Muath Alrammal
Department of Computer and Information
    Sciences
Higher Colleges of Technology
Abu Dhabi, United Arab Emirates
and
LACL (Laboratoire d'Algorithmique,
    Complexité et Logique)
University Paris-Est (UPEC)
Créteil, France

Hanêne Ben-Abdallah
Computer and Information Science
University of Pennsylvania
Philadelphia, PA

Ameur Bensefia
Department of Genie Informatique
University of Rouen Normandy
Laboratoire d'Informatique de Traitement de
    l'Information et des Systèmes (LITIS)
Rouen, France
and
Department of Computer and Information
    Sciences
Higher Colleges of Technology
Abu Dhabi, United Arab Emirates

Ghazala Bilquise
Department of Computer and Information
    Sciences
Higher Colleges of Technology
Abu Dhabi, United Arab Emirates

Muhammad Fahim
Department of Computer and Information
    Sciences
Higher Colleges of Technology
Abu Dhabi, United Arab Emirates

Thaeer Kobbaey
Department of Computer and Information
    Sciences
Higher Colleges of Technology
Abu Dhabi, United Arab Emirates

Christos Manolas
Department of Theatre, Film, Television and
    Interactive Media
The University of York
York, United Kingdom
and
Department of Media Works
Ravensbourne University London
London, United Kingdom

Munir Naveed
Department of Computer Science
University of Huddersfield
Huddersfield, United Kingdom
and
Department of Computer and Information
    Sciences
Higher Colleges of Technology
Abu Dhabi, United Arab Emirates

Sujni Paul
Department of Computer and Information
    Sciences
Higher Colleges of Technology
Abu Dhabi, United Arab Emirates

Han-I Wang
Department of Health Sciences
The University of York
York, United Kingdom

Dimitrios Xanthidis
School of Library, Archives, and Information
  Sciences
University College London
London, United Kingdom
and
Department of Computer and Information
  Sciences
Higher Colleges of Technology
Abu Dhabi, United Arab Emirates

Ourania K. Xanthidou
Department of Computer Science
Brunel University of London
Uxbridge, United Kingdom

1 Introduction

Dimitrios Xanthidis
University College London
Higher Colleges of Technology

Christos Manolas
The University of York
Ravensbourne University London

Ourania K. Xanthidou
Brunel University of London

Han-I Wang
The University of York

CONTENTS

1.1 Introduction 1
1.2 Audience 2
1.3 Getting Started with Jupyter Notebook 2
1.4 Creating Standalone, Executable Files 4
1.5 Structure of this Book 6
References 6

1.1 INTRODUCTION

Undoubtedly, at the time of writing, Python is among the most popular computer programming languages. Alongside other common languages like C# and Java, it belongs to the broader family of C/C++-based languages, from which it naturally borrows a large number of packages and modules. While Python is the youngest member in this family, it is widely adopted as the platform of choice by academic and corporate institutions and organizations on a global scale.

As a C++-based language, Python follows the *structured programming* paradigm, and the associated programming principles of *sequence*, *selection*, and *repetition*, as well as the concepts of *functions* and *arrays* (as *lists*). A thorough presentation of such concepts is both beyond the scope of this book and possibly unnecessary, as this was the subject of the seminal works of computer science giants like Knuth, Stroustrup, and Aho (Aho Alfred et al., 1983; Knuth, 1997; Stroustrup, 2013). Readers interested in an in-depth understanding of these concepts on a theoretical basis are encouraged to refer to such works that form the backbone of modern programming. As an *Object-Oriented Programming* (OOP) platform, it provides all the facilities and tools to support the OOP paradigm. Unlike its counterparts (i.e., C++, C#, and Java), Python does not provide a streamlined, centralized IDE to support GUI programming, but it does offer a significant number of related modules that cover most, if not all, of the various GUI requirements one may encounter. It includes a number of modules that allow for the implementation of *database programming*, *web development*, and *mobile development* projects, as well as platforms, modules, and methods that can be used for *machine* and *deep learning* applications and even *virtual and augmented reality* project development. Nevertheless, one of the main reasons that made Python such a popular option among computer science professionals and academics is the wealth of modules and packages it offers for *data science* tasks, including a large variety of libraries and tools specifically designed for *data analytics*, *data visualization*, and *statistical analysis* tasks.

Arguably, there is an abundance of online resources and tutorials and printed books that address most of the aforementioned topics in great detail. On the technical side, such resources may seem too complicated for someone who is currently studying the subject or approaches it without prior programming knowledge and experience. In other cases, resources may be structured more like *reference books* that may focus on particular topics without covering the introductory parts of computing with Python that some readers may find useful. This book aims at covering this gap by exploring how Python can be used to address various computational tasks of introductory to intermediate difficulty level, while also providing a basic theoretical introduction to the underlying concepts.

1.2 AUDIENCE

This book focuses on students of *computer science*, *information systems*, and *information technology*, or anyone who is involved in *computer programming*, *data analytics*, *HCI-GUI*, and related disciplines, at an entry-to-intermediate level. This book aims to provide a hands-on experience based on the underlying topics, and assist students and faculty members in developing their algorithmic thought process and programs for given computational problems. It can also be used by professionals who possess the necessary theoretical and computational thinking background but are presently making their transition to Python.

Considering the above, this book includes a wealth of examples and the associated Python code and output, presented in a context that also discusses the underlying concepts and their applications. It also provides key concepts in the form of quick access observations, so that the reader can skim through the various topics. Observations can be used as a reference and navigation tool, or as reminders for points for discussion and in-class presentation in the case of using this book as a teaching resource. Chapters are also accompanied by related exercises and case studies that can be used in this context, and their solutions are provided in the Appendix at the end of this book.

1.3 GETTING STARTED WITH JUPYTER NOTEBOOK

Ample information and support are available through online community channels and the official documentation and guides in terms of installing and running Python programming environments. Nevertheless, this section provides a brief and straightforward guide on how to use *Anaconda Navigator* and *Jupyter Notebook* in order to interpret and execute Python code, as the majority of examples in this book have been implemented and tested using this particular configuration.

Once Anaconda Navigator is launched, a number of different editors and environments are presented in the home page (Figure 1.1).

Launching the Jupyter Notebook (i.e., clicking the *Launch* button) initiates a web interface based on the file directory of the local machine (Figure 1.1). To create a new Python program, the user can select *New* from the top right corner and the *Python 3* notebook menu option (Figure 1.2). This action will launch a new Python file under Jupyter with a default name. This can be changed by clicking on the file name.

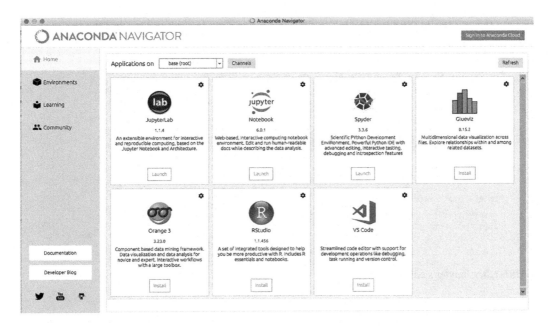

FIGURE 1.1 Anaconda IDE homepage.

FIGURE 1.2 Create a new Python file in Jupyter Notebook.

Jupyter editor is organized in cells. The user can add each line of code to a separate cell or add multiple lines to the same cell (Figure 1.3). The *Run* button in the main toolbar is used to execute the code in the selected cell. If the code is free from errors, the interpreter moves to the next cell; otherwise, an error message is displayed immediately after the cell where the error occurred (Figure 1.4).

FIGURE 1.3   Jupyter's editor.

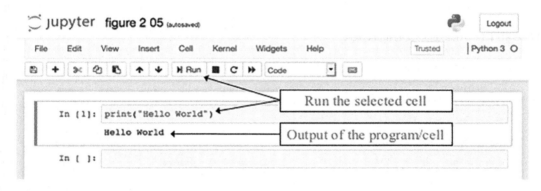

FIGURE 1.4   Run a Python program on Jupyter.

1.4   CREATING STANDALONE, EXECUTABLE FILES

With the exception of Chapter 12: Virtual Reality Application Development with Python that discusses applications that demand specific and highly specialized development platforms, the Python scripts and examples presented in this book were implemented and tested natively in the Anaconda Jupyter environment. In this context, the process of developing and testing software solutions is a rather straightforward and intuitive process. However, when it comes to the actual deployment of applications in more realistic scenarios, things become slightly more complex. This is mainly due to the fact that the Python code one develops is usually dependent on a number of *external libraries*, *packages*, and *files of various formats*. These are automatically provided in the background when working within the Anaconda environment, but this is not necessarily the case when scripts are exported as *standalone* files. The required libraries and resources may be located on numerous different places within the file structures of the computer and/or network systems used during development.

In the context of application deployment, references to such external files and objects are generally referred to as application *dependencies*. Dependencies form a crucial and essential part of the developed application, and the underlying files must be provided alongside the final deliverable program (e.g., a standalone, executable application), as their absence will prevent the program from running correctly in machines lacking the necessary libraries and file structures. Fortunately, the latter are automatically selected and packaged by special routines and processes during the *deployment* phase of the development cycle. This way, once the final deployment package is created, one can run the application on other computers, irrespectively of whether these include the necessary files and libraries or not.

Many *SDKs* and programming environments provide built-in routines (i.e., *wizards*) for the generation of the deployment packages and standalone executable files. In the case of Anaconda Jupyter, although there is no automated, built-in wizard for such tasks, one can resort to a number of external helper applications. A detailed, step-by-step tutorial of this process is beyond the scope of this book. However, some basic, introductory examples are provided below, in order to assist readers with minimal or no previous experience with command line environments in familiarizing with such tasks.

At the moment of writing, two of the most widely used third-party applications for generating standalone executable files from Python scripts are *PyInstaller* for Windows (PyInstaller Development Team, 2019) and *Py2app* for Windows/Mac OS (Oussoren & Ippolito, 2010). Both applications can handle dependencies and *linking*, and the decision on which one should be used comes down to the operating system at hand and personal preference. In broad terms, the steps one needs to follow when creating standalone executable files are summarized below:

- **Step 1:** Irrespectively of what program and procedure one choses to generate the standalone application, the original script(s) must be firstly exported from Anaconda Jupyter, as one or more Python.*py* file(s). This will be the file(s) used as input to the deployment application.
- **Step 2:** Another essential task is to ensure that the application is installed on the system. This can be achieved in a number of ways that are detailed in the numerous associated online guides and tutorials (Apple Inc, 2021; Cortesi, 2021; Microsoft, 2021a, 2021b; Oussoren & Ippolito, 2010; PyInstaller Development Team, 2019). For the purposes of this example, one possibility is to install PyInstaller using a *Command Prompt/PowerShell* window (Microsoft, 2021a, b) using the following command:
  - `pip install pyinstaller`
- **Step 3a (Windows):** Once PyInstaller is installed, and given that the associated files and the command line environment are set up appropriately, the generation of the standalone file could be as simple as the following command:
  - `pyinstaller yourprogram.py`
  Alternatively, the user can refer to the PyInstaller official documentation, in order to execute more specific and complex commands with appropriate *parameters* and *flags*, as necessary. For instance, using the same command with the `--onefile` flag would force the generated executable file to be packaged in a single file rather than in a folder structure containing multiple files:
  - `pyinstaller --onefile yourprogram.py`
- **Step 3b (Mac OS):** The same basic idea also applies when using the Py2app (Oussoren & Ippolito, 2010), although the procedure and commands may be slightly different. For instance, when used on a Mac OS system, Py2app generates *application bundles* instead of an executable file. As an example, users of Mac OS systems can use the *Terminal* window (Apple Inc, 2021) to firstly install Py2app:
  - `pip install -U py2app`
  Py2app can be then used to create a *setup* file:
  - `py2applet --make-setup yourprogram.py`
  Finally, the setup file can be used to generate the standalone application bundle:
  - `python setup.py py2app`
  In both cases, the standalone application is usually placed at a specified directory structure according to the settings and parameters used.

In order to be able to successfully execute the example commands provided here, the reader may have to execute a number of other necessary commands and set up tasks and navigate to the correct directories using the command line environment. Detailed information on how to use both PyInstaller and Py2app can be found on the official documentation pages (Cortesi, 2021; Oussoren & Ippolito, 2010) and on the large variety of associated online resources. It must be noted that the third-party applications mentioned here are just two of the tools one may choose to use for creating standalone executable files based on Python scripts, and they are not the only way of dealing with such tasks.

The development and deployment processes vary depending on the characteristics of the developed application, the chosen development platform, and the targeted operating system(s). As most chapters of this book utilize the Anaconda Jupyter environment, most of the examples and programming scripts can be developed and tested within the development platform (or even other platforms) without the need to generate standalone executable files. However, the information provided here can be used as a general guide for the deployment procedure and the necessary conversions, should the reader choose to create standalone versions of the various examples.

1.5   STRUCTURE OF THIS BOOK

This book is divided into three main parts, based on the knowledge field, character, and objective of the presented topics.

The first part (Chapters 2–5) covers classic computer programming topics like *introduction to programming*, *Object-Oriented Programming*, *Graphical User Interface (GUI) programming*, and *application development*. It is meant to assist readers with little or no prior programming experience to start learning computer programming using Python and the Anaconda Jupyter platform. The related concepts, techniques, and algorithms are discussed and explained with examples of the necessary code and the expected output.

The second part (Chapters 6–9) covers concepts related to *data structures and organization*, the *algorithms* used to manipulate these structures, *database programming (SQL)*, *data analysis* and *visualization*, and the basics of *statistical analysis*. These concepts cover most of the topics, algorithms, and applications that make up what is collectively referred to as *data science*. The structure of this part of this book provides a potential entry point for readers with no prior knowledge in data science, as well as a reference point for those who would like to focus on the implementation of specific data science tasks using Python.

The third part (Chapters 10–12) covers *machine* and *deep learning* concepts, while also providing a brief introduction to using Python in contexts not traditionally linked with the language like *virtual reality (VR)* application development. This part introduces concepts that are potentially more advanced from a contextual perspective, but not necessarily more challenging when it comes to their implementation using Python. For instance, while a deeper understanding of the principles and algorithms behind *machine and deep learning* may be out of scope for many of the readers of this book, the development of applications using the various related modules and methods provided by Python may be something that is of interest. Similarly, while video game and VR/AR application development is certainly a topic that falls outside the scope of a Python textbook in the strict sense, a basic understanding of how such applications could be developed using the Python language may provide a useful insight to the most adventurous of the readers.

All the scripts and case studies presented in this book, as well as the related data and files necessary for their execution, are included as supplementary material in Appendix A.

REFERENCES

Aho, A.V., Hopcroft, J.E., Ullman, J.D., Aho, A.V., Bracht, G.H., Hopkin, K.D., Stanley, J.C., Jean-Pierre, B., Samler, B.A., & Peter, B.A. (1983). *Data Structures and Algorithms*. USA: Addison-Wesley.

Apple Inc. (2021). *Terminal User Guide*. Support.Apple.Com. https://support.apple.com/en-gb/guide/terminal/welcome/mac/.

Cortesi, D. (2021). *PyInstaller Documentation*. PyInstaller 4.5. https://pyinstaller.readthedocs.io/_/downloads/en/stable/pdf/.

Knuth, D.E. (1997). *The Art of Computer Programming* (Vol. 3). Pearson Education.

Microsoft. (2021a). *Installing Windows PowerShell*. https://docs.microsoft.com/en-us/powershell/scripting/windows-powershell/install/installing-windows-powershell?view=powershell–7.1.

Microsoft. (2021b). *Windows Command Line*. https://www.microsoft.com/en-gb/p/windows-command-line/9nblggh4xtkq?activetab=pivot:overviewtab.

Oussoren, R., & Ippolito, B. (2010). *py2app – Create Standalone Mac OS X Applications with Python*. https://py2app.readthedocs.io/en/latest/.

PyInstaller Development Team. (2019). *PyInstaller Quickstart*. https://www.pyinstaller.org/.

Stroustrup, B. (2013). *The C++ Programming Language*. India: Pearson Education.

2 Introduction to Programming with Python

Ameur Bensefia
University of Rouen Normandy
Higher Colleges of Technology

Muath Alrammal
Higher Colleges of Technology
University Paris-Est (UPEC)

Ourania K. Xanthidou
Brunel University of London

CONTENTS

2.1	Introduction	10
2.2	Algorithm vs. Program	11
	2.2.1 Algorithm	11
	2.2.2 Program	12
2.3	Lexical Structure	12
	2.3.1 Case Sensitivity and Whitespace	13
	2.3.2 Comments	13
	2.3.3 Keywords	13
2.4	Punctuations and Variables	14
	2.4.1 Punctuations	14
	2.4.2 Variables	14
2.5	Data Types	15
	2.5.1 Primitive Data Types	15
	2.5.2 Non-Primitive Data Types	16
	2.5.3 Examples of Variables and Data Types Using Python Code	16
2.6	Statements, Expressions, and Operators	21
	2.6.1 Statements and Expressions	21
	2.6.2 Operators	21
	2.6.2.1 Arithmetic Operators	22
	2.6.2.2 Comparison Operators	23
	2.6.2.3 Logical Operators	24
	2.6.2.4 Assignment Operators	25
	2.6.2.5 Bitwise Operators	26
	2.6.2.6 Operators Precedence	28
2.7	Sequence: Input and Output Statements	29
2.8	Selection Structure	30
	2.8.1 The `if` Structure	30
	2.8.2 The `if...else` Structure	32
	2.8.3 The `if...elif...else` Structure	33
	2.8.4 Switch Case Structures	34

DOI: 10.1201/9781003139010-2

        2.8.5   Conditional Expressions 35
        2.8.6   Nested `if` Statements 35
2.9   Iteration Statements 36
        2.9.1   The `while` Loop 36
        2.9.2   The `for` Loop 40
        2.9.3   The Nested `for` Loop 42
        2.9.4   The `break` and `continue` Statement 45
        2.9.5   Using Loops with the Turtle Library 47
2.10   Functions 50
        2.10.1   Function Definition 50
        2.10.2   No Arguments, No Return 50
        2.10.3   With Arguments, No Return 51
        2.10.4   No Arguments, With Return 51
        2.10.5   With Arguments, With Return 52
        2.10.6   Function Parameter Passing 52
                 2.10.6.1   Call/Pass by Value 52
                 2.10.6.2   Call/Pass by Reference 53
2.11   Case Study 54
2.12   Exercises 55
        2.12.1   Sequence and Selection 55
        2.12.2   Iterations – `while` Loops 56
        2.12.3   Iterations – `for` Loops 56
        2.12.4   Methods 57
References 58

2.1   INTRODUCTION

It is hard to find a programming language that does not follow the norms of how a computer program should look like, as the underlying structures have been established for over 50 years. These norms, widely known as the basic programming principles, are broadly accepted by the academic, scientific and professional communities, something also reflected in the approaches of legendary figures in the field like (Dijkstra et al., 1976; Knuth, 1997; Stroustrup, 2013).

The three basic programming principles refer to the concepts of *sequence*, *selection*, and *repetition* or *iteration*. Sequence is the concept of executing instructions of computer programs from top to bottom, in a sequential form. Selection refers to the concept of deciding among different paths of execution that can be followed based on the evaluation of certain conditions. Repetition is the idea of repeating a particular block of instructions as long as a condition is evaluated to `True` (i.e., non-zero). The concept of computer programming in its most basic form can be defined as the integration of these programming principles with *variables* that store and manipulate data through programs and *methods* or *functions* that facilitate the fundamental idea of divide and conquer.

The aim of this chapter is not to propose any innovative ideas of how to change the above logic and structures. Nevertheless, although it is unlikely that these concepts can be changed or redefined in a major way, they can be fine-tuned and put into the context of new and developing programming languages. From this perspective, this chapter can be viewed as an effort to present how these fundamental principles of computer programming are applied to Python, one of the most popular and intuitive modern programming languages, in a comprehensive and structured way. To accomplish this, a number of related basic concepts are presented and discussed in detail in the various sections of this chapter:

1. Algorithms and Programs, Lexical Structures.
2. Variables & Data Types, Primitive and Non-primitive.

3. Statements, Expressions, Operators & Punctuations.
4. Sequence: Input, Basic Operations, and Output Statements.
5. Selection Structures: `if`, `if...else`, `if...elif...else`, Conditional Expressions.
6. Iteration structures: `for` Loops, `while` Loops, Nested Loops.
7. Functions.

It should be noted that this chapter introduces the *Turtle* library, which is used to demonstrate some of the uses of iteration structures.

2.2 ALGORITHM VS. PROGRAM

The demand for developing a program always originates from a problem that must be addressed by means of computer-based automation. However, an intermediate essential step exists between the problem and the actual program, namely the *algorithm*.

2.2.1 ALGORITHM

The term *algorithm* was firstly proposed by mathematician Mohamed Ibn Musa Al-Khwarizmi during the ninth century. It was defined as a set of ordered and finite mathematical operations designed to solve a specific problem. Nowadays, this term is being adopted in various fields and disciplines, most notably in Computer Science and Engineering, in which it is defined as *a set of ordered operations executed by a machine* (computer).

The first step in program development is where a problem is defined. At this point, a solution is formulated as a clear and unambiguous set of steps. This solution is the algorithm. The steps described in the algorithm are later translated into a program using a specific a *programming language* (Figure 2.1).

> **Observation 2.1 – Algorithm:** A set of ordered operations that can be executed by a machine (computer system).

The benefit of starting off with the formulation of an algorithm rather than directly implementing the actual program is that it allows the programmer to focus on how to solve the problem logically, free from any constraints or considerations related to the specifics of any given programming language. Indeed, algorithms are written in a format incorporating natural human language called *pseudo-code*, and follow particular formal rules. Ultimately, such approaches ensure a certain level of clarity and detail that reduces or eliminates ambiguity without having to deal with the technicalities of the implementation.

The examples below provide two cases of algorithms demonstrating the clarity and simplicity that should characterize the solution to the problem at hand before it comes to translating this solution into an actual program. Both algorithms are in the form of pseudo-code and, thus, independent of any particular programming languages used for the implementation of the solutions:

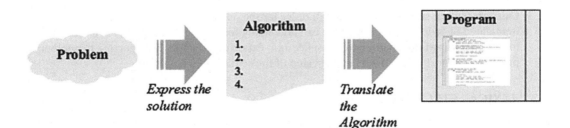

FIGURE 2.1 Phases of program development.

Algorithm 1: Calculate the Area of a Rectangle

```
Start
      Read the length of the rectangle
      Read the width of the rectangle
      Assign width*length to Area
      Display Area
End
```

Algorithm 2: Draw a Square of 50 Pixels Length

```
Start
      Draw a line of 50 pixels length
      Turn the pen right by 90 degrees
      Draw a line of 50 pixels length
      Turn the pen right by 90 degrees
      Draw a line of 50 pixels length
      Turn the pen right by 90 degrees
      Draw a line of 50 pixels length
      Turn the pen right by 90 degrees
      Display Area
End
```

2.2.2 PROGRAM

Once the algorithm is formed, the next step is to write the program in a specific programming language. Each programming language has its own rules and conventions. However, they all have a common core structure consisting of *inputs*, *processing*, and *outputs*. They are all implemented using some form of *code*, the format and structure of which could vary depending on the scope and purpose of each given language and program:

> **Observation 2.2 – Input, Processing, Output:** The basic structure of all programs irrespectively of the programming language used. Input represents any statement written to collect data from an external source. Output represents any statement that sends the outcome of the processing to a display unit, file, or another program.

1. **Input:** Statements dedicated to collecting data from external input sources (e.g., input from the user through the keyboard and mouse), opening and reading files, or accepting input from other programs. In most instances, input is managed at the beginning of the program execution, but this may vary between different languages and programs.
2. **Processing:** Processing lies at the core of the program and represents statements responsible for the manipulation of the information received at input. The length of this section can vary greatly, from a few simple statements to thousands of lines of code organized in numerous files and packages.
3. **Output:** Output statements are used in order for the outcome of the processing to be communicated outside the program. This can take many forms and includes, but is not limited to, sending visual information to a display unit, exporting to a file, or exporting to another program. In most cases, this is the last step of the sequence in a program.

2.3 LEXICAL STRUCTURE

Lexical structure refers to the basic conventions and restrictions in terms of the format and syntax of the text used in the programming environment, in this case Python. This is an important aspect of any programming language, as incorrect format or syntax may lead to compiling errors and code that is difficult to read and debug.

2.3.1 CASE SENSITIVITY AND WHITESPACE

Python is a *case-sensitive* programming language, which means that it distinguishes between keywords and variables written in capital and lower-case letters. Thus, `if` and `IF` are considered to be different words, with the first being recognized as a Python *keyword* and the second processed as a *variable* (see: Variables 2.4.2).

2.3.2 COMMENTS

A *program* is a set of instructions written in a specific language that can be translated and processed by a computer. In real life scenarios, programs can become quite sizable, with hundreds or even thousands of lines of code required. This can make it quite difficult for the programmer to remember the meaning, functionality, and purpose of each line of code. As such, good programming practice involves the use of comments in the program itself. Comments function as useful and intuitive reminders and descriptions to the programmer or anyone who may have direct access to the source code of the program. The comment is expressed in a natural human language and is ignored by the *interpreter* during *runtime*. Python allows the use of two main types of comments:

> **Observation 2.3 – Comments:** Natural language statements ignored by the *interpreter*, used to explain the purpose of the different parts of the code. Start a single line comment with #, or start and end a multiple line comment with """. Note that Python is *case-sensitive*.

- **Single Line Comment:** Starts with the # symbol and continues until the end of the current line:

```
# This statement displays the sentence Hello World
print ("Hello World")
```

- **Multiple Lines Comment:** Starts with the """ symbols and ends when the same symbol combination occurs again:

```
""" The statement below displays
the sentence Hello World """
print("Hello World")
```

2.3.3 KEYWORDS

Python reserves a number of keywords that are used by the interpreter to trigger specific actions when the code is compiled. As these keywords are *reserved*, the programmer is not allowed to use them as variable, function, method, or class names. A list of these keywords is provided in Table 2.1.

> **Observation 2.4 – Keywords:** Reserved words that cannot be used as names for *variables, functions, methods,* or *classes*.

TABLE 2.1
Python Keywords

and	continue	except	global	lambda	pass	while
as	def	False	if	None	raise	with
assert	del	finally	import	nonlocal	return	yield
break	elif	for	in	not	True	
class	else	from	is	or	try	

2.4 PUNCTUATIONS AND VARIABLES

Punctuations and *variables* are special types of symbols and text that dictate specific functionality. As such, when these symbols or text are encountered, the interpreter performs specific, predetermined tasks instead of treating them as common text.

2.4.1 PUNCTUATIONS

Python programs may contain punctuation characters that are combined with other symbols to denote specific functionality. These characters are divided into two main categories: *separators* and *operators* (Table 2.2).

2.4.2 VARIABLES

A *variable* describes a memory location used by a program to store data. Indeed, from a hardware standpoint, it is expressed as a *binary* or *hexadecimal* number that represents the *memory location* and another number that represents the actual *data* stored in it.

Observation 2.5 – Variable: Designated memory location used by the program to store values.

Since working directly with hexadecimal numbers is arguably impractical and counter-productive from a programming perspective, a variable is expressed as a combination of an *identifier* that replaces the actual memory location, a *data type* identifying the kind of data that can be stored in it, and a *value* that represents the actual data stored. Each programming language has its own rules when it comes to naming variables. In Python, a variable name has to conform to the following rules:

- It should start with a letter of the Latin alphabet ('a', 'b', ..., 'z', 'A', 'B', ..., 'Z').
- It may contain numbers.
- It may contain (or start with) the special character " _ ".
- It cannot contain any other character.
- It cannot be a Python keyword.

In line with the above, examples of allowed variable names include the following:

```
Salary, Name, Child1, Email_address, firstName, _ID
```

Similarly, examples of invalid variable names include the following:

```
print, 1Child, Email#address
```

TABLE 2.2
Separators and Operators in Python

Separators:	( ) { } [ ] : " ,						
Operators:	&	|	<	<=	>=	>	==
	−	+	*	**	/	//	%
	<>	!=	=	+=	—+	*=	/=
	%=	//=	**=	&=	|=	^=	>>=	<<=

2.5 DATA TYPES

As stated previously, the purpose of a variable is to hold a value of a specified type. This value can be a number (e.g., decimal, real, octal, hexadecimal), text (i.e., a string of characters), a single character, or a Boolean value (i.e., one out of two possible values: `True` or `False`). More complex structures that consist of any of the aforementioned types may be also used. In general, Python supports two main different data types of variables in this context: *primitive* and *non-primitive* (Figure 2.2).

> **Observation 2.6 – Data Types:** The type of the value stored in a variable could be *primitive* (i.e., integer, string, float, Boolean) or *non-primitive* (i.e., a collection of primitive data types).

2.5.1 PRIMITIVE DATA TYPES

There are four primitive data types that are used when the variable is to hold pure, simple values of data:

- **String or Text:** In Python, a string variable is declared with the `str` keyword. It can hold any set of characters, including letters, numbers, or other symbols, enclosed in double quotation marks:
  - `"This is a text."`
  - `"Do you accept the proposal (Yes/No)?."`
- **Numeric:** Since there are different types of numbers, Python provides variables suitable for different numerical formats and representations:
  - `int` represents integer number (e.g., +24509129)
  - `float` represents real numbers (e.g., −123.0968)
  - `complex` represents complex numbers (e.g., +45−33.6j)
  - `0o` represents octal numbers (e.g., 0o7652001)
  - `0x` represents hexadecimal numbers (e.g., 0x34EF1C3)
- **Boolean:** A Boolean variable is used to represent only two possible values: `True` or `False`.

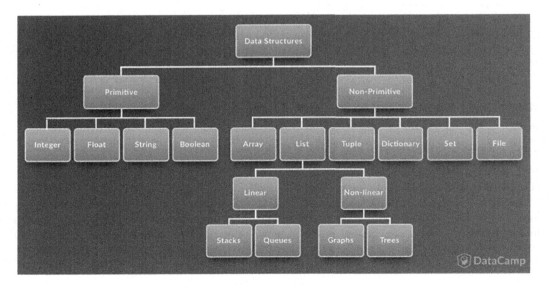

FIGURE 2.2 Python's data types. (See Jaiswal, 2017.)

2.5.2 NON-PRIMITIVE DATA TYPES

Non-primitive data types are complex types consisting of two or more other data types. Such structures are convenient when one needs to manipulate *collections of values* of different types. A list of non-primitive variables is provided below:

- **Sequence:** This type is suitable to use when different values have to be stored and grouped together. It can be further divided into the following categories:
  - **List:** This category represents a collection of any primitive data types where the elements of the list can be accessible through an index and can be modified (*mutable*).
  - **Tuple:** This category represents a collection of any primitive data types where the elements of the list can be accessible through an index but cannot be modified (*immutable*).
  - **Set:** This category represents a collection of distinct, unique objects. It is useful when creating lists that hold strictly unique values in the dataset, and are especially relevant when this dataset is large. The data is unordered and mutable.
  - **Range:** This category represents a series of numbers starting at 0 and ending at a specified number.

Examples:

```
["car", "bike", "truck"]      # This is a list of strings
[200, 6423, -709, 1205]       # This is a list of integers
("car", "bike", "truck")      # This is a tuple of strings
(20.1, +23, -1.9, 12.5)       # This is a tuple of floats
{'O', 'E', 'K', 'C', 'I'}     # This is a set of unique strings
range(5)                      # This will generate the numbers 0 1 2 3 4
range(3)                      # This will generate the numbers 0 1 2
```

- **Dictionary or Mapping:** In cases where it is necessary to associate a pair of data (commonly known as *key* and *value*), dictionary or mapping types can be used. These types are labeled as dict. The declaration begins with curly brackets, followed by the set of pairs separated by commas. Each pair is represented with the key and the value separated by a colon. To access any value, the key name should be provided between brackets:

```
{"name": "Steve", "age":20}      # This is a mapping variable
```

More information on this topic can be found in Chapter 6.

2.5.3 EXAMPLES OF VARIABLES AND DATA TYPES USING PYTHON CODE

This section includes a number of practical examples that demonstrate typical uses and structures of variables and data types in Python.

The first example is related to the *string/text* data type, one of the fundamental and most commonly used data types in computer programming. In this rather simple example, the reader can find a number of coding conventions and commands relating to this data type. For instance, the string values that are being passed to the firstName variable are enclosed in single quotes.

This is also the case when a string is used directly as an argument of the print() function, used to display the information of its arguments on screen. It must be also noted that good programming practice dictates that variables start with lower-case letters, (e.g., firstName instead of FirstName).

This example also highlights that, in addition to simple arguments like strings in quotation marks, functions like print() may accept multiple arguments of different types or formats, such as other variables, or calls to functions (e.g., .format(firstName)). The format() function takes a float value as an argument and loads it in the brackets {} of the preceding string (e.g., 'firstName is {}'.format(firstName)). Note the use of the type() function that returns the data type of the value stored in the provided variable (i.e., firstName).

In the *Jupyter Notebook* editor, if the output is text, it is provided immediately after the current code cell when the program is executed.

Last but not least, the reader should note that comments are included before every distinct piece of code that performs a particular task. While this is not a strict coding requirement, it is an important aspect of good programming practice.

```
1   # Declare a variable named firstName and assign its value to Steve
2   firstName = 'Steve'
3
4   # Print the value of variable firstName
5   print('firstName is {}'.format(firstName))
6
7   # Print the data type of variable firstName
8   print(type(firstName))
```

Output 2.5.3.a:

```
firstName is Steve
<class 'str'>
```

Variables of the *integer* data type are non-decimal numbers (e.g., numberOfStudents = 20):

```
1   # Declare a variable named numberOfStudents and assign its value to 20
2   numberOfStudents = 20
3
4   # Print the value of variable numberOfStudents
5   print('Number of students is {}'.format(numberOfStudents))
6
7   # Print the data type of variable numberOfStudents
8   print(type(numberOfStudents))
```

Output 2.5.3.b:

```
Number of students is 20
<class 'int'>
```

Variables of the *float* data type are floating-point numbers that require a decimal value. Note that the inclusion of the decimal value is mandatory even if it is zero:

```
1   # Declare a variable named salary and assign its value to 20000.0
2   salary = 20000.0
3
4   # Print the value of variable salary
5   print('Salary is {}'.format(salary))
6
7   # Print the data type of variable salary
8   print(type(salary))
```

Output 2.5.3.c:

```
Salary is 20000.0
<class 'float'>
```

Variables of the *complex* data type are in the form of an expression containing real and imaginary numbers, such as +x–y.j (e.g., complexNumber = +45–33.6j):

```
1   # Declare variable complexNumber; assing its value to +45-33.6j
2   complexNumber = +45–33.6J
3
4   # Print the value of variable complexNumber
5   print('complexNumber is {}'.format(complexNumber))
6
7   # Print the data type of variable complexNumber
8   print(type(complexNumber))
```

Output 2.5.3.d:

```
complexNumber is (45-33.6j)
<class 'complex'>
```

Values of the *octal* data type start with 0o (e.g., octalNumber = 0o7652001). In this particular example, the reader should also note the use of comments stretching across multiple lines. As mentioned, comments of this type start and end with three double quotation marks ("""):

```
1   # Declare a variable named octalNumber and assign its value to 0o7652001
2   octalNumber = 0o7652001
3
4   # Print the value of variable octalNumber
5   print('octalNumber is {}'.format(octalNumber))
6
7   """Print the data type of variable octalNumber: notice that the type
8   is octal integer; this is why a class int text appears in the result"""
9   print(type(octalNumber))
```

Output 2.5.3.e:

```
octalNumber is 2053121
<class 'int'>
```

Boolean variables can only take two different values: `True` or `False`. In the following code, variable `married` is `True`, but the only other possible value this variable could take would be `False`:

```
1    # Declare a variable named married and assign its value to True
2    married = True
3
4    # Print the value of variable married
5    print('married is {}'.format(married))
6
7    # Print the data type of variable married
8    print(type(married))
```

Output 2.5.3.f:

```
married is True
<class 'bool'>
```

Mapping variables are always enclosed in curly brackets (e.g., `mappingVariable = {'name': 'Steve', 'age': 20}`):

```
1    # Declare a variable named mappingVariable and assign its
2    # value to {'name':'Steve', 'age':20}
3    mappingVariable = {'name':'Steve', 'age':20}
4
5    # Print the value of variable mappingVariable
6    print('mappingVariable is {}'.format(mappingVariable))
7
8    # Print the data type of variable mappingVariable
9    print(type(mappingVariable))
```

Output 2.5.3.g:

```
mappingVariable is {'name': 'Steve', 'age': 20}
<class 'dict'>
```

List variables are enclosed in square brackets (e.g., `listVariable = [200, 6423, -709, 1205]`):

```
1    # Declare a variable named listVariable and assign
2    # its value to [200, 6423, -709, 1205]
3    listVariable = [200, 6423, -709, 1205]
4
5    # Print the value of variable listVariable
6    print('listVariable is {}'.format(listVariable))
7
8    # Print the data type of variable listVariable
9    print(type(listVariable))
```

Output 2.5.3.h:

```
listVariable is [200, 6423, -709, 1205]
<class 'list'>
```

Tuple variables are enclosed in parentheses (e.g., tupleVariable = ('car', 'bike', 'truck')):

```
1    # Declare a variable named tupleVariable and assign
2    # its value to ('car', 'bike', 'truck')
3    tupleVariable = ('car', 'bike', 'truck')
4
5    # Print the value of variable tupleVariable
6    print('tupleVariable is {}'.format(tupleVariable))
7
8    # Print the data type of variable tupleVariable
9    print(type(tupleVariable))
```

Output 2.5.3.i:

```
tupleVariable is ('car', 'bike', 'truck')
<class 'tuple'>
```

Range variables hold integers ranging from 0 up to a specified number (e.g., rangeVariable = range(5)). Note that the specified number is *not inclusive*, so rangeVariable in this example will hold values 0, 1, 2, 3, and 4:

```
1    # Declare a variable named rangeVariable and assign its value to a
2    # range of integers from 0 to 4 (i.e., 0 1 2 3 4)
3    rangeVariable = range(5)
4
5    # Print the value of variable rangeVariable
6    print('rangeVariable is {}'.format(rangeVariable))
7
8    # Print the data type of variable rangeVariable
9    print(type(rangeVariable))
```

Output 2.5.3.j:

```
rangeVariable is range(0, 5)
<class 'range'>
```

Set variables hold sets of unique values of primitive data types. In the following code, command set('cookie') allocates unique values 'i', 'c', 'o', 'e', 'k' to variable setVariable:

```
1    # Declare a variable named setVariable and assign its value to
2    # the set of unique letter in the word 'cookie'
3    setVariable = set('cookie')
4
5    # Print the value of variable setVariable
6    print('setVariable is {}'.format(setVariable))
7
8    # Print the data type of variable setVariable
9    print(type(setVariable))
```

Output 2.5.3.k:

```
setVariable is {'i', 'e', 'c', 'k', 'o'}
<class 'set'>
```

2.6 STATEMENTS, EXPRESSIONS, AND OPERATORS

Statements and *expressions* refer to specific syntactical structures that provide instructions to the *interpreter* in order to execute specific tasks. They can be simple structures executing a simple task, like printing a message on screen, or more complicated ones that perform a number of tasks and generate multiple threads of information and results.

Operators refer to special symbols that perform particular, pre-determined tasks, and can be used as building blocks for building logical statements and expressions. This section introduces basic concepts related to these fundamental programming elements.

> **Observation 2.7 – Statement:** A line of code that can be executed by the Python *interpreter*.

2.6.1 STATEMENTS AND EXPRESSIONS

A *statement* is a unit/line of code (i.e., an *instruction*) that the Python interpreter can execute. So far, two kinds of statements have been presented in this chapter, *assignment* and *print*:

> **Observation 2.8 – Expression:** Any combination of values, variables, operators, and/or calls to functions that result in an unambiguous value.

```
1   # Assignment statement produces no output
2   name = 'Steve'
3
4   # Print function
5   print('Name is:', name)
```

Output 2.6.1:

```
Name is: Steve
```

A *script* usually contains a sequence of statements. When there are more than one statements, the results appear one at a time, as each statement is executed.

An *expression* is a combination of values, variables, operators, and calls to functions resulting in a clear and unambiguous value upon execution.

2.6.2 OPERATORS

Operators are tokens/symbols that represent computations, such as addition, multiplication and division. The values an operator acts upon are called *operands*.

Let us consider the simple expression x = 3*2. The reader should note the following:

> **Observation 2.9 – Operators/Operands:** Operators are symbols representing computations like additions, multiplications, divisions. Operands are the values that the operators act upon.

- x is a variable.
- 3 and 2 are the operands.
- * is the multiplication operator.
- 3*2 is considered an expression since it results in a specific value.

TABLE 2.3

Python Arithmetic Operators

Operator	Example	Name	Description
+ (unary)	+a	Unary positive	a
+ (binary)	a + b	Addition	Sum of a and b. The + *operator* adds two numbers. It can be also used to *concatenate strings*. If either operand is a string, the other is converted to a string too.
- (unary)	-a	Unary negation	It converts a positive value to its negative equivalent and vice versa.
- (binary)	a - b	Subtraction	b subtracted from a.
*	a * b	Multiplication	Product of a and b.
/	a / b	Division	The division of a by b. The result is always of type float.
%	a % b	Modulo	The remainder when a is divided by b.
//	a // b	Floor division (also called integer division)	The division of a by b, rounded to the next smallest integer.
**	a ** b	Exponentiation	a raised to the power of b.

Python supports many operators for combining data into *expressions*. These can be divided into *arithmetic*, *comparison*, *logical*, *assignment*, and *bitwise*:

2.6.2.1 Arithmetic Operators

These operators can be used with integers, floating-point numbers, or even characters (i.e., they can be used with any primitive type other than Boolean). Table 2.3 lists

Observation 2.10 – Efficient Script Writing: Include expressions that display results inside the print function to avoid multiple instructions. Use a single statement to declare and assign values to multiple variables.

the arithmetic operators supported by Python, and the example that follows presents a script that applies a number of these operators. It is worth noting that the arithmetic expressions are not separate statements in the script. Instead, they appear as arguments in the `print()` function. Both options are correct, although it is advisable to follow a syntax similar to the script in order to write shorter, and thus more efficient, scripts.

```
1   a = 5
2   b = 4
3
4   # Addition expression
5   print('a+b=', a + b)
6
7   # Subtraction expression
8   print('a-b=', a - b)
9
10  # Multiplication expression
11  print('a*b=', a * b)
12
13  # Division expression
14  print('a/b=', a / b)
15
16  # Exponent expression
17  print('a raised to the power of b =', a ** b)
```

```
18
19  # Unary negation expression
20  print('a negated is =', - a)
21
22  # Modulus expression
23  print('The remainder of the integer division between a and b is:', a % b)
24
25  # Floor division
26  print('Floor division of a and b is:', a // b)
```

Output 2.6.2.a:

```
a+b= 9
a-b= 1
a*b= 20
a/b= 1.25
a raised to the power of b = 625
a negated is = -5
The remainder of the integer division between a and b is: 1
Floor division of a and b is: 1
```

2.6.2.2 Comparison Operators

These operators compare values for equality or inequality, (i.e., the relation between the two operands, be it numbers, characters, or strings). They yield a Boolean value as a result. The comparison operators are typically used with some type of conditional statement (see: 2.8 Selection Structures) or within an iteration structure (see: 2.9 Iteration Structures), determining the *branching* or *looping* directions to follow. Table 2.4 lists the comparison operators supported by Python, and the code that follows provides some relevant example cases using a Python script.

TABLE 2.4
Python Comparison Operators

Operator	Example	Name	Description
==	a == b	Equal to	True if the value of a is equal to that of b; False otherwise
!=	a != b	Not equal to	True if a is not equal to b; False otherwise
<	a < b	Less than	True if a is less than b; False otherwise
<=	a <= b	Less than or equal to	True if a is less than or equal to b; False otherwise
>	a > b	Greater than	True if a is greater than b; False otherwise
>=	a >= b	Greater than or equal to	True if a is greater than or equal to b; False otherwise

An interesting point about this particular script is that the variables are all declared and assigned with values in one statement separated by commas. The script also demonstrates the use of a mix of strings and arithmetic expressions as arguments of the print function, separated by commas:

```
1   a, b, c, d, e = 5, 4, 5, 'Dubai', 'Abu Dhabi'
2
3   # Test for equality and print directly the result of the expression
4   print(a == b, 'and', a == c)
```

```
5
6   # Test for inequality and print directly the result of the expression
7   print(a != b, 'and', a != c)
8
9   # Test for 'less than' and for 'less than' or 'equal to' and
10  # print directly the result of the expression
11  print(a < b, 'and', a <= b)
12
13  # Test for 'greater than' and for 'greater than or equal to' and
14  # print directly the result of the expression
15  print(a > b, 'and', a >= b)
16
17  # Test for equality and 'less than' between strings
18  print(d == e, 'and', d > e)
```

Output 2.6.2.b:

```
False and True
True and False
False and False
True and True
False and True
```

2.6.2.3 Logical Operators

As mentioned, comparison operators compare their operands and produce a Boolean output. This type of output is commonly used in branching and looping statements. Boolean operators are used to combine multiple comparison expressions into a more complex, singular expression. The Boolean operators require their operands to be Boolean values. Table 2.5 lists the logical operators supported by Python and the following script demonstrates some of their indicative applications:

```
1   # Apply the 'not' logical operator
2   x = 5
3   print(not (x < 10))
4   print(not (x < 3))
5
6   # Apply the 'or' logical operator
7   x, y = 5, 7
8   print((x > 3) or (y < 6))
9   print((x < 3) or (y < 6))
10
11  # Apply the 'and' logical operator
12  x, y = 5, 7
13  print((x > 3) and (y > 6))
14  print((x < 3) and (y > 6))
15
16  # Combine 'not', and 'and or' operators
17  x, y = 5, 7
18  print(not (x < 3) and (y > 6))
19  print((x < 3) or (y > 6) and (x < 10))
```

```
Output 2.6.2.c:

   False
   True
   True
   False
   True
   False
   True
   True
```

TABLE 2.5
Python Logical Operators

Operator	Example	Description
not	not a	True if a is `False`; False if a is `True`
or	a or b	True if either a or b is `True`; `False` otherwise
and	a and b	True if both a and b are `True`; `False` otherwise

TABLE 2.6
Python Assignment Operators

Operator	Example	Description
=	c = a + b	Assigns the result of the *expression* on the right side of the *assignment operator* to the *variable* on the left side.
+=, -=	c += a, c -= b	Equivalent to `c = c + a` or `c = c - a`
*=, /=	c *= a, c /= b	Equivalent to `c = c * a` or `c = c / b`
//=	c //= a	Equivalent to `c = c // a`
%=	c %= a	Equivalent to `c = c % a`
**=	c **= a	Equivalent to `c = c ** a`

2.6.2.4 Assignment Operators

These quite significant operators allow the manipulation of variables by saving or updating their values. Table 2.6 and the code that follows summarize the use of the different assignment operators in Python:

```
1   # Assign the result of the expression on the right side of
2   # the assignment operator to the variable on the left side
3   a, b = 12, 10
4   c = a + b
5   print('The value of c is:', c)
6
7   # Use +=, -+, *=, /= in assignments
8   a, c = 2, 12
9   c += a
10  print('The value of c is:', c)
11
12  a, c = 2, 12
13  c -= a
14  print('The value of c is:', c)
15
16  a, c = 2, 12
17  c *= a
18  print('The value of c is:', c)
19
20  a, c = 2, 12
21  c /= a
22  print('The value of c is:', c)
```

```
23
24  # Use the %= and **= in assignments
25  a, c = 4, 10
26  c %= a
27  print('The value of c is:', c)
28
29  a, c = 4, 10
30  c **= a
31  print('The value of c is:', c)
```

Output 2.6.2.d:

```
The value of c is: 22
The value of c is: 14
The value of c is: 10
The value of c is: 24
The value of c is: 6.0
The value of c is: 2
The value of c is: 10000
```

2.6.2.5 Bitwise Operators

These are considered to be *low-level* operators. They treat operands as sequences of binary digits and operate on them bit by bit. Table 2.7 details the bitwise operators supported by Python and the example that follows demonstrates their application within a script. The reader should note that when assigning values to variables in the binary system, the values must be preceded by 0b, followed by the value in the binary form. Likewise, when variable values must be displayed in the binary form, the form {:04b} must be used in order to display the binary value with four digits.

TABLE 2.7
Python Bitwise Operators

Operator	Example	Name	Description
&. |	a & b, a | b	bitwise AND, OR	Each bit position in the result is the logical AND (or OR) of the bits in the corresponding position of the operands; 1 if both are 1, otherwise 0 for AND; 1 if either is 1, otherwise 0.
~	~a	bitwise negation	Each bit position in the result is the logical negation of the bit in the corresponding position of the operand; 1 if 0, 0 if 1.
^	a ^ b	bitwise XOR (exclusive OR)	Each bit position in the result is the logical XOR of the bits in the corresponding position of the operands; 1 if the bits in the operands are different, 0 if they are the same.
>>, <<	a >> n, a << n	Shift right or left n places	Each bit is shifted right or left by n places.

```
1  # Bitwise 'and'
2  a, b = 0b1100, 0b1010
3  print('0b{:04b}'.format(a & b))
4
5  # Bitwise 'and'
6  a, b, c, = 12, 10, 0 # 12 = 0b1100, 10 = 0b1010
7  C = a & b # 8 = 0b1000
8  print('Value of c is', c)
9
```

```
10   # Bitwise 'or'
11   a, b = 0b1100, 0b1010
12   print('0b{:04b}'.format(a | b))
13
14   # Bitwise 'or'
15   a, b, c, = 12, 10, 0 # 10 = 0b1100, 12 = 0b1010
16   c = a | b # 14 = 0b1110
17   print('Value of c is', c)
18
19   # Bitwise negation
20   a = 0b1100
21   b = ~a
22   print('0b{:04b}'.format(b))
23
24   # Bitwise negation
25   a, b = 12, ~(a)   # 12 = 0b1100, -13 = 0b-1101
26   print('Value of b is', b)
27
28   # Bitwise XOR (exclusive OR)
29   a, b = 0b1100, 0b1010
30   print('0b{:04b}'.format(a ^ b))
31
32   # Bitwise XOR (exclusive OR)
33   a, b, c = 12, 10, a ^ b  # 12 = 0b1100, 10 = 0b1010, 6 = 0b0110
34   print ('Value of c is', c)
35
36   # Shift right 'n' places
37   a = 0b1100
38   print('0b{:04b}'.format(a >> 2))
39
40   # Shift right 'n' places
41   a, b, = 12, a >> 2 # 3 = 0b0011
42   print('Value of c is', b)
43
44   # Shift left 'n' places
45   a = 0b1100
46   print('0b{:04b}'.format(a << 2))
```

Output 2.6.2.e:

```
0b1000
Value of c is 8
0b1110
Value of c is 14
0b-1101
Value of b is -13
0b0110
Value of c is 6
0b0011
Value of c is 3
0b110000
```

2.6.2.6 Operators Precedence

Python, like other programming languages, uses the standard algebraic procedure to evaluate expressions. All operators are assigned a precedence:

- Operators with the highest precedence are applied first.
- Next, the results of their expression are used to determine those with the next highest precedence.
- In case of operators with equal precedence their application starts from left to right.
- This pattern continues until the full expression is calculated.

Observation 2.11 – Order of Precedence: The order of precedence of operator execution determines the result of complex expressions. Inconsistencies can lead to incorrect scripts.

Table 2.8 lists the operator precedence for Python, from lowest to highest. The code following this provides some examples of their application. It is essential for the reader to keep in mind the order of precedence of the various operators, since failure to do so will most certainly lead to inconsistencies in the way the complex expressions are calculated by the system:

TABLE 2.8
Python Precedence Operators

Precedence	Operator	Description
Lowest	or	Boolean OR
	and	Boolean AND
	not	Boolean NOT
	==, != , <, <=, >, >=, is, is not	Comparisons, identity
	|	Bitwise OR
	^	Bitwise XOR
	&	Bitwise AND
	<< , >>	Bit shifts
	+ , -	Addition, subtraction
	*, /, //, %	Multiplication, division, floor division, modulo
	+x, -x, ~x	Unary positive, unary negation, bitwise negation
Highest	**	Exponentiation

```
1    # The order of execution is exponentiation first,
2    # then multiplication: 2 * 2 = 4, then, 4 * 5 = 20
3    a = 5 * 2 ** 2
4    print('The value of a is:', a)
5
6    # The order of execution is multiplication first,
7    # then addition: 2 * 3 = 6, then 2 + 6 = 8
8    a = 2 + 2 * 3
9    print('The value of a is:', a)
10
11   # Parentheses have the highest precedence,
12   # then everything else: (2 + 2) = 4, then, 4 * 3 = 12
13   a = (2 + 2) * 3
14   print('The value of a is:', a)
15
```

```
16   # Addition and subtraction have the same precedence,
17   # hence, they are evaluated from left to right.
18   # This is also the case between arithmetic operators
19   # with equal precedence: 2 + 2 = 4, then 4 − 3 = 1
20   print('The value of a is:', a)
```

Output 2.6.2.f:

```
The value of a is: 20
The value of a is: 8
The value of a is: 12
The value of a is: 12
```

2.7   SEQUENCE: INPUT AND OUTPUT STATEMENTS

Similarly to most other contemporary programming languages, Python is organized around *functions*, reusable programming routines that can be attached to an *object* of a class or used as standalone pieces of code that perform specific tasks. Python has a quite extensive array of functions, both predefined ones that are inherently built in the core of the language itself, or as part of the various classes used by it.

An example of a Python function that has already appeared in several of the exercises presented in this chapter is the print() function. As the name suggests, this is a function used to display output on screen. To invoke it one simply has to call it with an argument (e.g., print(<argument>)).

Another frequently used Python function is input(), used to get input from the keyboard. This function prompts the user to provide input in the form of text. The function stops the program execution until the text input has been provided and resumes only when the user presses the designated key (i.e., *Enter* or *Return*). The following example demonstrates the use of both print() and input() in a single Python script:

> **Observation 2.12 – Input/Output:**
> Use the print() function to display output on screen. Output is passed to the function as an argument. Use the input() function to receive input from the keyboard. Ensure that input() is assigned to a variable, as Python may treat it as memory *garbage.*

```
1   # Call the 'input' function to accept the user's input from
2   # the keyboard and assign the provided data to a variable
3   fullName = input('Insert your full name\n')
4
5   # Print the contents of the variable fullName on screen
6   print('The name you entered is', fullName)
```

Output 2.7.a & 2.7.b:

```
Insert your full name
```

```
Insert your full name
Rania
The name you entered is Rania
```

It is important to point out the following in regard to this particular script:

- Any value received as input must be assigned to a suitable variable. If input data are unallocated, there is a serious risk that Python will treat them as memory *garbage*.
- *Escape* character \n should be used to force the display of the next output of the program to the next line.
- The input() function treats all input streams as text regardless of whether numeric values are provided. If an input stream is meant to be treated as a numerical value, further processing is required.

2.8   SELECTION STRUCTURE

One of the three principles of computer programming is to make a decision of the next block of statements to execute, based on the result of the evaluation of a certain condition. Such a condition, and the statements to execute based on it, is referred to as a *selection*. There are three main types of selection statements: if, if...else, and if...elif...else.

2.8.1   THE if STRUCTURE

The if structure is used to determine whether a certain statement or block of statements will be executed or not, based on a simple or complex condition. If the condition is True (or non-zero), then the block of statements is executed, otherwise it is not executed and the program flow continues from the next statement outside the if structure. This means that the evaluation of the condition must yield a Boolean or arithmetic (i.e., zero/non-zero) value. The syntax of the basic if statement is provided below:

> **Observation 2.13 – Condition:** A True/False or zero/non-zero value expression used to determine the flow of program execution.

```
if (condition):
    Block of statements to execute if condition is True
    Statements to execute outside the if statement
```

Similarly, Figure 2.3 illustrates a simple if statement in the form of a *flowchart*.

Most high-level programming languages, such as C++ or Java, use brackets {} to mark a block of statements. Since Python does not have any type of designated markers for such purpose, it uses *indentation* to identify these blocks. Under this scheme, the block starts with the indentation and ends at the first non-indented line of code. Consider the following script:

```
1    # Simple 'if' statement
2    a = int(input('Enter the first integer to continue: '))
3    b = int(input('Enter the second integer to continue: '))
4    if (a > b):
5        print("The first integer is larger than the second")
```

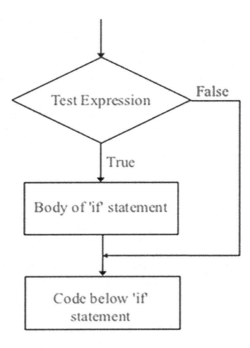

FIGURE 2.3 Flowchart of the `if` statement.

Output 2.8.1:

```
Enter the first integer to continue: 5
Enter the second integer to continue: 3
The first integer is larger than the second
```

In this example, the user is prompted to enter two integer values assigned to two different, corresponding variables. Next, the variables are compared based on their values. This is done with a simple `if` statement that, when `True`, displays a message on screen. Both the `input()` and `print()` functions are used in the script. The reader should note that, since the `input()` function treats every input as *text*, it is necessary to convert this value into a suitable *primitive type* for the required calculations or processing to take place. This is the idea behind *casting*. In this particular example, the input value is *cast* into an integer using the `int()` function. Also, the reader should note that it is possible to use one function call inside another, in this case the `input()` function call inside the `int()` cast call.

Observation 2.14 – `if` Statement: Used to determine whether a statement or block of statements will be executed or not, based on a simple or complex condition.

Observation 2.15 – Indentation: Use indentation to mark a block of statements.

Observation 2.16 – Casting: Convert input values to appropriate *primitive* data type, as required for calculations or processing.

2.8.2 THE if...else STRUCTURE

It is possible to write the if statement in a way that it executes a block of statements when the condition is True and another when it is not. This is the concept behind the if...else statement:

```
if (condition):
   Block of statements to execute if
   condition is True
else:
   Block of Statements to execute if
   condition is False
```

Figure 2.4 illustrates an if...else structure as a flowchart and the following code provides an example of its application. This particular script prompts the user to enter two integers (note that input is treated as *text* by default), converts the input to actual integers, compares the two values, and displays one of the two outputs, depending on the result of the comparison. In this example, there is only one statement to execute, as the condition of the if statement will be either True or False. However, the user can add multiple instructions within the block of statements, while it is also possible to have another if statement *nested* inside the block. Such cases are discussed at later sections of this chapter.

Observation 2.17 – Selection:

- Use the if statement for the execution of one block of statements if the condition is True.
- Use the if...else statement for the execution of either of two possible blocks of statements depending on a particular condition.
- Use the if...elif...else statement for the execution of multiple possible blocks of statements depending on a number of conditions.
- Use *dictionary/mapping* structures in place of the *switch* structure of C++, Java, etc.
- Use *conditional expression* in place of the conditional operator used in C++, Java, etc.
- Use nested if structures in more complex cases.

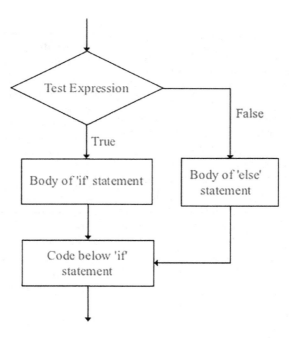

FIGURE 2.4 Flowchart of the if...else statement.

```
1    # The 'if...else...' statement
2    a = int(input('Enter the first integer to continue: '))
3    b = int(input('Enter the second integer to continue: '))
4    if (a > b):
5        print('First integer holds a value greater than the second')
6    else:
7        print('Second integer holds a value greater than the first')
```

Output 2.8.2:

```
Enter the first integer to continue: 13
Enter the second integer to continue: 20
Second integer holds a value greater than the first
```

2.8.3 THE **if...elif...else** STRUCTURE

Python allows the execution of more than two blocks of statements in a single if structure. If one of the conditions controlling the if structure is True, the block associated with that structure is executed. The remaining blocks are just ignored and the program execution continues at the first line after the if structure. If none of the conditions are True, then the else statement is executed. The syntax of the if...elif...else structure is provided below, and its flowchart can be found in Figure 2.5:

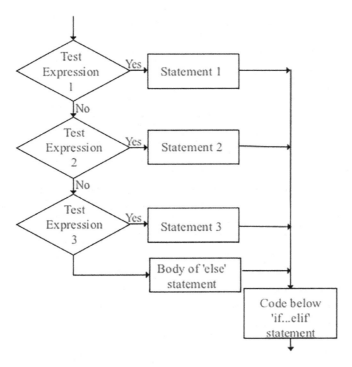

FIGURE 2.5 Flowchart of the if...elif...else statement.

```
        if (condition1):
                Block to execute if condition1 is True
        elif (condition2):
                Block to execute if condition2 is True
        ...
        else:
                Block to execute if none of the conditions are True
```

The following script demonstrates the application of an if...elif...else structure. The script prompts the user to enter an integer between 0 and 100. Depending on the input value, a particular block of code is executed based on the conditions of the various if...elif...else structures:

```
1    # The 'if...elif...else...' statement
2    a = int(input('Enter a grade between 0 and 100: '))
3
4    if (a < 60):
5        print('I am sorry but you failed the course.\n'\
6        'Please try harder next semester')
7    elif (a < 70):
8        print('Task completed! You passed the course')
9    elif (a < 80):
10       print('Well done! You did well in the course')
11   elif (a < 90):
12       print('Very good job. Keep up the good work')
13   elif (a < 100):
14       print('Excellent performance. Congratulations.')
15   else:
16       print('I am sorry but an integer between 0 and 100 was expected')
```

Output 2.8.3:

```
Enter a grade for the course between 0 and 100: 92
Excellent performance. Congratulations.
```

2.8.4   SWITCH CASE STRUCTURES

A *switch case* structure is used as an alternative to long if structures that compare a variable against several values. Unlike other programming languages, Python does not have a dedicated *switch case* statement. To get around the lack of such statements, programmers may use an if... elif...else structure, as described in the previous section. Alternatively, *dictionary/mapping* can be used as shown in the script below:

```
1    # Dictionary mapping used to check against a range of options
2    numberToTextSwitcher = {
3        1: 'One',
4        2: 'Two',
5        3: 'Three'
6    }
7
8    number = input('Insert 1, 2, or 3: ')
```

```
9    intNumber = int(number)
10   print('The string value of', intNumber, \
11          'is', numberToTextSwitcher.get(intNumber))
```

Output 2.8.4:

```
Insert 1, 2, or 3: 3
The string value of 3 is Three
```

The reader should note some interesting points in relation to this script:

- The *dictionary/mapping* variable type, in this example `numberToTextSwitcher`, can be used to substitute the functionality of the missing switch statement.
- When a statement is long and difficult to include in a single line, the programmer can use the \ symbol to inform the Python interpreter that the statement continues in the next line.
- Apply the `get()` function of the dictionary/mapping variable with the *key* (i.e., the first part of the pair) to get access to the *value* (i.e., the second part of the pair).

2.8.5 CONDITIONAL EXPRESSIONS

Another expression that can be used in Python instead of the missing *conditional operator* of C++ or Java, is what is often called the *conditional expression*. The syntax is the following:

```
Statement 1 if condition else Statement 2
```

In this case, the first part of the expression that is executed is the `if` condition. If this is `True`, the first statement is executed; otherwise, the second statement is executed. The following code provides an example of the application of the conditional expression:

```
1    # Use of 'conditional expression' instead of the 'if...else' statement
2    a = int(input('Enter the first integer (a): '))
3    b = int(input('Enter the second integer (b): '))
4
5    print('a is greater than b') if (a > b) else print('b is greater than a')
```

Output 2.8.5:

```
Enter the first integer (a): 3
Enter the second integer (b): 6
b is greater than a
```

2.8.6 NESTED `if` STATEMENTS

As already implied, it is possible to have an `if` structure nested inside another. In fact, such a practice could go to as much depth as the programmer wishes, although it is not advisable to go deeper than three levels since it will be difficult to conceptually control the resulting structure. A possible syntax for the nested `if` structure is presented below:

```
if (condition 1):
        if (condition 2):
                Block 1 executes
```

```
            else:
                    Block 2 executes
        else:
                Block 3 to execute if
                condition 1 is False
```

Block 1 will be executed if condition 2 is `True`. Condition 1 is not considered at this point, as it is `True` by default. Note that if this was not the case, the program flow would never reach the nested `if(<condition 2>)` statement. Also, the first `else` statement is an alternative to the `if(<condition 2>)` part of the structure and not to the `if(<condition 1>)` part. The latter is taken care of by the second `else` statement. The code that follows is an example of a nested `if`, based on a simple variation of a previously used script:

```
1    # A script with a basic nested 'if' structure
2    inputGrade = int(input('Enter your grade between 0 and 100: '))
3
4    if (inputGrade >= 80):
5        if (inputGrade >= 90):
6            print('Excellent performance')
7        else:
8            print('Very good. Keep up the good work')
9    else:
10       if (inputGrade >= 60):
11           print('You did well')
12       else:
13           print('Sorry, you failed the course')
```

Output 2.8.6:

```
Enter your grade between 0 and 100: 50
Sorry, you failed the course
```

2.9   ITERATION STATEMENTS

Application developers and programmers always look to optimize their programs using appropriate, efficient statements and minimizing the lines of code in order to create an easy to maintain program. A common way to reduce the lines of code is the concept of *iteration*. Indeed, *iteration*, alongside *sequence* (i.e., sequential execution of statements) and *selection* (see previous sections) constitute what is known in computer program-

Observation 2.18 – Loop: A block of statements that is executed repeatedly while a certain condition is `True`. There are three possible forms of loops: `while` loops, `for` loops, and nested loops.

ming as the *three basic principles of programming*. The iteration concept applies to cases where a block of statements has to be repeated several times. There are three possible iteration alternatives offered in Python: the `while` loop, the `for` loop, and the nested loops.

2.9.1   THE `while` LOOP

The `while` loop is suitable for cases where the number of iterations is unknown and depends on certain conditions. These conditions need to be specified explicitly, similarly to the various forms of *selection* statements. The block of statements inside the loop is repeated as long as the specified conditions are satisfied. Once the conditions become `False` the Python interpreter exits the loop and proceeds with the rest of the program. The block of statements within the loop structure needs to be *indented*. The syntax of the basic `while` loop and its flowchart (Figure 2.6) are provided below:

Observation 2.19 – while Loop: Repeatedly executes a block of statements while a certain condition is `True`. If the condition is never `True`, the block is never executed. If the condition never changes to `False`, the block is executed indefinitely, causing an *infinite* loop.

```
# while loop with one condition
while (condition):
        Block of statements
...


# while loop with two conditions;
# op can be any logical operator
while (condition) op (condition2):
        Block of statements
...
```

If the condition before the beginning of the loop is not met, the block of statements will not be executed and/or repeated. It is also possible that the conditions inside the `while` loop are not updated, in which case the block will be executed indefinitely resulting in an undesirable *infinite* loop. In order to avoid the latter, it is essential for the conditions to be updated inside the `while` loop.

The following script provides a basic example of the `while` loop. The program starts by prompting the user to decide whether the message should be displayed or not. This is done by entering either 'Y'/'y' or 'N'/'n'. Any other input is considered as not 'Y'/'y'. In this arrangement, the flow goes into the block that belongs to the `while` loop only when the user enters 'Y' or 'y'. Note that the same prompt for input is given to the user inside the loop. This is because it is necessary to change this value in order to determine the `while` condition. As mentioned, if this value is not modified inside the loop (i.e., if the statement showMessage = input ('Do you want to

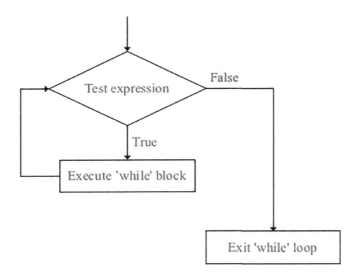

FIGURE 2.6 Flowchart of the `while` loop.

show the message again (Y/N)?)' is missing) the program execution would lead into an infinite loop. The program will continue to run as long as the user enters 'Y' or 'y':

```
1    # Use of 'while' loop to show the message 'Hello world'
2    # as long as the user enters 'Y' or 'y'
3    showMessage = input('Do you want to show the message again (Y/N)? ')
4
5    while (showMessage == 'Y' or showMessage == 'y'):
6        print('Hello world')
7        showMessage = input('Do you want to show the message again (Y/N)? ')
```

Output 2.9.1.a:

```
Do you want to show the message again (Y/N)? Y
Hello world
Do you want to show the message again (Y/N)? Y
Hello world
Do you want to show the message again (Y/N)? N
```

Another example of a while loop can be seen in the script below, which introduces the use of the end = '' clause in the print() function. This results in the program stopping and waiting for new output at the end of the same print without proceeding to the next line:

```
1    # Use the 'while' loop to display all integers
2    # between two values provided by the user
3
4    numberToShow = int(input('Enter the starting integer: '))
5    endInteger = int(input('Enter the ending integer: '))
6
7    while (numberToShow <= endInteger):
8        print(numberToShow, ' ', end = '')
9        numberToShow += 1
```

Output 2.9.1.b:

```
Enter the starting integer: 5
Enter the ending integer: 10
5  6  7  8  9  10
```

The next script is a classic example of adding together two integers, the values of which are entered by the user at runtime. The reader should note how the *loop control* variable (i.e., currentInteger) is being modified inside the block of statements. Also, it should be noted how the two print() functions are used and connected through the end = '' clause, in order to display the results in a single line:

```
1    # Use the 'while' loop to add all integers between two values
2    # provided by the user
3
4    currentInteger = int(input('Enter the starting integer:'))
5    endingInteger = int(input('Enter the ending integer:'))
```

```
6    sumOfValues = 0
7
8    while (currentInteger <= endingInteger):
9        print('currentInteger value is', currentInteger, end = '')
10       sumOfValues += currentInteger
11       currentInteger += 1
12       print(' and sumOfValues currently is', sumOfValues)
```

Output 2.9.1.c:

```
Enter the starting integer:1
Enter the ending integer:5
currentInteger value is 1 and sumOfValues currently is 1
currentInteger value is 2 and sumOfValues currently is 3
currentInteger value is 3 and sumOfValues currently is 6
currentInteger value is 4 and sumOfValues currently is 10
currentInteger value is 5 and sumOfValues currently is 15
```

In addition to the above, it is also possible to have an `if` structure of any type nested inside the `while` loop. The following code provides an example of a script that repeatedly accepts integers from the keyboard, and displays the integers plus a calculation of the even and odd numbers present. What is noteworthy in this script is the use of an `if...else` structure inside the `while` loop:

```
1    """ Use of the 'while' loop to count the number of even and
2    odd numbers from an input stream provided by the user.
3    Stop the loop and display the results when the user enters 0 """
4
5    # Declare the counters for even and odd numbers
6    countEven, countOdd = 0, 0
7
8    # Declare a variable to temporarily store current input value
9    userInput = int(input('Enter an integer, \
10   or 0 to display the results and exit: '))
11
12   # The 'while' loop that repeatedly executes the main block of code
13   while (userInput != 0):
14       if (userInput % 2 == 0):
15           countEven += 1
16       else:
17           countOdd += 1
18
19       # Repeatedly accept new input from the user until 0 is entered
20       userInput = int(input('Enter an integer, or 0 to display \
21           the results and exit: '))
22
23   # Display the results of the program
24   print('You entered', countEven,'even and', countOdd,'odd numbers')
```

Output 2.9.1.d:

```
Enter an integer, or 0 to display the results and exit: 2
Enter an integer, or 0 to display the results and exit: 3
Enter an integer, or 0 to display the results and exit: 4
Enter an integer, or 0 to display the results and exit: 5
Enter an integer, or 0 to display the results and exit: 6
Enter an integer, or 0 to display the results and exit: 0
You entered 3 even and 2 odd numbers
```

Programmers can also use a logically modified version of the while loop in place of the *do...until* (or *repeat...until*) loop, another classic programming language loop structure that is not directly available in Python. When using the while loop to replace the *do...until* functionality, the programmer should make sure that the while condition is True during the first iteration, and that its value is repeatedly updated at the end of the block of statements inside the loop.

2.9.2 THE for LOOP

The for loop structure allows for the execution of a block of statements for a *predefined* number of iterations. The loop controls the number of iterations using a counter (i.e., a variable declared locally in the loop), within a specific range defined by two numbers: *start* and *end*. The range can be also specified by just one *end* number, in which case the *start* will be considered to be 0 by default. Additionally, it is possible to include an incremental or decremental *step* inside the for header. Each repeated

Observation 2.20 – for Loop: Repeatedly executes a block of statements for a predefined number of times. The end of the loop must be defined, the start can be omitted, and the step can be specified in the header.

statement is placed within the block of statements, inside the for loop. The syntax for each of the three types of the for loop is provided below, while Figure 2.7 showcases the associated flowchart:

```
# Number of iterations is end-start
for counter in range (start, end):
    Block of statements

# Number of iterations is end and starts from 0
for counter in range (end):
    Block of statements

""" Number of iterations is (end-start)/step; counter increases/
decreases by step """
for counter in range (start, end, step):
    Block of statements
```

The next script showcases a script used to display the list of names stored in a *tuple*. The block of statements inside the for loop is executed four times with the i index starting at 0 and increasing up to 3 (inclusive):

```
1    # Declare a variable as a 'tuple' of immutable string elements
2    myFriends = ('John', 'Ali', 'Steven', 'Catherine')
3
4    # Use a 'for' loop to read the elements in the 'tuple', first to last
5    for i in range (0, 4):
6        print('Happy New Year:', myFriends[i])
7    print('Done.')
```

Output 2.9.2.a:

```
Happy New Year: John
Happy New Year: Ali
Happy New Year: Steven
Happy New Year: Catherine
Done.
```

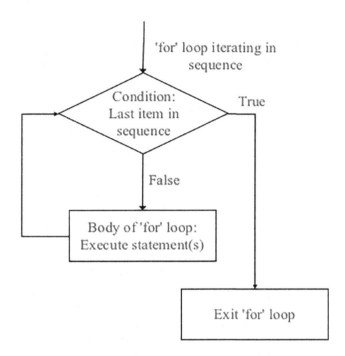

FIGURE 2.7   Flowchart of the for loop.

A similar example is provided in the following script, where instead of a *tuple* variable a *list* is used. The user is prompted to enter four names into the empty list, which are subsequently displayed on screen:

```
1    # Declare a 'list' variable that will accept names provided by the user
2    nameList = []
3
4    # Declare a 'dictionary' mapping numbers 1-4
5    # to text values 'first', 'second', 'third', 'fourth', respectively
6    numberToText = {
7        1: 'first',
8        2: 'second',
9        3: 'third',
10       4: 'fourth'
11   }
```

```
12
13  # Use 'for' loop to accept 4 names; store them in dictionary
14  for i in range (0, 4):
15      message = ('Enter the ' + str(numberToText.get(i + 1)) + \
16                      ' name to insert in the dictionary: ')
17      newName = input(message)
18      nameList.insert(i, newName)
19
20  # Use a 'for' loop to display the newly created name list
21  for i in range (4):
22      print(nameList[i])
23
24  print('Done.')
```

Output 2.9.2.b:

```
Enter the first name to insert in the dictionary: Hellen
Enter the second name to insert in the dictionary: Steven
Enter the third name to insert in the dictionary: Ahmed
Enter the fourth name to insert in the dictionary: Catherine
Hellen
Steven
Ahmed
Catherine
Done.
```

The reader should note the following:

- A *list* is declared using square brackets instead of the parentheses used for *tuples*. By leaving the square brackets empty, an empty list is created.
- Use a *dictionary mapping* to convert numeric values into the corresponding text (e.g., numberToText).
- Use the str() function to convert a numeric value into a string.
- Use the *concatenation* operator (+) to combine strings.
- Use the insert() function to populate the list. The first argument is the index of the new element and the second is the actual value.
- If the start number is omitted in the for loop header, zero is assumed as a default value.

2.9.3 THE NESTED for LOOP

As with if statements, it is possible to embed a for loop (i.e., *inner* loop) into another (i.e., *outer* loop) to create a nested for loop. This is particularly convenient when dealing with non-primitive data types of two or more dimensions, or with more complex problems. The syntax is provided below, and the associated flowchart is presented in Figure 2.8:

Observation 2.21 – Nested Loops: Use nested loops of any type to address complex situations like mathematical problems, drawing shapes, searching or shorting, or dealing with multi-dimensional non-primitive data types.

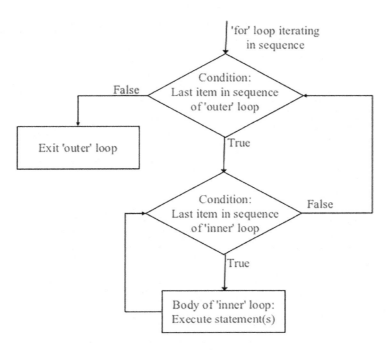

FIGURE 2.8 Flowchart of the nested for loop.

```
for counter1 in range (start1, end1):
    Block of statements 1
    ...
for counter2 in range (start2, end2):
    Block of statements 2
    ...
for counter3 in range (start3, end3):
    Block of statements 3
    ...
```

Nested loops are commonly used for the implementation of programs that deal with various types of non-primitive data types, such as *lists*, *tuples*, or *sets*. The following script provides an example of a nested for loop structure, in which a two-dimensional *list* variable (i.e., languages) is displayed on screen. This particular variable stores six different elements (i.e., names of programming languages) in two different dimensions (i.e., three elements on each dimension). The reader should note how the *counters* of the nested loops are used as *indices* for the displayed items of the list:

```
1   # Define a two-dimensional list with 3 programming languages
2   # as its elements (per dimension)
3   languages=[['Python','Java','C++'],['PhP','HTML','Java Script']]
4
5   # A nested 'for' loop prints the 2 different dimensions of the list
6   for i in range(2):
7       print(i, 'Set of programming languages:')
8       for j in range(3):
9           print('Happy new year:', languages[i][j])
10  print('All languages displayed')
```

Output 2.9.3.a:

```
0 Set of programming languages:
Happy new year: Python
Happy new year: Java
Happy new year: c++
1 Set of programming languages:
Happy new year: PhP
Happy new year: HTML
Happy new year: Java Script
All languages displayed
```

Another common use of nested loops relates to the implementation of various *sorting* or *searching algorithms* (see: Chapter 6). The following script provides another example of a nested `for` loop structure that implements a classic sorting algorithm referred to as the *Bubble Sort*. This script does the following:

- It declares two lists, one to accept the original list of integers and the other to store the sorted list.
- It runs a `for` loop that accepts a number of integers as input from the user and transfers them to the first list.
- It runs a second `for` loop that reads from the original list and transfers to into the second one (sorted list).
- It runs a nested `for` loop that utilizes the Bubble Sort algorithm.
- Finally, it runs two more `for` loops: one that displays the original list of integers and one that displays the sorted one.

It should be noted that the code presented in this script is not an example of the most efficient or complete sorting algorithm, but a more simplistic implementation of it, as the main purpose was to help the reader gain a better understanding of the use of nested loops:

```
1    originalList, sortedList = [], []
2
3    # The first 'for' loop accepts a number
4    # of integers and populate the 'originalList'
5    sizeOfList = int(input('Total number of integers in the list? '))
6    for i in range (sizeOfList):
7        tempValue = int(input('Add an integer to the list: '))
8        originalList.insert(i, tempValue)
9
10   # The second 'for' loop copies the 'originalList' into the
11   # 'sortedListed' in preparation for sorting the latter
12   for i in range (sizeOfList):
13       sortedList.insert(i, originalList[i])
14
15   # Use a nested 'for' loop to sort the 'originalList' into the
16   # 'sortedList' using the Bubble Sort algorithm
17   for i in range (sizeOfList - 1):
18       for j in range (sizeOfList):
19           if (sortedList[i] > sortedList [i + 1]):
```

```
20                    temp = sortedList[i]
21                    sortedList[i] = sortedList[i + 1]
22                    sortedList[i + 1] = temp
23
24  # Use two 'for' loops to successively display the two lists
25  print('The original list is: ', end = '')
26  for i in range (sizeOfList):
27      print(originalList[i], '', end = '')
28
29  print('\nThe sorted list is: ', end = '')
30  for i in range (sizeOfList):
31      print(sortedList[i], '', end = '')
```

Output 2.9.3.b:

```
Total number of integers in the list? 3
Add an integer to the list: 2
Add an integer to the list: 1
Add an integer to the list: 4
The original list is: 2 1 4
The sorted list is: 1 2 4
```

2.9.4 THE break AND continue STATEMENT

Another common use of nested loops is related to the implementation of algorithms for the solution of mathematical problems. The following script presents an implementation of a program calculating the prime numbers. In this particular case, the user is prompted to enter the last integer of the prime numbers list the program should calculate. Next, a for loop nested inside a while loop determines whether this integer is a prime number or not.

Observation 2.22 – break and continue: Use the break statement combined with a selection statement in a loop, to permanently interrupt loop execution. Use the continue statement combined with a selection statement in a loop to skip the current iteration.

The script introduces the break statement, which forces the interpreter to skip all the remaining statements and iterations, and exit the current iteration. As shown in the script, break is generally combined with a selection statement:

```
1   # Use a nested 'for' loop inside a 'while' loop to find primary numbers.
2   # Variable 'endInteger' stores the last integer of the sequence
3   endInteger = int(input('Enter the last integer \
4   of the sequence of primary numbers: '))
5
6   # Print default prime integers 1 and 2. This is subsequently followed
7   # by the rest of the sequence on the same line
8   print('1 2 ', end = '')
9
10  # The 'counter' variable is used to evaluate
11  # whether a number within the range is prime
12  counter, flag = 3, 'true'
13
```

```
14  # 'while' loop controls the counter variable used for evaluation
15  while (counter <= endInteger):
16
17      # 'for': check current 'counter' value against the integers
18      # in the list up to itself to determine if it is a prime number
19      for i in range (2, counter):
20          if ((counter % i) == 0):
21              flag = 'false'
22              break
23      if (flag == 'true'):
24          print(counter, '', end = '')
25      flag = 'true'
26      counter += 1
```

Output 2.9.4.a:

```
Enter the last integer of the sequence of primary numbers: 100
 1 2 3 5 7 11 13 17 19 23 29 31 37 41 43 47 53 59 61 67 71 73 79 83 89 97
```

The following example provides a more direct demonstration of how the break statement is used. The code instructs the interpreter to read from a non-primitive data type list, but *breaks* just after reading its first element:

```
1   # Declare variable 'myFriends' and populate with a list of names
2   myFriends = ('Ahmed', 'John', 'Emma', 'Hind')
3
4   # Use a 'for' loop to read the elements of the list
5   for i in range (4):
6       # Use an 'if' statement to stop reading the list once
7       # the second element (i.e., index 1) is reached
8       if (i == 1):
9           break
10      print('Happy new year:', myFriends[i])
11
12  print('Done')
```

Output 2.9.4.b:

```
Happy new year: Ahmed
Done
```

Another statement that is commonly used in loops, and particularly in nested loops, is the continue statement. It is used when there is a need to skip one or more particular iterations, and continue with the rest of the program. It is worth noting that this statement is frequently combined with *selection* statements. The main difference between the continue and the break statements is that the former stops the active iteration without completely interrupting the loop. The following script demonstrates the use of the continue statement:

```
1   # Declare variable 'myFriends' and populate with a list of names
2   myFriends = ('Ahmed', 'John', 'Emma', 'Rania')
3
4   # Use a 'for' loop to read the elements of the list
5   for i in range (4):
6       # Use an 'if' statement to skip the second element
7       # (i.e., the element with index 1)
8       if (i == 1):
9           continue
10      print('Happy new year:', myFriends[i])
11
12  print('Done.')
```

Output 2.9.4.c:

```
Happy new year: Ahmed
Happy new year: Emma
Happy new year: Rania
Done.
```

2.9.5   USING LOOPS WITH THE TURTLE LIBRARY

In addition to a multitude of other uses, loops are also convenient when using code for drawing shapes. Among the most important programming tools for such tasks is the Turtle library. The following script provides an example of how to draw a basic shape of four squares (100 pixels in length). The reader should note the use of the forward(length) function of the t object (turtle class), which draws a straight line of 100 pixels. Next, the script uses the left(degrees) function on the t object to turn the drawing pen 90 degrees left and repeat the 100-pixel drawing. At the end of the script it is necessary to use the mainloop() function on the t object to ensure that the drawing process is completed promptly. The output of this example shows the four squares drawn as a result of the for loop:

```
1   # Import the 'turtle' library
2   import turtle as t
3
4   # Use a 'for' loop to draw 4 squares with sides of 100 pixels
5   for i in range (4):
6       t.forward(100)
7       t.left(90)
8       t.forward(100)
9       t.left(90)
10      t.forward(100)
11      t.left(90)
12      t.forward(100)
13
14  # Use the mainloop() function of the 'turtle' class
15  t.mainloop()
```

Output 2.9.5.a:

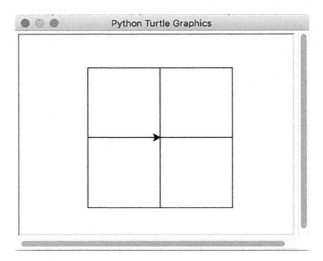

Nested loops can be also used with Turtle to draw more complex shapes. The following script demonstrates this by building on the previous example and forcing the drawing process to be repeated three more times with the use of a nested loop. In each repetition, the rectangular shape is rotated by 30 degrees to the left:

```
1    # Import the 'turtle' library
2    import turtle as t
3
4    # Nested 'for' to draw a complex of squares with sides of 100 pixels
5    for i in range (3):
6        for j in range (4):
7            t.forward(100)
8            t.left(90)
9            t.forward(100)
10           t.left(90)
11           t.forward(100)
12           t.left(90)
13           t.forward(100)
14       t.left(30)
15
16   # Use the mainloop() function of the 'turtle' class
17   t.mainloop()
```

Output 2.9.5.b:

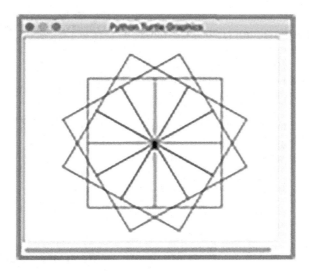

The Turtle library comes with a rich set of functions that support a large variety of drawing tasks. Table 2.9 provides a sample based on this set, including some of the most important of its functions.

TABLE 2.9
Methods Available in the Turtle Class

Method or Command	Required Parameters	Description
forward	Length in pixels	Moves the Turtle pen forward by the specified amount
backward	Length in pixels	Moves the Turtle pen backward by the specified amount
right	Angle in degrees	Turns the Turtle pen a number of degrees clockwise
left	Angle in degrees	Turns the Turtle pen a number of degrees counter-clockwise
penup	None	Picks up the Turtle pen
pendown	None	Puts down the Turtle pen to start drawing
pensize	Thickness of pen	The thickness of the Turtle pen
color, pencolor	Color name	Changes the color of the Turtle pen
fillcolor	Color name	Changes the fill color for the drawing
begin_fill, end_fill	None	Defines the start and the end of the application of the fillcolor() method
setposition	None	Set the current position
goto	x, y coordinates	Moves the Turtle pen to coordinate position x, y
shape	Shape name	Can accept values 'arrow', 'classic', 'turtle', or 'circle'.
speed	Time delay	Dictates the speed of the Turtle pen (i.e., slow (0) to fast (10+)).
circle	Radius, arc, steps	Draws a circle counter-clockwise with a pre-set *radius*. If *arc* is used, it will draw an arc from 0 up to a given number in degrees. If *steps* is used, it will draw the shape in pieces resembling a *polygon*.

2.10 FUNCTIONS

A *function* is a block of statements that performs a specific task. It allows the programmer to reuse parts of their code, promoting the concept of *modularity*. The main idea behind this approach is to divide a large block of code into smaller, and thus more manageable, sub-blocks. There are two types of functions in Python:

> **Observation 2.23 – Function:** A defined structure of statements that can be called repeatedly. It has a unique name, and may take arguments and/or return values to the caller.

- **Built-in:** The programmer can use these functions in the program without defining them. Several functions of this type were used in the previous sections (e.g., `print()` and `input()`).
- **User-defined:** Python allows programmers to create their own functions. The following section focuses on this particular function type.

2.10.1 FUNCTION DEFINITION

The main rules for defining functions in Python are the following:

> **Observation 2.24 – Four Types of Functions:**
>
> 1. No arguments, no return value.
> 2. With arguments, no return value.
> 3. No arguments, with return value.
> 4. With arguments, with return value.

- The function block begins with the keyword `def`, followed by the function name and parentheses. Note that, as Python is case-sensitive, the programmer must use `def` instead of `Def`.
- Similar to variable names, function names can include letters or numbers, but no spaces or special characters, and cannot begin with a number.
- Optional input parameters, called arguments, should be placed within the parentheses. It is also possible to define the parameters inside the parentheses.
- The block of statements within a function starts with a colon and is indented.
- A function that *returns* data must include the keyword `return` in its block of code.

The syntax for a function declaration is as follows:

```
def functionName (var1, var2, … etc.):
        Statements
```

Depending on the presence or absence of arguments, and on the presence of input and/or return values, functions can be classified under four possible types. These types are presented in detail in the following section.

2.10.2 NO ARGUMENTS, NO RETURN

This is a type in which the function does not accept variables as arguments, and does not return any data. This is demonstrated in the following script that merely prints a predefined string on screen. The reader should note that there are no arguments inside the parameters and no `return` statement inside the block of statements. The structure simply invokes the `print()` function displaying the desired message. Invoking such a function inside the main program is a rather simple and straightforward task:

```
1   # Define function that neither accepts arguments nor returns values
2   def printSomething():
3       print('Hello world')
4
5   # Call the function from the main program
6   printSomething()
```

Output 2.10.2:

```
Hello world
```

2.10.3   With Arguments, No Return

Another type of a function is one in which the function accepts variables as arguments, but does not return any data. In the following script, the function is invoked by declaring its name while also including a number of values in the parentheses. These values are *passed* to the main body of the function, and can be treated as normal variables:

```
1   # Define a function that accepts arguments but does not return values
2   def printMyName(fName, lName):
3       print('Your name is:', fName, lName)
4
5   # Prompt user to input their name
6   firstName = input('Enter your first name: ')
7   lastName = input('Enter your last name: ')
8
9   # Call the function from the main program
10  printMyName(firstName, lastName)
```

Output 2.10.3:

```
Enter your first name: Alex
Enter your last name: Fora
Your name is: Alex Fora
```

2.10.4   No Arguments, With Return

The third type involves a function that does not accept arguments, but returns data. It is important to remember that since this type of function returns a value to the calling code, this value must be assigned to a variable before being used or processed:

```
1   # Define a function that does not accept arguments but returns values
2   def returnFloatNumber():
3       inputFloat = float(input('Enter a real number ' \
4                                'to return to the main program: '))
5       return inputFloat
6
7   # Call the function from the main program to display the input
8   x = returnFloatNumber()
9   print('You entered:', x)
```

Output 2.10.4:

```
Enter a real number to return to the main program: 5.7
You entered: 5.7
```

2.10.5 WITH ARGUMENTS, WITH RETURN

The fourth type involves a function that both accepts arguments and returns values back to the calling code. The following script demonstrates this. In this case, the call of the function must include a list of arguments and assign the return value to a specific variable for later use:

```
1   # Function accepts arguments & returns values to the caller
2   def calculateSum(number1, number2):
3       print('Calculate the sum of the two numbers.')
4       return(number1 + number2)
5
6   # Accept two real numbers from the user
7   num1 = float(input('Enter the first number: '))
8   num2 = float(input('Enter the second number: '))
9
10  # Call the function to calculate the sum for the two numbers
11  addNumbers = calculateSum(num1, num2)
12
13  # Print the sum for the numbers
14  print('The sum for the two numbers is:', addNumbers)
```

Output 2.10.5:

```
Enter the first number: 3
Enter the second number: 5
Calculate the sum of the two numbers.
The sum for the two numbers is: 8.0
```

2.10.6 FUNCTION PARAMETER PASSING

There are two different ways to pass parameters to functions. Determining which of the two should be chosen depends on whether the value of the *original variables* should be changed within the function or not. These two ways for passing parameter values to a function are commonly referred to as *call/pass by value* and *call/pass by reference*.

2.10.6.1 Call/Pass by Value

In this case, the value of the argument (parameter) is processed as a *copy* of the original variable. Hence, the original variable in the caller's scope will be unchanged when program control returns to the caller. In Python, if *immutable* parameters (e.g., integers and strings) are passed to a function, the common practice is to call/pass parameters by value. The example below illustrates such a case by introducing the id() function. It accepts an object as a parameter (i.e., id(object)) and returns the identity of this particular object. The return value of

---

Observation 2.25 – Passing Values to Argument:

1. **By Value:** Argument is a copy of the original variable, which remains unchanged.
2. **By Reference:** Changes apply directly to the original variable, thus, changing its value.

`id()` is an integer, which is unique and permanent for this object during its lifetime. As shown in the example, the id of variable x before calling the `checkParamemterID` function is 4564813232. It should be noted the id of x is not changed within the function as long as the value of x is not updated. However, once the value is updated to 20, its corresponding id is changed to 4564813552. The most important thing to note is that the id of x does not change after calling the function, and its original value is maintained (4564813232). That means that the change of the value of x was applied on a *copy* of the variable, and not the original one within the caller's scope:

```
1    # Define function 'checkParameterID' that accepts a parameter (by value)
2    def checkParameterID(x):
3        print('The value of x inside checkParameterID',\
4        'before value change is', x, '\nand its id is', id(x))
5
6        # Change the value of parameter 'x' within the scope of the function
7        x = 20
8        print('The value of x inside checkParameterID',\
9        'after value change is', x, '\nand its id is', id(x))
10
11   # Declare variable 'x' in the main program and assign initial value
12   x = 10
13
14   print('The value of x before calling the function ',\
15        'checkParameterID is', x, '\nand its id is', id(x))
16
17   # Call function 'checkParameterID'
18   checkParameterID(x)
19
20   # Display info about 'x' in the main program after function call
21   print('The value of x after calling the function checkParameterID '\
22        'is', x, '\nand its id is', id(x))
```

Output 2.10.6.a:

```
The value of x before calling the method checkParameterID is 10
and its id is 140715021772880
The value of x inside checkParameterID before value change is 10
and its id is 140715021772880
The value of x inside checkParameterID after value change is 20
and its id is 140715021773200
The value of x after calling the method checkParameterID is 10
and its id is 140715021772880
```

2.10.6.2 Call/Pass by Reference

In this case, the function gets a *reference* to the argument (i.e., the original variable) rather than a copy of it. The value of the original variable in the caller's scope will be modified if a change occurs within the function. In Python, if mutable parameters (e.g., a *list*) are passed to a function, the call/pass is by reference. As shown below, `updateList` appends a value of 5 to the list named y. The fact that the value of the original mutable variable x changes demonstrates the functionality of argument call/pass by reference:

```
1    # Define function 'upDateList' that changes values within the list
2    def updateList(y):
3        y = y.append(5)
4        return y
5
6    # Declare list 'x' with 4 elements and assign values
7    x = [1, 2, 3, 4]
8    print('The content of x before calling the function updateList is:', x)
9
10   # Call function 'updateList'
11   print('Call the function updateList')
12   updateList(x)
13   print('The content of x after calling the function updateList is:', x)
```

Output 2.10.6.b:

```
The content of x before calling the method updateList is: [1, 2, 3, 4]
Call the method updateList
The content of x after calling the method updateList is: [1, 2, 3, 4, 5
```

2.11 CASE STUDY

Write a Python application that displays the following menu and runs the associated `functions` based on the user's input:

- Body mass index calculator.
- Check customer credit.
- Check a five-digit for palindrome.
- Convert an integer to the binary system.
- Initialize a list of integers and sort it.
- Exit.

Specifics on the components of the application:

- **Body Mass Index Calculator:** Read the user's weight in kilos and height in meters, and calculate and display the user's body mass index. The formula is: $BMI = (weightKilos)/(heightMeters \times heightMeters)$. If the BMI value is less than 18.5, display the message "Underweight: less than 18.5". If it is between 18.5 and 24.9, display the message "Normal: between 18.5 and 24.9". If it is between 25 and 29.9, display the message "Overweight: between 25 and 29.9". Finally, if it is more than 30, display the message "Obese: 30 or greater".
- **Check Department-Store Customer Balance:** Determine if a department-store customer has exceeded the credit limit on a charge account. For each customer, the following facts are to be entered by the user:
  - Account number.
  - Balance at the beginning of the month.
  - Total of all items charged by the customer this month.
  - Total of all credits applied to the customer's account this month.
  - Allowed credit limit.

The program should accept input for each of the above from as integers, calculate the new balance (= beginning balance + charges − deposits), display the new balance, and determine if the new balance exceeds the customer's credit limit. For customers whose credit limit is exceeded, the program should display the message "Credit limit exceeded".

- A palindrome is a number or a text phrase that reads the same backward as forward (e.g., 12321, 55555). Write an application that reads a five-digit integer and determines whether or not it is a palindrome. If the number is not five digits long, display an error message indicating the issue to the user. When the user dismisses the error dialog, allow them to enter a new value.
- **Convert Decimal to Binary:** Accept an integer between 0 and 99 and print its binary equivalent. Use the modulus and division operations, as necessary.
- **List Manipulation and Bubble Sort:** Write a script that does the following:
  a. Initialize a list of integers of a maximum size, where the maximum value is entered by the user.
  b. Prompt the user to select between automatic or manual entry of integers to the list.
  c. Fill the list with values either automatically or manually, depending on the user's selection.
  d. Sort the list using Bubble Sort.
  e. Display the list if it has less than 100 elements.

The above should be implemented using a single Python script. Avoid adding statements in the main body of the script unless necessary. Try to use `functions` to run the various tasks of the application. Have the application/menu run continuously until the user enters the value associated with exiting.

2.12   EXERCISES

2.12.1   SEQUENCE AND SELECTION

1. Write a script that displays numbers 1–4 on the same line and in one output, separated by one space.
2. Write a script that accepts three integers and calculates and displays their sum, average, product, lowest, and highest.
3. Write a script that accepts five integers and prints how many of them are odd and even. (Hint: An even number leaves a remainder of zero when divided by 2. Use the modulus operator.)
4. Write a script that accepts five numbers and calculates and prints the number of negatives, positives, and zeros.
5. Write a script that accepts two integers and determines and prints whether the first is a multiple of the second.
6 Write a script that accepts one number consisting of five digits, separates the number into the individual digits, and prints each digit separated by three spaces from each other. (Hint: use both division and modulus operations to break down the number.)
7. Write a script that accepts the radius of a circle as an integer and prints the circle's diameter, circumference, and area. (Hint: Use the constant value *3.1459* for π. Calculate the diameter as *radius*2*, the circumference as *2π*radius*, and the area as *π*radius²*.)
8. Write a script that accepts the first and the last name from the user as two separate inputs, concatenates them separated by one space character, and displays the result.
9. Write a script that accepts a character and displays it in the ASCII format. (Hint: use the `ord()` function.)
10. Write a script that accepts an ASCII value between 50 and 255 and displays its character. (Hint: use the `chr()` function.)

2.12.2 ITERATIONS – `while` LOOPS

1. Drivers are concerned with the accumulated mileage of their automobiles. One particular driver has been monitoring trips by recording miles driven and petrol gallons used. Write a script that uses a `while` statement to accept the miles and petrol gallons used for each trip. The script should calculate and display the miles per gallon obtained for each trip, and the combined, total miles per gallon obtained up to date.
2. Write a script that accepts integers within the range of 1–30. For each number entry, the script should print a line containing adjacent asterisks of the same number (e.g., for number 7 it should display: "7: *******"). The script should run until the user enters a predefined exit value.
3. A company pays its employees partially based on commissions. The employees receive $200 per week, plus 9% of their gross sales for the week. Write a script that accepts the items sold for a week by a single employee and calculates and displays their earnings. There is no limit to the number of items that can be sold by an employee.
4. Write a script that uses a `while` statement to determine and print the largest number entered by the user. The user is allowed to enter numbers until a predefined exit value is entered.
5. Write a script that uses a `while` statement and the tab escape sequence (\t) to print the tabular form of: a number, its multiple by 2, its multiple by 10, the square, and its cube number.
6. *Armstrong* numbers represent the sum of their digits to the power of the total number of digits. Therefore, for a three-digit Armstrong number, the sum of the cube roots of each digit should equal to the number itself (e.g., 153 = 1 ^ 3 + 5 ^ 3 + 3 ^ 3 = 1 + 125 + 27 = 153). Based on the above, write a script that displays all three-digit Armstrong numbers between 130 and 140, as well as their breakdown.
7. The factorial of a non-negative integer is written as *n!* and is defined as *n! = n*(n–1)*(n–2)*...*1* for values of n greater than or equal to 1, and as *n! = 1* for *n = 0*. Write a script that accepts a non-negative integer and computes and prints its factorial.
8. Write a script that converts Celsius temperatures to Fahrenheit. The program should print a table displaying all the Celsius temperatures and their Fahrenheit equivalents. (Hint: the formula for the conversion is: *F = 9/5C + 32*.)
9. A company wants to send data over the Internet and has requested a script that will encrypt this data. The desired encryption function is the following: each digit should be replaced by a value calculated by adding 7 to it and getting the remainder after dividing the new value by 10. Next, the first digit should be swapped with the third and the second with the fourth. The program should print the resulting encrypted integer.
10. Write a script that reads an encrypted four-digit integer, decrypts it by reversing the encryption scheme of the previous exercise, and prints the result.

2.12.3 ITERATIONS – `for` LOOPS

1. Write a script that uses a `for` statement to display the following patterns:

(a)	(b)	(c)	(d)
*	*********	*********	*
**	********	********	**
***	*******	*******	***
****	******	******	****
*****	*****	*****	*****
******	****	****	******
*******	***	***	*******
********	**	**	********

2. Write a script that prompts the user to enter a number of integer values and calculate their average. Use a `for` statement to receive and add up to the sequence of integers, based on user input.

3. A mail order house sells five different products with the following codes and retail prices: *001 = $2.98, 002 = $4.50, 003 = $9.98, 004 = $4.49,* and *005 = $6.87.* Write a script that accepts the following two values from the user: *product number* and *quantity sold.* This process must be repeated as long as the user enters a valid code. The script should use a mapping technique to determine the retail price for each product. Finally, the script should calculate and display the total value of all products sold.

2.12.4 METHODS

1. Write a script that uses methods to do the following: (a) continuously accept integers into a two-dimensional list of integers until the user enters an exit value (e.g., 0), (b) find and display the min value for each row and/or column of the list and of the whole list, (c) find and display the max value for each row and/or column of the list and of the whole list, and (d) find and display the average value for each row and/or column of the list and of the whole list.

2. Write a script that uses methods to continuously accept the following details for a series of books: ISBN number, title, author, publication date, and publication company. The details of each book must be stored in five lists associated with the book information categories. The script should accept books until the user enters an ISBN number of 0. Before exiting, the script must print the details of the books.

3. Write a script that uses different methods to print a box, an oval, an arrow, and a diamond on screen. Use the Turtle library for this purpose.

4. Using the Olympic Games logo as a reference, write a Python script that uses the Turtle library and appropriate methods to draw the logo rings, matching the color order and position.

5. Using only the Turtle library methods `fillcolor()`, `begin _ color()`, `end _ color()`, `color()`, `penup()`, `pendown()`, and `goto()`, write a Python script that uses various methods to draw Figure Exercise 5.

6. Write a Python script that uses appropriate methods and the Turtle library to draw a regular polygon of N sides. The script should use a method to prompt the user to enter the number of sides (N). (Hint: a regular polygon of N sides is the combination of N equilateral triangles.) The figure drawn should look like Figure Exercise 6.

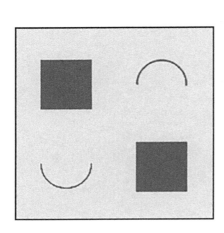

Figure Exercise 5.

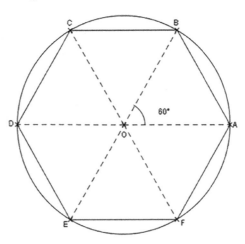

Figure Exercise 6.

REFERENCES

Dijkstra, E. W., Dijkstra, E. W., Dijkstra, E. W., & Dijkstra, E. W. (1976). *A Discipline of Programming* (Vol. 613924118). New Jersey: Prentice-Hall Englewood Cliffs.

Jaiswal, S. (2017). *Python Data Structures Tutorial*. DataCamp. https://www.datacamp.com/community/tutorials/data-structures-python.

Knuth, D. E. (1997). *The Art of Computer Programming* (Vol. 3). Pearson Education.

Stroustrup, B. (2013). *The C++ Programming Language*. India: Pearson Education.

3 Object-Oriented Programming in Python

Ghazala Bilquise and Thaeer Kobbaey
Higher Colleges of Technology

Ourania K. Xanthidou
Brunel University London

CONTENTS

3.1 Introduction 60
3.2 Classes and Objects in Python.. 62
    3.2.1 Instantiating Objects 63
    3.2.2 Object Data (Attributes).. 63
        3.2.2.1 Instance Attributes 63
        3.2.2.2 Class Attributes... 64
    3.2.3 Object Behavior (Methods).. 66
        3.2.3.1 Instance Methods 66
        3.2.3.2 Constructor Methods 68
        3.2.3.3 Destructor Method 71
3.3 Encapsulation 72
    3.3.1 Access Modifiers in Python... 72
    3.3.2 Getters and Setters... 72
    3.3.3 Validating Inputs before Setting 73
    3.3.4 Creating Read-Only Attributes.. 75
    3.3.5 The property() Method 76
    3.3.6 The `@property` Decorator 77
3.4 Inheritance.. 78
    3.4.1 Inheritance in Python 78
        3.4.1.1 Customizing the Sub Class 79
    3.4.2 Method Overriding 81
        3.4.2.1 Overriding the Constructor Method 82
    3.4.3 Multiple Inheritance 83
3.5 Polymorphism – Method Overloading 85
    3.5.1 Method Overloading through Optional Parameters in Python 86
3.6 Overloading Operators 87
    3.6.1 Overloading Built-In Methods... 90
3.7 Abstract Classes and Interfaces in Python 91
    3.7.1 Interfaces 94
3.8 Modules and Packages in Python 94
    3.8.1 The import Statement 95
    3.8.2 The from...import Statement.. 95
    3.8.3 Packages.. 96
    3.8.4 Using Modules to Store Abstract Classes.. 97
3.9 Exception Handling 98

DOI: 10.1201/9781003139010-3

    3.9.1   Handling Exceptions in Python...98
            3.9.1.1   Handling Specific Exceptions...100
    3.9.2   Raising Exceptions 101
    3.9.3   User-Defined Exceptions in Python..102
3.10 Case Study 103
3.11 Exercises 104

3.1   INTRODUCTION

The *Object-Oriented Programming* (*OOP*) paradigm is a powerful approach that involves problem solving by means of programming components called classes, and the associated programming objects contained in these classes. This approach aims at the creation of an environment that reflects method structures from the real world. Within the OOP paradigm, variables, and the associated data and methods (see: Chapter 2), are logically grouped into reusable objects belonging to a parent class. This enables a modular approach to programming. Some of the most significant benefits of developing software using this paradigm is that it is easier to implement, interpret, and maintain.

OOP is developed around two fundamental pillars of programming, and four basic principles of how these could be used efficiently. The two pillars are the *class* and its *objects*. The four principles are the concepts of *encapsulation*, *abstraction*, *inheritance*, and *polymorphism*. Although it is true that various other programming techniques and approaches are also applied within the OOP paradigm, they all share the above core components and concepts.

A real-life analogy that demonstrates the class and object relationship is that of a recipe of a cake. The recipe provides information about the ingredients and the method of how to bake it. Using the recipe, several cakes may be baked. In this context, the recipe represents the class, and each cake that is baked using the recipe represents the object. Similarly, in software development, if it is required to store the data of numerous employees, a class that describes the general specifications of an employee is created. This class defines what *types* of data are required for employees (*class properties*) and what *actions* can be performed on the data (*class methods*). New employees are then created using the class. What is important to note is that the class does not hold any data. It is simply a template used as a model for the container of employees of the same kind, alongside any related actions that can be performed on the data. The relation between these two fundamental elements (i.e., class and objects) is illustrated in Figure 3.1.

In OOP terminology, the process of creating an object based on a specific class is known as *instantiation*. During instantiation, the created object *inherits* the properties described in the class. For example, an object named `car1` may have properties like `make`, `model`, and `color`, while

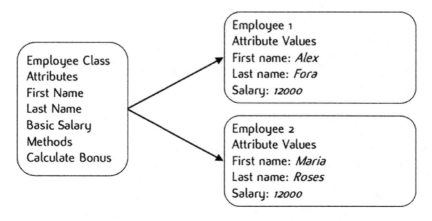

FIGURE 3.1   Using class Employee to generate the objects Employee1 and Employee2.

book1 may have ISBN, title, price and publication _ year. Similarly, the *methods* of the object are the actions or tasks it can perform. Using the same object examples, a car may perform actions like startEngine(), stopEngine() and moveCar(), and a book update-Price() and calculateDiscount().

In terms of communicating complex OOP structures and ideas, programmers use the *Unified Modelling Language* (*UML*), a tool that allows them to draw standardized diagrams that visualize the structure of programs independently of the programming language used for the implementation. The basic building block of UML is the *class diagram*, a graphical representation of a class as a rectangle with three sections, namely the class name, the class attributes, and the class methods. The basic structure of a class diagram is illustrated in Figure 3.2, and a related example is provided in Figure 3.3.

The top section of the class diagram contains the class name, which should adhere to the following naming conventions:

- It must be a noun.
- It must be written in singular form.
- It must start with an upper-case letter (upper camel case should be used for multiple words in the class name).

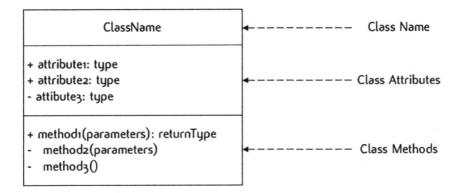

FIGURE 3.2 Syntax of a class diagram.

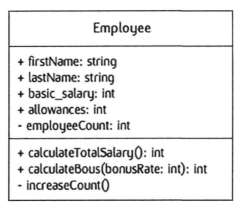

FIGURE 3.3 A simple class with its attributes and methods.

The middle section of the class diagram consists of the class *attributes*. These should be written using lowercase letters, with compound words separated by an underscore. Optionally, the data type of each attribute can be specified after its name, separated by a colon.

> **Observation 3.1 – Camel Case:** The practice of starting each word of a sentence in capital.

The last section of the class diagram contains the *operations* or *methods* of the class. Method names should be verbs and follow the *lower camel case* naming convention (i.e., the first word is in lower case and the first letters of all subsequent words are in upper case). Similar to attributes, the input and output parameters of the method can be specified. The input parameters are written within the parentheses following the method name. The output parameters are specified at the end of the method, separated by a colon.

Finally, *access modifiers*, represented with a *plus* or *minus* symbol, are used to specify the *scope of access* of an attribute or method. The plus symbol indicates that the attribute or method is *public* and can be accessed by any object of any class outside the current one, whereas the minus symbol indicates that the method or attribute is *private* and can only be accessed from within the current class or its objects.

This chapter covers basic concepts related to the usage of classes and objects, and the four main principles of OOP, namely:

- **Encapsulation:** The process of wrapping the attributes and methods of the objects of a class in one unit, and managing the access to these attributes and methods.
- **Abstraction:** The technique used to hide the implementation details of a class, by providing a more abstract view. This allows for the development of a simpler interface, by focusing on what the object does rather than how it does it.
- **Inheritance:** The mechanism used for the creation of a parent-child relationship between classes, where the child (or *sub*) class acquires the attributes and the methods of the parent (or *super*) class, thus, eliminating redundant code and facilitating reusability and maintainability.
- **Polymorphism:** A feature of OOP languages that enables methods to perform different tasks based on the context of the variables used. This is achieved through designated processes like method overriding and overloading.

3.2 CLASSES AND OBJECTS IN PYTHON

Contextualizing the concepts of classes and methods and their relationship is frequently easier through the use of working examples. Consider the common case of developing a simple application that must store employees' data. Every employee is likely to have an *employee ID*, a *first name*, a *last name*, a *basic salary*, and *allowances*. The first step toward the implementation of such an application in OOP would be to define a class that holds the appropriate, general specification for all employees. This will be used as a blueprint to create a record for each employee in the application.

> **Observation 3.2 – `pass`:** The `pass` keyword is a line of code that does nothing. It is necessary when defining an empty class since it is required that every class has at least one line of code.

In Python, a class is created simply by using the `class` keyword followed by the name of the class. The name must follow the same naming rules that also apply to variables. However, for clarity purposes, it is recommended that the name of the class is capitalized using the *CapWords* notation (i.e., the first letter of each word in the class name should be capitalized).

> **Observation 3.3 – `class` keyword:** Create a class simply by using the `class` keyword followed by the name of the class. The class name must adhere to the naming conventions of Python for variables and should have the first letter in capital.

The example below creates an empty class with no attributes or methods, and thus no functionality:

```
1    # Define a class with no functionality
2    class Employee:
3          Pass
```

3.2.1 INSTANTIATING OBJECTS

To *instantiate* an object means to create a new object using a class as a template. An object is instantiated by passing the class name (followed by parentheses) to a variable. In the script example provided below emp1 and emp2 are instances of the Employee class. Note that, in the output of the script, each object reserves a different memory location, as the attributes of the two employees will be stored *separately*:

Observation 3.4 – Creating/ Instantiating Objects: An object is created by using the name of the class it belongs followed by parentheses.

```
1    # Define the class
2    class Employee:
3          Pass
4
5    # Create two instances/objects based on the class
6    emp1 = Employee()
7    emp2 = Employee()
8
9    # Print the memory address of instances 'emp1' and 'emp2'
10   print(emp1)
11   print(emp2)
```

Output 3.2.1:

```
<__main__.Employee object at 0x0000026242C487F0>
<__main__.Employee object at 0x0000026242C483D0>
```

3.2.2 OBJECT DATA (ATTRIBUTES)

Object data, also known as *attributes*, are stored in variables. There are two types of attributes in a class, namely *instance* and *class* attributes.

Observation 3.5 – Object Data (Attributes): Data that is associated with each instantiated object and is unique to that object. Use the dot notation syntax to call it (e.g., obj. attribute = value).

3.2.2.1 Instance Attributes

An *instance attribute* contains data associated with each instantiated object, and is therefore unique to that object. Instance attributes are created using the *dot notation* syntax (obj.attribute = value) and are only accessible by the object associated with them. In the example below, class Employee is used to instantiate objects emp1 and emp2. These objects will store the first and last names, the basic salary, and the allowance of two different employees.

The reader should note the use of the dot notation to assign values to the instance/object attributes, and how the `print()` method is used to show the first and last names of the two `Employee` instances/objects:

```
1   # Define the class
2   class Employee:
3         Pass
4
5   # Create two instances/objects based on the class
6   emp1 = Employee()
7   emp2 = Employee()
8
9   # Provide attributes and assign values to the instances
10  emp1.firstName = "Maria"
11  emp1.lastName = "Rena"
12  emp1.basicSalary = 12000
13  emp1.allowance = 5000
14  emp2.firstName = "Alex"
15  emp2.lastName = "Flora"
16  emp2.basicSalary = 15000
17  emp2.allowance = 5000
18
19  # Print the objects and their attributes
20  print(emp1.firstName, emp1.lastName)
21  print(emp2.firstName, emp2.lastName)
```

Output 3.2.2.1:

```
Maria Rena
Alex Flora
```

3.2.2.2 Class Attributes

While *instance* attributes are specific to each individual object, *class* attributes belong to the class itself, and are thus shared among all instances of the class. In the following example, the class attribute `bonusPercent` is defined within the scope of the `Employee` class. Unlike instance attributes `firstName` and `lastName`, which take unique values for each of the two employees (i.e., emp1 and emp2), class attribute `bonusPercent` is common to both employees:

Observation 3.6 – Class Attribute: Data that belongs to the class and has its values shared among each object instantiated through the class. Define it the same way as a simple variable.

Observation 3.7: It is recommended to use lower-case letters when naming attributes. If an attribute name has more than one word, use lower case for the first word and capital first letters for the rest, all combined in one word.

```
1   class Employee:
2   # Define the class attribute
3         bonusPercent = 0.2
4
5   # Define and create the 'emp1' instance
6   emp1 = Employee()
7   emp1.firstName = "Maria"
8   emp1.lastName = "Rena"
9
10  # Define and create the 'emp2' instance
11  emp2 = Employee()
12  emp2.firstName = "Alex"
13  emp2.lastName = "Flora"
14
15  # Print class attribute
16  print(Employee.bonusPercent)
17  # Each instance is associated with the same class attribute value
18  print(emp1.firstName, emp1.lastName, emp1.bonusPercent)
19  print(emp2.firstName, emp2.lastName, emp2.bonusPercent)
20
21  # Accessing the class attribute by using the class name
22  Employee.bonusPercent = 0.3
23  print(Employee.bonusPercent)
24  # Accessing the class attribute by using the instance name
25  print(emp1.bonusPercent)
26  print(emp2.bonusPercent)
27
28  # Accessing the dictionary of the class and its objects
29  print(emp1.__dict__)
30  print(emp2.__dict__)
31  print(Employee.__dict__)
```

Output 3.2.2.2:

```
0.2
Maria Rena 0.2
Alex Flora 0.2
0.3
0.3
0.3
{'firstName': 'Maria', 'lastName': 'Rena'}
{'firstName': 'Alex', 'lastName': 'Flora'}
{'__module__': '__main__', 'bonusPercent': 0.3, '__dict__': <attribute '__dict__' of 'Employe
e' objects>, '__weakref__': <attribute '__weakref__' of 'Employee' objects>, '__doc__': None}
```

In terms of declaration and value assignments, a class attribute is treated as any other regular variable within the class, in contrast to instance attributes where the dot notation is used. It is accessed by using the name of the class to which it belongs followed by the attribute name:

```
<className>.<attribute_name> = value
```

When a class attribute is associated with an instantiated object name, Python firstly checks if that attribute is available in that particular object, and if not, whether it is available in the associated class or any super class the object inherits from (see Section: 3.4.1 Inheritance in Python).

There is a simple way to determine whether an attribute belongs to an object or to the class used to instantiate it. Every Python object contains a special attribute called __dict__ (i.e., *dictionary*), which includes references to all the attributes within this object. Using the previous example, if __dict__ is called for emp1 and

> **Observation 3.8:** Call the __dict__ attribute on any object to find the attributes that belong to that particular object.

emp2 it will not include the bonusPercent class attribute. On the contrary, this will be the case if it is called for the Employee class.

3.2.3 OBJECT BEHAVIOR (METHODS)

A *method* is a structured block of code that is associated with an object. It is defined in a class and contains code that performs specific tasks using data from either the class itself or the instantiated objects inheriting from the class. Methods must have a distinct name, and may or may not take parameters or return values. All methods in a class must include an essential parameter, usually named self, that references the current object instance. It is important to note that self is not a reserved word. Any variable name may be used to reference the object, as long as it follows the Python variable naming rules.

3.2.3.1 Instance Methods

An *instance* method, just like an instance attribute, is specific to a particular object rather than the class used to instantiate it. It is, thus, invoked for each separate object, and uses the data of the object that invoked it. Instance methods are defined within a class and include the mandatory self parameter. However, passing the

> **Observation 3.9 – *Instance Method*:** Defined as any other method but includes the *self* parameter as one of its arguments.

self parameter to the method is not required when calling the method.

In the following Python example, instance method printDetails(self) is defined in the Employee class and called twice to print each of the two employees' data (i.e., firstName, lastName, and salary). It does not accept any arguments and it displays the required information utilizing the attributes of the particular object it is associated with. Instance method calculateBonus(self, bonusPercent) collects data from the attribute of the associated object, calculates the bonus for the employee, and displays the result. The reader should note that defining and calling instance and class methods is similar, with the exception of the use of dot notation to associate the instance method with the super class:

```
1    # Define the class
2    class Employee:
3
4          # Define the 'printDetails' method
5          def printDetails(self):
6                print("Employee Name", self.firstName, self.lastName,
7                      "earns", self.salary)
8
9          # Define the 'calculateBonus' method
10         def calculateBonus(self, bonusPercent):
11               return self.salary * bonusPercent
12   # Create the two objects and print their attributes
13   emp1 = Employee()
14   emp1.firstName = "Maria"
```

```
15  emp1.lastName = "Rena"
16  emp1.salary = 15000
17  emp1.printDetails()
18  print("Bonus amount is", emp1.calculateBonus(0.2))
19
20  emp2 = Employee()
21  emp2.firstName = "Alex"
22  emp2.lastName = "Flora"
23  emp2.salary = 18000
24  emp2.printDetails()
25  print("Bonus amount is", emp1.calculateBonus(0.2))
```

Output 3.2.3.1.a:

```
Employee Name Maria Rena earns 15000
Bonus amount is 3000.0
Employee Name Alex Flora earns 18000
Bonus amount is 3000.0
```

From a structural and logical viewpoint, class and instance methods can be used strategically to further improve the efficiency and clarity of the code. For instance, the class used in the previous examples can be further improved by introducing the following change. Since bonusPercent is the same for both employees, its value can be stored in a class attribute and be shared among all the instances of the class. In this case, calling the instance method is simplified, as it is no longer necessary to pass any parameters as method arguments. Instead, instance or class attributes can be accessed directly, as shown in the example below:

```
1   # Define the class
2   class Employee:
3
4          # Define a class attribute common for all objects
5          bonusPercent = 0.2
6          # Define an instance method that takes no arguments
7          def calculateBonus(self):
8                  return self.salary * Employee.bonusPercent
9
10  # Create two objects and an instance attribute
11  emp1 = Employee()
12  emp1.salary = 15000
13  emp2 = Employee()
14  emp2.salary = 18000
15
16  # Print using the instance method and the class attribute
17  print("Bonus amount is", emp1.calculateBonus(),
18        "calculated at", Employee.bonusPercent)
19  print("Bonus amount is", emp2.calculateBonus(),
20        "calculated at", Employee.bonusPercent)
21
22  # Change the value of the class attribute
```

```
23  Employee.bonusPercent = 0.3
24
25  # Print again using the instance method and the changed class attribute
26  print("Bonus amount is", emp1.calculateBonus(),
27        "calculated at", Employee.bonusPercent)
28  print("Bonus amount is", emp1.calculateBonus(),
29        "calculated at", Employee.bonusPercent)
```

Output 3.2.3.1.b:

```
Bonus amount is 3000.0 calculated at 0.2
Bonus amount is 3600.0 calculated at 0.2
Bonus amount is 4500.0 calculated at 0.3
Bonus amount is 4500.0 calculated at 0.3
```

3.2.3.2 Constructor Methods

A *constructor* is a special method used to *initialize* the data of an object. In Python, constructors are implemented using the __init__() method. This method is automatically invoked whenever a new instance of the class is created. If not explicitly defined, the compiler assumes a *default* constructor with no implementation details. It is important to note that a constructor does not return any value.

The programmer can optionally define constructors other than the default one. A user-defined constructor is created by defining the __init__() method within the class. Like all methods in a class, it takes a self argument that references the current object. The syntax of the __init__() method is the following:

> **Observation 3.10 – Constructor Method:** Defined either automatically or by using the __init__() method. It is invoked automatically when a new instance of a class is created. It can be used to initialize the data of the new object or to perform any other task necessary. It can take arguments with or without default values. It does not return any value.

```
def __init__ (self [, arguments])
```

User-defined constructors can be one out of three different types, depending on whether they take arguments or not. The first is the simple constructor, which takes no arguments. The following Python script presents such a case, where the constructor takes no arguments and prints a default text message. Notice that every time a new object is instantiated the message is displayed:

```
1   # Define the class
2   class Employee:
3
4       # Default constructor takes no arguments, prints message
5       def __init__ (self):
6           print("Object created")
7
8   # Every time a new object is created the constructor is called and
9   # the message is displayed
10  emp1 = Employee()
11  emp3 = Employee()
```

Output 3.2.3.2.a:

```
Object created
Object created
```

The default constructor may be also used to initialize instance attributes with default values. In the following example, when a new Employee object is created, instance attributes salary and allowances are set to a default value of 0:

```
1   # Define the class
2   class Employee:
3
4       """ Define the default constructor that takes no arguments
5       but initializes the values of the instance attributes """
6       def __init__ (self):
7               self.salary = 0
8               self.allowances = 0
9
10  """ Every time a new object is created the constructor is called
11  and the instance attributes are set to 0 """
12  emp1 = Employee()
13  emp1.salary = 15000
14
15  """ Print the instance attributes of the objects. The default
16  allowances value is printed """
17  print(emp1.salary, emp1.allowances)
18
19  # Change the value of the allowances attribute
20  emp1.allowances = 3000
21
22  # Print the instance attribute of the object after the value
23  # of allowances is changed
24  print(emp1.salary, emp1.allowances)
```

Output 3.2.3.2.b:

```
15000 0
15000 3000
```

The second constructor type accepts parameters as arguments. It is used when initialization of the attributes of the new object involves the assignment of specific values rather than the default ones. To highlight this, in the following example, a list of the arguments used to initialize the attributes of the object is provided after the default self attribute:

```
1   # Define the class
2   class Employee:
3
4       # Define the constructor with four arguments
5       def __init__ (self, first, last, salary, allowances):
6               # Initialize instance attributes: use values of arguments
```

```
7                      self.firstName = first
8                      self.lastName = last
9                      self.salary = salary
10                     self.allowances = allowances
11
12   # Create a new object with specific instance attribute values
13   emp1 = Employee("Maria", "Rena", 15000, 3000)
14
15   # Print the object's attributes
16   print(emp1.firstName, emp1.lastName, emp1.salary, emp1.allowances)
```

Output 3.2.3.2.c:

```
Maria Rena 15000 3000
```

For simplicity reasons, Python does not support method *overloading* and, thus, the definition of multiple constructors is not allowed. Additionally, if a user-defined constructor is provided, it is no longer possible to use the default constructor in order to create a new object with no parameters. This limitation can be overcome by means of the third constructor type, which is used to accept arguments with default values. This allows the programmer to initialize the associated object with or without values. This constructor type is illustrated in the following example. When emp1 is instantiated, the constructor is invoked without any parameter values. In contrast, in the case of emp2, it is invoked with predefined parameter values, which are assigned to the respective instance attributes. Once both objects are instantiated, the instance attributes of both emp1 and emp2 are accessed and printed using regular dot notation:

```
1    # Define the class
2    class Employee:
3
4          """ Define a constructor that takes four arguments with
5          default empty values (None) if no values are passed """
6          def __init__ (self, first = None, last = None, salary = None,
7                        allowances = None):
8              if first!= None and last!= None and salary!= None \
9                          and allowances!= None:
10                     self.firstName = first
11                     self.lastName = last
12                     self.salary = salary
13                     self.allowances = allowances
14                     print("Object initialized with supplied values")
15             else:
16                     self.salary = 0
17                     self.allowances = 0
18                     print("Object initialized with default values")
19
20   # Create a new object invoking the constructor with no parameters
21   emp1 = Employee()
```

```
22   emp1.firstName = "Alex"
23   emp1.lastName = "Flora"
24   print(emp1.firstName, emp1.lastName, emp1.salary, emp1.allowances)
25   # Create a new object invoking the constructor with parameters
26   emp2 = Employee("Maria", "Rena", 15000, 5000)
27   print(emp2.firstName, emp2.lastName, emp2.salary, emp2.allowances)
28   # Change and reprint the value of instance attribute of 'emp2'
29   emp2.salary = 20000
30   print(emp2.firstName, emp2.lastName, emp2.salary, emp2.allowances)
```

Output 3.2.3.2.d:

```
Object initialized with default values
Alex Flora 0 0
Object initialized with supplied values
Maria Rena 15000 5000
Maria Rena 20000 5000
```

3.2.3.3   Destructor Method

Destructors are special methods invoked at the end of the lifecycle of objects, when they must be deleted. In Python, destructors are implemented using the __del__() method, and are invoked when all references to an object have been deleted. The following Python script provides an example of two objects (i.e., emp1 and emp2) firstly being created and then destroyed:

Observation 3.11 – Destructor Method: Defined by using the __del__() method. It is used to delete an instance/object when it is not needed anymore. The method takes no arguments, and returns no values.

```
1    # Define the class
2    class Employee:
3
4         # Define the default constructor that only prints a message
5         def __init__(self):
6              print("Employee created")
7
8         # Destructor deletes the object and prints a message
9         def __del__(self):
10             print("Employee deleted")
11
12   # Constructor automatically invoked to create 'emp1' and 'emp2'
13   emp1 = Employee()
14   emp2 = Employee()
15
16   # Destroy objects 'emp1' and 'emp2'. Destructor method is called
17   del emp1
18   del emp2
```

Output 3.2.3.3:

```
Employee created
Employee created
Employee deleted
Employee deleted
```

3.3   ENCAPSULATION

Encapsulation is one of the pillars of Object-Oriented Programming. It is based on the idea of wrapping up the attributes and methods in a class and controlling access when instantiating new objects/instances. Instead, *access modifiers* are used to dictate and control how the instance attributes can be accessed.

Observation 3.12 – Encapsulation: Wrapping up the attributes and methods in a class and controlling access when instantiating new objects/ instances.

3.3.1   ACCESS MODIFIERS IN PYTHON

As mentioned, objects store data in attributes. Appropriate protective measures ensure that this data is accessed and modified in a controlled way. In general, OOP languages provide access modifiers that specify how an attribute or method can be accessed. There are three main types of access modifiers:

Observation 3.13 – Access Modifiers: Access modifiers control how the instance attributes can be accessed. Access modifiers can be *public* with no special notation needed, *private* denoted by double underscore (__), or *protected* denoted by single underscore (_).

- **Public:** Attribute/method can be accessed by any class or program without any restrictions.
- **Private:** Attribute/method can be accessed only within the container class.
- **Protected:** Attribute/method can be accessed within the container class and its sub-classes.

By default, all attributes and methods in Python are *public*. Instead of using special keywords to specify whether an attribute is *public*, *private*, or *protected*, Python uses a special naming convention to control access. An attribute with an *underscore* prefix (_) denotes a protected attribute, while a double underscore prefix (__) a private attribute. As mentioned, the absence of a prefix denotes the default, *public* modifier.

3.3.2   GETTERS AND SETTERS

When defining a class, it is good programming practice to control the access to instance attributes by means of two special types of methods commonly referred to as *getters* and *setters*. Many OOP languages use such methods to implement the principle of encapsulation. A getter is a method that reads (*gets*) the value of an attribute, while a setter writes (*sets*) it. Using getters and set-

Observation 3.14 – Getters and Setters: Used to implement encapsulation. Setters are used to store data into private instance attributes whereas getters are used to read that data.

ters to access object attributes ensures that the data is protected (i.e., encapsulated). The benefits of using these special methods are the following:

- Ensuring validation when reading or writing attribute data.
- Setting different access levels for the class attributes.
- Preventing direct manipulation of the attribute data.

In the Python example below, the `Employee` class uses `setFirstName()`, a setter method, to store data in a protected attribute of the object (denoted by the double underscore symbol), while getter method `getFirstName()` is used to read and print the employee's first name. As the attribute is protected, it is accessible using the methods within the class, and within the object created using the class. Getter and setter methods should be used for all instance attributes defined in the class. In other words, for every instance attribute, it is recommended that the associated getter and setter methods are provided. The reader should also notice the use of the `self` parameter with all methods, as it provides the reference to the current object being used:

In this context, if the `print(emp1.getFirstName())` command is replaced by `print(emp1.__first)` in an attempt to access the private instance attribute directly, an error will occur:

```
1    # Define the class
2    class Employee:
3
4          # Define the getter method to read private attribute __first
5          def getFirstName(self):
6                  return self.__first
7
8          # Setter method writes to private attribute __first
9          def setFirstName(self, value):
10                 self.__first = value
11
12   # Create object emp1
13   emp1 = Employee()
14
15   # Use the setter to store new data in the private attribute
16   emp1.setFirstName("George")
17
18   # Getter reads the data from the private attribute and prints it
19   print(emp1.getFirstName())
```

Output 3.3.2:

```
George
```

3.3.3   VALIDATING INPUTS BEFORE SETTING

As discussed, getter and setter methods shield the data values of private instance attributes. In addition, they also provide *data validation* functionality. As an example, if the value of private instance attribute __firstName should not exceed 15 characters in length, and __salary should be a

number between 0 and 20,000, the associated *validation code* can be added to the setter methods of the attributes. Similarly, if it is necessary to format the output in
a particular way, the associated code could be added to
the getter methods. The following script provides a class
example demonstrating this concept:

Observation 3.15 – Validating Data:
Use getters and setters to validate data
stored in the private attributes and format data appropriately before used as
output.

```
1   # Define the class
2   class Employee:
3
4       # Define a setter for private attribute '__firstName'.
5       # Check the attribute value and store it if it is lower than 15
6       def setFirstName(self, value):
7           if len(value) < 15:
8               emp1.__firstName = value
9
10      # Define a getter for private attribute '__firstName'.
11      # Print the data with an appropriate message
12      def getFirstName(self):
13          return "The first name is :", self.__firstName
14
15      # Define a setter for private attribute '__salary'.
16      # Check attribute value; store it if it is between 0 and 20000
17      def setSalary(self, value):
18          if (value > 0 and value < 20000):
19              emp1.__salary = value
20
21      # Define a getter for private attribute '__salary'.
22      # Print the data with an appropriate message
23      def getSalary(self):
24          return "The salary is ", self.__salary
25
26  # Create a new object and call its setters
27  # to validate and store values in its attributes
28  emp1 = Employee()
29  emp1.setFirstName("John")
30  emp1.setSalary(17000)
31
32  # Attribute getters print stored values and associated messages
33  print(emp1.getFirstName(), emp1.getSalary())
34
35  # Repeat the previous tasks with an invalid first name entry.
36  # Notice: no change takes place in the '__firstName' attribute
37  emp1.setFirstName("Check to see if more than 15 characters are stored")
38  emp1.setSalary(19000)
39  print(emp1.getFirstName(), emp1.getSalary())
```

```
40
41  # Repeat the previous tasks with invalid salary entry.
42  # Notice: there is no change taking place in the '__salary' attribute
43  emp1.setFirstName("George")
44  emp1.setSalary(21000)
45  print(emp1.getFirstName(), emp1.getSalary())
```

Output 3.3.3:

```
('The first name is :', 'John') ('The salary is ', 17000)
('The first name is :', 'John') ('The salary is ', 19000)
('The first name is :', 'George') ('The salary is ', 19000)
```

3.3.4 CREATING READ-ONLY ATTRIBUTES

Getter and setter methods may be also used to control *read-only* or *write-only* attributes. For example, attribute age may be designated as read only, since it should be calculated using the value of attribute dateOfBirth. In this case, age will require a getter but no setter method, allowing thus the user to *read* the age value but not to *update* it.

In the following example, class Employee defines instance attributes for employees' first and last names, and the corresponding getter and setter methods. The class also defines attributes for the employees' emails and full names, which as read-only attributes do not have setter methods. In this case, the values of these attributes are constructed when they are being read using the getter method:

Observation 3.16 – Creating Read-Only Attributes: Use getters with no setters to create and output the values of read-only attributes, whose data are calculated using private attributes.

```
1   # Define the class
2   class Employee:
3
4         # The getter and setter methods for the first name
5         def getFirstName(self):
6               return self.__first
7
8         def setFirstName(self, value):
9               self.__first = value
10
11        # The getter and setter methods for the last name
12        def getLastName(self):
13              return self.__last
14        def setLastName(self, value):
15              self.__last = value
16
17        # Read-only attributes with only a getter method
18        def getEmail(self):
19              return self.__first + "." + self.__last + "@company.com"
20        def getFullName(self):
21              return self.__first + " " + self.__last
```

```
22
23   # Create a new 'Employee' object
24   emp1 = Employee()
25
26   # Setter stores value to the '__private' instance attributes
27   emp1.setFirstName("George")
28   emp1.setLastName("Davies")
29
30   # Print the read-only attributes
31   print(emp1.getFullName(), emp1.getEmail())
```

Output 3.3.4:

```
George Davies George.Davies@company.com
```

3.3.5   THE PROPERTY() METHOD

In the example presented below, methods getFirst-
Name() and setLastName() are used to read from,
and write to, private attribute __first. In order to
make this particular example more user-friendly, the
getter and setter methods could be automatically called
when accessing the attribute, using the dot notation (i.e.,
<obj>.<property>). The property() method pro-
vides the necessary interface by encapsulating the getter

**Observation 3.17 – Property
Method:** Use it to encapsulate the
getter and setter methods in a single
interface that facilitates access to a
private attribute using simply the dot
notation.

and setter methods, which are invoked when reading from, or writing to it. The method syntax is
the following:

```
property_name = property(gettermethod, settermethod)
```

After defining the property method, the attribute is accessed using the dot notation on the property
name (<obj>.<property>) instead of invoking the getter and setter methods directly:

```
1    # Define the class
2    class Employee:
3
4         # Define the getter method
5         def getFirstName(self):
6              return self.__first
7
8         # Define the setter method
9         def setFirstName(self, value):
10             self.__first = value
11
12        """ Use the property method to encapsulate the getter and setter
13        in a single method interface """
14        firstName = property(getFirstName, setFirstName)
15
```

```
16   # Create the 'emp1' object
17   emp1 = Employee()
18
19   """ Use dot notation to invoke the setter and getter methods through
20   the property interface """
21   emp1.firstName = "George"
22   print(emp1.firstName)
```

Output 3.3.5:

```
George
```

3.3.6 THE @property DECORATOR

Another way to define attributes in Python is to use the @property *decorator*, which is built in the property() method. In the example below, @property defines the firstName attribute by using two different methods with the property name. The firstName(self)

Observation 3.18 – The @property **Decorator:** It allows the extension of the property method in a similar way.

method is decorated with the @property decorator, indicating that the method is a getter. Accordingly, the firstName(self, value) method is decorated with @firstName.setter, indicating that this is a setter. With this structure in place, the attribute can be accessed by using its property name with the dot notation, without explicitly calling the getter and setter methods:

```
1    # Define the class
2    class Employee:
3
4        # Use the property decorator to define the getter method
5        @property
6        def firstName(self):
7            return self.__first
8
9        # Use the property decorator to define the setter method
10       @firstName.setter
11       def firstName(self, value):
12           self.__first = value
13
14   # Create the 'emp1' object
15   emp1 = Employee()
16
17   # Access private attribute '__first' through property name 'firstName'
18   emp1.firstName = "George"
19   print(emp1.firstName)
```

Output 3.3.6:

```
George
```

3.4   INHERITANCE

Inheritance is one of the four main principles of OOP. It allows the programmer to extend the functionality of a class by creating a *parent-child* relationship between classes. In such a relationship, the *child* (also called *sub* or *derived* class) inherits from the *parent* (also called *super* or *base* class). The reader should note that these terms may be used interchangeably in this chapter, based

> **Observation 3.19 – Inheritance:** Allows the extension of the functionality of a *parent/super/base* class, by creating a *child/sub/derived* class that inherits its attributes and behavior.

on the context of each discussion. Inheritance is extremely useful, as it facilitates *code reusability*, thus minimizing code and making it easier to maintain. An important concept relating to child classes is that they may have their own new attributes and methods, and can optionally *override* the functionality of the respective parent class.

3.4.1   INHERITANCE IN PYTHON

The Python syntax for implementing the concept of inheritance is the following:

```
Class Parent:
Parent class definition
Class Child(Parent):
Child class definition
```

As a practical example of inheritance, the reader can consider two classes, a super class named Employee and a sub class named SalesEmployee (Figure 3.4). Instead of creating the general attributes of SalesEmployee (e.g., *first name, last name, salary, or allowances*) from scratch, they can be inherited from Employee. Accordingly, the sub class can also inherit the setters and getters, and generally all the *functionality* of the Employee class. Additional attributes that may be unique to SalesEmployee (e.g., *commission rate*) can be also added to the inherited ones, as required.

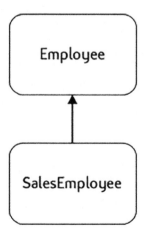

FIGURE 3.4   Parent-child relationship between classes.

The implementation of this particular example of super class Employee and sub class SalesEmployee is presented in the Python script examples below. In the first script, Employee class is defined with private attributes __first, __last, __salary, and __allowances, and class method getTotalSalary(). In the second, SalesEmployee class is created as an empty class, hence the use of the pass keyword. Private attributes and the method are inherited from the Employee class. Note that the name of super class Employee is passed to SalesEmployee as an argument:

```
1    # Define class 'Employee' and its private attributes and method
2    class Employee():
3
4         def __init__(self, first, last, salary, allowances):
5              self.__first = first
6              self.__last = last
7              self.__salary = salary
8              self.__allowances = allowances
9
10        def getTotalSalary(self):
11             return self.__salary + self.__allowances
12
13   # Create object 'emp1' and print the total salary of the current employee
14   emp1 = Employee("George", "White", 16000, 5200)
15   print(emp1.getTotalSalary())
```

Output 3.4.1.a:

```
21200
```

```
1    # Define sub class 'SalesEmployee' based on super class 'Employee'
2    class salesEmployee(Employee):
3         pass
4
5    """ Create a new object of the sub class that inherits
6    attributes and behavior from the super class """
7    semp1 = salesEmployee("Alex", "Flora", 12000, 4000)
8    print(semp1.getTotalSalary()) # Method of the superclass is invoked
```

Output 3.4.1.b:

```
16000
```

When the `semp1` object is instantiated, Python scans `SalesEmployee` for an initialization method (i.e., `__init__()`). If this is not found, it scans and executes the initialization method of the super class (i.e., `Employee`), with the parameters associated with the current object. Similarly, when `getTotalSalary()` is invoked for object `semp1`, the method is called from the super class, since it does not exist in the sub class. The same order of resolution is followed for all methods and attributes in the sub class.

3.4.1.1 Customizing the Sub Class

As mentioned, sub classes can be further customized by adding new attributes and methods. For instance, in the case of sub class `SalesEmployee` this can be done by adding attribute `commission _ percent`. The reader should note that attempting to use the added attribute for an object that belongs to the `Employee` class will raise an error. This is because there is no such

Observation 3.20 – Customize Sub Classes: Add attributes and/or methods to sub classes to extend their behavior beyond that of the super class. Using the added behavior on objects of the super class will raise an error. Attributes of the super class that will be used in the sub class need to be declared as *protected*.

attribute or method in the super class. It is also worth noting that in order to be able to use super class attributes `salary` and `allowances`, they must be declared as *protected* instead of *private*. The following scripts demonstrate these concepts:

```python
1   # Define class 'Employee'
2   class Employee():
3
4       """ Define the constructor of the class with parameters.
5       Define the attributes of the class """
6       def __init__(self, first, last, salary, allowances):
7           self.__first = first
8           self.__last = last
9           self._salary = salary
10          self._allowances = allowances
11
12      # Define a derived attribute
13      def getTotalSalary(self):
14          return self._salary + self._allowances
15
16  # Define the 'SalesEmployee' sub class
17  class salesEmployee(Employee):
18
19      # Use the property decorator to define the getter method
20      @property
21      def commissionPercent(self):
22          return self.__comm
23
24      # Use the property decorator to define the setter method
25      @commissionPercent.setter
26      def commissionPercent(self, value):
27          self.__comm = value
28
29  # Create and use object 'emp1' based on super class 'Employee'
30  emp1 = Employee("Maria", "Rena", 15000, 5000)
31  print(emp1.getTotalSalary())
32
33  # Create and use object 'semp1' based on sub class 'SalesEmployee'
34  semp1 = salesEmployee("Alex", "Flora", 16000, 6000)
35  # The attribute is set in the sub class
36  semp1.commissionPercent = 0.05
37
38  print(semp1.commissionPercent)
39
40  """ The next line generates an error since its
41  attribute only exists in the sub class """
42  print(emp1.commissionPercent)
43
44  # Print the attributes of objects 'emp1' and 'semp1'
45  print(semp1.__dict)
46  print(emp1.__dict)
```

Output 3.4.1.1:

```
20000
0.05
```

```
--------------------------------------------------------------------------------
AttributeError                           Traceback (most recent call last)
<ipython-input-9-0e8e58d5eaf8> in <module>
     40 """ The next line generates an error since its
     41 attribute only exists in the sub class """
---> 42 print(empl.commissionPercent)
     43
     44 # Print the attributes of objects 'empl' and 'sempl'

AttributeError: 'Employee' object has no attribute 'commissionPercent'
```

3.4.2 METHOD OVERRIDING

Method overriding is another important programming feature that is common in OOP languages. It allows a sub class to contain a method with a different implementation than the one inherited from the super class. In the context of the previous examples, the programmer may wish to compute the total salary of a sales employee by adding *commissions* to their salary and allowances. In this case, sub class method `getTotalSalary()` must be implemented differently to the original one inherited from `Employee`. As shown in the following example, super class method `getTotalSalary()` has to be called in the implementation of sub class method `getTotalSalary()`:

```
1    # Define class 'Employee'
2    class Employee():
3
4        # Define the constructor and the attributes of the super class
5        def __init__(self, first, last, salary, allowances):
6            self.__first = first
7            self.__last = last
8            self._salary = salary
9            self._allowances = allowances
10
11       # Define 'getTotalSalary'
12       def getTotalSalary(self):
13           return self._salary + self._allowances
14
15   # Define sub class 'salesEmployee'
16   class salesEmployee(Employee):
17
18       # Use the property decorator to define the getter method
19       @property
20       def commissionPercent(self):
21           return self.__comm
22
23       # Use the property decorator to define the setter method
24       @commissionPercent.setter
25       def commissionPercent(self, value):
26           self.__comm = value
27
```

```
28        # Super class getter overrides the parent class method
29        def getTotalSalary(self):
30            return super().getTotalSalary() + (super().getTotalSalary()
31   *self.__comm)
32
33   # Create and use object 'emp1' based on super class 'Employee'
34   emp1 = Employee("Maria", "Rena", 15000, 5000)
35   print(emp1.getTotalSalary())
36
37   # Create and use object 'semp1' based on sub class 'salesEmployee'
38   semp1 = salesEmployee("Alex", "Flora", 16000, 6000)
39
40   # Set the attribute in the sub class
41   semp1.commissionPercent = 0.05
42
43   # Invoke the overridden getter method from the sub class
44   print(semp1.getTotalSalary())
```

Output 3.4.2:

```
20000
23100.0
```

3.4.2.1 Overriding the Constructor Method

The concept of method overriding is also used to create customized constructors in the sub class. In this case, the super() method is used to invoke the __init__() method of the super class, as shown in the following script:

> **Observation 3.21 – Constructor Overriding:** Call the __init__() method of the super class to access the constructor and add attributes to extend it.

```
1    # Define class 'Employee'
2    class Employee():
3
4        # Define the constructor of the super class and its attributes
5        def __init__(self, first, last, salary, allowances):
6            self.__first = first
7            self.__last = last
8            self._salary = salary # Protected attribute
9            self.__allowances = allowances
10
11       # Define the getter of the class
12       def getTotalSalary(self):
13           return self._salary + self.__allowances
14
15   # Define sub class 'salesEmployee'
16   class salesEmployee(Employee):
17
18       """ Define the constructor of the sub class adding the 'comm'
19       attribute. Call the 'init' method of the super class """
20       def __init__(self, first, last, salary, allowances, comm):
```

```
21                 super().__init__(first, last, salary, allowances)
22                 self.__comm = comm
23
24         # Access protected attribute '_salary' from the sub class
25         def getTotalSalary(self):
26                 return super().getTotalSalary() + (self._salary *
27 self.__comm)
28
29 # Create and use object 'emp1' based on the super class
30 emp1 = Employee("Maria", "Rena", 15000, 5000)
31 print(emp1.getTotalSalary())
32
33 # Create and use object 'semp1' based on the sub class
34 semp1 = salesEmployee("Alex", "Flora", 16000, 6000, 0.05)
35 print(semp1.getTotalSalary()) # Method of the child class is invoked
```

Output 3.4.2.1:

```
20000
22800.0
```

3.4.3 MULTIPLE INHERITANCE

Sub classes can inherit attributes and methods from multiple super classes, a concept known as *multiple inheritance*. In Python, this can be implemented using the following syntax:

> **Observation 3.22 – Multiple Inheritance:** The concept of having a sub class inheriting from more than one super classes.

```
class Parent1
    pass
class Parent2
    pass
class Child (Parent1, Parent2):
    pass
```

As an example of multiple inheritance, Figure 3.5 presents a structure consisting of two super classes (Person and Employee) and one sub class (Manager) that inherits from both super classes.

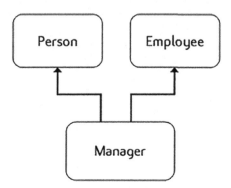

FIGURE 3.5 A representation of multiple inheritance between three classes.

The following Python scripts implement this structure. The reader should note that the constructor in the Manager class calls the respective constructors of both super classes during initialization. Methods getFullName and getContact are inherited from super class Person, while getAnnualSalary and getDepartment are inherited from Employee:

```
1    # Define the first super class ('Person')
2    class Person():
3
4         # Define class constructor and attributes
5         def __init__(self, firstName, lastName, contact):
6              self.__firstName = firstName
7              self.__lastName = lastName
8              self.__contact = contact
9
10        # Getter for the first & last name of the first super class
11        def getFullName(self):
12             return "Employee name is: " + self.__firstName +" " \
13   + self.__lastName
14
15        # Define the getter for the contact of the first parent
16        def getContact(self):
17             return "Contact number is: " + self.__contact
18
19   # Define the second Parent base class Employee
20   class Employee():
21        # The constructor & the attributes of the second super class
22        def __init__(self, salary, dept):
23             self.__salary = salary
24             self.__dept = dept
25
26        # Define the getter for the salary of the second super class
27        def getAnnualSalary(self):
28             return "The annual salary is: " + str(self.__salary * 12)
29
30        # The getter for the department of the 2nd super class
31        def getDepartment(self):
32             return "The employee belongs to the department: " +\
33                  self.__dept
34   # Define subclass 'Manager' inheriting from both 'Person' and 'Employee'
35   class Manager(Person, Employee):
36        def __init__(self, firstName, lastName, contact, salary, dept):
37             Person.__init__(self, firstName, lastName, contact)
38             Employee.__init__(self, salary, dept)
```

```
39
40   # Create and use a new instance of the 'Manager' class
41   mgr1 = Manager("Maria", "Rena", "0123456789", 14500, "Marketing")
42
43   # Call inherited behaviour from super class 'Person'
44   print(mgr1.getFullName())
45   print(mgr1.getContact())
46
47   # Call inherited behaviour from super class 'Employee'
48   print(mgr1.getAnnualSalary())
49   print(mgr1.getDepartment())
```

Output 3.4.3:

```
Employee name is: Maria Rena
Contact number is: 0123456789
The annual salary is: 174000
The employee belongs to the department: Marketing
```

3.5   POLYMORPHISM – METHOD OVERLOADING

Another powerful feature of OOP languages is the support of *method overloading*. This is a fundamental element of *polymorphism*, the option of defining and using two or more methods with the same name but different parameter lists or *signatures*. Overloading a method improves code readability and maintainability, as implementation is divided into multiple methods instead of being concentrated into a single, complex one.

Observation 3.23 – Polymorphism/ Method Overloading: The concept of using method overloading to implement two or more methods with the same name but different signatures.

While method overloading is a prominent feature in many OOP languages, such as Java and C++, it is not entirely supported in Python. Python is a *dynamically typed* language and *datatype binding* occurs at runtime. This is known as *late binding* and it differs from the *static binding* used in languages like Java and C++, in which overloaded methods are invoked at *compile time* based on the arguments they are supplied with. In Python, when multiple methods with the same name are defined, the last definition overrides all previous ones. As an example, consider method calculateTotal-Salary() in the Employee class. The method computes the annual salary of the employee without the bonus. A second method that calculates the total salary plus the bonus can be implemented with the same name, thus, overloading calculateTotalSalary(). In this case, the first method will be ignored and any reference to it will raise an error, as shown in the following example:

```
1    # Define class 'Employee'
2    class Employee:
3
4        # Define method 'calculateTotalSalary'
5        def calculateTotalSalary(self):
6            return(self.salary + self.allowances)
7        # Define a method overloading 'calculateTotalSalary'
8        def calculateTotalSalary (self, bonus):
9            return(self.salary + self.allowances) + bonus
```

```
10
11  # Create and use the 'emp1' object
12  emp1 = Employee()
13  emp1.salary = 15000
14  emp1.allowances = 5000
15  print("Total salary is ", emp1.calculateTotalSalary(2000))
16
17  # Create and use the 'emp2' object
18  emp2 = Employee()
19  emp2.salary = 18000
20  emp2.allowances = 4000
21  # This method call will generate an error
22  print("Total salary is ", emp2.calculateTotalSalary())
```

Output 3.5:

```
Total salary is  22000

-----------------------------------------------------------------------
TypeError                              Traceback (most recent call last)
<ipython-input-8-517b173547e9> in <module>
     22
     23 # This method call will generate an error
---> 24 print("Total salary is ", emp2.calculateTotalSalary())

TypeError: calculateTotalSalary() missing 1 required positional argument: 'bonus'
```

3.5.1 METHOD OVERLOADING THROUGH OPTIONAL PARAMETERS IN PYTHON

Although Python does not directly support method overloading in the same form as other OOP languages, it offers an alternative approach to achieve the same functionality. Instead of resorting to the creation of multiple methods, it allows methods to take *optional parameters* with *default values*. When a method is invoked in the code, the programmer can choose whether to provide the parameter values or not. This, in turn, dictates which

Observation 3.24 – Method Overloading in Python: In Python, use optional method parameters to emulate the method overloading feature available in other OOP languages.

block of statements would be executed within the method. Commonly, the None value is used to assign a default null value to the attribute.

In the example below, constructor method calculateTotalSalary() is defined with optional parameter bonus. The implementation subsequently returns different values, depending on whether a new value has been assigned to the optional parameter. If this is not the case, the default None value is used.

```
1   class Employee:
2
3        def calculateTotalSalary(self, bonus = None):
4        # None statement supports both 'is' and '=='comparison operators
5            if bonus is None:
6                return(self.salary + self.allowances)
7            else:
```

```
8                          return(self.salary + self.allowances) + bonus
9
10  emp1 = Employee()
11  emp1.salary = 15000
12  emp1.allowances = 5000
13  emp2 = Employee()
14  emp2.salary = 18000
15  emp2.allowances = 4000
16
17  print("Total salary is ", emp2.calculateTotalSalary(2000))
18  print("Total salary is ", emp1.calculateTotalSalary())
```

Output 3.5.1:

```
Total salary is 24000
Total salary is 20000
```

3.6   OVERLOADING OPERATORS

Operator overloading refers to the process of changing the default behavior of an operator based on the *operands* being used. A classic case of operator overloading in Python is the modification of the behavior of the addition (+) and multiplication (*) operators based on the input type. For instance, when the addition operator is used on two numbers it performs regular numerical

Observation 3.25 – Operator Overloading: Apply the + and * operators on operands of different primitive data types to yield different results.

addition, but when it is used with strings it *concatenates* them. Similarly, when the multiplication operator is used on numbers it multiplies them, while when it is used on a string and an integer *it repeats the string*. The reader should note that this fundamental operator overloading functionality works on operands of primitive data types, like in the following example:

```
1  a = 1
2  b = 2
3  print(a + b) # Adds the two numbers
4  print(a * b) # Multiplies the two numbers
5
6  a = 'Python'
7  b = ' is fun'
8  print(a + b) # Concatenates the two strings
9  print(a + b * 3) # Concatenates and repeats the string
```

Output 3.6.a:

```
3
2
Python is fun
Python is fun is fun is fun
```

If the addition operator is used on user-defined objects it raises a `TypeError`, since it does not support the instance type, as shown below:

```
1   # Define class 'Employee'
2   class Employee:
3
4        salary = 0
5
6   # Create and use two objects of the 'Employee' class
7   emp1 = Employee()
8   emp1.salary = 15000
9   emp2 = Employee()
10  emp2.salary = 22000
11
12  # Attempting the following print will generate a TypeError
13  print(emp1 + emp2)
```

Output 3.6.b:

```
---------------------------------------------------------------------------
TypeError                                 Traceback (most recent call last)
<ipython-input-11-527139aab026> in <module>
     11
     12 # Attempting the following print will generate a TypeError
---> 13 print(emp1 + emp2)

TypeError: unsupported operand type(s) for +: 'Employee' and 'Employee'
```

This issue can be bypassed by utilizing the built-in *magic* or *dunder* methods, which can be invoked by means of the respective operators. For instance, in the case of the addition operator the associated __add__() method is firstly extended in terms of its functionality and, subsequently, invoked as shown in the following script:

Observation 3.26 – Magic or Dunder Methods: Special methods invoked when a basic operator is called, with a double underscore as a prefix and a suffix. They are used to overload operators with the object type of operands.

```
1   # Define class 'Employee'
2   class Employee:
3
4        # Overload the + operator to add the 'salary' of two objects
5        def __add__(self, other):
6             return self.salary + other.salary
7
8   # Create the two objects of the 'Employee' class
9   emp1 = Employee()
10  emp1.salary = 15000
11  emp2 = Employee()
12  emp2.salary = 22000
13
14  # Invoke the overloaded + operator by extending the '__add__' method
15  print(emp1 + emp2)
```

Output 3.6.c:

```
37000
```

In order to implement operator overloading, the programmer has to define the appropriate magic method according to the operator in the class definition.

Tables 3.1–3.4 provide a list of magic methods corresponding to the respective *binary*, *comparison*, *unary*, and *assignment* operators. Changing the implementation of the magic method associated with the respective operator can provide a different meaning to that particular operator. For example, the plus (+) operator can be used with the Employee objects to add their salaries (i.e., emp1 + emp2). Similarly, the *less than* (<) operator can be used to compare which employee was hired first, or which is older. Conceptually, the idea is to use operator overloading in order to define and implement the functionality of operators in a way that is logical and appropriate in the context of the overall program structure and requirements.

TABLE 3.1
List of Binary Operators and Their Corresponding Magic Method

Operator	Magic Method
+	__add__(self, other)
–	__sub__(self, other)
*	__mul__(self, other)
//	__floordiv__(self, other)
/	__div__(self, other)
%	__mod__(self, other)
**	__pow__(self, other)
<<	__lshift__(self, other)
>>	__rshift__(self, other)
&	__and__(self, other)
^	__xor__(self, other)
|	__or__(self, other)

TABLE 3.2
List of Comparison Operators and Their Corresponding Magic Method

Operator	Magic Method
<	__lt__(self, other)
>	__gt__(self, other)
<=	__le__(self, other)
>=	__ge__(self, other)
==	__eq__(self, other)
!=	__ne__(self, other)

TABLE 3.3

List of Unary Operators and Their Corresponding Magic Method

Operator	Magic Method
–	__neg__(self, other)
+	__pos__(self, other)
~	__invert__(self, other)

TABLE 3.4

List of Assignment Operators and Their Corresponding Magic Method

Operator	Magic Method
+=	__iadd__(self, other)
–=	__isub__(self, other)
*=	__imul__(self, other)
/=	__ifloordiv__(self, other)
//=	__idiv__(self, other)
%=	__imod__(self, other)
**=	__ipow__(self, other)
<<=	__ilshift__(self, other)
>>=	__irshift__(self, other)
&=	__iand__(self, other)
^=	__ixor__(self, other)
|=	__ior__(self, other)

3.6.1 Overloading Built-In Methods

While Python does not support overloading of custom methods in a class, it does so for *built-in* methods. This allows the programmer to change the default behavior of an existing method within the context of a class. For example, in the case of the print() method, the default behavior is to print a string if the input is text or an *object reference* if the argument is an object, as shown in the following example:

> **Observation 3.27 – Overloading Built-In Methods:** It is possible to overload built-in methods (e.g., print, len, bool) by extending the functionality of their respective magic methods.

```
1   # Define class 'Employee'
2   class Employee:
3        Pass
4
5   # Create a new 'emp1' object based on the class
6   emp1 = Employee()
7   emp1.firstName = "George"
8   emp1.lastName = "Comma"
9
10  # Use the print method to show the object's reference
11  print(emp1)
```

Output 3.6.1.a:

```
<__main__.Employee object at 0x000002A2140033D0>
```

Nevertheless, when an object is used as an argument, it can be overloaded. Using the usual Employee example, overloading the appropriate magic method, in this particular instance __str__(), allows the program to print the respective employee's details (e.g., firstName, lastName) instead of the object reference as in the following example:

```
1    # Define class 'Employee'
2    class Employee:
3
4            # Define and extend the constructor of the class
5            def __init__(self, first, last, salary):
6                    self.firstName = first
7                    self.lastName = last
8                    self.salary = salary
9
10           # Overload print: extend the functionality of '__str__'
11           def __str__(self):
12                   return "Employe name: " + self.firstName + " " + \
13                           self.lastName + " Salary: " = str(self.salary)
14
15    # Create and use the 'emp1' object based on the 'Employee' class
16    emp1 = Employee("George", "Comma", 15000)
17
18    # Use the overloaded print method
19    print(emp1)
```

Output 3.6.1.b:

```
Employee name: George Comma Salary: 15000
```

3.7 ABSTRACT CLASSES AND INTERFACES IN PYTHON

An *abstract* class is a class that cannot be instantiated. It serves as a blueprint or template for creating sub classes, but it cannot be used to create objects. An abstract class contains declarations of abstract methods. Declarations of this type include the names and parameter lists of the methods, but no implementation. The latter must be defined in the corresponding sub class.

Observation 3.28 – Abstract Class: A class that cannot be instantiated, but serves as a template for sub classes. Abstract classes contain declarations of abstract methods (i.e., methods whose implementation must be defined in the sub classes or non-abstract methods).

In order to create abstract classes and methods, modules ABC and abstractmethod must be imported to the program. The syntax for doing so is the following:

```
from abc import ABC, abstractmethod
```

ABC stands for *Abstract Base Classes*. Newly created abstract classes inherit from ABC and must include at least one abstract method using the @abstractmethod built-in decorator, with no implementation. The following script provides an example of an abstract class (i.e., Employee) with one abstract method (i.e., getTotalSalary()). Running this script raises an error, since abstract classes cannot instantiate objects:

```
1    # Import ABC
2    from abc import ABC, abstractmethod
3
4    # Define abstract class 'Employee'
5    class Employee(ABC):
6
7          # Define abstract method 'getTotalSalary', which must be empty
8          @abstractmethod
9          def getTotalSalary(self):
10               Pass
11
12   # Abstract classes cannot instantiate objects
13   emp1 = Employee()
```

Output 3.7.a:

```
----------------------------------------------------------------------
TypeError                                     Traceback (most recent call last)
<ipython-input-16-47be1b52dd97> in <module>
      11
      12 # Abstract classes cannot instantiate objects
---> 13 emp1 = Employee()

TypeError: can't instantiate abstract class Employee with abstract methods getTotalSalary
```

Once the abstract class is implemented, it can be used as a super class for deriving sub classes. Sub classes of this type must implement the abstract method of the abstract class as a minimum requirement. In this context, as shown in the first of the following scripts, sub class FullTimeEmployee will raise an error, since it does not implement the abstract method (i.e., getTotalSalary()) of its super abstract class (i.e., Employee). On the contrary, the second script presents the implementation of abstract method getTotalSalary() that resolves this issue:

```
1    # Import ABC
2    from abc import ABC, abstractmethod
3
4    # Define abstract class 'Employee'
5    class Employee(ABC):
6
7          # Define abstract method 'getTotalSalary'
8          @abstractmethod
9          def getTotalSalary(self):
10               Pass
11
12   # Define class 'fullTimeEmployee' based on the abstract class
13   class fullTimeEmployee(Employee):
14
15         # Define the constructor of the sub class and its attributes
16         def __init__(self, first, last, salary, allowances):
```

```
17              self.__first = first
18              self._last = last
19              self.__salary = salary
20              self.__allowances = allowances
21
22  # Error will be raised as the sub class does not implement
23  # the abstract method
24  ftl = fullTimeEmployee("Maria", "Rena", 15000, 6000)
```

Output 3.7.b:

```
------------------------------------------------------------
TypeError                               Traceback (most recent call last)
<ipython-input-12-7e5c51df1210> in <module>
     21
     22 # Error will be  raised as the sub class does not implement the abstract method
---> 23 ftl = fullTimeEmployee("Maria", "Rena", 15000, 6000)

TypeError: Can't instantiate abstract class fullTimeEmployee with abstract methods getTotalSalary
```

```
1   # Import ABC
2   from abc import ABC, abstractmethod
3
4   # Define abstract class 'Employee'
5   class Employee(ABC):
6
7           # Define abstract method 'getTotalSalary'
8           @abstractmethod
9           def getTotalSalary(self):
10                  Pass
11
12  # Define class 'fullTimeEmployee' based on the abstract class
13  class fullTimeEmployee(Employee):
14
15          # Define the constructor of the sub class and its attributes
16          def __init__(self, first, last, salary, allowances):
17                  self.__first = first
18                  self._last = last
19                  self.__salary = salary
20                  self.__allowances = allowances
21
22          # Implement the abstract method of the abstract class
23          def getTotalSalary(self):
24                  return self.__salary + self.__allowances
25
26  # Create and use a new 'fullTimeEmployee' object
27  ftl = fullTimeEmployee("Maria", "Rena", 15000, 6000)
28  print(ftl.getTotalSalary())
```

Output 3.7.c:

21000

Abstract classes may include both abstract and non-abstract methods with implementations. Sub classes that inherit from the abstract class also inherit the implemented methods. If required, the latter can be overridden, but in all cases, implementations must include the abstract method.

3.7.1 INTERFACES

In OOP, an *interface* refers to a class that serves as a template for the creation of other classes. Its main purpose is to improve the organization and efficiency of the code by providing blueprints for prospective classes. As such, interfaces describe the behavior of inherited classes, similarly to abstract classes. However, contrary to the latter, they cannot contain non-abstract methods. Python does not support the explicit creation of interfaces. However, since it does support multiple inheritance, the programmer can mimic the interface functionality by utilizing abstract class inheritance, limited to the exclusive use of *abstract methods*.

> **Observation 3.29 – Interface:** A class that cannot be instantiated but serves as a template for sub classes. Unlike abstract classes, interfaces cannot have non-abstract methods.

3.8 MODULES AND PACKAGES IN PYTHON

Modules and *packages* refer to structures used for organizing code in Python. Modules are files containing Python code structures (e.g., classes, methods, attributes, or simple variables) signified by the *.py* file extension. Instead of rewriting particular blocks of code, modules can be imported into other Python files or applications, thus allowing for a *modular* programming approach based on *reusable* code.

> **Observation 3.30 – Module:** A module provides a way of organizing code in Python. Modules can host classes, methods, attributes, or even simple variables that can be imported and reused in other classes. Modules are commonly used with abstract classes.

Abstract classes and interfaces are two of the programming structures commonly stored in modules, from where they can be imported on demand. In the example provided in the following script, the entire definition of class Employee is stored in a module named *employee.py*:

```
1   # 'Employee' module saved in 'employee.py' file
2   class Employee:
3
4       # Define the constructor and private attributes of the class
5       def __init__(self, first, last, salary):
6           self.__firstName = first
7           self.__lastName = last
8           self.__salary = salary
9
10      # Define the getter for annual salary
11      def getAnnualSalary(self):
12          return self.__salary * 12
13
14      # Define the getter for fullName
15      def getFullName(self):
16          return self.__firstName + " " + self.__lastName
```

3.8.1 THE IMPORT STATEMENT

Python module files are imported using the `import` statement. The statement may include one or more modules. The syntax is the following:

```
import module1, [module2, module3...]
```

Once a module is imported, its classes and methods can be referenced using its name as a prefix (i.e., `module.classname`). The following example imports the `Employee` class from the associated *employee.py* module, and accesses its attributes and methods from the main body of the program:

> **Observation 3.31 – The** `import` **Statement:** Used to import either specific methods and attributes or entire classes stored in modules.

```
1    # Import the 'employee.py' file as a module
2    import employee
3
4    # Use the module to create and use a new object
5    emp1 = employee.Employee("Maria", "Rena", 15000)
6    print(emp1.getFullName())
7    print()
```

Output 3.8.1:

```
Maria Rena
```

3.8.2 THE FROM...IMPORT STATEMENT

A Python module may contain several classes, methods, attributes, or variables. The `from...import` statement allows the programmer to selectively import specific components from a module. The syntax is the following:

```
from module import name1, [name2, name3...]
```

Note that the names used in this example (e.g., name1, name2, name3) represent names of classes, methods, or attributes.

To import all objects from a module the following syntax can be used:

```
from module import *
```

The reader should note that if a specific class is imported *explicitly*, it can be referenced without a prefix, like in the next example:

```
1    # Import class 'Employee' from 'employee' module in 'employee.py'
2    from employee import Employee
3
4    # Use the imported class to create and use a new object
5    emp1 = Employee("Alex", "Flora", 18000)
6    print(emp1.getFullName())
7    print(emp1.getAnnualSalary())
```

Output 3.8.2:

```
Alex Flora
216000
```

3.8.3 PACKAGES

A *package* is a collection of modules grouped together in a common folder. The package folder must contain a file with the designated name *__init__.py*, which indicates that the folder is a package. The *__init__.py* file can be empty, but it must be always present in the package folder. Once the package structure is created,

Observation 3.32 – Package: A mechanism used to store a number of different modules in the same folder for better code organization.

Python modules can be added as required. The example in Figure 3.6 illustrates the structure of a package named *hr*, containing the mandatory *__init__.py* file, and a module named *employee.py*.

Modules contained in packages can be imported to an application using the package name as a prefix in the import statement, as shown in the following scripts:

```
1   # Import the employee module from the 'hr' package
2   import hr.employee
3
4   # Use 'Employee' class stored in the module to create & use an object
5   emp1 = hr.employee.Employee("Alex", "Flora", 16000)
6   print(emp1.getFullName())
7   print(emp1.getAnnualSalary())
```

Output 3.8.3.a:

```
Alex Flora
216000
```

```
1   # Import 'Employee' class in the employee module from 'hr' package
2   from hr.employee import Employee
3
4   # Use the 'Employee' class of the module to create and use an object
5   emp2 = Employee ("Alex", "Flora", 15000)
6   print(emp2.getFullName())
7   print(emp2.getAnnualSalary())
```

FIGURE 3.6 Package hr contains the __init__.py file and the employee.py module.

Output 3.8.3.b:

```
Alex Flora
180000
```

3.8.4 USING MODULES TO STORE ABSTRACT CLASSES

Modules may be also used to store abstract classes or interfaces. In the following example, abstract class IEmployee is stored in module *employee.py*, which is contained in the *hr* package named:

```
1   # Use 'abc' module to create an abstract class: store it as a module
2   # ('employee.py') in the hr package
3   from abc import ABC, abstractmethod
4
5   # Define abstract class 'IEmployee' and its behavior
6   class IEmployee(ABC):
7       @abstractmethod
8       def getTotalSalary(self):
9           Pass
10      @abstractmethod
11      def getFullName(self):
12          Pass
```

The following script demonstrates how the programmer can import the IEmployee class to the application, and use it to create a sub class (FullTimeEmployee):

```
1   # Import the 'IEmployee' class from the employee module ('hr' package)
2   from hr.employee import IEmployee
3
4   # Define a new sub class inheriting from the 'IEmployee' super class
5   class fullTimeEmployee(IEmployee):
6
7       # The constructor, attributes & behavior of the sub class
8       def __init__(self, first, last, salary, allowances):
9           self.__first = first
10          self.__last = last
11          self.__salary = salary
12          self.__allowances = allowances
13
14      def getTotalSalary(self):
15          return self.__salary + self.__allowances
16
17      def getFullName(self):
18          return self.__first + " " + self.__last
19
20  # Create and use a new object
21  ftl = fullTimeEmployee("Maria", "Rena", 15000, 6000)
22  print(ftl.getFullName())
23  print(ftl.getTotalSalary())
```

Output 3.8.4:

```
Maria Rena
21000
```

3.9 EXCEPTION HANDLING

When writing programs in Python, or in any other programming language for that matter, the code may include errors. Depending on their nature and significance, these errors may lead to a number of issues, such as preventing the program from executing, generating incorrect output, or causing the program to crash. It is, thus, the responsibility of the programmer to provide error identification and handling solutions, whenever possible. Errors can be classified into three main categories:

- **Compile Time Errors:** They occur due to incorrect syntax, datatype use, or parameters in a method call among others. Whenever the compiler encounters a compile error in the program it will stop execution. Compile time errors are the easiest to handle and can be fixed easily by correcting the problematic code line(s).
- **Logical Errors:** They occur due to incorrect program logic. A program containing logical errors may run normally without crashing, but will generate incorrect output. Logical errors are handled by testing the application with various different input values, and making corrections to the program logic as necessary.
- **Runtime Errors:** They occur during the execution of a program, due to external factors not necessarily related to the code. For example, a user may provide an invalid input that the application is not expecting, or the code is attempting to read a file that does not exist in the system. In Python, these types of errors raise exceptions and cause the program to crash and terminate abruptly. To prevent this, the programmer should *catch* these exceptions by adding appropriate error handling code to the program.

> **Observation 3.33 – Types of Errors:** There are three types of errors that may be encountered:
>
> 1. Compile Time: This is due to incorrect syntax and will not allow the program to execute.
> 2. Logical: This error type will allow execution of the program but may produce incorrect output.
> 3. Runtime: Raised because of unexpected external issues, wrong input, or wrong expressions. This error type will cause the program to crash.

3.9.1 HANDLING EXCEPTIONS IN PYTHON

In Python, when a runtime error occurs, the program crashes and a *built-in exception* is raised. The exception provides information about the error. For example, running the following script will cause a ZeroDivisionError exception as it attempts to divide a value by 0. The exception provides information about the nature of the issue (i.e., division by zero).

> **Observation 3.34 – Handling Exception:** Use the try...exception... [else:]...[finally] syntax to identify possible errors that might be encountered during execution and handle them appropriately, avoiding abnormal termination of the program.

```
1   a = 10
2   b = 0
3   print(a / b)
```

Output 3.9.1.a:

```
-------------------------------------------------------------------------
ZeroDivisionError                         Traceback (most recent call last)
<ipython-input-2-dd04aeeae314> in <module>
      1 a = 10
      2 b = 0
----> 3 print(a / b)

ZeroDivisionError: division by zero
```

Exceptions can be handled using a `try`/`except` block of statements. As the name suggests, this structure consists of two distinct blocks: try and `except`. The `try` block includes critical statements that are most likely to cause an exception. When the exception occurs within the `try` block, the execution of the program jumps to the `except` block. This part contains code that handles the exception appropriately. For example, it may display a related user message, close an open file, or log the error to a file. If no exception is raised in the `try` block, the program skips the `except` block and execution continues as normal.

Two optional blocks may also be added to the exception handling code, namely `else` and `finally`. The `else` block contains statements that are executed in case no exception occurs. The `finally` block contains code that must be executed irrespectively of whether an exception occurs or not, and is mainly used for releasing external resources, such as closing an open file.

> **Observation 3.35 – Raising Exceptions:** Instead of using built-in exceptions, it is possible to define user-defined exception to address specific errors in the program execution.

The main Python syntax for catching exceptions is shown below:

```
try:
        critical statement
except [ExceptionClass as err]:
        exception handling statements
[else:
        statements to execute when exception has not occurred
finally:
        statements to execute whether an exception has occurred or not]
```

The `ExceptionClass` is optional, and refers to the type of exception being handled. If omitted, all types of exceptions are handled by the except block.

The following example is an improved version of the code used in previous examples, since in this occasion the program will not crash abruptly. Instead, it will terminate with a user-friendly error message:

```
1    # Declare variables 'a' and 'b'
2    a, b = 10, 0
3
4    """ Try to divide the variables and if an exception is raised
5    execute the alternative statement in the 'except' block """
6    try:
7            print(a / b)
8    except:
9            print("An error has occurred")
```

Output 3.9.1.b:

```
An error has occurred
```

3.9.1.1   Handling Specific Exceptions

Trying to catch all types of errors within a single `try/except` block is not considered good programming practice, as it does not allow the programmer to handle exceptions on a case-by-case basis. Python provides various different built-in exception classes that are raised automatically, according to the type of error being encountered. These specific exceptions can be utilized by referring to their designated names. Table 3.5 lists a number of common built-in exception classes in Python.

The example presented below demonstrates how a specific error can be handled using the `ZeroDivisionError` exception class:

```
1    # Declare variables 'a' and 'b'
2    a, b = 10, 0
3
4    # Attempt to print the result of the division of 'a' by 'b'
5         try:
6     print(a / b)
7
8    # If a specific 'ZeroDivisionError' occurs print a relevant message
9    except ZeroDivisionError as err:
10         print("An error has occurred")
11         print(err)
```

Output 3.9.1.1:

```
An error has occurred
division by zero
```

A `try` block may also contain multiple `except` blocks. This is useful when the programmer wants to handle various different types of errors. However, only one of these blocks will be executed when

---

TABLE 3.5
Common Exception Classes in Python

Exception Class	Description
ArithmeticError	Raised when arithmetic operations fail. Includes the following exception sub classes: `OverflowError`, `ZeroDivisionError`, `FloatingPointError`
OverflowError	The result of an arithmetic operation is out of range
ZeroDivisionError	Attempting to divide by zero
FloatingPointError	Floating-point operation failure
IndexError	An array index is invalid
AttributeError	A non-existing attribute is referenced for an instance
TypeError	An operator or method is applied to an inappropriate type of object
FileNotFoundError	A file is not found
ValueError	The parameter of a method is of an inappropriate type

an exception occurs. When multiple `except` blocks are used, the code structure must start with the more specific exception classes and end with the more generic ones. In this case, the latter are used as an added measure of trying to handle unexpected errors that are not accounted for explicitly. The syntax of a multiple exceptions block is provided below:

```
try:
        # critical statements
        pass
except FileNotFoundError:
        # handle FileNotFound exception
        pass
except (IndexError, ArithmeticError):
        # except block with multiple exceptions
        # index out of range in an array and arithmetic error
        pass
except:
        # must be placed at end. Handles all other errors
        pass
```

3.9.2 RAISING EXCEPTIONS

In Python, built-in exceptions are raised automatically when a corresponding runtime error occurs. However, it also allows raising exceptions defined by the programmer. This is achieved by using the `raise` keyword followed by the exception name. When raising user-defined exceptions, it is also possible to provide a string parameter that describes the reason for raising the exception. The next example demonstrates such a case, where if the user input (i.e., user's age) is less than 18, a user-defined exception (i.e., `ValueError`) is raised:

```
1   # Accepts the user's age
2   age = int(input("Enter your age: "))
3
4   # If the input is an integer less than 18 raise an error
5   if age < 18:
6           raise ValueError("Age cannot be below 18")
```

Output 3.9.2.a:

```
Enter your age: 17
---------------------------------------------------------------------------
ValueError                                Traceback (most recent call last)
<ipython-input-6-de16dc8d8553> in <module>
      4 # If the input is an integer less than 18 raise an error
      5 if age < 18:
----> 6     raise ValueError("Age cannot be below 18")

ValueError: Age cannot be below 18
```

In the example below, built-in exception `AttributeError` is raised when the value of private attribut*e* __first is invalid.

```
1    # Define class 'Employee'
2    class Employee:
3
4            # Define the getter method
5            def getFirstName(self):
6                    return self.__first
7
8            # Define the setter method
9            def setFirstName(self, value):
10             if len(value) < 15:
11                     self.__first = value
12             else: # Raise error if the input exceeds 14 characters
13                     raise AttributeError("First name must be less than 15 \
14   characters")
15
16   # Attempt to create a new object and set the first name
17   try:
18           emp1 = Employee()
19           emp1.setFirstName("Maria Rena White") # Exception raised
20
21   # Raise the 'AttributeError' exception if the first name exceeds 14
22   # characters
23   except AttributeError as err:
24           print(err)
25   except:
26           print("An error has occurred")
```

Output 3.9.2.b:

```
First name must be less than 15 characters
```

Raising exceptions is also a convenient way of handling invalid values passed to an attribute setter method. However, in this case, instead of raising built-in exceptions, it is preferable to create custom, in-class ones.

3.9.3 User-Defined Exceptions in Python

As mentioned, Python raises built-in exceptions whenever a runtime error occurs. However, for *custom errors*, Python also allows the creation of *custom exceptions* that can be raised from within the code. For example, instead of raising built-in exception `AttributeError`, the programmer can create a user-defined exception by deriving a new class from the `Exception` base class, as shown below:

```
class NewExceptionName (Exception):
    pass
```

In the following script, user-defined exception `FirstNameException` is created and subsequently raised in the setter method, when the length of the first name exceeds the limit of 14 characters:

```
1    # Define the new exception class based on the built-in exceptions
2    class FirstNameException(Exception):
3
4          def __init__(self, message):
5                super().__init__(message)
6
7    # Define class 'Employee'
8    class Employee:
9
10         # Getter method
11         def getFirstName(self):
12               return self.__first
13
14         # Setter method
15         def setFirstName(self, value):
16           if len(value) < 15:
17                 self.__first = value
18           else:
19                 # Raise an extended exception 'FirstNameException' if
20                 # the first name exceeds 14 chars
21                 raise FirstNameException(
22                 # Raise error
23                 "First name should be less than 15 characters")
24
25   # Create and use the new object handling possible user-defined
26   # exceptions
27   try:
28     emp1 = Employee()
29         emp1.setFirstName("Maria Rena White") # Exception raised
30   except FirstNameException as err:
31         print(err)
32   except:
33         print("An error has occurred")
```

Output 3.9.3:

```
First name should be less than 15 characters
```

3.10   CASE STUDY

Sherwood real estate requires an application to manage properties. There are two types of properties: apartments and houses. Each property may be available for rent or sale.

Both types of properties are described using a reference number, address, built-up area, number of bedrooms, number of bathrooms, number of parking slots, pool availability, and gym availability. A house requires extra attributes such as the number of floors, plot size and house type (villa or townhouse). An apartment requires additional attributes such as floor and number of balconies.

Each type of property (house or apartment) may be available for rent or sale.

A rental property should include attributes such as deposit amount, yearly rent, furnished (yes or no), and maids' room (yes or no). A property available for sale has attributes such as sale price and estimated annual service charge.

All properties include a fixed agent commission of 2%. Both types of sale properties have a fixed tax of 4%.

All properties require a method to display the details of the property.

All properties should include a method to compute the agent commission. For rental properties, agent commission is calculated by using the yearly rental amount, whereas for purchase properties it is calculated using the sale price.

Both types of purchase properties should include a method to compute the tax amount. Tax amount is computed based on the sale price.

Design and implement a Python application that creates the four types of properties (e.g., RentalApartment, RentalHouse, SaleApartment, SaleHouse) by using multiple inheritance and abstract classes. Implement class attributes and instance attributes using encapsulation. All numeric attributes, such as price, should be validated for inputs with a suitable minimum and maximum price.

Define the methods in the abstract class and implement it in the respective classes. Override the print method to display each property details.

Test your application by creating new properties of each type and calling the respective methods.

3.11 EXERCISES

1. Using the diagram shown below, write Python code for the following:

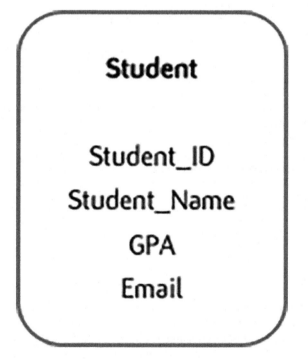

a. Create a class named Student.
b. Create appropriate getters and setters using the @property decorator for Student_ Name and GPA attributes. The Student_ID and Email attributes are read only. Create only getter methods for these attributes.
c. Add a private class attribute named MAX_ID and set it to 0.

d. Add a default constructor method to the Student class. The default constructor should initialize the GPA attribute to 0 and Student_ID to MAX_ID + 1.

e. Add an overloaded constructor that takes Student_Name and GPA as arguments and initializes private data variables with the values provide. In addition, it should set the Student_ID to MAX_ID + 1 and the email attribute to first_name.last_name@university.edu.

f. Modify the setter method of the GPA attribute to check if the provided value is between 0 and 4 before storing it.

g. Add a destructor method to the Student class. The method should print the message "All student records destroyed".

h. Instantiate two new objects called std1 and std2, using the default and the overloaded constructors, respectively.

i. Print the data values stored in each object's attributes.

j. Delete objects std1 and std2.

4 Graphical User Interface Programming with Python

Ourania K. Xanthidou
Brunel University London

Dimitrios Xanthidis
University College London
Higher Colleges of Technology

Sujni Paul
Higher Colleges of Technology

CONTENTS

4.1 Introduction 108
    4.1.1 Python's GUI Modules 109
    4.1.2 Python IDE (Anaconda) and Chapter Scope 109
4.2 Basic Widgets in Tkinter 109
    4.2.1 Empty Frame 110
    4.2.2 The Label Widget 111
    4.2.3 The Button Widget 119
    4.2.4 The Entry Widget 120
    4.2.5 Integrating the Basic Widgets 121
4.3 Enhancing the GUI Experience 126
    4.3.1 The Spinbox and Scale Widgets inside Individual Frames 126
    4.3.2 The Listbox and Combobox Widgets inside LabelFrames................. 131
    4.3.3 GUIs with CheckButtons, RadioButtons and SimpleMessages 138
4.4 Basic Automation and User Input Control... 146
    4.4.1 Traffic Lights Version 1 – Basic Functionality................................. 146
    4.4.2 Traffic Lights Version 2 – Creating a Basic Illusion 148
    4.4.3 Traffic Lights Version 3 – Creating a Primitive Automation 149
    4.4.4 Traffic Lights Version 4 – A Primitive Screen Saver with a Progress Bar.............. 151
    4.4.5 Traffic Lights Version 5 – Suggesting a Primitive Screen Saver............................ 156
4.5 Case Studies... 159
4.6 Exercises 159

DOI: 10.1201/9781003139010-4

4.1　INTRODUCTION

In modern day software development, creating an application with an intuitive Windows style *Graphical User Interface* (*GUI*) is a must in order to make it attractive for the user. There are four essential concepts related to this, and the associated programming tools:

- **Widgets:** The different components used to create an application GUI. These are relatively simple, pre-defined objects available through Python libraries. In this chapter, the libraries and modules used include *tkinter* and *PIL*, providing *visual attributes* that supply the necessary windows object aesthetic. The associated objects can be as simple as *labels*, *texts*, and *buttons* or as complex as *frames* and *grids*.

  > **Observation 4.1 – Widget:** A graphical component used to create the interface of the Python application. This is provided as a pre-defined class of the *tkinter* or *PIL* packages.

- **Options:** Characteristics or attributes of a widget/object that dictate the way the latter looks and behaves (e.g., the object color, text, position, or alignment). Value changes, usually integrated with interactions between the user and the GUI, control aspects like the visual appearance or format of the application and its behavior.

  > **Observation 4.2 – Option:** An attribute of the widget that controls its look and behavior.

- **Methods:** Pre-defined or newly developed snippets of Python code, aiming to affect the widgets by changing the values of their properties/attributes. There is a wealth of method in the various packages offered by Python, such as *tkinter* and *PIL*. They can be as simple or complex as the developer intends.

  > **Observation 4.3 – Method:** A specific structure of code that changes the value of an option of a particular widget. It can be either pre-defined or newly developed.

- **Events:** The interaction between the user of a GUI-based Windows style application and the various widgets of the application is expressed through the various available *events* that trigger the execution of particular commands or blocks of code. There are numerous such events offered by Python, some of them applicable to several different widgets. Examples are the *click* or *double-click* of a mouse, pressing the *enter key* in the keyboard, *hovering* over a widget, or changing the text of a text widget.

  > **Observation 4.4 – Event:** An interaction between the user and an object that causes a change in terms of the object's appearance and/or value. Many types of interactions are available.

  > **Observation 4.5 – Event-Driven (or Visual) Programming:** The concept of handling events, through the use of methods in order to change the options of an object and, thus, their look and actions.

Event-driven (or visual) programming is the process during which one or more of the properties/attributes of a widget/object changes state or value. This is done through the use of specific *methods* and is triggered through interactions between the user and the widget/object, caught by the associated *event*.

The focus of this chapter is to introduce the concept of event-driven (or visual) programming by presenting some of the most popular widgets and the associated methods and properties/attributes/options, and the most commonly used *events* for the creation of a GUI experience.

4.1.1   PYTHON'S GUI MODULES

Python provides a rather complete set of widgets (pre-sented as classes) to create objects for user-friendly applications, a comprehensive and developer-friendly set of methods available through these widgets, a rich set of attributes of these widgets, and an adequate number of well-defined programmable events that can be triggered through user interactions. There are two basic modules that define the components and functionality of these widgets, namely the *tkinter* and the *PIL* modules.

Observation 4.6 – Python GUI Modules: The most important and frequently used modules for GUI programming in Python are *Tk/Tcl*, *Tkinter.Tix*, and *tkinter.ttk*.

The *tkinter* module provides a number of classes, including the fundamental *Tk* class, as well as numerous other classes associated with GUIs. It consists of the following:

- **Tk/Tcl:** A toolkit that includes widgets for GUI applications.
- **Tkinter.Tix:** An extension of *tkinter* including more advanced GUI widgets (e.g., spin boxes, trees).
- **tkinter.ttk:** a collection of widgets, some of which are part of the original *tkinter* module (e.g., combo boxes, progress bars).

Although it is not possible to describe all the widgets, methods, properties, and events available through all these modules in detail in this chapter, an effort is made to present the most commonly used ones and provide examples of their application. This chapter gradually moves from simpler to more sophisticated cases of increasing complexity.

4.1.2   PYTHON IDE (ANACONDA) AND CHAPTER SCOPE

In line with the approach taken in previous chapters, the *Jupyter Notebook* (*Anaconda*) is the platform of choice for the code developed in this chapter. Detailed download and installation instructions are provided in the introductory Chapter 1.

It is worth noting that when writing programs in Python, or any other language indeed, it is useful following good programming practices. It is a good habit and a helpful strategy in the long run to use *pseudocode* in the form of comments before lines or blocks of code that are written to accomplish a specific and well-defined task. This allows the reader or the owner of the program to understand the underlying algorithm, making the program more readable and user-friendly.

It is beyond the scope of this chapter to write "highly intelligent" Python programs that create complex and sophisticated GUI applications, as this would make this chapter content difficult to digest. Instead, this chapter aims at presenting the tools and their suggested uses for the creation of common tasks and applications, without trying to offer the most efficient or optimal solution for such tasks.

4.2   BASIC WIDGETS IN TKINTER

Arguably, when creating a GUI, there are four basic widgets that intuitively come to mind. These are the actual *frame*, and the *label*, the *button*, and the *entry* widgets (the latter is commonly referred to as *textbox* in other programming languages). In this section, these particular widgets will be presented and utilized to create simple GUI applications.

Observation 4.7 – Basic Widgets: The basic widgets of any GUI in Python are the *form*, and the *label*, the *button*, and the *entry* widgets.

4.2.1 EMPTY FRAME

The basic *frame* is the initial *parent* object that a Python GUI application requires in order to support the GUI interface and functionality. The following Python code creates a basic, empty frame titled "Python Basic Window Frame":

```
1    # Import the necessary library
2    import tkinter as tk
3
4    # Create the frame using the tk class
5    winFrame = tk.Tk()
6    winFrame.title("Python Basic Window Frame")
7
8    winFrame.mainloop()
```

Output 4.2.1.a:

A few things are worth noting in this example:

- Every frame is an object of the tk class, initiated by the Tk() *constructor*. The object *must have a name*.
- It is common practice to give a title to every frame using the title() method.
- The mainloop() method runs the frame and puts tkinter in a *wait state*, which internally monitors *user-generated events*, such as keyboard and mouse activity.

By default, the basic frame is resizable and its size is determined automatically. If there is a requirement for specifically defining and controlling whether it should be resizable, two methods can be used, namely: resizable() and geometry(). If it is preferred to have a non-resizable frame, one can just pass Boolean value False to both parameters of the resizable() method. Accordingly, passing True would result in a resizable frame. The geometry() method is used to pass the initial size of the frame as a string. It is also possible to define the maximum and minimum sizes of the window frame, as well as its background color. The aforementioned methods and their application are demonstrated in the following example:

> **Observation 4.8 – The mainloop() Method:** Use the mainloop() method to monitor and control any type of interaction between the user and the application.

> **Observation 4.9 – Frame Methods:** Use the title(), resizable(), geometry(), maxsize(), minsize(), config() methods to configure the basic content, size, geometry, flexibility, and look of the main window frame.

```
1    # Import the necessary library
2    import tkinter as tk
3
4    # Create the frame using the tk object
5    winFrame = tk.Tk()
6
7    # Provide a title for the frame
8    winFrame.title("Python Controlled Frame")
9
10   # The frame is resizable if the method parameters are set
11   # to True or non-zero; if set to False, it is not resizable
12   winFrame.resizable(True, True)
13
14   # The frame will have initial dimensions of 500 by 200
15   winFrame.geometry('500x200')
16
17   # The frame can be resized up to a maximum of 1500 by 600
18   winFrame.maxsize(1500, 600)
19
20   # The frame can be resized down to a minimum of 250 by 100
21   winFrame.minsize(250, 100)
22
23   # The background colour of the frame can be changed with
24   # the use of the configure method and the bg option
25   winFrame.configure(bg = 'dark grey')
26   winFrame.mainloop()
```

Output 4.2.1.b:

Once the basic frame is set, the actual GUI can be created by adding the desired widgets.

4.2.2   THE LABEL WIDGET

The *label* widget is a basic widget class from the *tkinter* module. It is used to display a message or image on screen. As it does not accept input from the keyboard its value *cannot be changed directly during runtime*, but this can be done indirectly through the code. The widget comes with several methods and the associated parameters and options that can be used to change its

Observation 4.10 – Labels: Basic widgets used to display a message or an image. They do not accept input and, thus, their value cannot be changed directly by the user. Label widgets must be attached to a frame or window through the pack() or grid() methods.

appearance and functionality. The following script is an example showcasing the use of some of the available options:

```
1    # Import the tkinter library
2    import tkinter as tk
3
4    # Define the parent frame
5    winFrame = tk.Tk()
6    winFrame.title("Labels in Python")
7    winFrame.resizable(True, True)
8    winFrame.geometry('300x100')
9
10   # Create a label object based on the tk.Label class
11   winLabel = tk.Label(winFrame, text = "Hello Python programmer")
12
13   # Associate the label object with the parent frame
14   winLabel.pack()
15
16   # Run the interface
17   winFrame.mainloop()
```

Output 4.2.2.a:

The script creates a window frame containing a basic label widget, used to display a text message. The label widget (`winLabel`) is derived from the `tk.Label` class, by means of the related `tk.Label()` constructor. This call takes a minimum of two parameters, namely the parent frame (`winFrame`) and the *text* that assigns the label with a message to display. The label widget is tied to the parent frame through the `pack()` method. Finally, the `mainloop()` method activates the application.

An extension of this basic use of the label widget could involve the use of the `grid()` method, in order to control its placement within the parent frame more efficiently:

```
1    # Import the tkinter library
2    import tkinter as tk
3
4    # Define the parent frame
5    winFrame = tk.Tk()
6    winFrame.title("Python Label using the Grid")
7
8    # Create a label and place it in the Grid
9    winLabel = tk.Label(winFrame, text = \
10       "Use the Grid method to \nplace the label in a static position")
11   # Specify the row and column the label
```

```
12   # is to be placed, regardless of the size of the parent frame
13   winLabel.grid(column = 0, row = 0)
14
15   winFrame.mainloop()
```

Output 4.2.2.b:

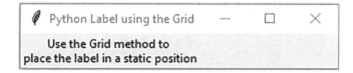

A couple of things are noteworthy in this case:

- For clarity purposes, if the statement is lengthy, it can be broken by inserting the *backslash special character* ("\"). This character informs Python that the statement continues on the next line.
- Using the `grid()` method instead of `pack()` ensures that the label widget will be placed in the respective grid cell, in this case in the first row (`row = 0`) and first column (`column = 0`), and that its position *will not be directly adjusted based on the size of the frame or parent widget*.

> **Observation 4.11 – The Backslash Special Character ("\"):** Use the backslash special character ("\") to break a lengthy line.

> **Observation 4.12 – *expand, foreground, background, font, anchor*:** Use the *expand, foreground, background, font*, and *anchor* options to improve the appearance of widgets.

It is possible to further enhance the appearance of a label by changing its *foreground* and *background* colors, its *alignment*, and its *expandability*, as shown in the following script. This example demonstrates the behavior of the alignment of labels before and after resizing the window frame:

```
1    # Import the relevant library
2    import tkinter as tk
3
4    # The basic frame with the tk.Tk() constructor and provide a title
5    winFrame = tk.Tk()
6    winFrame.title('More options for label widgets')
7
8    # Create the 1st label and place it in the middle of the parent window
9    winLabel1 = tk.Label(winFrame, fg = 'green', font = "Arial 24",
10       text = 'A green label of Arial 24, that does not expand')
11   winLabel1.pack(expand = 'N')
12
13   # The second label that expands vertically when the frame is resized
14   winLabel2 = tk.Label(winFrame, bg = 'red', fg = 'white',
15       text = 'A label in red background that expands only vertically')
16   winLabel2.pack(expand = 1, fill = tk.Y)
17
18   # The third label that expands horizontally when the frame is resized
19   winLabel3 = tk.Label(winFrame, bg = 'blue', fg = 'yellow',
20       text = 'A label in blue background that expands only horizontally')
```

```
21  winLabel3.pack(expand = 1, fill = tk.X)
22
23  # The fourth label 'anchored' (i.e., align always to the right/east)
24  winLabel4 = tk.Label(winFrame, anchor = 'e', bg = 'green',
25      text = 'A right, i.e., east, aligned label')
26  winLabel4.pack(expand = 1, fill = tk.BOTH)
27
28  winFrame.mainloop()
```

Output 4.2.2.c:

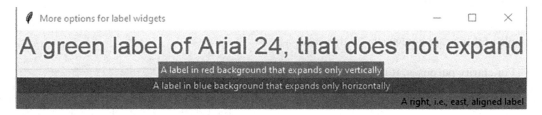

A number of key observations can be made based on this example:

1. The expand option can be used to control whether a label widget will expand in line with its parent widget. If the value is 0 or "N", the label *will not expand.*
2. If the expand option is set to 'Y' or non-zero, the label widget can expand in line with its parent widget. It can be also specified whether the expansion will be horizontal, vertical, or both. In this case, one can use the fill option with the following arguments: X for horizontal expansion only; Y for vertical expansion only, and BOTH for a simultaneous expansion in both directions.
3. The fg and bg options can be used to define the color of the foreground and background of the label widget, respectively.
4. The font option can be used to set up the font name and size of the text in the label widget.
5. The anchor option can be used to ensure that the label widget will not relocate if the parent widget does.

Ultimately, label widgets can provide additional functionality and can be further enhanced in terms of their appearance. Indeed, they can be loaded with *image objects* with or without associated text, and can function as *buttons* (covered in a later section of this chapter). If images are to be used, the *PIL* module must be imported, as it provides the necessary methods to support such processes. The following Python program uses image objects as buttons that change the text-related properties of the main label:

```
1  # Import the relevant library
2  import tkinter as tk
3  # Import the necessary image processing classes from PIL
4  from PIL import Image, ImageTk
5
6  global photo1, photo2, photo3, photo4, photo5, photo6
7
8  # Declare the methods to control the click events from each of the
9  # labels and change the settings of the main label
```

```
10   def changeBorders(a, b):
11       winLabel5.config(relief = a, borderwidth = b)
12   def changeText(a):
13       winLabel5.config(text = a)
14   def changeAlignment(a):
15       winLabel5.config(anchor = a)
16
17   # Declare the method that will open the various images
18   def photos():
19       global photo1, photo2, photo3, photo4, photo5, photo6
20
21       image1 = Image.open('LabelsDynamicWithImageGoodMorning.gif')
22       image1 = image1.resize((100, 50), Image.ANTIALIAS)
23       photo1 = ImageTk.PhotoImage(image1)
24
25       image2 = Image.open('LabelsDynamicWithImageGoodAfternoon.gif')
26       image2 = image2.resize((100, 50), Image.ANTIALIAS)
27       photo2 = ImageTk.PhotoImage(image2)
28
29       image3 = Image.open('LabelsDynamicWithImageGoodEvening.gif')
30       image3 = image3.resize((100, 50), Image.ANTIALIAS)
31       photo3 = ImageTk.PhotoImage(image3)
32
33       image4 = Image.open('LabelsDynamicWithImageAlignLeft.gif')
34       image4 = image4.resize((100, 50), Image.ANTIALIAS)
35       photo4 = ImageTk.PhotoImage(image4)
36
37       image5 = Image.open('LabelsDynamicWithImageAlignRight.gif')
38       image5 = image5.resize((100, 50), Image.ANTIALIAS)
39       photo5 = ImageTk.PhotoImage(image5)
40
41       image6 = Image.open('LabelsDynamicWithImageAlignCenter.gif')
42       image6 = image6.resize((100, 50), Image.ANTIALIAS)
43       photo6 = ImageTk.PhotoImage(image6)
44
45   # Declare the method that will create the first row of labels
46   # that will shape the main label
47   def firstRow():
48       winLabel1a = tk.Label(winFrame, text = "Left click to \
49           \n change to raised label \nwith border width of 4",
50           relief = "raised")
51       winLabel1a.grid(column = 1, row = 0)
52       winLabel1a.bind("<Button-1>", lambda event, a = "raised",
53           b = 4: changeBorders(a, b))
54       winLabel1b = tk.Label(winFrame, text = "Left click to \n change \
55           to sunken label \nwith border width of 6", relief = "raised")
56       winLabel1b.grid(column = 2, row = 0)
57       winLabel1b.bind("<Button-1>", lambda event, a = "sunken",
58           b = 6: changeBorders(a, b))
59       winLabel1c=tk.Label(winFrame, text = "Left click to \n change \
60           to flat label \nwith border width of 8", relief = "raised")
61       winLabel1c.grid(column = 3, row = 0)
```

```
62      winLabel1c.bind("<Button-1>", lambda event, a = "flat",
63          b = 8: changeBorders(a, b))
64
65  # Declare the method that will create the second row of labels
66  # that will shape the main border
67  def secondRow():
68      winLabel2a = tk.Label(winFrame, text = "Left click to \n change \
69          to ridge label \nwith border width of 10", relief = "raised")
70      winLabel2a.grid(column = 1, row = 4); winLabel2a.bind("<Button-1>",
71          lambda event, a = "ridge", b = 10: changeBorders(a, b))
72      winLabel2b = tk.Label(winFrame, text="Left click to \nchange to \
73          solid label \nwith border width of 12", relief = "raised")
74      winLabel2b.grid(column = 2, row = 4); winLabel2b.bind("<Button-1>",
75          lambda event, a = "solid", b = 12: changeBorders(a, b))
76      winLabel2c = tk.Label(winFrame, text="Left click to \n change to \
77          groove label \nwith border width of 14", relief = "raised")
78      winLabel2c.grid(column = 3, row = 4); winLabel2c.bind("<Button-1>",
79          lambda event, a = "groove", b = 14: changeBorders(a, b))
80
81  # Declare the method that will create the third row of labels
82  # that will change the text of the main label
83  def thirdRow():
84      global photo1, photo2, photo3, photo4, photo5, photo6
85
86      winLabel3a = tk.Label(winFrame,
87          text="Double left click to\n change to",
88          image = photo1, compound = 'left', relief = "raised")
89      winLabel3a.grid(column = 0, row = 1)
90      winLabel3a.bind("<Double-Button-1>", lambda event,
91          a = "Good morning": changeText(a))
92      winLabel3b = tk.Label(winFrame, image = photo2, relief = "raised")
93      winLabel3b.grid(column = 0, row = 2)
94      winLabel3b.bind("<Double-Button-1>", lambda event,
95          a = "Good afternoon": changeText(a))
96      winLabel3c=tk.Label(winFrame, image=photo3, compound="center",
97          text="Double click to\n change the text to", relief="raised")
98      winLabel3c.grid(column = 0, row = 3)
99      winLabel3c.bind("<Double-Button-1>", lambda event,
100         a = "Good evening": changeText(a))
101
102 # Declare the method that will create the fourth row of labels
103 # that will adjust the alignments of the text of the main label
104 def fourthRow():
105     winLabel4a = tk.Label(winFrame, image = photo4,
106         text = "Right click to \n left align the text\nof the label",
107         compound = "center", relief = "raised")
108     winLabel4a.grid(column = 4, row = 1)
109     winLabel4a.bind("<Button-3>", lambda event,
110         a = "w": changeAlignment(a))
111     winLabel4b = tk.Label(winFrame, image = photo5, relief = "raised",
112         text = "Right click to \nright align the text\nof the label")
113     winLabel4b.grid(column = 4, row = 2)
```

```
114      winLabel4b.bind("<Button-3>", lambda event,
115          a = "e": changeAlignment(a))
116      winLabel4c = tk.Label(winFrame, image = photo6, compound = "right",
117          text = "Right click to \ncenter align the text\nof the label",
118          relief = "raised")
119      winLabel4c.grid(column = 4, row = 3)
120      winLabel4c.bind("<Button-3>", lambda event, a = "center":
121          changeAlignment(a))
122
123 # The basic frame with the tk.Tk() constructor and provide a title
124 winFrame = tk.Tk()
125 winFrame.title("Playing with Label options at runtime")
126
127 photos()
128 firstRow()
129 secondRow()
130 thirdRow()
131 fourthRow()
132
133 # Create the main label
134 winLabel5=tk.Label(winFrame, text = "...", font= "Arial 18", width= 30)
135 winLabel5.grid(column = 2, row = 2)
136
137 winFrame.mainloop()
```

Output 4.2.2.d:

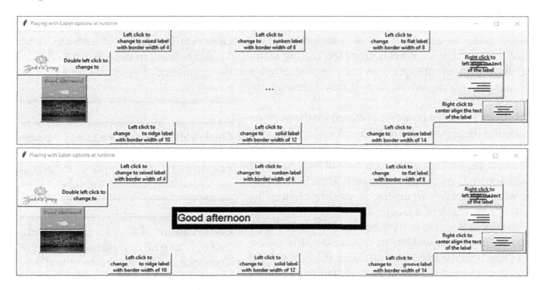

As mentioned, the PIL module provides the necessary classes to support processes related with images, in this case Image and ImageTk.

The photos() method includes six sets of three lines/steps, and deals with the opening and reading of the images, as well as their preparation in order to be loaded to the respective labels. In the first step (i.e., the first line of each set) the Image class and the open() method are used to read the images and create an image object. Next, the script uses the resize() method with the preferred dimensions for the image and the ANTIALIAS option in order to ensure that quality is maintained when downsizing an image to fit the label. This applies to all six cases. During the final step, a new image object is created based on the previously processed image. This is accomplished by using PhotoImage method from the ImageTk class for each of the six cases. It is worth noting that this process applies to images with a *gif* file type. The reader should check the Python documentation to find the exact classes, methods, and options that should be used when working with other types of images, as well as the exact process that must be followed. Nevertheless, the latter should not differ significantly from the process presented above.

The next part of the script involves the use of four methods (firstRow(), secondRow(), thirdRow(), and fourthRow()) to create the twelve labels of the application (i.e., three labels for each row). For each label, three statements are used. The first statement creates the label widget and sets its text property to show the associated message, and the relief property to enhance the widget appearance to raised. The second statement places the label in the desired position within the grid of the current frame. The third statement calls the bind method in order to associate the particular widget with an *event*.

There are a number of events that can be associated with the various widgets. This example involves three basic events, namely: <Button-1> that is triggered when the user *left-clicks* on the parent widget (label in this case), <Button-3> that is triggered when the user *right-clicks*, and <Double-Button-1> that is triggered when the parent widget is *double left-clicked*.

Whenever an event is triggered, a method is usually called in order to execute a set of statements. If the method is to accept arguments from the calling statement, the *lambda* event expression must be also called in order *to define the arguments before they are passed to the method*.

There are a number of options offered for the purpose of changing the appearance of the border of a label widget. These include options such as raised, sunken, flat, ridge, solid, and groove and have to be set through the relief property. Property borderwidth, used with an integer argument, is used to change the default border width of a label.

Finally, it is possible to have both a text and an image appearing in a label widget. In such cases, it is necessary to combine the two elements using the compound expression. The expression accepts different alignment values, namely left when the image is to be placed before the text, right when the image is to be placed after the text, and center when both objects are to be placed at the same position, one over the other.

Observation 4.13 – resize(), ANTIALIAS: Use the resize() method to set the preferred dimensions of the image, and the ANTIALIAS option to ensure that the highest quality is maintained when resizing an image.

Observation 4.14 – <button-1>, <button-3>, <Double-Button-1>: Use the <Button-1>, <Button-3>, and <Double-Button-1> events to catch when the parent widget is *left-clicked*, *right-clicked* or *double-left-clicked*.

Observation 4.15 – lambda: Use the *lambda* event expression to define the arguments passed by an event to a method.

Observation 4.16 – relief, borderwidth: Use the relief and borderwidth properties to adjust the visual attributes of the label.

Observation 4.17 – compound, left, right, center: Use the compound filter to combine text and image objects in a label. Options include left, right, and center.

4.2.3 THE BUTTON WIDGET

As mentioned previously, the label widget is not meant to be used to trigger events initiated by the user interaction with the GUI. In such cases, the *button* widget can be used instead. This widget also belongs to the *tkinter* module, although it can be also found in the *ttk* module, where button objects can be created by defining the *button* class. The following script demonstrates the possible output of five different user interactions through the use of a simple button widget. The script also provides user feedback depending on the type of interaction, by displaying relevant messages through a label widget:

Observation 4.18 – The Button Widget: Use the *button* widget to create objects that are responsive to various types of events (e.g., *click*, *double-click*, *right-click*), and the corresponding options or properties to modify its appearance.

```
1    # Import the relevant library
2    import tkinter as tk
3
4    # Define the method that controls the mouse click events
5    def changeText(a):
6        winLabel.config(text = a)
7
8    # The basic frame with the tk.Tk() constructor and provide a title
9    winFrame = tk.Tk()
10   winFrame.title("A simple button and label application")
11
12   # Create the label
13   winLabel = tk.Label(winFrame, text = "...")
14   winLabel.grid(column = 1, row = 0)
15
16   # Create the button widget and bind it with the associated events
17   winButton=tk.Button(winFrame, text="Left, right, or double left Click "\
18       "\nto change the text of the label", font = "Arial 16", fg = "red")
19   winButton.grid(column = 0, row = 0)
20   winButton.bind("<Button-1>", lambda event, \
21       a = "You left clicked on the button": changeText(a))
22   winButton.bind("<Button-3>", lambda event, \
23       a = "You right clicked on the button": changeText(a))
24   winButton.bind("<Double-Button-1>", lambda event, \
25       a = "You double left clicked on the button": changeText(a))
26   winButton.bind("<Enter>", lambda event, \
27       a = "You are hovering above the button": changeText(a))
28   winButton.bind("<Leave>", lambda event, \
29       a = "You left the button widget": changeText(a))
30
31   winFrame.mainloop()
```

Output 4.2.3:

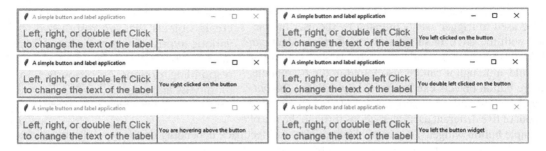

As shown, the process of creating a button widget object and assigning values to its basic options or properties (e.g., text, font, fg) is not different to the one used in the case of the label widget. Accordingly, binding the button widget to an event and calling a method (with or without arguments) is also following the same syntax and logic as in the label widget case.

4.2.4   THE ENTRY WIDGET

The *entry* widget is a basic widget from the *ttk* module (*tkinter* package), which allows input from the keyboard *as a single line*. The widget offers several methods and options that allow the control of its appearance and/or functionality. The widget must be placed in a *parent* widget, usually the current frame, through the .pack() or .grid() methods. The following script introduces the basic use of the entry widget, and its output:

> **Observation 4.19 – Entry/Text:** Use the *entry* and/or *text* widgets from the *ttk* module (*tkinter* package) to allow the user to enter text as a single line or multiple lines respectively. When using the text widget, specify the number of text lines through the height = <number of lines> option.

```
1    # Import the necessary library
2    import tkinter as tk
3    from tkinter import ttk
4
5    # Create the frame using the tk object
6    winFrame = tk.Tk()
7    winFrame.title("Python GUI with text")
8
9    # Create a StringVar object to accept user input from the keyboard
10   textVar = tk.StringVar()
11
12   # Set the initial text for the StringVar
13   textVar.set('Enter text here')
14
15   # Create an entry widget and associate it to the StringVar object
16   winText = ttk.Entry(winFrame, textvariable = textVar, width = 40)
17   winText.grid(column = 1, row = 0)
18
19   winFrame.mainloop()
```

Output 4.2.4:

In line with common GUI development practice, the frame is created first and any child objects (in this case the entry widget) are created and placed in it subsequently. Finally, the `mainloop()` method is called to run the application and monitor its interactions. The `width` property specifies the number of characters the widget can display. The reader should note that this is not necessarily the total number of *accepted* characters, rather the number of *displayed* characters. It must be also noted that if it is necessary to have multiple lines entered, it would be preferable to use the text widget (*tk* module, *tkinter* library) and specify the number of lines through the `height = <number of lines>` option.

The script also introduces a method that helps the programmer monitor the execution of the application: the `StringVar()` constructor from the *tk* class. When associated with relevant widgets, such as the entry widget, its functionality is to create objects that accept text input. Once such an object is created it can have its content set through the `.set()` method. If no content is set, the object will remain empty until the user provides input through the associated widget. The entry widget and the `StringVar` object are associated via the `textvariable`.

4.2.5 INTEGRATING THE BASIC WIDGETS

Having introduced the syntax and functionality of the basic Python widgets included in the *tkinter*, *PIL*, and *ttk* modules/libraries, it would be useful to attempt to create an interface that integrates all of them in one application. The following Python script displays a message to the user, accepts a text input from the keyboard, and uses a number of buttons to change the various attributes of the text, through the integration of *label*, *entry*, and *button* widgets:

```
1    # Import the necessary library
2    import tkinter as tk
3    from tkinter import ttk
4
5    # The tempText variable will store the contents of the entry widget
6    global tempText
7    # The textVar object will associate the entry widget with the input
8    global textVar
9    # Define the winText widget
10   global winText
11
12   # ===================================================================
13   # Declare the methods that will run the application
14   def showHideLabelEntry(a):
15       if (a == 's'):
16           winText.grid()
17       elif (a == 'h'):
```

```
18              winText.grid_remove()
19
20  def showHideEntryContent(a):
21      global tempText
22      global textVar
23      if (a == 's'):
24          if (tempText!= ''):
25              textVar.set(tempText)
26      if (a == 'h'):
27          tempText = textVar.get()
28          textVar.set('')
29
30  def enableLockDisableEntryWidget(a):
31      if (a == 'e'):
32          winText.config(state = 'normal')
33      elif (a == 'l'):
34          winText.config(state = 'disabled')
35
36  def boldContentsOfEntryWidget(a):
37      if (a == 'b'):
38          winText.config(font = 'Arial 14 bold')
39      elif (a == 'n'):
40          winText.config(font = 'Arial 14')
41
42  def passwordEntryWidget(a):
43      if (a == 'p'):
44          winText.config(show = '*')
45      elif (a == 'n'):
46          winText.config(show = '')
47
48  # ====================================================================
49  # Declare the method that will create the application GUI
50  def createGUI():
51      createLabelEntry()
52      showHideButton()
53      showHideContent()
54      enableDisable()
55      boldOnOff()
56      passwordOnOff()
57
58  # Create a label and an entry widget to prompt for input and
59  # associate it with a StringVar object
60  def createLabelEntry():
61      global textVar
62      global winText
63
64      winLabel = tk.Label(winFrame, text = 'Enter text:', bg = 'yellow',
65          font = 'Arial 14 bold', relief = 'ridge', fg = 'red', bd = 8)
```

```
66      winLabel.grid(column = 0, row = 0)
67
68      # A StringVar object to accept user input from the keyboard
69      textVar = tk.StringVar()
70      winText = ttk.Entry(winFrame, textvariable = textVar, width = 20)
71      winText.grid(column = 1, row = 0)
72
73  # Create two button widgets to show/hide the label and entry widgets
74  def showHideButton():
75      winButtonShow = tk.Button(winFrame, font='Arial 14 bold',
76          text = 'Show the\nentry widget', fg='red',
77          borderwidth=8, height=3, width=20)
78      winButtonShow.grid(column = 0, row = 1)
79      winButtonShow.bind('<Button-1>',lambda event,
80          a = 's': showHideLabelEntry(a))
81      winButtonHide = tk.Button(winFrame, font = 'Arial 14 bold',
82          text = 'Hide the\nentry widget',
83          fg = 'red', borderwidth = 8, height = 3, width = 20)
84
85      winButtonHide.grid(column = 1, row = 1)
86      winButtonHide.bind('<Button-1>', lambda event, \
87                      a = 'h': showHideLabelEntry(a))
88
89  # Two button widgets to show/hide the contents of the entry widget
90  def showHideContent():
91      winButtonContentShow = tk.Button(winFrame, font = 'Arial 14 bold',
92          text = 'Show the contents\nof the entry widget',
93          fg = 'blue', borderwidth = 8, height = 3, width = 20)
94      winButtonContentShow.grid(column = 0, row = 2)
95      winButtonContentShow.bind('<Button-1>', lambda event,
96          a = 's': showHideEntryContent(a))
97      winButtonContentHide = tk.Button (winFrame,
98      text = 'Hide the contents\nof the entry widget',
99          font = 'Arial 14 bold', fg = 'blue', borderwidth = 8,
100         height = 3, width = 20)
101     winButtonContentHide.grid (column = 1, row = 2)
102     winButtonContentHide.bind ('<Button-1>', lambda event,
103         a = 'h': showHideEntryContent(a))
104
105 # Button widgets to enable/disable & lock/unlock the entry widget
106 def enableDisable():
107     winButtonEnableEntryWidget = tk.Button(winFrame,
108         text = 'Enable the\nentry widget', font = 'Arial 14 bold',
109         fg = 'green', borderwidth = 8, height = 3, width = 20)
110     winButtonEnableEntryWidget.grid(column = 0, row = 3)
111     winButtonEnableEntryWidget.bind('<Button-1>', lambda event,
112         a = 'e': enableLockDisableEntryWidget(a))
113     winButtonDisableEntryWidget = tk.Button(winFrame,
```

```
114        text = 'Lock the\nentry widget', font = 'Arial 14 bold',
115        fg = 'green', borderwidth = 8, height = 3, width = 20)
116    winButtonDisableEntryWidget.grid(column = 1, row = 3)
117    winButtonDisableEntryWidget.bind('<Button-1>', lambda event,
118        a = 'l': enableLockDisableEntryWidget(a))
119
120 # Create two button widgets to switch the "bold" property
121 # of the entry widget content on or off
122 def boldOnOff():
123    winButtonBoldEntryWidget = tk.Button (winFrame,
124        text = 'Bold contents of\nthe entry widget',
125        font = 'Arial 14 bold',
126        fg = 'brown', borderwidth = 8, height = 3, width = 20)
127    winButtonBoldEntryWidget.grid (column = 0, row = 4)
128    winButtonBoldEntryWidget.bind ('<Button-1>', lambda event,
129        a = 'b': boldContentsOfEntryWidget(a))
130    winButtonNoBoldEntryWidget = tk.Button (winFrame,
131        text = 'No bold contents of \nthe entry widget',
132        font = 'Arial 14 bold', fg = 'brown', borderwidth = 8,
133        height = 3, width = 20)
134    winButtonNoBoldEntryWidget.grid (column = 1, row = 4)
135    winButtonNoBoldEntryWidget.bind ('<Button-1>', lambda event,
136        a = 'n': boldContentsOfEntryWidget(a))
137
138 # Button widgets to convert the entry widget text to a password
139 def passwordOnOff():
140    winButtonPasswordEntryWidget = tk.Button(winFrame,
141        text ='Show entry widget \ncontent as password', borderwidth=8,
142        font = 'Arial 14 bold', fg = 'grey', height = 3, width = 20)
143    winButtonPasswordEntryWidget.grid(column = 0, row = 5)
144    winButtonPasswordEntryWidget.bind('<Button-1>', lambda event,
145        a = 'p': passwordEntryWidget(a))
146    winButtonNormalEntryWidget = tk.Button(winFrame,
147        font = 'Arial 14 bold',
148        text = 'Show entry widget \ncontent as normal text',
149        fg = 'grey', borderwidth = 8, height = 3, width = 20)
150    winButtonNormalEntryWidget.grid(column = 1, row = 5)
151    winButtonNormalEntryWidget.bind('<Button-1>', lambda event,
152        a = 'n': passwordEntryWidget(a))
153
154 # =====================================================================
155 # Create the frame using the tk object and run the application
156 winFrame = tk.Tk()
157 winFrame.title("Wrap up the basic widgets")
158 createGUI()
159 winFrame.mainloop()
```

Output 4.2.5.a–4.2.5.f:

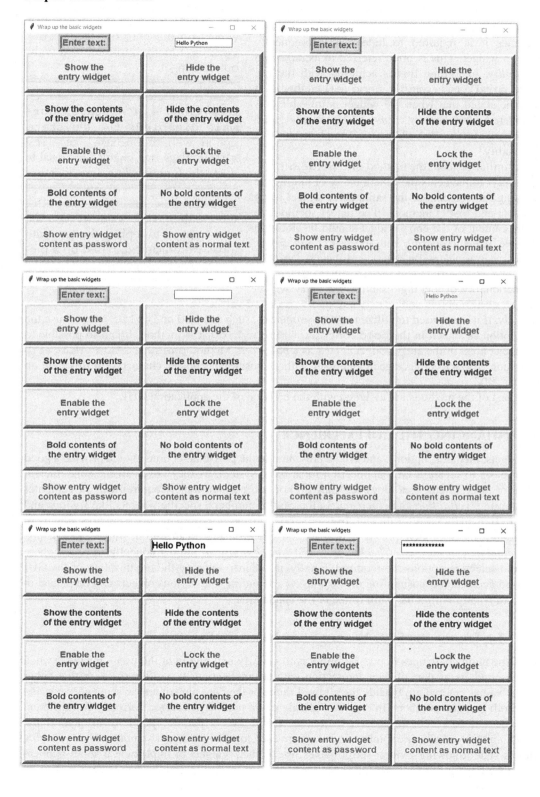

There are some noteworthy ideas presented in this script, relating to the need to *hide*, *disable*, and *lock* the text of a widget, or make it *appear as a password*. For example, sometimes it is required to hide, and subsequently unhide, a widget. This is often referred to as *adjusting its visibility*. In Python this is achieved with the use of the `grid()` and `grid_remove()` methods. It should be stated that when the widget is invisible *it is not deleted, but merely removed from the grid*. Method `showHideLabelEntry()` implements this functionality.

In a similar fashion, the method `showHideEntry-Content()` implements the functionality of hiding and displaying the contents of the same entry widget using the `set()` and `get()` methods. The reader should note that the content of the entry widget should be stored in a variable, since tampering with the `set()` and `get()` methods may accidentally delete it. Likewise, method `enableLockDisableEntryWidget()` implements the functionality of locking/disabling the entry widget using the `state` option and its `normal` and `disabled` values.

Observation 4.20 – `grid()`: Use the `grid()` method to position a widget on the grid; use the `grid_remove()` method to remove it without deleting it.

Observation 4.21 – state, normal, disabled: Use the `state` option with the `normal` or `disabled` flags to enable or disable (lock) the functionality of a widget.

Observation 4.22 – show: Use the `show` option to replace the text with a password-like text, based on a preferred character/symbol.

Finally, if it is required to utilize text font properties, such as `bold` or `italic`, one can use the `font` option as shown in the `boldContentsOfEntryWidget()` method. It is also possible to make the content of the entry widget appear as a *password*. Method `passwordEntryWidget()` uses option `show` to replace each character with a chosen placeholder character, in this case an asterisk ("*").

The rest of the methods are assigned with the creation of the application GUI.

4.3   ENHANCING THE GUI EXPERIENCE

The widgets, methods, options, and events presented in the previous sections should provide a good enough basis to create a GUI application for a basic system, as they cover all the fundamental aspects of basic interaction. However, they do not address two major requirements in computer programming: *validation* and *efficiency*. In the case of numbers, specific widgets like *spinbox* and *scale* are frequently used for the purposes of validation and improvement of visual appearance. In the case of text, for tasks requiring optimized and synchronized organization, widgets like *listbox* and *combobox* can be used. *Checkbuttons* and *radiobuttons* are used frequently in cases where improved selection options are required. Finally, in order to improve the organization of the GUI and avoid accidental repositioning of the widgets at runtime, the various objects can be placed in *individual frames* within the main frame of the application.

4.3.1   THE SPINBOX AND SCALE WIDGETS INSIDE INDIVIDUAL FRAMES

One of the main challenges in programming is to identify and highlight the user's mistakes when entering numbers as part of their interaction with an application. It is often the case that either numeric values entered are outside the allowed range or they are alphanumeric sequences consisting of both text and numbers. In order to validate that a number is entered correctly two different approaches are followed: (a) code is written to ensure the correct, acceptable form of the input number, and (b) widgets like spinbox and scale are used to restrict the user's options when *selecting* numbers. The following Python script makes use of such widgets to implement a small application in which the user may enter the *speed limit*, the *current speed*, and the *fine per km/h over the speed limit.* Once these numbers are entered, the fine is calculated based on the following formula: *fine = (current speed − speed limit)×fine per km/h.* For improving the organization of the GUI, the script uses a frame widget, which the various other widgets are placed upon:

```
1    # Import the necessary modules
2    import tkinter as tk
3    from tkinter import ttk
4
5    # Declare and initialise the global variables and widgets
6    # and define the associated methods
7    currentSpeedValue, speedLimitValue, finePerKmValue = 0, 0, 0
8    global speedLimitSpinbox
9    global finePerKmScale
10   global currentSpeedScale
11   global fine
12
13   # ================================================================
14   # Define the methods to run the control speed application
15   # Define the method to control the Current Speed Scale widget change
16   def onScale(val):
17       global currentSpeedValue
18       v = float(val)
19       currentSpeedValue.set(v)
20       calculateFine()
21
22   # Define the method to control the Speed Limit Spinbox widget change
23   def getSpeedLimit():
24       global speedLimitValue
25       v = float(speedLimitSpinbox.get())
26       speedLimitValue.set(v)
27       calculateFine()
28
29   # Define the method to control the Fine per Km Spinbox widget change
30   def getFinePerKm(val):
31       global finePerKmValue
32       v = int(float(val))
33       finePerKmValue.set(v)
34       calculateFine()
35
36   # Define the method to calculate the Fine given the 3 user parameters
37   def calculateFine():
38       global currentSpeedValue, speedLimitValue, finePerKmValue
39       global fine
40       diff = float(currentSpeedValue.get())-float(speedLimitValue.get())
41       finePerKm = float(finePerKmValue.get())
42       if (diff <= 0):
43           fine.config(text = 'No fine')
44       else:
45           fine.config(text = 'Fine in USD: '+ str(diff * finePerKm))
46
47   # ================================================================
```

```
48   # Define the methods that will create the interface of the application
49   def createGUI():
50       currentSpeedFrame()
51       speedLimitFrame()
52       finePerKmFrame()
53       fineFrame()
54
55   # Create the frame to include the Current Speed widgets
56   def currentSpeedFrame():
57       global currentSpeedValue
58
59       CurrentSpeedFrame = tk.Frame (winFrame, bg = 'light grey', bd = 2,
60           relief = 'sunken')
61       CurrentSpeedFrame.pack()
62       CurrentSpeedFrame.place(relx = 0.05, rely = 0.05)
63       currentSpeed = tk.Label(CurrentSpeedFrame, text = 'Current speed:',
                                                           width = 24)
64       currentSpeed.config(bg = 'light blue', fg = 'red', bd = 2,
65           font = 'Arial 14 bold')
66       currentSpeed.grid(column = 0, row = 0)
67
68       # Create Scale widget; define variable to connect to scale widget
69       currentSpeedValue = tk.DoubleVar()
70       currentSpeedScale = tk.Scale (CurrentSpeedFrame, length = 200,
71           from_ = 0, to = 360)
72       currentSpeedScale.config(resolution = 0.5,
73           activebackground = 'dark blue', orient = 'horizontal')
74       currentSpeedScale.config(bg = 'light blue', fg = 'red',
75           troughcolor = 'cyan', command = onScale)
76       currentSpeedScale.grid(column = 1, row = 0)
77       currentSpeedSelected = tk.Label(CurrentSpeedFrame, text = '...',
78           textvariable = currentSpeedValue)
79       currentSpeedSelected.grid(column = 2, row = 0)
80
81   # Create the frame to include the Speed Limit widgets
82   def speedLimitFrame():
83       global speedLimitValue
84       global speedLimitSpinbox
85
86       SpeedLimitFrame = tk.Frame(winFrame, bg = 'light yellow', bd = 4,
87           relief = 'sunken')
88       SpeedLimitFrame.pack()
89       SpeedLimitFrame.place(relx = 0.05, rely = 0.30)
90       # Create the prompt label on the Speed Limit frame
91       speedLimit=tk.Label(SpeedLimitFrame, text='Speed limit:', width=24)
92       speedLimit.config(bg = 'light blue', fg = 'yellow', bd = 2,
93           font = 'Arial 14 bold')
94       speedLimit.grid(column = 0, row = 0)
95       # Create the Spinbox widget; define variable to connect to Spinbox
96       speedLimitValue = tk.DoubleVar()
```

```
97       speedLimitSpinbox = ttk.Spinbox(SpeedLimitFrame,
98           from_ = 0, to = 360, command = getSpeedLimit)
99       speedLimitSpinbox.grid(column = 1, row = 0)
100      speedLimitSelected = tk.Label(SpeedLimitFrame, text = '...',
101          textvariable = speedLimitValue)
102      speedLimitSelected.grid(column = 2, row = 0)
103
104  # Create the frame to include the Fine per Km widgets
105  def finePerKmFrame():
106      global finePerKmValue
107
108      FinePerKmFrame = tk.Frame(winFrame, bg = 'light blue',
109          bd = 4, relief = 'sunken')
110      FinePerKmFrame.pack()
111      FinePerKmFrame.place (relx = 0.05, rely = 0.55)
112      # Create the prompt label on the Fine per Km frame
113      finePerKm=tk.Label(FinePerKmFrame, text='Fine/Km overspeed (USD):',
114          width = 24)
115      finePerKm.config(bg = 'light blue', fg = 'brown', bd = 2,
116          font = 'Arial 14 bold')
117      finePerKm.grid(column = 0, row = 0)
118      # Create Scale widget; define variable to connect to Scale widget
119      finePerKmValue = tk.IntVar()
120      finePerKmScale = ttk.Scale(FinePerKmFrame, orient = 'horizontal',
121          length = 200, from_ = 0, to = 100, command = getFinePerKm)
122      finePerKmScale.grid(column = 1, row = 0)
123      finePerKmSelected = tk.Label(FinePerKmFrame, text = '...',
124          textvariable = finePerKmValue)
125      finePerKmSelected.grid(column = 2, row = 0)
126
127  # Create the frame to include the Fine for speeding
128  def fineFrame():
129      global fine
130
131      FineFrame = tk.Frame(winFrame, bg='yellow', bd=4, relief='raised')
132      FineFrame.pack()
133      FineFrame.place(relx = 0.05, rely = 0.80)
134      # Create the label that will display the fine on the Fine frame
135      fine = tk.Label(FineFrame, text = 'Fine in USD:...', fg = 'blue')
136      fine.grid(column = 0, row = 0)
137  # ===================================================================
138  # Create the main frame for the application and run it
139  winFrame = tk.Tk()
140  winFrame.title("Control speed")
141  winFrame.config(bg = 'light grey')
142  winFrame.resizable(False, False)
143  winFrame.geometry('500x170')
144  createGUI()
145  winFrame.mainloop()
```

Output 4.3.1:

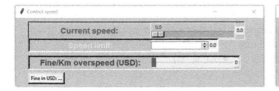 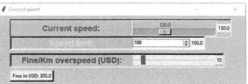

Conceptually, the script may be divided into three parts. The first part involves the declaration of the global variables and their initialization, so that they can be used in runtime when the user interacts with the program (line 7). This is important since the methods implementing the interaction will be using the same variables *dynamically*. At this stage, the main frame is also initialized and formed (lines 139–145), although this is done outside the initial phase. Eventually, a frame is created with a single label placed in it, with the sole purpose of displaying the calculated fine for speeding (lines 128–137).

The second part includes the creation of the four different frames inside the main frame, and the placement of the relevant widgets in each of them. These frames are created by means of a call to the relevant methods, through the createGUI() method (lines 49–54).

In the first case, (lines 56–80), the frame is placed inside the main window frame in a particular position (relx and rely options). Next, a label and a scale widget are placed in the frame. The reader should note the use of the config() method that defines the background (bg), foreground (fg), borderwidth (bd), and font name and size (font) of the label. It must be also noted that the label is placed in column 0 and row 0 of the *current frame*, and not of the *main window frame*.

In addition to the label, the scale widget is also placed in the frame. It is set to have a length (length) of 200 pixels, and its values are restricted within a lower boundary of 0 and upper boundary of 360. The reader should

Observation 4.23 – frames, relx, rely: Use *frames* for improved control of the interface. Contain the various widgets of the interface in the relevant frames. Use options relx and rely to place the frames in specified positions, relative to the main window.

Observation 4.24 – scale: Use the *scale* widget to create a controlled mechanism that will accept numerical user input. The *tkinter* widget has more visual options than the *ttk* alternative.

Observation 4.25 – Options: Use the required options, such as active-background, troughcolor, bg, fg, to modify the visual attributes of the widget. Use the resolution option to specify the *increment* and *decrement* steps. Use the orient option to specify its orientation (i.e., horizontal or vertical). Use the from _ = and to = options to set the numerical boundaries of the widget.

also observe the use of the config() method that sets the resolution option of the widget, allowing for user-defined increments (including decimals) of the values, the activebackground option that sets the color of the widget when it is active, and the orientation (orient) that can take one of two values: horizontal or vertical. For clarity reasons, the config() method is used for a second time to set some more options for the widget, such as the background (bg), the foreground (fg), and the troughcolor that sets the color of the *trough*. Additionally, another label is placed in the frame in order to display the current value of the *scale* widget, as an optional visual aid.

The second frame and the associated label introduce the spinbox widget (lines 82–103). This is also used to control user input when entering numeric values. It is very similar to the scale widget, allowing for the setting of the lower and upper boundaries of the accepted values, with two main differences: (a) it is visually different, and (b) the user may directly enter a value to the textual part of the widget, and/or control it with the increase/decrease arrows. As in the previous case, another label is added to the frame as an extra visual aid.

The third frame introduces another scale widget (lines 105–126). This is different to the one used in the first frame in that (a) it is visually different and restricted as to its visual attributes (i.e., it is not offering several of the *tk* widget options), and (b) it belongs to the *ttk* class/library instead of *tk*. The reader should notice the distinctly different visual results of the two scale widgets.

> **Observation 4.26:** Use the *spinbox* widget to create a controlled mechanism that will accept numerical user input, while also allowing direct input.

The third part defines the four methods used to control the interaction between the user and the application (lines 16–46). The reader should note that three of the methods (i.e., onScale(val), getSpeedLimit(), and getFinePerKm(val)) are directly associated with widgets currentSpeedScale, speedLimitSpinbox, and finePerKmScale, respectively. This is done through the command option. More specifically, when the user interacts with a particular widget, the resulting values are captured and the respective methods are called for the calculation of the fine. In the case of the scale widget, the value is passed with the call to the method. This is the case for both *tk* and *ttk*. The reader should observe (a) the use of the set and get methods applied to the objects of the widgets in order to tamper with the widget values, (b) the use of the casing operators (i.e., float(), int(float())) to control the type of numerical values used in the calculation, and (c) the declaration of the *global* variables that must be called and used in the methods. At the end of each of these methods the calculateFine() method is called to perform the associated calculation.

4.3.2 THE LISTBOX AND COMBOBOX WIDGETS INSIDE LABELFRAMES

Two of the most well-known widgets used in programming are the *listbox* and the *combobox*. These widgets are used to present the user with lines of text as a list, with the purpose of allowing them to make a selection. This selection can be also used to synchronize the contents between multiple instances of different widgets. The programmer can be creative as to the appearance of the widgets, as it is possible to manipulate their visual attributes, despite the fact that the basic form cannot be modified. The main difference between the two widgets is that the former provides an open list whereas the latter is a collapsed list that opens upon the user's click. Another widget which can help further enhancing the appearance of an application is the *labelframe* widget. This widget is similar to the frame widget, but it allows for a label to be specified on the frame itself, thus, removing the need for the creation of an extra label widget into the frame. Some of the visual attributes of this widget (including those related to the label font) can be manipulated.

> **Observation 4.27 – listbox, combobox:** Use the *listbox* and *combobox* widgets to display lists of lines of text, select one or more of these lines and, synchronize their contents as necessary.

> **Observation 4.28 – labelframe:** As with the *frame* widget, one can use the *labelframe* widget without the need to create an extra label for descriptions. The same options as with the *frame* and *label* widgets apply.

> **Observation 4.29 – randint():** Use the randint() method of the *random* library to generate random numbers within a specified range.

In this section, two additional libraries are introduced: *random* and *time*. The former is introduced in order to use method randint() that generates random numbers, and the latter in order to use process_time() that records the starting and/or ending time of a particular process.

> **Observation 4.30 – process_time():** Use the process_time() method of the *time* library to mark a particular moment in time and use it to count the time elapsed for a given process.

The following Python script allows the user to select a number of randomly generated integers in order to populate a listbox. Subsequently, it sorts this list into a second listbox before displaying the *size of the list*, the *sum of the numbers* and *their average*, and the *processing time for completing the sorting process*:

```
1    # Import the necessary modules
2    import tkinter as tk
3    from tkinter import ttk
4    from tkinter import *
5    import random
6    import time
7
8    # Initialise various lists used by the listboxes, comboboxes, & methods
9    unsortedL = []; sortedL = []; statisticsData = [];
10   sizes = [5, 20, 100, 1000, 10000, 20000]
11   global UnsortedList, SortedList
12   global startTime, endTime, ListSizeSelection, size
13   global UnsortedListScrollBar, SortedListScrollBar
14   global EntryFrame, UnsortedFrame, SortedFrame
15
16   # Populate the unsorted list with random numbers and
17   # the unsorted listbox
18   def populateUnsortedList():
19       global size
20       global UnsortedListScrollBar
21       global UnsortedList
22       global ListSizeSelection
23
24       # Read the number of elements as they are selected from the combobox
25       size = int(ListSizeSelection.get())
26
27       # randint() method of the random class generates random integers
28       for i in range (size):
29           n = random.randint(-100, 100)
30
31           # Enter the generated random integer to the relevant place in the
32           # unsorted list
33           unsortedL.insert(i, n)
34
35       # Populate the listbox with the elements of the unsorted list
36       for i in range (0, size):
37           UnsortedList.insert(i, unsortedL[i])
38       UnsortedListScrollBar.config(command = UnsortedList.yview)
39
40   # Use Bubble sort to sort the list & record the statistics for later use
41   def sortToSortedList():
42       global size, startTime, endTime
43       global SortedListScrollBar
44       global SortedList
45
46       # Load the unsorted list and listbox to the sorted list and listbox
47       for i in range (0, size):
48           sortedL.insert(i, unsortedL[i])
```

```
49
50      # Start the timer
51      startTime = time.process_time()
52
53      # The Bubble sort algorithm
54      for i in range (0, size-1):
55          for j in range (0, size-1):
56              if (sortedL[j] > sortedL[j+1]):
57                  temp = sortedL[j]
58                  sortedL[j] = sortedL[j+1]
59                  sortedL[j+1] = temp
60
61      # End the timer
62      endTime = time.process_time()
63
64      # Load the sorted list to the relevant listbox
65      for i in range (0, size):
66          SortedList.insert(i, sortedL[i])
67          SortedListScrollBar.config(command = SortedList.yview)
68
69  # Clear all lists, listboxes, & comboboxes, & the global size variable
70  def clearLists():
71      global size
72      sortedL.clear()
73      unsortedL.clear()
74      UnsortedList.delete('0', 'end')
75      SortedList.delete('0', 'end')
76      statisticsData.clear()
77      StatisticsCombo.delete('0', 'end')
78
79  # Calculate and report the statistics from the sorting process
80  def statistics():
81      global size, startTime, endTime
82      statisticsData.clear()
83      statisticsData.insert(1, 'The size of the lists is ' + str(size))
84      statisticsData.insert(2,'The sum of the lists is '+str(sum(sortedL)))
85      statisticsData.insert(3, 'The time passed to sort the list was ' \
86          + str(round(endTime - startTime, 5)))
87      statisticsData.insert(4, 'The average of the sorted list is: ' \
88          + str(round(sum(sortedL) / size, 2)))
89      StatisticsCombo['values'] = statisticsData
90
91  # =====================================================================
92  # Define the methods that will create the GUI of the application
93  def createGUI():
94      unsortedFrame()
95      entryFrame()
96      entryButton()
97      sortButton()
98      sortedFrame()
99      clearButton()
100     statisticsButton()
```

```
101        statisticsSelection()
102
103 # Create the labelframe & place the Unsorted Array Listbox widgets in it
104 def unsortedFrame():
105        global unsortedList
106        global UnsortedListScrollBar
107        global UnsortedList
108        global winFrame
109        global UnsortedFrame
110
111        UnsortedFrame = tk.LabelFrame (winFrame, text = 'Unsorted Array')
112        UnsortedFrame.config(bg='light grey',fg='blue',bd=2, relief='sunken')
113        # Create a scrollbar widget to attach to the UnsortedList
114        UnsortedListScrollBar = Scrollbar (UnsortedFrame, orient = VERTICAL)
115        UnsortedListScrollBar.pack(side = RIGHT, fill = Y)
116        # Create the listbox in the Unsorted Array frame
117        UnsortedList = tk.Listbox(UnsortedFrame, bg='cyan', width=13, bd=0,
118            height = 12, yscrollcommand = UnsortedListScrollBar.set)
119        UnsortedList.pack(side = LEFT, fill = BOTH)
120        # Associate the scrollbar command with its parent widget,
121        # i.e., the UnsortedList yview
122        UnsortedListScrollBar.config(command = UnsortedList.yview)
123        # Place the Unsorted frame and its parts into the interface
124        UnsortedFrame.pack(); UnsortedFrame.place(relx = 0.02, rely = 0.05)
125
126 # Create the labelframe to include the Entry widget
127 def entryFrame():
128        global unsortedList
129        global UnsortedListScrollBar
130        global ListSizeSelection
131        global EntryFrame
132        global winFrame
133
134        EntryFrame = tk.LabelFrame(winFrame, text = 'Actions')
135        EntryFrame.config(bg='light grey', fg='red', bd=2, relief = 'sunken')
136        EntryFrame.pack(); EntryFrame.place(relx = 0.25, rely = 0.05)
137        # Create the label in the Entry frame
138        EntryLabel = tk.Label(EntryFrame,
139            text='How many integers\nin the list', width = 16)
140        EntryLabel.config(bg = 'light grey', fg='red', bd = 3,
141            relief = 'flat', font = 'Arial 14 bold')
142        EntryLabel.grid(column = 0, row = 0)
143        # Create the combobox to select the number of elements in the lists
144        ListSizeSelection = tk.IntVar()
145        ListSizeCombo = ttk.Combobox(EntryFrame,
146            textvariable=ListSizeSelection, width = 10)
147        ListSizeCombo['values'] = sizes
148        ListSizeCombo.current(0)
149        ListSizeCombo.grid(column = 1, row = 0)
150
151 # Create button to insert new entries into the unsorted array & listbox
152 def entryButton():
```

```
153        global EntryFrame
154
155        EntryButton = tk.Button(EntryFrame, text = 'Populate\nUnsorted list',
156            relief = 'raised', width = 16)
157        EntryButton.bind('<Button-1>', lambda event: populateUnsortedList())
158        EntryButton.grid(column = 0, row = 2)
159
160 # Create the button that will sort the numbers and display them
161 # in the sorted array and listbox
162 def sortButton():
163        global EntryFrame
164
165        SortButton=tk.Button(EntryFrame,text='Sort numbers\nwith BubbleSort',
166        relief = 'raised', width = 16)
167        SortButton.bind('<Button-1>', lambda event: sortToSortedList())
168        SortButton.grid(column = 1, row = 2)
169
170 # Create the labelframe to include the Sorted Array Listbox widgets
171 def sortedFrame():
172        global sortedList
173        global SortedListScrollBar
174        global SortedList
175        global winFrame
176        global SortedFrame
177
178        SortedFrame = tk.LabelFrame(winFrame, text = 'Sorted Array')
179        SortedFrame.config(bg='light grey', fg='blue', bd=2, relief='sunken')
180        # Create a scrollbar widget to attach to the SortedList
181        SortedListScrollBar = Scrollbar (SortedFrame)
182        SortedListScrollBar.pack(side = RIGHT, fill = Y)
183        # Create the listbox in the Sorted Array frame
184        SortedList = tk.Listbox (SortedFrame, bg='cyan', width=13, height=12,
185            yscrollcommand = SortedListScrollBar.set, bd = 0)
186        SortedList.pack(side = LEFT, fill = BOTH)
187        # Associate the scrollbar command with its parent widget,
188        # i.e., the SortedList yview
189        SortedListScrollBar.config(command = SortedList.yview)
190        # Place the Unsorted frame and its parts into the interface
191        SortedFrame.pack(); SortedFrame.place(relx = 0.75, rely = 0.05)
192
193 # Create the button that will clear the two listboxes and the two lists
194 def clearButton():
195        global EntryFrame
196
197        ClearButton = tk.Button(EntryFrame, text = 'Clear lists',
198            relief = 'raised', width = 16)
199        ClearButton.bind('<Button-1>', lambda event: clearLists())
200        ClearButton.grid(column = 0, row = 3)
201
202 # Create the button that will display the statistics for the sorting
203 def statisticsButton():
204        global EntryFrame
```

```
205
206     StatisticsButton = tk.Button(EntryFrame, text = 'Show statistics',
207         relief = 'raised', width = 16)
208     StatisticsButton.bind('<Button-1>', lambda event: statistics())
209     StatisticsButton.grid(column = 1, row = 3)
210
211 # Create the option menu that will show the statistical results
212 # from the sorting process
213 def statisticsSelection():
214     global EntryFrame
215     global StatisticsCombo
216
217     StatisticsSelection = tk.StringVar()
218     statisticsData = ['The statistics will appear here']
219     StatisticsSelection.set(statisticsData[0])
220     StatisticsCombo = ttk.Combobox(EntryFrame, width = 30,
221         textvariable = StatisticsSelection)
222     StatisticsCombo['values'] = statisticsData
223     StatisticsCombo.grid(column = 0, columnspan = 2, row = 4)
224 # ===================================================================
225
226 # Create the main frame for the application
227 winFrame = tk.Tk()
228 winFrame.title("Bubble Sort"); winFrame.config(bg = 'light grey')
229 winFrame.resizable(True, True); winFrame.geometry('650x300')
230
231 createGUI()
232
233 winFrame.mainloop()
```

Output 4.3.2:

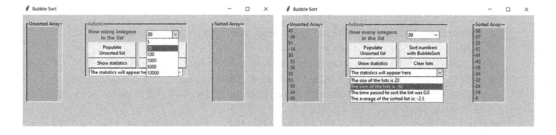

Initially, the necessary libraries are imported (i.e., *tkinter*, *time*, and *random*, lines 2–6). Next, the various lists, variables, and listboxes are initialized (lines 9–14). Note that the lists are not defined as *global*, since they are accessed *by reference* by all methods in the script by default. It must be also noted that different types of objects and/or variables must be declared as global in separate lines, since declaring them together may raise errors. After initialization, the main frame is created and configured in lines 227–229.

The next step is to create the application interface. In this case, the interface consists of two distinct parts. The first includes two listboxes created and placed inside the associated labelframes (lines 103–124 and 170–191). The use of labelframes makes the creation of additional labels obsolete. The visual properties of the listboxes can be configured through their options, which are almost identical to those of an entry widget. The listboxes can be populated at run time using the `insert(index, value)` method, and cleared at run time using the `delete(index, index)` method. Likewise, the properties/options of the labelframes are similar to those of regular frames and labels.

The second part is to create the labelframe that hosts the comboboxes and the buttons required in the application. The purpose of the first combobox is to display the number of random integers in the unsorted list. The second one displays basic statistics related to the sorting process, the size of the lists, the sum and average of the integers, and the time required to sort the list. There are three notable observations related to the creation and use of the comboboxes (lines 143–149 and 211–223). Firstly, they must include a `["values"]` list which will take its values from an associated list. The latter can be initially empty or populated. Secondly, their selection value (e.g., `textvariable`), must be associated with an object of the `IntVar()` class (or any similar alternative) that will store it for further use, since the selected combobox value is not directly accessible. Thirdly, the currently selected value must be defined through the `current(index)` method.

The last step is to create the interaction between the user and the application. For this purpose, four buttons are created and bound with click events to trigger the respective methods. This populates, sorts, and clears the relevant lists, and displays the basic statistics. The `populateUnsortedList()` method uses the `randint()` method to generate random integers, and the `insert()` method populates the unsorted list (lines 16–38). It is worth noting the declaration of global variable `size`, and the use of the `get()` method to read the value from the *private* attribute of the `ListSizeSelection` object (line 25). The `sortToSortedList()` method (lines 40–67) declares global variables `size`, `startTime` and `endTime`, uses the `process_time()` method to mark the start and end of the sort process, and utilizes a common *Bubble Sort* algorithm to sort the list and populate the `sortedList`. The `clearLists()` method uses methods `clear()` to clear the values of the lists and `delete()` to delete the values of the listboxes (lines 69–77). Finally, the `statistics()` method uses methods `sum()` and `round()` to produce the basic statistics that will be displayed (lines 79–89).

Observation 4.31 – `insert()`, `delete()`: Use the `insert()` and `delete()` methods to populate or clear a listbox.

Observation 4.32 – ["values"]: Use the `["values"]` property to populate a combobox with an initial list of values.

Observation 4.33 – `textvariable`: Use the `textvariable` option of the combobox to associate it with an `IntVar()` object that will store the selected value.

Observation 4.34 – `current()`: Use the `current()` method to define the currently selected value of the combobox.

Observation 4.35 – `get()`: It is necessary to use the `get()` method to read from the `IntVar()` object, as it is *private* and, hence, not directly accessible.

Observation 4.36 – `clear()`: Use the `clear()` method to clear the values of the lists.

Observation 4.37 – `xview`, `yview`, `xscrollcommand`, `yscrollcommand`: Use the *scrollbar* widget to attach a scrollbar to the associated widget (usually a listbox). Use `xview` or `yview` to control its orientation (i.e., *horizontal* or *vertical*). Use the `xscrollcommand` or the `yscrollcommand` to activate it.

The reader should observe the use of the *scrollbar* widget introduced in this script. The idea behind, and the use of, this particular widget is intuitive and quite straightforward. Firstly, the labelframe inside which the scrollbar operates is created. Next, the scrollbar is created and connected (*packed*) to the parent widget (i.e., in this case the associated labelframe), specifying its orientation and positioning. Lastly, the widget/object that will make use of the scrollbar is created and associated with the scrollbar through either `yscrollcommand` or `xscrollcommand` (depending on whether the scrollbar orientation is *vertical* or *horizontal* respectively), and configured to scroll the contents of the attached widget (lines 38, 120–124, and 67, 187–191).

4.3.3 GUIs with CheckButtons, RadioButtons and SimpleMessages

In addition to listboxes and comboboxes, there are two more widgets that users of windows-based applications are familiar with, namely *checkbuttons* and *radiobuttons*. These widgets allow the user to make one or more selections from a set of different available options/actions. Their main difference is that while in the case of checkbuttons the user may select more than one option at any given time, radiobuttons only allow a single selection from the set of available options. Finally, another handy widget available in Python the reader should be familiar with is the *message* widget. In this section the most basic form of this widget will be introduced and explained.

The following script implements an interface that includes two listboxes with associated, attached vertical scrollbars. The listboxes are populated with the names of various countries and their capital cities. It also includes two entry boxes for accepting new entries to the listboxes. Insertions are triggered using the associated button-click events. The contents of all listboxes are synchronized with the user's click *on any listbox*. The interface also includes four buttons that handle the interaction between the application and the user, allowing for the insertion and deletion of particular entries, the clearance of all entries from all three containers, and exiting the application. Finally, two checkbuttons control whether the relevant containers are enabled or not, and two radiobuttons whether they are visible:

```
1    import tkinter as tk
2    from tkinter import *
3    from tkinter import ttk
4    from tkinter import messagebox
5
6    countries = ['E.U.', 'U.S.A.', 'Russia', 'China', 'India', 'Brazil']
7    Capital = ['Brussels', 'Washinghton', 'Moscow', 'Beijing', 'New Delhi',
                                                                'Brazilia']
8
9    global newCountry, newCapital
10   global CountriesFrame, CapitalFrame
11   global checkButton1, checkButton2
12   global radioButton
13   global CountriesList, CapitalList
14   global CountriesScrollBar, CapitalScrollBar
15
16   # Create the interface for the listboxes
17   def drawListBoxes():
18       global CountriesList, CapitalList
19       global CountriesFrame, CapitalFrame
20       global CountriesScrollBar, CapitalScrollBar
21
22       # Create CountriesFrame labelframe; place CountriesList widget in it
23       CountriesFrame = tk.LabelFrame(winFrame, text = 'Countries')
24       CountriesFrame.config(bg = 'light grey', fg = 'blue', bd = 2,
```

```
25              width = 15, relief = 'sunken')
26          # Create a scrollbar widget to attach to the CountriesList
27          CountriesScrollBar = Scrollbar(CountriesFrame, orient = VERTICAL)
28          CountriesScrollBar.pack(side = RIGHT, fill = Y)
29          # Create the listbox in the CountriesFrame
30          CountriesList = tk.Listbox(CountriesFrame, bg = 'cyan', width = 15,
31              height = 8, yscrollcommand = CountriesScrollBar)
32          CountriesList.pack(side = LEFT, fill = BOTH)
33          # Associate the scrollbar command with its parent widget,
34          # (i.e., the CountriesList yview)
35          CountriesScrollBar.config(command = CountriesList.yview)
36          # Place the Countries frame and its parts on the interface
37          CountriesFrame.pack(); CountriesFrame.place(relx = 0.03, rely = 0.05)
38          CountriesList.bind('<Double-Button-1>',
39              lambda event: alignList('countries'))
40          # Create the CapitalFrame labelframe; place CapitalList widget on it
41          CapitalFrame = tk.LabelFrame(winFrame, text = 'Countries Capital')
42          CapitalFrame.config(bg = 'light grey', fg = 'blue', bd = 2,
43              width = 13, relief = 'sunken')
44          # Create a scrollbar widget to attach to the CapitalFrame
45          CapitalScrollBar = Scrollbar(CapitalFrame, orient = VERTICAL)
46          CapitalScrollBar.pack(side = RIGHT, fill = Y)
47          # Create the listbox in the CapitalFrame
48          CapitalList = tk.Listbox(CapitalFrame, bg = 'cyan',
49              yscrollcommand = CapitalScrollBar, width = 16, height = 8, bd = 0)
50          CapitalList.pack(side = LEFT, fill = BOTH)
51          # Associate the scrollbar command with its parent widget,
52          # (i.e., the CapitalList yview)
53          CapitalFrame.pack(); CapitalFrame.place(relx = 0.70, rely = 0.05)
54          CapitalList.bind('<Double-Button-1>',
55              lambda event: alignList('capital'))
56  # Create the interface for the new entries
57  def drawNewEntries():
58      global newCountry, newCapital
59
60          # Create the labelframe and place the newCountry entry widget on it
61          NewCountryFrame = tk.LabelFrame(winFrame, text = 'New Country')
62          NewCountryFrame.config(bg = 'light grey', fg = 'blue', bd = 2,
63              width = 13, relief = 'sunken')
64          NewCountryFrame.pack(); NewCountryFrame.place(relx= 0.03, rely = 0.75)
65          newCountry = tk.StringVar(); newCountry.set('')
66          NewCountryEntry = tk.Entry(NewCountryFrame, textvariable = newCountry,
67              width = 15)
68          NewCountryEntry.config(bg= 'dark grey', fg = 'red', relief = 'sunken')
69          NewCountryEntry.grid(row = 0, column = 0)
70
71          # Create the labelframe and place the newCapital entry widget on it
72          NewCapitalFrame = tk.LabelFrame(winFrame, text = 'New Capital')
73          NewCapitalFrame.config(bg = 'light grey', fg = 'blue', bd = 2,
74              width = 13, relief = 'sunken')
```

```
75      NewCapitalFrame.pack(); NewCapitalFrame.place(relx= 0.70, rely = 0.75)
76      newCapital = tk.StringVar(); newCapital.set('')
77      NewCapitalEntry = tk.Entry(NewCapitalFrame, textvariable = newCapital,
78          width = 15)
79      NewCapitalEntry.config(bg= 'dark grey', fg = 'red', relief = 'sunken')
80      NewCapitalEntry.grid(row = 0, column = 0)
81
82  # Create the interface for the action buttons
83  def drawButtons():
84      # Create the labelframe that will host the buttons
85      ButtonsFrame = tk.Frame(winFrame)
86      ButtonsFrame.config(bg= 'light grey', bd=2, width=14, relief='sunken')
87      ButtonsFrame.pack(); ButtonsFrame.place(relx = 0.30, rely = 0.07)
88
89      newRecordButton = tk.Button(ButtonsFrame, text = 'Insert\nnew record',
90          width = 11, height = 2)
91      newRecordButton.grid(row = 0, column = 0)
92      newRecordButton.bind('<Button-1>', lambda event,
93          a = 'insertRecord': buttonsClicked(a))
94
95      deleteRecordButton = tk.Button (ButtonsFrame,
96          text = 'Delete\n record', width = 11, height = 2)
97      deleteRecordButton.grid (row = 0, column = 1)
98      deleteRecordButton.bind('<Button-1>', lambda event,
99          a = 'deleteRecord': buttonsClicked(a))
100
101     clearRecordsButton = tk.Button (ButtonsFrame,
102         text = 'Clear\n records', width = 11, height = 2)
103     clearRecordsButton.grid (row = 1, column = 0)
104     clearRecordsButton.bind('<Button-1>', lambda event,
105         a = 'clearAllRecords': buttonsClicked(a))
106
107     exitButton = tk.Button(ButtonsFrame, text='Exit', width=11, height=2)
108     exitButton.grid (row = 1, column = 1)
109     exitButton.bind('<Button-1>', lambda event : winFrame.destroy())
110     exit()
111
112  # Create the interface for the checkbuttons
113  def drawCheckButtons():
114     global checkButton1, checkButton2
115
116     # Create the labelframe that will host the checkbuttons
117     CheckButtonsFrame = tk.Frame(winFrame)
118     CheckButtonsFrame.config(bg = 'light grey', bd = 2, relief = 'sunken')
119     CheckButtonsFrame.pack();CheckButtonsFrame.place(relx=0.34, rely=0.43)
120
121     checkButton1 = IntVar(value = 1)
122     CountriesCheckButton = tk.Checkbutton (CheckButtonsFrame,
123         variable = checkButton1, text = 'Countries \nenabled/disabled',
124         bg = 'light blue', onvalue = 1, offvalue = 0, width = 15,
```

```
125            height = 2, command = checkClicked).grid(row = 0, column = 0)
126
127        checkButton2 = IntVar(value = 1)
128        CapitalCheckButton = tk.Checkbutton (CheckButtonsFrame,
129            variable = checkButton2, onvalue = 1, offvalue = 0,
130            text = 'Capitals \nenabled/disabled', width = 15, height = 2,
131            bg = 'light blue', command = checkClicked).grid (row=1, column=0)
132
133 # Create the interface for the radiobuttons
134 def drawRadioButtons():
135     global radioButton
136
137 # Create the labelframe that will host the radiobuttons
138     RadioButtonsFrame = tk.Frame(winFrame)
139     RadioButtonsFrame.config(bg = 'light grey', bd = 2, relief = 'sunken')
140     RadioButtonsFrame.pack();RadioButtonsFrame.place(relx=0.31, rely=0.78)
141
142     radioButton = IntVar()
143     visibleRadioButton = tk.Radiobutton (RadioButtonsFrame,
144         text = 'Containers \nvisible', width = 8, height = 2,
145         bg = 'light green', variable = radioButton, value = 1,
146         command = radioClicked).grid(row = 0, column = 0)
147
148     invisibleRadioButton = tk.Radiobutton (RadioButtonsFrame,
149         text = 'Containers \ninvisible', width = 8, height = 2,
150         bg = 'light green', variable = radioButton, value = 2,
151         command = radioClicked).grid(row = 0, column = 1)
152
153     radioButton.set(1)
154
155 # Define method alignList that will identify the selected row
156 # in any of the listboxes and align it with the corresponding row others
157 def alignList(a):
158     global CountriesList, CapitalList
159     global selectedIndex
160
161     if (a == 'countries'):
162         selectedIndex = int(CountriesList.curselection()[0])
163         CapitalList.selection_set(selectedIndex)
164
165     if (a == 'capital'):
166         selectedIndex = int(CapitalList.curselection()[0])
167         CountriesList.selection_set(selectedIndex)
168
169 # Define checkClicked method to control the state of the containers
170 def checkClicked():
171     global checkButton1, checkButton2
172
173     # Control the state of the containers as NORMAL or DISABLED
174     # based on the state of the checkbuttons
```

```
175        if (checkButton1.get() == 1):
176            CountriesList.config(state = NORMAL)
177        else:
178            CountriesList.config(state = DISABLED)
179
180        if (checkButton2.get() == 1):
181            CapitalList.config(state = NORMAL)
182        else:
183            CapitalList.config(state = DISABLED)
184
185 # Define the radioClicked method that will display or hide the frames
186 # of the containers
187 def radioClicked():
188        global CountriesFrame, CapitalFrame
189        global radioButton
190
191        # Use the destroy() method to destroy the frames of the containers.
192        # The lists are not destroyed
193        CountriesFrame.destroy()
194        CapitalFrame.destroy()
195
196        if (radioButton.get() == 1):
197            drawListBoxes()
198            populate()
199
200 # Populate the listboxes
201 def populate():
202        global CountryList, CapitalList
203        global selectedIndex
204
205        for i in range (int(len(countries))):
206            CountriesList.insert(i, countries[i])
207
208        for i in range (int(len(capital))):
209            CapitalList.insert(i, capital[i])
210
211 # Define method buttonsClicked that will trigger the corresponding code
212 # when any of the buttons is clicked
213 def buttonsClicked(a):
214        global CountriesList, PopulationCombo, CapitalList
215        global newCountry, newPopulation, newCapital, populationSelection
216        global selectedIndex
217
218        if (a == "insertRecord"):
219            if (newCountry!= '' and newCapital!= ''):
220                countries.append(newCountry.get()); CountriesList.delete('0',
                                                                          'end')
221                capital.append(newCapital.get());CapitalList.delete('0','end')
222                # Call method populate() to re-populate the containers
```

```
223                     # with the renewed lists
224                     populate()
225
226       if (a == 'deleteRecord'):
227           # Use messagebox.askyesno() to pop a confirmation message
228           # for deleting the elements
229           deleteElementOrNot=messagebox.askokcancel(title="Delete element",
230               message="Are you ready to delete the elements?", icon='info')
231       if (deleteElementOrNot == True):
232           # Use the pop() method to remove selected elements from the lists
233           countries.pop(selectedIndex); capital.pop(selectedIndex)
234           CountriesList.delete('0', 'end'); CapitalList.delete('0', 'end')
235           # Call method populate() to re-populate the containers
236           # with the renewed lists
237           populate()
238
239       if (a == 'clearAllRecords'):
240           # Use messagebox.askyesno() to pop a confirmation message
241           # for clearing the lists
242           clearListsOrNot=messagebox.askokcancel(title="Clear all elements",
243               message = "Are you ready to clear the lists?", icon = 'info')
244       if (clearListsOrNot == True):
245           countries.clear(); capital.clear()
246           CountriesList.delete('0', 'end'); CapitalList.delete('0', 'end')
247           # Call method populate() to re-populate the containers
248           # with the renewed lists
249           populate()
250
251 # Create the frame for the Countries program and configure its size
252 # and background color
253 winFrame = tk.Tk()
254 winFrame.title ('Countries')
255 winFrame.geometry("500x250")
256 winFrame.config (bg = 'light grey')
257 winFrame.resizable(False, False)
258
259 # Create the Graphical User Interface
260 drawListBoxes()
261 drawNewEntries()
262 drawButtons()
263 drawCheckButtons()
264 drawRadioButtons()
265
266 # Call populate()to populate the listboxes and comboboxes
267 populate()
268
269 winFrame.mainloop()
```

Output 4.3.3:

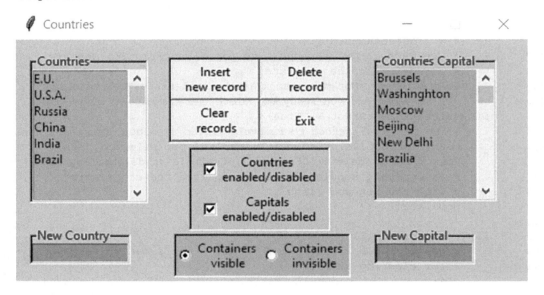

As in previous examples, the first part of the application deals with drawing the interface. In this particular case this task is assigned to methods `drawListBoxes()`, `drawNewEntries()`, `drawButtons()`, `drawCheckButtons()`, and `drawRadioButtons()`. Method `drawListBoxes()` (lines 16–55) creates the relevant frames and containers. The reader should note the call to method `alignList()` that causes the contents of the two containers to be aligned, and the use of the `relx` and `rely` options that position the respective frames in the appropriate places within the interface. The `drawNewEntries()` method (lines 56–80) creates the entry widgets that will accept the user's input for new entries. Observe how the entry widgets are associated with the respective `StringVar()` objects that allow the use of the input through the appropriate `set()` and `get()` methods. Similarly, the `drawButtons()` method (lines 82–110) creates the frame and places the buttons that perform the basic actions of the application (i.e., insert a new entry, delete a selected entry, clear all contents of the containers, and exit the application). In the case of the *Exit* button in particular, one should note the use of the `destroy()` method that destroys the interface of the main window, and the `exit()` method that exits the application.

The `drawCheckButtons()` method (lines 112–131) creates the frame for the checkbutton widgets.

Observation 4.38 – destroy(), exit(): Use methods `destroy()` to destroy the interface (i.e., the widgets of the particular frame it applies) and `exit()` to exit the application.

Observation 4.39 – checkbutton, onvalue, offvalue: Use the *checkbutton* widget to offer selection options. Each option is represented by a separate widget. If an option is selected, the widget is given an `onvalue`, otherwise it is given an `offvalue` through the associated `IntVar()` object.

Observation 4.40 – radiobutton: Use the *radiobutton* widget to offer a number of mutually exclusive options. Each option is represented by a different widget. If an option is selected, the widgets are given a particular value through the associated `IntVar()` object.

Notice how each of the checkbuttons is associated (*bound*) with a separate `IntVar()` object to monitor its state (i.e., `onvalue = 1` if it is *checked* or `offvalue = 0` if it is *unchecked*). The reader should also notice that when the user checks/unchecks the checkbutton the checkClicked() method is triggered through the command option. This is in order to control the appearance of the respective container. Likewise, in the case of drawRadiobuttons() (lines 133–153), two of them are placed in the relevant frame and trigger the radioClicked() method through the command option. This controls the appearance of the containers as a whole. It is important to note that in such cases where multiple radiobuttons are associated/bound with the same IntVar() object, *only one can be selected*.

The second part of the application deals with the interactions that take place between the interface and the user and their results, through the use of methods alignList(), checkClicked(), radioClicked(), populate(), and buttonsClicked(). In the case of alignList() (lines 155–167), the curselection() method is applied to the relevant container (listbox) to identify the element of the container that was selected. Since the method results to a tuple, it is necessary to limit the result to the first element of the tuple (i.e., the [0] value). Once the element of the container is identified through its index, the selection _ set() method is executed. This allows the other container to align the two listboxes based on the selections. Ultimately, this process synchronizes the two containers.

In the case of the checkClicked() method (lines 169–183) the reader should note the following:

> **Observation 4.41 – command, checkbutton, radiobutton:** Use the command option to trigger a particular action when any of the checkButton or radioButton widgets are selected.

> **Observation 4.42 – curselection():** Use methods curselection() to identify the selected element from a listbox and selection _ set() to select a particular indexed element.

> **Observation 4.43 – state, NORMAL, DISABLED:** Use the state option to determine whether a particular listbox is enabled (NORMAL) or disabled (DISABLED).

- The use of the state option and its two possible values (i.e., NORMAL and DISABLED), which determine whether the associated widget will be enabled or not. More specifically, NORMAL dictates that the user is *allowed to click in the relevant container and select one or more of its elements* and DISABLED the opposite.
- The use of the get() method to access the value of objects checkButton1 and checkButton2. The reader is reminded that accessing the values of these objects is only possible through such methods, since the objects and their values are private. The checkButton1 and checkButton2 widgets are declared as *global* to ensure that they are used *by reference*, taking their values from *the original objects in the main application*.

In the case of the radioClicked() method (lines 185–198), frames CountriesFrame and CapitalFrame are destroyed alongside their containers/listboxes (i.e., CountriesList and CapitalList) and are only recreated and repopulated if the user selects the appropriate visibleRadioButton from the interface (i.e., assigning a value of 1 to the radioButton object).

> **Observation 4.44 – append(), delete(), clear():** Use methods append() to append a list (i.e., insert a new element at the end of the list), delete() to delete a selected element from a list, and clear() to clear all the elements of a list.

Finally, the buttonsClicked() method (lines 211–249) has three main tasks. Firstly, it inserts a new element in each of the listboxes when the user clicks the *Insert* button. In this case, the values of the newCountry and newCapital entry widgets are checked and, if not empty, used to *append* the relevant lists. Notice that it is preferable to *append the lists* and *not the listboxes*, as the former host the actual values. The listboxes are repopulated only after this task is completed.

Secondly, the method has the task of deleting the selected elements from the listboxes when the user clicks the *Delete* button. In this case, as long as an element of the listboxes is selected, a simple messagebox pops up to confirm the user's choice. Notice that the `askyesno()` method provides one of the simplest available forms of messages, and results in either `True` or `False`. The programmer can use these values to determine further actions. The reader should note that the *messagebox* module is part of the *tkinter* library. It is also noteworthy that the `delete()` method is used in the code to initially clear the listboxes from their contents, and subsequently re-populate them with the refreshed, appended lists. This particular method accepts the first and the last index in the range of elements that should be deleted from the lists as arguments. Similarly, a third task is to completely clear the listboxes from their contents. For this purpose, the `clear()` method is applied to both lists (but *not the listboxes*), given that confirmation is provided by the user through another simple messagebox interaction.

In all the cases discussed above, the `populate()` method (lines 200–209) is responsible for reading the lists and using their contents to populate the listboxes.

> **Observation 4.45 – `askyesno()`:**
> Use the appropriate *messagebox* module method (e.g., `askyesno()`) to confirm the user's choice.

4.4 BASIC AUTOMATION AND USER INPUT CONTROL

A common characteristic of visual programming is the creation of the illusion that the application objects/widgets change shape, content, or status, either automatically or based on the user's input or automatically. If an object/widget is to be activated and put in operation *automatically*, the programmer needs to associate it with a respective *time-controlled event*. The latter enables the programmer to change the properties of the object/widget at run time, through the activation and execution of appropriate blocks of code that are based on the time-controlled event.

In this section, the reader will have the opportunity to get some exposure to the creation of applications that manipulate objects/widgets without the user's input, or with interactions of a different type than direct written input or button-click events. Throughout the section, a basic *Traffic Lights* application is gradually developed toward a primitive, but informative, *automated* user experience.

4.4.1 TRAFFIC LIGHTS VERSION 1 – BASIC FUNCTIONALITY

The *Traffic Lights* sample project can start by creating a very basic application that uses three images (loaded in labels) displaying a green, a yellow, and a red traffic light, respectively. The three images can be programmed to appear and disappear based on user's selection. The following Python script creates this interface and implements the related interactions:

```
1   # Import libraries
2   import tkinter as tk
3   from tkinter import *
4   # Import the necessary image processing classes
5   from PIL import Image, ImageTk
6
7   global radioButton
8   global image1, image2, image3
9   global photo1, photo2, photo3
10  global winLabel1, winLabel2, winLabel3
11  global winFrame
12
13  # Create the main frame
14  winFrame = tk.Tk()
```

```
15  winFrame.title("Traffic Lights v1")
16
17  # Create the interface with the images and labels
18  def photos():
19      global radioButton
20      global image1, image2, image3
21      global photo1, photo2, photo3
22
23      image1 = Image.open("TrafficLightsGreen.gif")
24      image1 = image1.resize((50, 100), Image.ANTIALIAS)
25      photo1 = ImageTk.PhotoImage(image1)
26
27      winLabel1=tk.Label(winFrame,text='', image=photo1, compound='left')
28      winLabel1.grid(row = 0, column = 0)
29
30      image2 = Image.open("TrafficLightsYellow.gif")
31      image2 = image2.resize((50, 100), Image.ANTIALIAS)
32      photo2 = ImageTk.PhotoImage(image2)
33
34      winLabel2 = tk.Label(winFrame,text='',image=photo2,compound='left')
35      winLabel2.grid(row = 0, column = 1)
36
37      image3 = Image.open("TrafficLightsRed.gif")
38      image3 = image3.resize((50, 100), Image.ANTIALIAS)
39      photo3 = ImageTk.PhotoImage(image3)
40
41      winLabel3 = tk.Label(winFrame,text='',image=photo3,compound='left')
42      winLabel3.grid(row = 0, column = 2)
43
44      # Control active traffic lights based on the radio button selection
45      if (radioButton.get() == 1):
46          winLabel2.destroy()
47          winLabel3.destroy()
48
49      if (radioButton.get() == 2):
50          winLabel1.destroy()
51          winLabel3.destroy()
52
53      if (radioButton.get() == 3):
54          winLabel1.destroy()
55          winLabel2.destroy()
56
57  # Create the radio button interface
58  def drawRadioButtons():
59      global radioButton
60
61      visibleGreenRadioButton = tk.Radiobutton (winFrame, text = 'Green',
62          width=17, height=1, bg = 'light grey', variable = radioButton,
63          value = 1, command = photos).grid(row = 1, column = 0)
64
65      visibleYellowRadioButton = tk.Radiobutton(winFrame, text='Yellow',
66          width= 17, height= 1, bg= 'light grey', variable = radioButton,
```

```
67            value = 2, command = photos).grid(row = 1, column = 1)
68
69       visibleRedRadioButton = tk.Radiobutton (winFrame, text = 'Red',
70            width= 17, height= 1, bg= 'light grey', variable = radioButton,
71            value = 3, command = photos).grid(row = 1, column = 2)
72
73   radioButton = IntVar()
74   photos()
75   drawRadioButtons()
76
77   winFrame.mainloop()
```

Output 4.4.1:

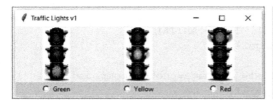

The output demonstrates the two main parts of the application. In the first part, the `photos()` method loads the three images and controls their visibility within the interface (lines 17–55). The reader will notice that part of the method is the destruction of two of the images, in order to leave only one on display (lines 44–56). For this task, the reader might also consider to use the `grid_remove()` method (covered in previous sections), which will have the same result.

The second part controls which of the three images will be displayed. Once the desired radiobutton has been clicked upon, the corresponding image stays on display and the other two are hidden (lines 57–71). It is worth noting that all three radio buttons are associated with the same variable. This is reflected on the fact that they cancel each other when selected, as the value of the common associated object is altered.

4.4.2 TRAFFIC LIGHTS VERSION 2 – CREATING A BASIC ILLUSION

Taking things one step further, the application is changed in such a way as to make only one image appearing instead of three. The impression that there is only one image is of course illusory, as it is essentially caused by manipulating the visual properties of the associated widget and/or its position in the interface. In this case, the traffic images are stacked upon each other using the same grid coordinates, and, subsequently, two of them are being removed from the interface.

This version is almost identical to the original one, with the exception of the positioning of the widgets and the slightly modified title. The proposed modification only requires the replacement of lines 15, 35, 42, 62, 66–67, and 70–71 with the ones provided below, which are only different in terms of their grid coordinates and width:

```
15       winFrame.title ("Traffic Lights v2"); winFrame.geometry("200x180")
    [...]
35       winLabel2.grid(row = 0, column = 0)
    [...]
42       winLabel3.grid(row = 0, column = 0)
    [...]
```

```
62      width = 20, height = 1, bg = 'light grey', variable = radioButton,
    [...]
66      width = 20, height = 1, bg = 'light grey', variable = radioButton,
67          value = 2, command = photos).grid(row = 2, column = 0)
    [...]
70      width = 20, height = 1, bg = 'light grey', variable = radioButton,
71          value = 3, command = photos).grid(row = 3, column = 0)
```

Output 4.4.2

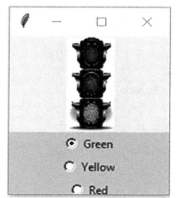 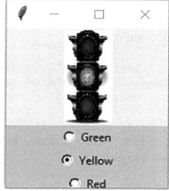 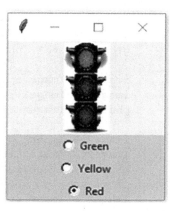

4.4.3   TRAFFIC LIGHTS VERSION 3 – CREATING A PRIMITIVE AUTOMATION

In this version of the sample application, there is no need for the user to click on the respective radio buttons in order to cause the traffic light images to appear/disappear. The change happens automatically after 5 seconds from the time one of the images is turned *on* (and the other two turned *off*). In order to enable timed functionality, in addition to the libraries used in the previous versions, the *time* library must be imported to the script.

This version differs from the previous ones in a number of ways:

- The radio buttons that were dealing with the interaction are removed, and a new `manage-Labels()` function is introduced to control the *automated* process of traffic light changes.
- Every time there is a change of the displayed image, the `time.sleep()` function (*time* library) is used to *freeze the execution of the application* for a given period of time (in this case 3 seconds).
- Since there are no radiobuttons, the application uses another object (`trafficLight`), to control which image is displayed. This is accomplished by setting its value through the `set()` method.
- The `update()` function is applied to the main frame in order to refresh the interface based on the latest status update.

The complete script is provided below:

```
1   # Import libraries
2   import tkinter as tk
3   from tkinter import *
4   # Import the necessary image processing classes
```

```
5    from PIL import Image, ImageTk
6    # Import the timer threading library
7    import time
8
9    global image1, image2, image3
10   global photo1, photo2, photo3
11   global winLabel1, winLabel2
12   global winFrame
13   global trafficLight
14
15   # Open the traffic images and create the relevant pointers
16   def photos():
17       global image1, image2, image3
18       global photo1, photo2, photo3
19
20       image1 = Image.open("TrafficLightsGreen.gif")
21       image1 = image1.resize((50, 100), Image.ANTIALIAS)
22       photo1 = ImageTk.PhotoImage(image1)
23
24       image2 = Image.open("TrafficLightsYellow.gif")
25       image2 = image2.resize((50, 100), Image.ANTIALIAS)
26       photo2 = ImageTk.PhotoImage(image2)
27
28       image3 = Image.open("TrafficLightsRed.gif")
29       image3 = image3.resize((50, 100), Image.ANTIALIAS)
30       photo3 = ImageTk.PhotoImage(image3)
31
32   # Manage label visibility based on time.
33   def manageLabels():
34       global winLabel1, winLabel2
35       global Photo1, Photo2, Photo3
36       global winFrame
37       global trafficLight
38
39       if (trafficLight.get() == 1):
40           winLabel1.config(image = photo1)
41           winLabel1.grid(row = 0, column = 0)
42           winLabel2.config(text = 'Green')
43           time.sleep(3)
44
45       if (trafficLight.get() == 2):
46           winLabel1.config(image = photo2)
47           winLabel1.grid(row = 0, column = 0)
48           winLabel2.config(text = 'Yellow')
49           time.sleep(3)
50
51       if (trafficLight.get() == 3):
52           winLabel1.config(image = photo3)
53           winLabel1.grid(row = 0, column = 0)
54           winLabel2.config(text = 'Red')
55           time.sleep(3)
56
```

```
57        winFrame.update()
58
59  # Create the main frame
60  winFrame = tk.Tk()
61  winFrame.title ("Traffic Lights v3"); winFrame.geometry("200x180")
62
63  photos()
64
65  winLabel1 = tk.Label(winFrame, text='', image=photo1, compound='left')
66  winLabel1.grid(row = 0, column = 0)
67  winLabel2=tk.Label(winFrame,text='...'); winLabel2.grid(row=1,column=0)
68
69  trafficLight = IntVar()
70  trafficLight.set(1)
71
72  while (True):
73        if (trafficLight.get() == 1):
74              trafficLight.set(2)
75        elif (trafficLight.get() == 2):
76              trafficLight.set(3)
77        elif (trafficLight.get() == 3):
78              trafficLight.set(1)
79        manageLabels()
80
81  winFrame.mainloop()
```

Output 4.4.3:

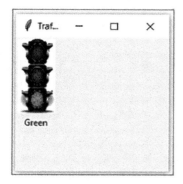 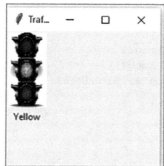 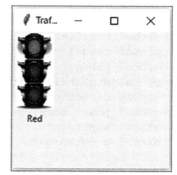

4.4.4 TRAFFIC LIGHTS VERSION 4 – A PRIMITIVE SCREEN SAVER WITH A PROGRESS BAR

Having introduced the concept of timed events and how they can be used to control the flow of events in an application, it is rather straightforward to expand the same idea to *the creation of an illusory movement of particular objects* inside a frame. A good example of this is the creation of a primitive screen saver using the existing *Traffic Lights* application as a basis.

In addition to the existing widgets, an additional widget that can be used in this scenario is the *progressbar* widget. This will assist in making the screen saver a bit more informative, by providing clues about the elapsed and remaining time in any particular condition (i.e., *green*, *yellow*, and *red* traffic light). The widget belongs to the *ttk* library and can take several parameters that control its appearance and functionality, with the most important ones being length, orient, and mode. Length determines the size (i.e., length) of the progress bar, orient the orientation of the widget (i.e., VERTICAL or HORIZONTAL), and mode if the displayed value is *predetermined* ("determinate") or *indetermined* ("intederminate"). In the case of the former, the bar will appear moving toward one end of the widget until the specified value is reached, while in the case of the latter the bar will appear moving continuously from one end to the other and back.

The following script implements a related implementation example, where the three traffic lights are controlling the movement of a *car* image (embedded in a label widget). When the green light is on, the car is moving at a particular speed and when yellow is on at half that speed. Similarly, when the red light is on, the car appears to stop and the progressbar appears to be *loading* to reflect the elapsed time in this particular condition (i.e., red light) and remaining time until the next condition is triggered (i.e., green light). The car image appears to be bouncing across the frame, moving toward a different direction every time it reaches the edges of the parent frame. The movement of the car image is always diagonal, and follows four different directions. The program stops when the user interrupts (closes) the application. The associated Python script is provided below:

```
1    # Import libraries
2    import tkinter as tk
3    from tkinter import ttk
4    from tkinter import *
5    # Import the necessary image processing classes
6    from PIL import Image, ImageTk
7    # Import threading libary for the timer threading
8    import time
9
10   global trafficLight
11   global image1, image2, image3
12   global photo1, photo2, photo3
13   global winLabel1, winLabel2, winLabel3
14   global direction
15   global posx, posy
16   global winFrame
17   global progressBar
18
19   # Open the traffic and car images and create the relevant pointers
20   def photos():
21       global image1, image2, image3, image4
22       global photo1, photo2, photo3, photo4
23
24       image1 = Image.open("TrafficLightsGreen.gif")
25       image1 = image1.resize((50, 100), Image.ANTIALIAS)
26       photo1 = ImageTk.PhotoImage(image1)
27
28       image2 = Image.open("TrafficLightsYellow.gif")
29       image2 = image2.resize((50, 100), Image.ANTIALIAS)
30       photo2 = ImageTk.PhotoImage(image2)
31
```

```
32        image3 = Image.open("TrafficLightsRed.gif")
33        image3 = image3.resize((50, 100), Image.ANTIALIAS)
34        photo3 = ImageTk.PhotoImage(image3)
35
36        image4 = Image.open("Car.gif")
37        image4 = image4.resize((30, 15), Image.ANTIALIAS)
38        photo4 = ImageTk.PhotoImage(image4)
39
40  # Manage label visibility based on time
41  def manageLabels():
42        global trafficLight
43        global winLabel1, winLabel2
44        global Photo1, Photo2, Photo3
45        global winFrame
46
47        if (trafficLight.get() == 1):
48          winLabel1.config(image=photo1)
49          winLabel2.config(text='Green'); a=1
50        elif (trafficLight.get() == 2):
51           winLabel1.config(image=photo2)
52           winLabel2.config(text='Yellow'); a=2
53        elif (trafficLight.get() == 3):
54           winLabel1.config(image = photo3)
55           winLabel2.config(text = 'Red'); a = 3
56
57        winLabel1.pack(); winLabel1.place(x = 1, y = 1)
58        winLabel2.pack(); winLabel2.place(x = 1, y = 100)
59        winFrame.update
60
61        # Call method moveCar()to move the image within the interface
62        moveCar(a)
63
64  # Control the direction of the movement
65  def checkDirection():
66        global direction
67        global posx, posy
68
69        if (posx >= 400 and direction == 1):
70            direction = 2
71        elif (posx >= 400 and direction == 4):
72            direction = 3
73        elif (posx <= 0 and direction == 2):
74            direction = 1
75        elif (posx <= 0 and direction == 3):
76            direction = 4
77        elif (posy <= 0 and direction == 3):
78            direction = 2
79        elif (posy <= 0 and direction == 4):
80            direction = 1
81        elif (posy >= 200 and direction == 1):
82            direction = 4
83        elif (posy >= 200 and direction == 2):
```

```
84            direction = 3
85
86   # Manage the movement of the car
87   def moveCar(a):
88       global direction
89       global posx, posy
90       global winLabel3
91       global winFrame
92       global progressBar
93
94       progressBar['value'] = 0
95
96       for i in range(10):
97           # Call checkDirection() to control the movement direction
98           checkDirection()
99
100      if (a == 1):
101          move = 10
102      elif (a == 2):
103          move = 5
104      else:
105          move = 0
106          progressBar['value'] = int((i/(10 - 1)) * 100)
107
108      if (direction == 1):
109          posy += move; posx += move
110      elif (direction == 2):
111          posy += move; posx -= move
112      elif (direction == 3):
113          posy -= move; posx -= move
114      elif (direction == 4):
115          posy -= move; posx += move
116          winLabel3.pack(); winLabel3.place(x = posx, y = posy)
117
118          winFrame.update()
119          time.sleep(0.3)
120
121  # Create the main frame
122  winFrame = tk.Tk()
123  winFrame.title ("Traffic Lights v4"); winFrame.geometry("400x200")
124
125  photos()
126
127  winLabel1 = tk.Label(winFrame, text='', image=photo1, compound='left')
128  winLabel1.pack(); winLabel1.place(x = 1, y = 1)
129
130  winLabel2 = tk.Label(winFrame, text = '...')
131  winLabel2.pack(); winLabel2.place(x = 1, y = 100)
132
133  winLabel3 = tk.Label(winFrame, text='', image=photo4, compound='left')
134  winLabel3.pack(); winLabel3.place(x = 1, y = 1)
135  posx = 0; posy = 0
```

```
136
137  progressBar = ttk.Progressbar(winFrame, length=100, orient = VERTICAL,
138          mode = 'determinate')
139  progressBar.place(relx = 0.13, rely = 0.02)
140
141  trafficLight = IntVar()
142  trafficLight.set(3)
143  direction = 1
144
145  while (True):
146      winFrame.update_idletasks()
147      if (trafficLight.get() == 1):
148          trafficLight.set(2)
149      elif (trafficLight.get() == 2):
150          trafficLight.set(3)
151      elif (trafficLight.get() == 3):
152          trafficLight.set(1)
153      manageLabels()
154
155  winFrame.mainloop()
```

Output 4.4.4:

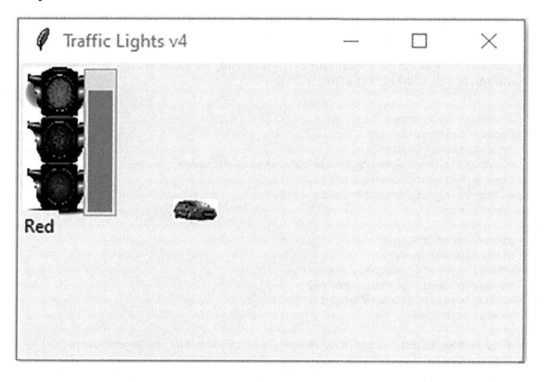

A number of new methods, options and computational ideas are introduced in this script. First, the reader will notice the use of the update_idletasks() method, which ensures that objects or methods not being currently used *are still updated every time the* while *loop is executed* (line 132). This safeguards from unwanted *garbage collection* processes that might occur for the, seemingly unused, objects. Second, it is worth noting the use of absolute coordinates x and y to continuously position the relevant widgets on the interface, instead of the relative ones (relx and rely) used in previous examples. This is especially relevant in the case of the moving car in order to trace and handle its movement when reaching the edges of the interface.

In terms of the actual movement of the car, the computational idea is quite simple. For instance, when it reaches the east edge of the interface, (a) if it is moving *southeast* (i.e., direction=1) it should bounce toward the *southwest* (i.e., direction=2), and (b) if it is moving *northeast* (i.e., direction=4) it should bounce toward the *northwest* (i.e., direction=3). The checkDirection() method (lines 59–79) takes care of the rest of the movements of the car. Once the step and directions are set, the actual movement takes place in method movecar() (lines 81–109). The method recalculates the current placement coordinates of the car based on the actual coordinates, given both the intended direction and the state of the traffic light.

> **Observation 4.47 – update_idle-tasks():** Use the update _ idle-tasks() method to ensure that idle widgets/objects are not being destroyed when not being used for extended periods of time.

> **Observation 4.48 – x, y coordinates:** It is often preferable to use the x and y coordinates when placing a widget on an interface, in order to ensure its absolute placement in pixels instead of the relative positions (i.e., using relx and rely).

4.4.5 Traffic Lights Version 5 – Suggesting a Primitive Screen Saver

As a conclusion of this automation-related series of scripts based on the Traffic Lights sample application, it is useful to introduce the idea of using designated keyboard input commands to achieve a certain level of control over the automated events. The following script introduces functionality that allows the user to move the car dynamically at run time using the *up*, *down*, *left*, and *right* keys on the keyboard, as well as the *esc* key to exit:

```
1    # Import libraries
2    import tkinter as tk
3    from tkinter import *
4    # Import the necessary image processing classes
5    from PIL import Image, ImageTk
6    # Import the timer threading libary
7    import time
8
9    global trafficLight
10   global posx, posy
11   global image1, image2, image3
12   global photo1, photo2, photo3
13   global winLabel1, winLabel2
14   global winFrame
15
16   # Open the traffic and car images and create the relevant pointers
17   def photos():
18       global image1, image2, image3, image4
19       global photo1, photo2, photo3, photo4
20
21       image1 = Image.open("TrafficLightsGreen.gif")
```

```
22        image1 = image1.resize((50, 100), Image.ANTIALIAS)
23        photo1 = ImageTk.PhotoImage(image1)
24
25        image2 = Image.open("TrafficLightsYellow.gif")
26        image2 = image2.resize((50, 100), Image.ANTIALIAS)
27        photo2 = ImageTk.PhotoImage(image2)
28
29        image3 = Image.open("TrafficLightsRed.gif")
30        image3 = image3.resize((50, 100), Image.ANTIALIAS)
31        photo3 = ImageTk.PhotoImage(image3)
32
33        image4 = Image.open("Car.gif")
34        image4 = image4.resize((30, 15), Image.ANTIALIAS)
35        photo4 = ImageTk.PhotoImage(image4)
36
37    # Manage the movement based on the traffic light
38    def keyPressed (event):
39        global trafficLight
40        global posx, posy
41        global winFrame
42        global winLabel3
43
44        # Set the moving step based on the traffic light
45        if (trafficLight == 1):
46            move = 10
47        elif (trafficLight == 2):
48            move = 5
49        elif (trafficLight == 3):
50            move = 0
51
52        print(event.keycode)
53
54        # Prepare the moving step (up, down, left, right, esc)
55        # Mac codes: (8320768,8255233, 8124162, 8189699, 3473435)
56        # The user pressed 'up'. Move the car accordingly
57        if (event.keycode == 38):
58            if (move == 10 and posy >= 20):
59                posy -= 10
60            elif (move == 5 and posy >=20):
61                posy -= 5
62        # The user pressed 'down'. Move the car accordingly
63        elif (event.keycode == 40):
64            if (move == 10 and posy <= 270):
65                posy += 10
66            elif (move == 5 and posy <= 270):
67                posy += 5
68        # The user pressed 'right'. Move the car accordingly
69        elif (event.keycode == 39):
70            if (move == 10 and posx <= 570):
71                posx += 10
72            elif (move == 5 and posx <= 570):
```

```
73              posx += 5
74          # The user pressed 'left'. Move the car accordingly
75          elif (event.keycode == 37):
76              if (move == 10 and posx >= 20):
77                  posx -= 10
78              elif (move == 5 and posx >= 20):
79                  posx -= 5
80          # The user pressed 'escape'. Close the program
81          elif (event.keycode == 27):
82              winFrame.destroy()
83              exit()
84
85          winLabel2.pack(); winLabel2.place(x = posx, y = posy)
86          winFrame.update()
87
88      def trafficLightsLoop():
89          global trafficLight
90          global winFrame
91          global winLabel1
92
93          winFrame.update_idletasks()
94          if (trafficLight == 1):
95              trafficLight = 2; winLabel1.config(image = photo2)
96          elif (trafficLight == 2):
97              trafficLight = 3; winLabel1.config(image = photo3)
98          elif (trafficLight == 3):
99              trafficLight = 1; winLabel1.config(image = photo1)
100
101         winLabel1.pack(); winLabel1.place(x = 1, y = 1)
102         winFrame.update
103
104         winFrame.after(3000, trafficLightsLoop)
105
106     # Create the main frame
107     winFrame = tk.Tk()
108     winFrame.title ("Traffic Lights v5"); winFrame.geometry("600x300")
109     winFrame.bind('<Key>', keyPressed)
110
111     photos()
112
113     winLabel1 = tk.Label(winFrame, text='', image=photo1, compound='left')
114     winLabel1.pack(); winLabel1.place(x = 1, y = 1)
115
116     winLabel2 = tk.Label(winFrame, text='', image=photo4, compound='left')
117     winLabel2.pack(); winLabel2.place(x = 1, y = 1)
118     trafficLight = 1; posx = 0; posy = 0
119
120     winFrame.after(3000, trafficLightsLoop)
121
122     winFrame.mainloop()
```

Output 4.4.5:

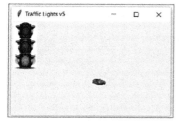  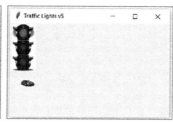

The script introduces some new ideas and techniques aiming to make the user experience more engaging, and to encourage further enhancements. Firstly, it must be noted that, in the main program, the main frame is *bound* to the keypressed() method through the <Key> event (line 102). It must be stressed that the naming of the event is important and that any deviations (e.g., <key>) may not be translated correctly by Python. The use of the binding results in the user being able to press any of the *up*, *down*, *left*, and *right* directional keys in order to move the car to the relevant direction. This is achieved by checking the values of the event.keycode produced based on the user's input. It is worth noting that these values may vary between different systems, so the code should include appropriate controls and solutions for such variations (lines 37–73).

Secondly, the reader should note the *avoidance of a loop* and its replacement by the after() method, which is applied to the main frame (winFrame). The reason for this decision was that since the program activates the monitoring of the <Key> event, the presence of a second monitoring event like a loop *would cause conflicts in the internal threading of the application*. The after() method serves the purpose of creating a loop-like behavior without causing such a conflict (lines 97, 113). Finally, the reader should note the use of the esc code in the keypressed() method (line 76) to exit the application in a controlled way.

> **Observation 4.49 – <key>, event. keycode:** Use the <Key> event to bind a particular frame or widget to a *key press* event. Once the key input is captured, use event.keycode to determine the appropriate action.

> **Observation 4.50 – after():** Use the after() method to call a method or execute a command after a predetermined number of seconds has elapsed since the initiation of the current method. This can be used as an alternative to for or while loops.

4.5 CASE STUDIES

Enhance the *Countries* application in order to include the following functionality:

- Add one more listbox to display more content for each country (e.g., size, population, etc.).
- Add a combobox to allow the user to select the *font name* of the contents of the listboxes.
- Add a combobox to allow the user to select the *font size* of the contents of the listboxes.
- Add a combobox to change the *background color* of the content in the listboxes.

4.6 EXERCISES

Enrich the *Traffic Lights* application by including one more car. The new car must be controlled by another set of keys on the keyboard, using the same traffic lights as those on the original application.

5 Application Development with Python

Dimitrios Xanthidis
University College London
Higher Colleges of Technology

Christos Manolas
The University of York
Ravensbourne University London

Hanêne Ben-Abdallah
University of Pennsylvania

CONTENTS

5.1 Introduction 161
5.2 Messages, Common Dialogs, and Splash Screens in Python 162
    5.2.1 Simple Message Boxes 162
    5.2.2 Message Boxes with Options 164
    5.2.3 Message Boxes with User Input 166
    5.2.4 Splash Screen/About Forms 168
    5.2.5 Common Dialogs 169
5.3 Menus 171
    5.3.1 Simple Menus with Shortcuts 171
    5.3.2 Toolbar Menus with Tooltips 175
    5.3.3 Popup Menus with Embedded Icons 178
5.4 Enhancing the GUI Experience 181
    5.4.1 Notebooks and Tabbed Interfaces 181
    5.4.2 Threaded Applications 185
    5.4.3 Combining Multiple Concepts and Applications in a Multithreaded System 190
5.5 Wrap Up 199
5.6 Case Study 205

5.1 INTRODUCTION

Application development can be viewed as a process that is both scientific and creative. Scientific because it follows the systematic process of the software development *life-cycle*. This covers all development steps, from requirement analysis and implementation to deployment and maintenance. Creative as it calls for the creativity of the developer to design a system that incorporates features that make it suitable and efficient for the task at hand, while also being attractive to the end user.

The previous chapter introduced and discussed some of the key objects for the development of an appealing user interface. In this chapter, the concept of application development is examined more

DOI: 10.1201/9781003139010-5

thoroughly, by introducing ideas and tools that call for the integration of *multiple functions within a single application*. These include:

- **Dialogs, Messages, and the Splash Screen:** Simple and intuitive objects that most users of Windows style applications are quite familiar with. Each of these objects serves a particular function and is part of the *Python API* (*Application Programming Interface*), thus, requiring only minimal coding.
- **Menus, Toolbar Menus, Popup Menus:** Variations of the well-known *menu* object allowing the user to select different functions available in the application. Menus are usually accompanied by extra functionality options like *hot keys*, *shortcuts*, and *tooltips*, in order to enhance their attractiveness and efficiency.
- **Tabs:** Tabs provide an effective way to optimize the use of the *real estate* of the running interface, allowing the inclusion of more than one application in the same space. This idea is simple, but intuitive and effective. Tabs are commonly used to separate a single *notebook* into various sections and load various independent applications.
- **Threads:** Threading involves the simultaneous execution of code relating to multiple instances of the same process, class or application. Different threads can be executed simultaneously, either in parallel or in explicitly defined time slots. Each thread can have its own widgets (if it is GUI based) and attributes. Threaded objects do not necessarily communicate with each other, although this is possible and can be implemented when and if necessary.

The focus of this chapter is on discussing and illustrating key underlying concepts and mechanisms associated with these tools and structures.

5.2   MESSAGES, COMMON DIALOGS, AND SPLASH SCREENS IN PYTHON

Messageboxes, *common dialogs*, and *splash screens* are some of the most understated, but useful objects that can help in enhancing the functionality of an application without adding lengthy code to it. They are user-friendly and multifunctional, and provide instant, and strictly restricted and managed input from the user during the execution of an application. Several types of these components are available with varied and diverse functions, such as the display of user messages, the creation of menus of options/choices, the acceptance and verification of user input, the management of display parameters and options (e.g., colors), and the management of files, file structures and directories. Each of the above can be called and implemented with relatively simple Python code commands, as described in the following sections.

5.2.1   SIMPLE MESSAGE BOXES

The simple *message box* displays a message to the user and stays on display until the corresponding (OK) button is clicked, at which point the application resumes execution. As there is no input to be received, the user reaction to the message is irrelevant and the only possible choice is to click the OK button. The object has three distinct forms represented by methods showinfo(), showerror(), and showwarning(), which are embedded in the messagebox object (tkinter library). These

> **Observation 5.1 – Simple Message Box:** Methods showinfo(), showerror(), and showwarning() (members of the messagebox object, tkinter library) are used to display a simple message box with a respective *info*, *error*, or *warning* icon.

methods do not change any fundamental aspects of the message box, but modify the icon that accompanies it according to the type of information provided to the user. The following Python script presents a basic example of the use of each of the three methods:

```
1    # Import libraries
2    import tkinter as tk
3    from tkinter import messagebox
4
5    # Declare simpleMessage() function, invoked upon button click
6    def simpleMessage(a):
7        if (a == 1):
8            messagebox.showinfo("Simple Info Message",
9                "You clicked for the info message")
10       elif (a == 2):
11           messagebox.showerror("Simple Error Message",
12               "You clicked for the error message")
13       elif (a == 3):
14           messagebox.showwarning("Simple Warning Message",
15               "You clicked for the warning message")
16
17   # Create a non-resizable Windows frame using the tk object
18   winFrame = tk.Tk()
19   winFrame.title("Simple Messageboxes")
20   winFrame.resizable(False, False)
21   winFrame.geometry('290x180')
22   winFrame.configure(bg = 'dark grey')
23
24   # Create button that triggers an info message
25   winButton1 = tk.Button(winFrame, width = 25,
26       text = "Click to display \na simple info messagebox")
27   winButton1.pack(); winButton1.place(x = 50, y = 20)
28   winButton1.bind('<Button-1>', lambda event: simpleMessage(1))
29
30   # Create button that triggers an error message
31   winButton2 = tk.Button(winFrame, width = 25,
32       text = "Click to display \na simple error messagebox")
33   winButton2.pack(); winButton2.place(x = 50, y = 70)
34   winButton2.bind('<Button-1>', lambda event: simpleMessage(2))
35
36   # Create button that triggers a warning message
37   winButton3 = tk.Button(winFrame, width = 25,
38       text = "Click to display \na simple warning messagebox")
39   winButton3.pack(); winButton3.place(x = 50, y = 120)
40   winButton3.bind('<Button-1>', lambda event: simpleMessage(3))
41
42   winFrame.mainloop()
```

Output 5.2.1:

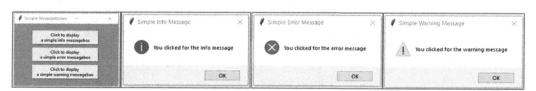

The reader should note that the first parameter passed to the message box is the *title*, whereas the second is the *content*. The program output provided above illustrates the resulting messages for each of the three simple types of message boxes.

5.2.2   MESSAGE BOXES WITH OPTIONS

Message boxes are commonly used to receive user confirmation for processes that take place at run-time. In such cases, instead of merely displaying information, the object must prompt the user to confirm their approval (or lack of) regarding the execution of particular processes. As in the case of simple messages, several options are available for message boxes with options, depending on the type of confirmation that is requested. However, there are two major differences between the two types of messages. Firstly, in the case of messages with options, the user makes a choice that may alter the execution order of the processes that follow, in contrast to the simple message box. The type and format of the input

> **Observation 5.2 – Message Box with Options:** Methods askokcancel(), askretrycancel(), askyesno(), and askquestion() (members of the messagebox object, tkinter library) are used to display a message, while also requesting some sort of confirmation from the user. The responses can be True or False for the first three and 'Yes' or 'No' for the last one.

depends on the type of the message (e.g., *OK-Cancel, Retry-Cancel, Yes-No*). Secondly, the user's choice has a tangible value that can be stored in a variable and checked against other pre-defined values to determine the flow of execution. These values are True or False (no quotes and case-sensitive) in the case of OK-Cancel, Retry-Cancel, and Yes-No, and 'Yes' or 'No' (in single quotation marks and case-sensitive) in the case of a question message box.

The following Python script provides a simple example that integrates all four different types of messages with options. The script also makes use of the showinfo() and showerror() methods of the simple message box:

```
1    # Import libraries
2    import tkinter as tk
3    from tkinter import messagebox
4
5    # Declare optionMessage() function, invoked upon button click
6    def optionMessage(a):
7        if (a == 1):
8            response = messagebox.askokcancel(title = "ok-cancel Message",
9                message = "Clicked the OK-Cancel message", icon = 'info')
10           if (response == True):
11               messagebox.showinfo("Info Message", "Clicked OK")
12           elif (response == False):
13               messagebox.showerror("Error Message", "Clicked Cancel")
14       elif (a == 2):
15           response = messagebox.askquestion(title = "question Message",
16               message = "Clicked the question message", icon = 'info')
17           if (response == 'yes'):
18               messagebox.showinfo("Info Message", "Clicked Yes")
19           elif (response == 'no'):
```

```
20              messagebox.showerror("Error Message", "Clicked No")
21       elif (a == 3):
22            response=messagebox.askretrycancel(title="retry-cancel Message",
23                message = "Clicked the Retry-Cancel message", icon = 'info')
24            if (response == True):
25               messagebox.showinfo("Info Message", "Clicked Retry")
26            elif (response == False):
27               messagebox.showerror("Error Message", "Clicked Cancel")
28       elif (a == 4):
29            response = messagebox.askyesno(title = "yes-no Message",
30                message = "Clicked the Yes-No message", icon = 'info')
31            if (response == True):
32               messagebox.showinfo("Info Message", "Clicked Yes")
33            elif (response == False):
34               messagebox.showerror("Error Message", "Clicked No")
35
36   # Create a non-resizable Windows frame using the tk object
37   winFrame = tk.Tk()
38   winFrame.title("Messageboxes with options")
39   winFrame.resizable(False, False)
40   winFrame.geometry('320x220')
41   winFrame.configure(bg = 'grey')
42
43   # Create button that triggers an OK-Cancel message
44   winButton1 = tk.Button(winFrame, width = 20,
45       text = "Click to display \na OK-Cancel messagebox")
46   winButton1.pack(); winButton1.place(x = 85, y = 20)
47   winButton1.bind('<Button-1>', lambda event: optionMessage(1))
48
49   # Create button that triggers a question message
50   winButton2 = tk.Button(winFrame, width = 20,
51       text = "Click to display \na Question messagebox")
52   winButton2.pack(); winButton2.place(x = 85, y = 70)
53   winButton2.bind('<Button-1>', lambda event: optionMessage(2))
54
55   # Create button that triggers a Retry-Cancel message
56   winButton3 = tk.Button(winFrame, width = 20,
57       text = "Click to display \na Retry-Cancel messagebox")
58   winButton3.pack(); winButton3.place(x = 85, y = 120)
59   winButton3.bind('<Button-1>', lambda event: optionMessage(3))
60
61   # Create button that triggers a Yes-No message
62   winButton3 = tk.Button(winFrame, width = 20,
63       text = "Click to display \na Yes-No messagebox")
64   winButton3.pack(); winButton3.place(x = 85, y = 170)
65   winButton3.bind('<Button-1>', lambda event: optionMessage(4))
66
67   winFrame.mainloop()
```

Output 5.2.2:

5.2.3 MESSAGE BOXES WITH USER INPUT

Occasionally, message boxes are used instead of regular entry or text widgets, to prompt user input of various different data types (i.e., string, integer, float). This is a viable choice when the interface is heavily loaded or when the use of widgets is not desirable. When message boxes are used for this purpose, the following methods can be used: (a) `askstring()` for string input, (b) `askinteger()` for integer numbers input, and (c) `askfloat()` for float numbers (real numbers) input. These methods

Observation 5.3 – Message Box with User Input: Methods `askstring()`, `askinteger()`, and `askfloat()` (members of the `simpledialog` object, `tkinter` library) are used to display a message requesting input of a specific data type from the user.

are members of the simpledialog class of the `tkinter` library. As they return a particular data type value, it must be stored in a suitable variable declared for this purpose.

As shown in the following Python script, the title and the message of the message box must be also specified:

```
1   # Import libraries
2   import tkinter as tk
3   from tkinter import simpledialog
4   from tkinter import messagebox
5
6   global name; global birthyear; global gpa
7
8   # Declare optionMessage() function, invoked upon button click
9   def inputMessage(a):
10      global name; global birthyear; global gpa
11
12      # Accept student name, year of birth, and GPA
13      # and display it through a simple message box
14      if (a == 1):
15          name = simpledialog.askstring("Name", "What is your name?")
16      elif (a == 2):
17          birthyear = simpledialog.askinteger("Year of birth",
```

```
18                 "What is the year of your birth?")
19      elif (a == 3):
20          gpa = simpledialog.askfloat("GPA",
21                 "What is your GPA (out of 4 with one decimal)?")
22      elif (a == 4):
23          message="Student's name: "+name+"\nStudent's year of birth: "+\
24              str(birthyear) + "\nStudent's GPA: " + str(gpa)
25          messagebox.showinfo("Student's info", message)
26
27 # Create a non-resizable Windows frame using the tk object
28 winFrame = tk.Tk()
29 winFrame.title("Inputboxes")
30 winFrame.resizable(False, False)
31 winFrame.geometry('260x220')
32 winFrame.configure(bg = 'grey')
33
34 # Create buttons that will trigger the associated messages
35 winButton1 = tk.Button(winFrame,
36      text = "Click to ask \nthe student's name", width = 20)
37 winButton1.pack(); winButton1.place(x = 30, y = 20)
38 winButton1.bind('<Button-1>', lambda event: inputMessage(1))
39 winButton2 = tk.Button(winFrame, width = 20,
40      text = "Click to ask \nthe student's year of birth")
41 winButton2.pack(); winButton2.place(x = 30, y = 70)
42 winButton2.bind('<Button-1>', lambda event: inputMessage(2))
43 winButton3 = tk.Button(winFrame,
44      text = "Click to ask \nthe student's GPA", width = 20)
45 winButton3.pack(); winButton3.place(x = 30, y = 120)
46 winButton3.bind('<Button-1>', lambda event: inputMessage(3))
47 winButton4 = tk.Button(winFrame,
48      text = "Click to show \nthe student's info", width = 20)
49 winButton4.pack(); winButton4.place(x = 30, y = 170)
50 winButton4.bind('<Button-1>', lambda event: inputMessage(4))
51
52 name = ""; birthyear = 0; gpa = 0.0
53
54 winFrame.mainloop()
```

Output 5.2.3:

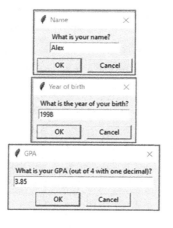

5.2.4 SPLASH SCREEN/ABOUT FORMS

A frequently underestimated type of object is the so-called *splash screen* or *about* form. It is most commonly used to provide information about application execution and processes, development details and dates, copyrights, and contacting the development team. The object does not follow a formal design and, therefore, it is not offered as a template by most well-known programming languages.

Observation 5.4 – *Splash screen*: A *splash screen* can be used in cases of excessive loading times of a window/widget or when there is a need to display information related to the application.

Among its various uses, the splash screen/about form can be used to give time to the main application to load its components. This is especially relevant if significant amounts of data need to be loaded, such as sizable databases or graphics, and heavy objects in general. The following script is a basic example of a splash screen with no apparent functionality. The form disappears after 8 seconds to give its place to the main application window:

```
1   # Import libraries
2   import tkinter as tk
3   import time
4
5   global winSplash
6
7   # Create the Splash screen
8   def splash():
9       global winSplash
10
11      winSplash = tk.Tk()
12      winSplash.title("Splash screen")
13      winSplash.resizable(False, False)
14      winSplash.geometry('250x100')
15      winSplash.configure (bg = 'dark grey')
16      winLabel1 = tk.Label(winSplash,
17          text = "Display the Splash screen \nfor 8 seconds")
18      winLabel1.grid(row = 0, column = 0)
19
20      # Use the update function to display the splash screen
21      # before the mainloop (main window) takes over
22      winSplash.update()
23
24  # Call the splash screen for 8 seconds
25  splash()
26  time.sleep(8)
27
28  # Destroy the splash screen before the mainloop
29  winSplash.destroy()
30
31  # Create the main window
32  winFrame = tk.Tk()
33  winFrame.title("Main Window")
34  winFrame.resizable(False, False)
```

```
35  winFrame.geometry('250x100')
36  winFrame.configure(bg = 'grey')
37
38  winLabel2 = tk.Label(winFrame, text = "Entered the main window")
39  winLabel2.grid(row = 0, column = 0)
40
41  winFrame.mainloop()
```

Output 5.2.4:

The user should note the use of the time.sleep() method after the splash() method is invoked. This delays the splash screen before the main window (winFrame) is loaded. It is also worth noting the use of the update() method on the winSplash object. This method ensures that the widget is displayed, although it is not the main window and, thus, the mainloop() method *cannot be used with it.*

5.2.5 COMMON DIALOGS

It is frequently the case that the programmer needs to utilize the *API* (*Application Programming Interface*) of the operating system in order to avoid writing code that is already provided as pre-packaged, essential functionality. Some of the most important GUI-related API elements can be found under the broader category of *dialogs*. Different versions of dialogs exist, such as *Color, Open File, Save File, Directory, Font Dialog*, and *Print*. These dialogs allow programmers to circumvent extensive GUI programming by offering instant access to basic, repetitive functional tasks. These are the common dialog objects that appear in various types of widely used GUI applications (e.g., MS Office or Adobe Creative Suite).

With the exception of the color dialog (askcolor), which is included in the colorchooser library, the aforementioned dialogs are all included in the filedialog library under the associated keywords (e.g., filedialog.askopenfile(), filedialog.asksaveasfile(), filedialog.askdirectory()). The syntax for invoking these API methods is simple and rather intuitive, and it allows a two-way communication with the user in order to obtain their selection. In the case of askcolor(), one should note that the result is a set of two values: an *rgb* (*red, green, blue*) value and a particular color selection. The color values selected

> **Observation 5.5 – API methods:** The API methods offered by Python can be used to perform basic repetitive tasks across many platforms and operating systems. These methods include askcolor() from the colorchooser library and asksavesasfile(), askopenfile(), and askdirectory() from the filedialog library.

can be stored in a variable for further use. The following Python script illustrates the use of the four API methods mentioned above:

```
1   # Import libraries
2   import tkinter as tk
3   from tkinter import filedialog
4   from tkinter import colorchooser
5
6   # Define openDialogs()  function, invoked upon button click
7   def openDialogs(a):
8       if (a == 1):
9           # Assign user color selection to a set of variables
10          (rgbSelected, colorSelected) = colorchooser.askcolor()
11          # Use the color element from the variable set to change
12          # the color of the form
13          winFrame.config(background = colorSelected)
14      elif (a == 2):
15          filedialog.askopenfile(title = "Open File Dialog")
16      elif (a == 3):
17          filedialog.askdirectory(title = "Directory Dialog")
18      elif (a == 4):
19          filedialog.asksaveasfilename(title = "Save As Dialog")
20
21  # Create a non-resizable Windows frame using the tk object
22  winFrame = tk.Tk()
23  winFrame.title("Common Dialogs")
24  winFrame.resizable(False, False)
25  winFrame.geometry('280x220')
26  winFrame.configure(bg = 'grey')
27
28  # Create button that triggers the Color dialog
29  winButton1 = tk.Button(winFrame,
30      text = "Click to open \nthe Color dialog", width = 20)
31  winButton1.pack(); winButton1.place(x = 60, y = 20)
32  winButton1.bind('<Button-1>', lambda event: openDialogs(1))
33
34  # Create button that triggers the Open File dialog
35  winButton2 = tk.Button(winFrame,
36      text = "Click to open \nthe File Dialog", width = 20)
37  winButton2.pack(); winButton2.place(x = 60, y = 70)
38  winButton2.bind('<Button-1>', lambda event: openDialogs(2))
39
40  # Create button that triggers the Directory dialog
41  winButton3=tk.Button(winFrame,
42      text="Click to open \nthe Directory Dialog", width = 20)
43  winButton3.pack(); winButton3.place(x = 60, y = 120)
44  winButton3.bind('<Button-1>', lambda event: openDialogs(3))
45
46  # Create button that triggers the Save As dialog
47  winButton3=tk.Button(winFrame,
48      text = "Click to open \nthe Save As Dialog", width = 20)
49  winButton3.pack(); winButton3.place(x = 60, y = 170)
50  winButton3.bind('<Button-1>', lambda event: openDialogs(4))
51
52  winFrame.mainloop()
```

Output 5.2.5:

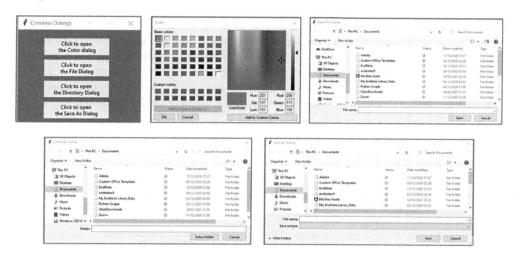

5.3 MENUS

It is quite rare for a desktop or mobile application to offer singular functionality. Developers usually create systems capable of performing numerous tasks and functions. An example of this are the scripts developed in the previous sections, where multiple, although quite simplistic, tasks were performed using a series of corresponding buttons. In reality, in most cases, access to different functions within an application is provided through *menus*. These can take different forms, such as *simple menus*, *single-layered menus*, *menus with nested sub-menus*, *toolbars*, and *pop-up menus*. These types of menus can be used in isolation, but are also frequently used in conjunction. This section covers basic menu concepts, as well as a number of particular options that can be used to further enhance menu functionality.

5.3.1 SIMPLE MENUS WITH SHORTCUTS

In all windows style applications, simple menus follow the same basic, but rather intuitive, style. They include a top-level list of items, usually displayed just below the title of the application. This top-level menu layer sits on top of sub-menus that are hidden in subsequent layers. Such basic menus are created using the constructor of the Menu class from the `tkinter` library. The idea is quite straightforward indeed. Firstly, the menu object is created using the Menu() constructor. Additional menu objects can be also created and attached to the main menu object, as necessary. Next, any required sub-menus can be added to the main menu. This can be accomplished with the `add _ command()` method for simple items or the `add _ checkbutton()` and `add _ radiobutton()` methods for check button and radio button items, respectively. For nested menus, these steps can be repeated as many times as necessary, although one should avoid going deeper than two levels of menus for clarity reasons. Finally, the `add _ cascade()` method is used to tie together the various menu pieces and activate the menu system.

> **Observation 5.6 – Menu class:** Use the constructor of the Menu class to create a menu object. The main menu choices can be added using the constructor (Menu()), while simple menu items can be added using the `add _ command()` method and radio and check buttons using the `add _ checkbutton()` and `add _ radiobutton()`methods, respectively. Use `add _ cascade()` to put all pieces of the menu together and display them on the menu bar.

In addition to creating the basic menu structure, developers often choose to extend its functionality by means of *menu shortcuts*. This can take the form of either *hot letters* using the *underline* option, or combinations of special keys (e.g., the *control* key) and letters through the *accelerator* option. In both cases, it is essential to remember that while these options may appear on the menu, they do not automatically trigger the relevant functionality. For this purpose, the main window form should be *bound* to the relevant event in order to trigger the respective functionality. This is achieved with the bind() method. The following application uses the functionality of the previous section, but with the implementation of a two-level deep basic menu instead of buttons:

```
1   # Import libraries
2   import tkinter as tk
3   from tkinter import filedialog
4   from tkinter import colorchooser
5   from tkinter import messagebox
6   from tkinter import Menu
7
8   # Define functions colorDialog, openDialog, saveAsDialog, quit, askyesno
9   # and askokcancel, invoking the relevant dialogs or message boxes
10  def colorDialog():
11      # Assign user color selection to a set of variables
12      (rgbSelected, colorSelected) = colorchooser.askcolor()
13      # Change the form color; use the color element from the variable set
14      winFrame.config(background = colorSelected)
15
16  def openDialog():
17      filedialog.askopenfile(title = "Open File Dialog")
18
19  def saveAsDialog():
20      filedialog.asksaveasfilename(title = "Save As Dialog")
21
22  def quit():
23      winFrame.destroy()
24      exit()
25
26  def askyesno():
27      messagebox.askyesno("YesNo message",
28          "Click on Yes or No to continue")
29
30  def askokcancel():
31      messagebox.askokcancel("OKCancel message",
32          "Click on OK or Cancel to continue")
33
34  # Define keypressedEvent() function that will invoke
35  # the associated function based on key press
36  def keypressedEvent(event):
37      if (event.keycode == 67 or event.keycode == 99):
38          colorDialog()
39      if (event.keycode == 70 or event.keycode == 102):
40          openDialog()
41      if (event.keycode == 83 or event.keycode == 115):
42          saveAsDialog()
43
```

```python
44  # Create non-resizable Windows frame using the tk object
45  winFrame = tk.Tk()
46  winFrame.title("Menus")
47  winFrame.resizable(False, False)
48  winFrame.geometry('260x220')
49
50  # Create the menu widget on the main window
51  menubar = tk.Menu(winFrame)
52
53  # Create the first series of sub-menus with dialogs
54  # and underline the shortcut letters
55  dialogs = tk.Menu(menubar, tearoff = 0)
56  dialogs.add_command(label = "Color dialog", command = colorDialog,
57      underline = 0)
58  dialogs.add_command(label = "Open File dialog", command = openDialog,
59      underline = 5)
60  dialogs.add_command(label = "Save As dialog", command = saveAsDialog,
61      underline = 0)
62  menubar.add_cascade(label = "Dialogs", menu = dialogs)
63
64  # Create the second series of sub-menus with messages
65  mssgs = tk.Menu(menubar, tearoff = 0)
66
67  # Create sub-menu inside the Yes/No, OK/Cancel message
68  mssgs1 = tk.Menu(mssgs, tearoff = 0)
69  mssgs1.add_command(label = "Yes/No Message", command = askyesno,
70      accelerator = 'Ctrl-Y')
71  mssgs1.add_command(label = "OK/Cancel Message", command = askokcancel,
72      accelerator = 'Ctrl-O')
73  mssgs.add_cascade(label = "Yes/No, OK/Cancel", menu = mssgs1)
74
75  mssgs.add_separator()
76  mssgs.add_command(label= "Exit", command = quit, accelerator = 'Ctrl-X')
77  menubar.add_cascade(label = "Messages", menu = mssgs)
78
79  # Create the third series of menus with check buttons and radio buttons
80  buttonmenus = tk.Menu(menubar, tearoff = 0)
81  buttonmenus.add_checkbutton(label = "Checkmenu1", onvalue=1, offvalue=0)
82  buttonmenus.add_checkbutton(label = "Checkmenu2", onvalue=1, offvalue=0)
83  buttonmenus.add_separator()
84  buttonmenus.add_radiobutton(label = "Radiomenu1")
85  buttonmenus.add_radiobutton(label = "Radiomenu2")
86  menubar.add_cascade(label = "Button menus", menu = buttonmenus)
87
88  # Bind the main window frame with the event/shortcut that will trigger
89  # the relevant function
90  winFrame.bind('<Key>', lambda event: keypressedEvent(event))
91  winFrame.bind('<Control-Y>', lambda event: askyesno())
92  winFrame.bind('<Control-O>', lambda event: askokcancel())
93  winFrame.bind('<Control-X>', lambda event: quit())
94
95  winFrame.config(menu = menubar)
96  winFrame.mainloop()
```

Output 5.3.1:

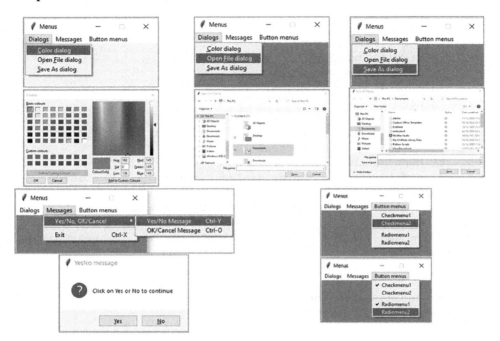

In addition to the necessary library calls, the script is split into three main parts. In the first part, the main window frame is created and configured (lines 44–48). Next, a menu object (`menubar`) is created (lines 50–51) and two main menu items (`dialogs` and `mssgs`) are attached to it (lines 55, 65). Notice the `tearoff` option, which prevents the menu from being detached from the main menu bar. Once the main menu components are in place, the various sub-menu items are created and associated with their parent menu item through the `add _ command()` method (lines 55–62

> **Observation 5.7 – add_separa-tor(), underline, accelerator:** Use the `add _ separator()` method to add a line separating the various items of a menu. Use the `underline` option to create *hot keys*, or the `accelerator` option to create *ctrl-*, *shift-*, or *alt-keys*, and to associate them with the desired functionality and events.

and 69–73). The `command` option binds particular menu items with the relevant methods. The `underline` option accepts the *index* of the text of the underlying object (starting at 0) and displays the associated character as a *hot key*. As in the case of hot keys in previous menu item examples, this is not enough by itself to trigger the relevant method or command, so a relevant event must be *bound* to the hot key character (lines 55–62 and 69–73). This is unlike the case of the `command` option.

When sub-menus are required as part of a menu item, the same process can be utilized. The only difference in this case would be that the referenced object should be the menu item instead of the main menu item (line 68). If it is preferred to use combinations of *special keys* (i.e., *Control, Shift*, or *Alt*) and characters, one can use the `accelerator` option instead of `underline` (lines 69–72, 76). As with `underline`, additional code should be written in order to trigger the function, method, or command associated with the menu item.

In cases where *check* or *radio* buttons are required instead of simple menu items, one can use methods `add _ checkbutton()` and `add _ radiobutton()`, respectively. These methods are used as alternatives to the `add _ command()` method (lines 81–82 and 84–85). When there is a need to separate the various menu items in groups, one can use the `add _ separator()` method (line 83). As mentioned, the `add _ cascade()` method ties together and activates the various items of the menu system.

In the second part of the script, the bindings between the menu item shortcuts (*hot keys* or *control characters*) and the associated commands are established (lines 90–93 and 36–42). The third part of the script involves the methods that perform the various functionality tasks (lines 8–32). Should the reader experience difficulties to follow through this example, the main coding concepts and commands used in the script are discussed in more detail in previous sections and/or chapters.

It is important to note that there is a difference in terms of how a menu is displayed in *Windows* (the menu bar is inside the running application window) and in *Mac OS* (the menu is displayed at the main system menu bar, detached from the running application window).

5.3.2   TOOLBAR MENUS WITH TOOLTIPS

An alternative form of presenting menu options to the user is the *toolbar* menu. It could either supplement the simple menu system or be used as a stand-alone component. The idea is rather straightforward: creating a collection of buttons (on a frame) and attaching it to the main window frame. The buttons are then bound to the respective commands.

> **Observation 5.8 – *toolbar menu*:** Use a *toolbar* menu system in addition to (or instead of) simple menus, to improve the GUI of a multi-functional application.

Buttons can display either images or text, or a combination of both. In order to improve clarity and make the interface more user-friendly, button text is often replaced by appropriate tooltips. The following Python script provides the same functionality as the one in the previous section, but is using a toolbar instead of a menu. The implementation also embeds tooltips to the toolbar buttons:

```
1    # Import libraries
2    import tkinter as tk
3    from tkinter import filedialog
4    from tkinter import colorchooser
5    from tkinter import Menu
6    from tkinter import *
7    # Import the necessary image processing classes from PIL
8    from PIL import Image, ImageTk
9
10   global openFileToolTip, saveAsToolTip, colorsDialogToolTip, exitToolTip
11   global photo1, photo2, photo3, photo4
12   global openFileButton, saveAsButton, colorsButton, exitButton
13
14   #-------------------------------------------------------------------------
15   # Open and resize images - load images to buttons
16   def images():
17       global photo1, photo2, photo3, photo4
18
19       image1 = Image.open("OpenFile.gif")
20       image1 = image1.resize((24, 24), Image.ANTIALIAS)
21       photo1 = ImageTk.PhotoImage(image1)
22       image2 = Image.open("SaveAs.gif")
23       image2 = image2.resize((24, 24), Image.ANTIALIAS)
24       photo2 = ImageTk.PhotoImage(image2)
```

```
25      image3 = Image.open("ColorsDialog.gif")
26      image3 = image3.resize((24, 24), Image.ANTIALIAS)
27      photo3 = ImageTk.PhotoImage(image3)
28      image4 = Image.open("Exit.gif")
29      image4 = image4.resize((24, 24), Image.ANTIALIAS)
30      photo4 = ImageTk.PhotoImage(image4)
31  #-----------------------------------------------------------------------
32  # Define the colorDialog, openDialog, saveAsDialog, and quit functions
33  # that will invoke the relevant dialogs or quit the application
34  def colorDialog():
35      # Assign user color selection to a set of variables
36      (rgbSelected, colorSelected) = colorchooser.askcolor()
37      # Change the form color; use the color element from set variable
38      winFrame.config(background = colorSelected)
39
40  def openDialog():
41      filedialog.askopenfile(title = "Open File Dialog")
42
43  def saveAsDialog():
44      filedialog.asksaveasfilename(title = "Save As Dialog")
45
46  def quit():
47      winFrame.destroy()
48      exit()
49  #-----------------------------------------------------------------------
50  # showToolTips function displays relevant message when hovering over a
51  # button; hideToolTips() function destroys/hides the tooltip
52  def showToolTips(a):
53      global openFileToolTip, saveAsToolTip
54      global colorsDialogToolTip, exitToolTip
55      if (a == 1):
56          openFileToolTip = tk.Label(winFrame, relief = FLAT,
57              text = "Open the Open File dialog", background = 'cyan')
58          openFileToolTip.place(x = 25, y = 30)
59      if (a == 2):
60          saveAsToolTip = tk.Label(winFrame, bd = 2, relief = FLAT,
61              text = "Open the Save As Dialog", background = 'cyan')
62          saveAsToolTip.place(x = 50, y = 30)
63      if (a == 3):
64          colorsDialogToolTip = tk.Label(winFrame, bd = 2, relief = FLAT,
65              text = "Open the Colors Dialog", background = 'cyan')
66          colorsDialogToolTip.place(x = 75, y = 30)
67      if (a == 4):
68          exitToolTip = tk.Label(winFrame, bd = 2, relief = FLAT,
69              text = "Click to exit the application", background = 'cyan')
70          exitToolTip.place(x = 100, y = 30)
71
72  def hideToolTips(a):
73      global openFileToolTip, saveAsToolTip
74      global colorsDialogToolTip, exitToolTip
75      if (a == 1):
```

```
76              openFileToolTip.destroy()
77      if (a == 2):
78              saveAsToolTip.destroy()
79      if (a == 3):
80              colorsDialogToolTip.destroy()
81      if (a == 4):
82              exitToolTip.destroy()
83  #------------------------------------------------------------------------
84  # Defing the bindButtons function to bind the buttons with the
85  # various events
86  def bindButtons():
87      global openFileButton, saveAsButton, colorsButton, exitButton
88
89      openFileButton.bind('<Button-1>', lambda event: openDialog())
90      openFileButton.bind('<Enter>', lambda event: showToolTips(1))
91      openFileButton.bind('<Leave>', lambda event: hideToolTips(1))
92      saveAsButton.bind('<Button-1>', lambda event: saveAsDialog())
93      saveAsButton.bind('<Enter>', lambda event: showToolTips(2))
94      saveAsButton.bind('<Leave>', lambda event: hideToolTips(2))
95      colorsButton.bind('<Button-1>', lambda event: colorDialog())
96      colorsButton.bind('<Enter>', lambda event: showToolTips(3))
97      colorsButton.bind('<Leave>', lambda event: hideToolTips(3))
98      exitButton.bind('<Button-1>', lambda event: quit())
99      exitButton.bind('<Enter>', lambda event: showToolTips(4))
100     exitButton.bind('<Leave>', lambda event: hideToolTips(4))
101 #------------------------------------------------------------------------
102 # Create non-resizable Windows frame using the tk object
103 winFrame = tk.Tk()
104 winFrame.title("Menus")
105 winFrame.resizable(False, False)
106 winFrame.geometry('260x220')
107
108 # Invoke the images function
109 images()
110
111 # Create toolbar with images and bind to related click event
112 toolbar = tk.Frame(winFrame, bd = 1, relief = RAISED)
113 toolbar.pack(side=TOP, fill=X)
114 # Create the toolbar buttons and invoke the bindButton function to bind
115 # them with the relevant events
116 openFileButton = tk.Button(toolbar, image = photo1, relief = FLAT)
117 saveAsButton = tk.Button(toolbar, image = photo2, relief = FLAT)
118 colorsButton = tk.Button(toolbar, image = photo3, relief = FLAT)
119 exitButton = tk.Button(toolbar, image = photo4, relief = FLAT)
120 bindButtons()
121 openFileButton.pack(side=LEFT, padx=0, pady=0)
122 saveAsButton.pack(side=LEFT, padx=0, pady=0)
123 colorsButton.pack(side=LEFT, padx=0, pady=0)
124 exitButton.pack(side=LEFT, padx=0, pady=0)
125
126 winFrame.mainloop()
```

Output 5.3.2:

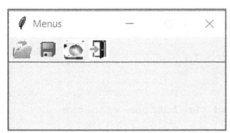 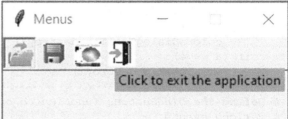

The script is similar to the previous versions in structure but with some notable differences. Firstly, a *toolbar* frame is created and populated with four buttons instead of creating a menu structure. Images are added to the buttons (lines 16–30) and activated through the associated `pack()` method calls (lines 121–124). Secondly, the buttons are associated with three events, namely

Observation 5.9 – Enter, Leave: Use the Enter and Leave events to trigger the desired actions when the mouse hovers over or moves away from an object.

`Button-1`, `Enter`, and `Leave` (lines 86–100, 120). `Button-1` is triggered when the left mouse button is pressed, `Enter` when the mouse pointer hovers over the button, and `Leave` when the mouse pointer exits the boundaries of the button.

Another key point in this script is the way tooltips are created and triggered. At the time of writing, Python did not provide an automatic method to create and trigger a tooltip. As such, developers wishing to use a tooltip should implement this functionality through coding. Nevertheless, the concept for doing so is rather simple: creating a label object that is displayed when the mouse

Observation 5.10 – *tooltip*: To add a *tooltip* to a particular object, associate a label with it and display or hide the label as the mouse hovers over or moves away from an object.

hovers over the button. This can be accomplished by creating separate labels for each button or by creating a single label and changing its text and location coordinates depending on the mouse pointer position. As mentioned, once the mouse pointer exits the boundaries of the button, the label can be hidden (destroyed). This implementation of tooltip functionality is illustrated in methods `showToolTips()` and `hideToolTips()` (lines 52–82).

5.3.3 Popup Menus with Embedded Icons

A third way to create menus in Python is through *pop-up* menus. Pop-up menus are quite similar to simple menus, with the difference that they are not attached to any particular, pre-defined position, but are floating on top of the application window. The creation and configuration of pop-up menus follow the same structure as simple menus; however, they are triggered in a slightly different way (e.g., left or right click on a designated space within the application window). Pop-up menus, similarly to simple menus, can include items of various

Observation 5.11 – *pop-up*: Use a *pop-up* menu to provide menu functionality without having to permanently display the menu within the application. Pop-up menus can be used as stand-alone menu options or in combination with simple menus and/or toolbars.

forms like text, images, combinations of both text and images, or shortcuts. They are often used in combination with menus of other types, like simple menus and toolbars, in order to improve application efficiency and make it more appealing to the user.

The following script implements the same functionality as the previous two examples, but uses pop-up menus instead of simple menus and/or toolbars. In this example, menu items include combinations of images and text:

```
1   # Import libraries
2   import tkinter as tk
3   from tkinter import filedialog
4   from tkinter import colorchooser
5   from tkinter import Menu
6   from tkinter import *
7   # Import the necessary image processing classes from PIL
8   from PIL import Image, ImageTk
9
10  global photo1, photo2, photo3, photo4
11  global popupmenu
12
13  # Open and resize images - load images to the buttons
14  def images():
15      global photo1, photo2, photo3, photo4
16
17      image1 = Image.open("OpenFile.gif")
18      image1 = image1.resize((24, 24), Image.ANTIALIAS)
19      photo1 = ImageTk.PhotoImage(image1)
20      image2 = Image.open("SaveAs.gif")
21      image2 = image2.resize((24, 24), Image.ANTIALIAS)
22      photo2 = ImageTk.PhotoImage(image2)
23      image3 = Image.open("ColorsDialog.gif")
24      image3 = image3.resize((24, 24), Image.ANTIALIAS)
25      photo3 = ImageTk.PhotoImage(image3)
26      image4 = Image.open("Exit.gif")
27      image4 = image4.resize((24, 24), Image.ANTIALIAS)
28      photo4 = ImageTk.PhotoImage(image4)
29
30  # Define the colorDialog, openDialog, saveAsDialog, and quit functions
31  # to invoke the relevant dialogs or quit the application
32  def colorDialog():
33      # Assign the user's selection of the color to a set of variables
34      (rgbSelected, colorSelected) = colorchooser.askcolor()
35      # Change the form color using the color part of the set of variables
36      winFrame.config(background = colorSelected)
37
38  def openDialog():
39      filedialog.askopenfile(title = "Open File Dialog")
40
41  def saveAsDialog():
42      filedialog.asksaveasfilename(title = "Save As Dialog")
43
44  def quit():
45      winFrame.destroy()
46      exit()
```

```
47
48  def popupMenu(event):
49      global popupmenu
50      popupmenu.tk_popup(event.x_root, event.y_root)
51  #------------------------------------------------------------------------
52  # Create non-resizable Windows frame using the tk objec,
53  winFrame = tk.Tk()
54  winFrame.title("Menus")
55  winFrame.resizable(False, False)
56  winFrame.geometry('260x220')
57
58  # Invoke the images function
59  images()
60
61  # Create the popup menu
62  popupmenu = tk.Menu(winFrame, tearoff = 0)
63  popupmenu.add_command(label = "Color dialog", image = photo1,
64      compound = LEFT, command = colorDialog)
65  popupmenu.add_command(label = "Exit", image = photo4, compound = LEFT,
66      command = quit)
67  popupmenu.add_separator()
68  popupmenu.add_command(label = "Open File dialog", image = photo2,
69      compound = LEFT, command = openDialog)
70  popupmenu.add_command(label = "Save As dialog", image = photo3,
71      compound = LEFT, command = saveAsDialog)
72
73  winFrame.bind('<Button-1>', lambda event: popupMenu(event))
74
75  winFrame.mainloop()
```

Output 5.3.3:

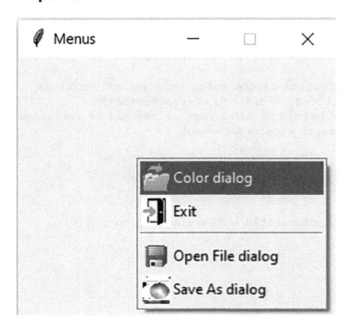

The reader should pay attention to two particular aspects of this script. Firstly, the `add_cascade()` method that was used in previous scripts to tie together the various menu items to the main menu system is missing. In this instance, the `tk_popup()` method is used instead. The method is called as a member of the popupmenu object (i.e., inside the `popupmenu(event)` method), and casts the pop-up menu at the current position of the mouse cursor (line 50). Secondly, it must be noted how *the text and the picture are combined on the menu items*. Hot keys and other types of shortcuts can be also used, as described in previous sections (lines 63–71).

> **Observation 5.12 – tk_popup(), add_cascade():** Use the `tk_popup(event.x_root, event.y_root)` method to display the pop-up menu at the current mouse location. Note that the `add_cascade()` method should not be used in this occasion, in contrast to the creation of simple menus.

> **Observation 5.13:** Use combinations of text, images, and hot keys to make the pop-up menu items more appealing and self-explanatory.

5.4   ENHANCING THE GUI EXPERIENCE

Three additional concepts can be utilized in order to further enhance the GUI experience. What these concepts have in common is that they can be used to improve the efficiency of *real estate* and memory usage of an application. Ultimately, good programming practice supports the creation of separate, autonomous GUIs and their ability to be reused in various programs by simple calls from the corresponding objects. This section examines these three concepts and provides some examples of their application.

5.4.1   NOTEBOOKS AND TABBED INTERFACES

As information systems grow larger in size, the management of real estate of the related applications (i.e., the creation of space that will host and display these applications) becomes increasingly important. The idea of using a menu system in its various different forms was introduced and explained in detail in previous sections. Menus offer a quite efficient way of addressing the management of real estate. An alternative way of doing so is through the use of *tabbed interfaces*. This approach is based on the creation of *separate sub-sections inside a single window* (i.e., *tabs*). Tabs are opened and run separately, but at the same time, they are parts of the same GUI structure. Tab-based implementations are commonly used in web browsers, where the various different web pages can be opened in separate tabs.

> **Observation 5.14 – Notebook(), Frame():** Use the `Notebook()` constructor (`ttk` module) to create the main object of a *tabbed interface*. Use the `Frame()` constructor (`ttk` module) to create each tab separately and to add them to the main object. Finally, `pack()` all the pieces together and load the applications in the respective tabs.

The following script combines two of the scripts covered in Chapter 4 (i.e., *Buttons and Text* and *Speed Control*) in a single application, utilizing a tab-based implementation:

```
1    # Import libraries
2    import tkinter as tk
3    from tkinter import ttk
4
5    # Declare and initialise the global variables and widgets
6    # for use with the functions
7    currentSpeedValue, speedLimitValue, finePerKmValue = 0, 0, 0
8    global speedLimitSpinbox
```

```
9    global finePerKmScale
10   global currentSpeedScale
11   global fine
12   global tab1, tab2
13   global winLabel
14   global winButton
15
16   # ============================================================
17   # Functions related to the tab2 application of Speed Control
18   # ============================================================
19   # Define the functions that will create the application interface
20   def createGUITab2():
21       currentSpeedFrame()
22       speedLimitFrame()
23       finePerKmFrame()
24       fineFrame()
25
26   # Define function to control changes in the Current Speed Scale widget
27   def onScale(val):
28       global currentSpeedValue
29       currentSpeedValue.set(float(val))
30       calculateFine()
31
32   # Define function to control changes in the Speed Limit Spinbox widget
33   def getSpeedLimit():
34       global speedLimitValue
35       speedLimitValue.set(float(speedLimitSpinbox.get()))
36       calculateFine()
37
38   # Define function to control changes in the Fine per Km Spinbox widget
39   def getFinePerKm(val):
40       global finePerKmValue
41       finePerKmValue.set(int(float(val)))
42       calculateFine()
43
44   # Define function to calculate Fine based on user input
45   def calculateFine():
46       global currentSpeedValue, speedLimitValue, finePerKmValue
47       global fine
48       diff = float(currentSpeedValue.get()) - float(speedLimitValue.get())
49       finePerKm = float(finePerKmValue.get())
50       if (diff <= 0):
51           fine.config(text = 'No fine')
52       else:
53           fine.config(text = 'Fine in USD: '+ str(diff * finePerKm))
54
55   # Add the Current Speed widgets to tab2
56   def currentSpeedFrame():
57       global currentSpeedValue
58
59       # Create the prompt label for the Current Speed tab
60       currentSpeed = tk.Label(tab2, text = 'Current speed:', width = 24)
```

```
61      currentSpeed.config(bg = 'light blue', fg = 'red', bd = 2,
62          font = 'Arial 14 bold')
63      currentSpeed.grid(column = 0, row = 0)
64      # Create Scale widget and define connection variable
65      currentSpeedValue = tk.DoubleVar()
66      currentSpeedScale=tk.Scale (tab2, length = 200, from_ = 0, to = 360)
67      currentSpeedScale.config(resolution = 0.5,
68      activebackground = 'dark blue', orient = 'horizontal')
69      currentSpeedScale.config(bg = 'light blue', fg = 'red',
70          troughcolor = 'cyan', command = onScale)
71      currentSpeedScale.grid(column = 1, row = 0)
72      currentSpeedSelected = tk.Label(tab2, text = '...',
73          textvariable = currentSpeedValue)
74      currentSpeedSelected.grid(column = 2, row = 0)
75
76  # Add the Speed Limit widgets to tab2
77  def speedLimitFrame():
78      global speedLimitValue
79      global speedLimitSpinbox
80
81      # Create the prompt label for the Speed Limit tab
82      speedLimit = tk.Label (tab2, text = 'Speed Limit:', width = 24)
83      speedLimit.config(bg = 'light blue', fg = 'yellow', bd = 2,
84          font = 'Arial 14 bold')
85      speedLimit.grid(column = 0, row = 1)
86      # Create the Spinbox widget and define variable to connect
87      # to Spinbox widget
88      speedLimitValue = tk.DoubleVar()
89      speedLimitSpinbox = ttk.Spinbox(tab2, from_ = 0, to = 360,
90          command = getSpeedLimit)
91      speedLimitSpinbox.grid(column = 1, row = 1)
92      speedLimitSelected = tk.Label(tab2, text = '...',
93          textvariable = speedLimitValue)
94      speedLimitSelected.grid(column = 2, row = 1)
95
96  # Add the Fine per Km widgets to tab2
97  def finePerKmFrame():
98      global finePerKmValue
99
100     # Create the prompt label for the Fine per Km tab
101     finePerKm=tk.Label(tab2, text='Fine/Km overspeed (USD):', width=24)
102     finePerKm.config(bg = 'light blue', fg = 'brown', bd = 2,
103         font = 'Arial 14 bold')
104     finePerKm.grid(column = 0, row = 2)
105     # Create Scale widget and define variable to connect to Scale widget
106     finePerKmValue = tk.IntVar()
107     finePerKmScale=ttk.Scale(tab2, orient = 'horizontal', length = 200,
108         from_ = 0, to = 100, command = getFinePerKm)
109     finePerKmScale.grid(column = 1, row = 2)
110     finePerKmSelected = tk.Label(tab2, text = '...',
111                                 textvariable = finePerKmValue)
111     finePerKmSelected.grid(column = 2, row = 2)
```

```
112
113  # Add the Fine for speeding label to tab2
114  def fineFrame():
115      global fine
116
117      # Create the label that will display the fine on the Fine tab
118      fine = tk.Label(tab2, text = 'Fine in USD:...', fg = 'blue')
119      fine.grid(column = 0, row = 3)
120  # ==============================================================
121  # The functions related to the tab1 application (button and text)
122  # ==============================================================
123  # Define the function that will control the mouse click events
124  def changeText(a):
125      global winLabel
126      winLabel.config(text = a)
127
128  # Define the function that will create the GUI for the tab1
129  def createGUITab1():
130      global winButton
131      global winLabel
132
133      winLabel = tk.Label(tab1, text = "...")
134      winLabel.grid(column = 1, row = 0)
135
136      # Create the button widget and bind it with the associated events
137      winButton=tk.Button(tab1, text="Left, right, or double left Click "
138          "\nto change the text of the label", font="Arial 16", fg="red")
139      winButton.grid(column = 0, row = 0)
140      winButton.bind("<Button-1>", lambda event, \
141          a = "You left clicked on the button": changeText(a))
142      winButton.bind("<Button-2>", lambda event, \
143          a = "You right clicked on the button": changeText(a))
144      winButton.bind("<Double-Button-1>", lambda event, \
145          a = "You double left clicked on the button": changeText(a))
146      winButton.bind("<Enter>", lambda event, \
147          a = "You are hovering above the button": changeText(a))
148      winButton.bind("<Leave>", lambda event, \
149          a = "You left the button widget": changeText(a))
150  # ==============================================================
151  # Create non-resizable Windows frame using the tk object
152  winFrame = tk.Tk()
153  winFrame.title("Tabs")
154  winFrame.resizable(True, True)
155  winFrame.geometry('500x180')
156
157  # Create notebook with tab pages
158  tabbedInterface = ttk.Notebook(winFrame)
159  tab1 = ttk.Frame(tabbedInterface)
160  tabbedInterface.add(tab1, text = "Buttons and Text")
161  tab2 = ttk.Frame(tabbedInterface)
162  tabbedInterface.add(tab2, text = "Speed control")
163  tabbedInterface.pack()
```

```
164
165 # Invoke the 2 functions to create the different GUIs for the 2 tabs
166 createGUITab1()
167 createGUITab2()
168
169 winFrame.mainloop()
```

Output 5.4.1:

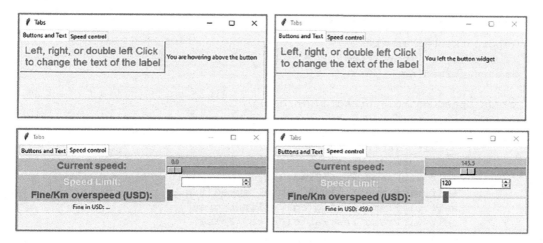

As shown in the output, the application implements an interface with two tabs, one hosting the *Buttons and Text* application and the other the *Speed Control* application. In this example, it is worth to raise some key points. Firstly, the tabs allow for a more efficient use of the real estate, since the two separate applications *run simultaneously* in a single window, but are displayed independently from each other. Secondly, the creation of the tabbed interface is through the Notebook() constructor of the ttk module (line 158). The two tabs are created using the Frame() constructor of the ttk module (lines 159 and 161) and are associated with the main notebook object by being added to it (lines 160 and 162). All the components are packed together in line 163. Ultimately, the tabs are created by means of the relevant GUI calls in lines 166 and 167.

There are two main differences between the way the applications are used in this example and in the original implementations presented in Chapter 4. The first is that, in both cases, the applications are converted to a completely procedural format, making full use of methods for all the required functionality and without any statements being added to the main body of the program. The second is that the *Speed Control* application is somewhat simplified, as the control variables associated with the Scale and Spinbox widgets and their respective labels are removed in order to avoid possible referencing issues between the various methods.

5.4.2 THREADED APPLICATIONS

One of the most important concepts in programming, and arguably among the most effective tools when creating real-life applications, is that of *threads* and *threading*. The idea behind threads is rather straightforward: multiple instances of an application can be run as independent processes. One way to conceptualize threads is to view them as *different objects of the same class*. Indeed, this is a rather accurate description, with the additional element of utilizing different processes of the operating system. One of the main characteristics of threaded applications is that they are meant to run *in parallel*. In reality, even in the case of using *multi-core* computer systems, this is not entirely feasible, but this is a rather specialized computer architecture consideration that exceeds the scope of this book.

In the following example, the *SpeedControl* application from Chapter 4 is converted to a class, for the purpose of demonstrating the implementation of threads. The script creates two objects of the *SpeedControl* class, and runs them separately on two different threads:

> **Observation 5.15 – *threads*:** Create different *threads* of the same objects of a class. Threads are separate and independent, and can run in parallel or sequentially. They use separate processes and allocated memory space.

```
1    # Import modules tk and ttk
2    import tkinter as tk
3    from tkinter import ttk
4    import threading
5
6    class SpeedControl(threading.Thread):
7
8        # Create and run the main window frame for the application
9        def __init__(self, winFrame):
10           super(SpeedControl, self).__init__()
11           self.winFrame = winFrame
12           self.winFrame.title("Control speed")
13           self.winFrame.config(bg = 'light grey')
14           self.winFrame.resizable(False, False)
15           self.winFrame.geometry('500x170')
16
17           # Create the frame, label and scale widgets for currentSpeed
18           self.currentSpeedFrame = tk.Frame (self.winFrame,
19               bg = 'light grey', bd = 2, relief = 'sunken')
20           self.currentSpeedFrame.pack()
21           self.currentSpeedFrame.place(relx = 0.05, rely = 0.05)
22           self.currentSpeed = tk.Label(self.currentSpeedFrame,
23               text = 'Current speed:', width = 24)
24           self.currentSpeed.config(bg = 'light blue', fg = 'red', bd = 2,
25               font = 'Arial 14 bold')
26           self.currentSpeed.grid(column = 0, row = 0)
27           self.currentSpeedScale = tk.Scale (self.currentSpeedFrame,
28               length = 200, from_ = 0, to = 360)
29           self.currentSpeedScale.config(resolution = 1,
30               orient = 'horizontal', activebackground = 'dark blue')
31           self.currentSpeedScale.config(bg = 'light blue', fg = 'red',
32               troughcolor = 'cyan', command = self.onScale)
33           self.currentSpeedScale.grid(column = 1, row = 0)
34           self.currentSpeedSel = tk.Label(self.currentSpeedFrame,
35               text='...')
36           self.currentSpeedSel.grid(column = 2, row = 0)
37           # Create the frame, label, & spinbox widget for the speedLimit
38           self.speedLimitFrame = tk.Frame(self.winFrame,
39               bg = 'light yellow', bd = 4, relief = 'sunken')
40           self.speedLimitFrame.pack()
41           self.speedLimitFrame.place(relx = 0.05, rely = 0.30)
42           self.speedLimit = tk.Label (self.speedLimitFrame,
43               text = 'Speed limit:', width = 24)
```

```
44            self.speedLimit.config(bg= 'light blue', fg = 'yellow', bd = 2,
45                font = 'Arial 14 bold')
46            self.speedLimit.grid(column = 0, row = 0)
47            self.speedLimitSpinbox = ttk.Spinbox(self.speedLimitFrame,
48                from_ = 0, to = 360, command = self.getSpeedLimit)
49            self.speedLimitSpinbox.grid(column = 1, row = 0)
50            self.speedLimitSel=tk.Label(self.speedLimitFrame, text='...')
51            self.speedLimitSel.grid(column = 2, row = 0)
52
53            # Create the frame, label, and scale widget for finePerKm
54            self.finePerKmFrame = tk.Frame(self.winFrame,
55                bg = 'light grey', bd = 2, relief = 'sunken')
56            self.finePerKmFrame.pack()
57            self.finePerKmFrame.place (relx = 0.05, rely = 0.55)
58            self.finePerKm = tk.Label(self.finePerKmFrame,
59                text = 'Fine/Km overspeed (USD):', width = 24)
60            self.finePerKm.config(bg = 'light blue', fg = 'red', bd = 2,
61                font = 'Arial 14 bold')
62            self.finePerKm.grid(column = 0, row = 0)
63            self.finePerKmScale = tk.Scale(self.finePerKmFrame,
64                length = 200, from_ = 0, to = 100)
65            self.finePerKmScale.config(resolution = 1,
66                activebackground = 'dark blue', orient = 'horizontal')
67            self.finePerKmScale.config(bg = 'light cyan', fg = 'red',
68                troughcolor = 'light blue', command = self.getFinePerKm)
69            self.finePerKmScale.grid(column = 1, row = 0)
70            self.finePerKmSel = tk.Label(self.finePerKmFrame, text='...')
71            self.finePerKmSel.grid(column = 2, row = 0)
72
73            # Create the frame for the fine and the related label
74            self.fineFrame = tk.Frame(self.winFrame, bg = 'yellow', bd = 4,
75                relief = 'raised')
76            self.fineFrame.pack()
77            self.fineFrame.place(relx = 0.05, rely = 0.80)
78            self.fine = tk.Label(self.fineFrame, text = 'Fine in USD:...',
79                fg = 'blue')
80            self.fine.grid(column = 0, row = 0)
81        # Define function to control changes in Current Speed Scale widget
82        def onScale(self, val):
83            v = int(float(val))
84            self.currentSpeedSel.config(text = v)
85            self.calculateFine()
86
87        # Define function to control changes in Speed Limit Spinbox widget
88        def getSpeedLimit(self):
89            v = self.speedLimitSpinbox.get()
90            self.speedLimitSel.config(text = v)
91            self.calculateFine()
92
93        # Define function to control changes in Fine per Km Spinbox widget
94        def getFinePerKm(self, val):
95            v = int(float(val))
96            self.finePerKmSel.config(text = v)
```

```
97              self.calculateFine()
98
99      # Define function to calculate the Fine based on user input
100     def calculateFine(self):
101         currentSpeed, speedLimit, finePerKm = 0, 0.0, 0
102
103         # Ensure relevant objects are initiated & assigned with values
104         if (self.currentSpeedScale.get()!= ''
105             and self.speedLimitSpinbox.get()!= ''
106             and self.finePerKmScale.get()!= ''):
107                 currentSpeed = self.currentSpeedScale.get()
108                 speedLimit = float(self.speedLimitSpinbox.get())
109                 finePerKm = self.finePerKmScale.get()
110         else:
111             currentSpeed, finePerkKm = 0, 0; speedLimit = 0.0
112
113         # Calculate the fine and display it on the associated label
114         diff = currentSpeed - speedLimit
115         if (diff <= 0):
116             self.fine.config(text = 'No fine')
117         else:
118             self.fine.config(text='Fine in USD: '+str(diff*finePerKm))
119
120 # Create two different GUI frames
121 winFrame1 = tk.Tk()
122 winFrame2 = tk.Tk()
123
124 # Create two different threads - one for each GUI frame
125 speedControl1 = SpeedControl(winFrame1)
126 speedControl2 = SpeedControl(winFrame2)
127
128 # Start each thread/frame and run it separately
129 speedControl1.start()
130 winFrame1.mainloop()
131
132 speedControl2.start()
133 winFrame2.mainloop()
```

Output 5.4.2:

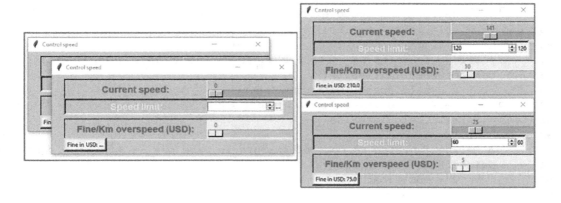

The output illustrates how this particular application runs the two different objects in separate threads. It must be noted that the threads are running simultaneously. The term *in parallel* should be avoided in this context, as it is uncertain whether the threads are indeed running in parallel. This is also something that can be affected by the operating system, the hardware and software settings, and the associated behaviors. Nevertheless, from the perspective of the user, this is of purely academic interest. As shown in the example above, the two threaded objects appear to run in parallel indeed, but at the same time they function independently and use different inputs as if they were run sequentially.

The order of statements between lines 121 and 133 is also important. Firstly, the two GUI window frames are created as normal. If the first GUI frame was to be created directly followed by the first threaded object, and before the second GUI frame and threaded object, the user would only get access to the first window frame. The second window frame would only appear once the first one was closed and stopped. The reader should also notice that the threading module needs to be inserted before the calls to the start methods of the threaded objects (i.e., speedControl1 and speedControl2).

> **Observation 5.16 – Threads:** Use the Thread class from the threading module to create threaded objects. Use the start() method to start the threads and the stop()method to stop them. Always use the self parameter on all widgets and attributes to refer to the specific object they belong to.

> **Observation 5.17:** Avoid using control variables (e.g., IntVar()) in threaded objects.

It must be noted that each threaded object is assigned to a separate window frame and has a dedicated main-loop() method to monitor its GUI and the associated events (lines 125–130 and 126–133). This assignment is taking place in lines 125–126, where the window frame for each threaded object is called as a parameter, and used on the specific, independent GUI for the underlying object.

> **Observation 5.18:** In cases of GUI-based threaded objects, use the mainloop() method for monitoring *each object.*

Another notable aspect of the script is the explicit definition of the __ init __ (self, win-Frame): super(SpeedControl, self). __ init __ () that loads the GUI widgets onto the window frames of each of the threaded objects. The reader should be reminded here that the __ init __ () method is provided by Python to automatically initialize basic and necessary widgets and attributes in preparation of launching the object. The self parameter is necessary in order for the Python interpreter to distinguish which object is running and what widgets and attributes belong to it. This is the reason why each widget and attribute, and even simple variables, are preceded by the self parameter.

Another key point in this particular script is that, since the object that is being created is threaded, it inherits from the Thread class of the threading module (line 6) *and is implemented on that class* (line 10). These two lines that essentially create the threaded object are called each time a new threaded object is initiated.

Finally, the reader should note that the control variables (e.g., IntVar()) are missing from this version of the code. This was done on purpose, as their inclusion could cause unnecessary conflicts between the threaded objects and the cross-method operations within any single threaded object, without offering any particular benefits to the application. In general, it is advisable that control variables on widgets are avoided, especially when implementing object-oriented and/or threaded object applications.

5.4.3 COMBINING MULTIPLE CONCEPTS AND APPLICATIONS IN A MULTITHREADED SYSTEM

Chapters 2–5 of this book provide a gradual progression from basic programming skills to more advanced application development concepts. Although there are certainly many more concepts and layers of depth to be explored when it comes to programming in Python, Chapters 2–5 should provide a solid basis for the aspiring programmer, as they cover the necessary building blocks required to make functional and well-structured applications. As a conclusion to this conceptual sub-section of this book, it was deemed necessary to provide an overview of how the concepts, mechanisms, and practices presented so far can be integrated into a coherent, centralized solution. Ultimately, this should provide an idea of how a multithreaded and multi-functional information system can be built, resembling the scenarios and challenges one may face in real life. The example presented below combines two of the applications developed earlier (Speed Control and Bubble Sort) into a multithreaded system that can be launched and operated as a single, unified platform. In order for this to be possible, two changes are required:

a. Each of the two individual applications (*Speed Control* and *Bubble Sort*) must be adjusted according to the object-oriented paradigm. This is done by separating and extracting the main code that is responsible for the GUI creation and all related methods, and save the remaining code as separate text files in Jupyter. By doing so, the original applications cannot be run separately, as there is no actual object being created in the remaining code. Instead of creating the object within the main body of each application, this is done through a call from another application, which now functions as the *main application*.

b. The code that was extracted from the original applications must be imported to this newly created application.

The code examples presented and discussed in the following pages provide a practical illustration of these changes:

Chapter5SpeedControl.py

```
1    # Import modules tk and ttk
2    import tkinter as tk
3    from tkinter import ttk
4    import threading
5
6    class SpeedControl(threading.Thread):
7
8        # Create and run the main window frame for the application
9        def __init__(self, winFrame):
10           super(SpeedControl, self).__init__()
11           self.winFrame = winFrame
12           self.winFrame.title("Control speed")
13           self.winFrame.config(bg = 'light grey')
14           self.winFrame.resizable(False, False)
15           self.winFrame.geometry('500x170')
16
17           # Create frame for currentSpeed & its label and scale widgets
18           self.currentSpeedFrame = tk.Frame(self.winFrame,
```

```
19              bg = 'light grey', bd = 2, relief = 'sunken')
20          self.currentSpeedFrame.pack()
21          self.currentSpeedFrame.place(relx = 0.05, rely = 0.05)
22          self.currentSpeed = tk.Label(self.currentSpeedFrame,
23              text = 'Current speed:', width = 24)
24          self.currentSpeed.config(bg = 'light blue', fg = 'red', bd = 2,
25              font = 'Arial 14 bold')
26          self.currentSpeed.grid(column = 0, row = 0)
27          self.currentSpeedScale = tk.Scale(self.currentSpeedFrame,
28              length = 200, from_ = 0, to = 360)
29          self.currentSpeedScale.config(resolution = 1,
30              activebackground = 'dark blue', orient = 'horizontal')
31          self.currentSpeedScale.config(bg = 'light blue', fg = 'red',
32              troughcolor = 'cyan', command = self.onScale)
33          self.currentSpeedScale.grid(column = 1, row = 0)
34          self.currentSpeedSel = tk.Label(self.currentSpeedFrame,
35              text = '...')
36          self.currentSpeedSel.grid(column = 2, row = 0)
37          # Create frame for speedLimit & its label and spinbox widgets
38          self.speedLimitFrame = tk.Frame(self.winFrame,
39              bg = 'light yellow', bd = 4, relief = 'sunken')
40          self.speedLimitFrame.pack()
41          self.speedLimitFrame.place(relx = 0.05, rely = 0.30)
42          self.speedLimit = tk.Label(self.speedLimitFrame,
43              text = 'Speed limit:', width = 24)
44          self.speedLimit.config(bg = 'light blue', fg = 'yellow',
45              bd = 2, font = 'Arial 14 bold')
46          self.speedLimit.grid(column = 0, row = 0)
47          self.speedLimitSpinbox = ttk.Spinbox(self.speedLimitFrame,
48              from_ = 0, to = 360, command = self.getSpeedLimit)
49          self.speedLimitSpinbox.grid(column = 1, row = 0)
50          self.speedLimitSel=tk.Label(self.speedLimitFrame,text='...')
51          self.speedLimitSel.grid(column = 2, row = 0)
52
53          # Create frame for finePerKm and its label and scale widgets
54          self.finePerKmFrame = tk.Frame(self.winFrame,
55              bg = 'light grey', bd = 2, relief = 'sunken')
56          self.finePerKmFrame.pack()
57          self.finePerKmFrame.place(relx = 0.05, rely = 0.55)
58          self.finePerKm = tk.Label(self.finePerKmFrame,
59              text = 'Fine/Km overspeed (USD):', width = 24)
60          self.finePerKm.config(bg = 'light blue', fg = 'red', bd = 2,
61              font = 'Arial 14 bold')
62          self.finePerKm.grid(column = 0, row = 0)
63          self.finePerKmScale = tk.Scale(self.finePerKmFrame,
64              length = 200, from_ = 0, to = 100)
65          self.finePerKmScale.config(resolution = 1,
66          activebackground = 'dark blue', orient = 'horizontal')
67          self.finePerKmScale.config(bg = 'light cyan', fg = 'red',
68              troughcolor = 'light blue', command = self.getFinePerKm)
```

```
69          self.finePerKmScale.grid(column = 1, row = 0)
70          self.finePerKmSel=tk.Label(self.finePerKmFrame, text = '...')
71          self.finePerKmSel.grid(column = 2, row = 0)
72
73          # Create the frame for Fine and its label
74          self.fineFrame = tk.Frame(self.winFrame, bg = 'yellow', bd = 4,
75             relief = 'raised')
76          self.fineFrame.pack()
77          self.fineFrame.place(relx = 0.05, rely = 0.80)
78          self.fine = tk.Label(self.fineFrame, text = 'Fine in USD:...',
79             fg = 'blue')
80          self.fine.grid(column = 0, row = 0)
81
82      # Define function to control changes in CurrentSpeedScale widget
83      def onScale(self, val):
84          v = int(float(val))
85          self.currentSpeedSel.config(text = v)
86          self.calculateFine()
87
88      # Define function to control changes in SpeedLimitSpinbox widget
89      def getSpeedLimit(self):
90          v = self.speedLimitSpinbox.get()
91          self.speedLimitSel.config(text = v)
92          self.calculateFine()
93
94      # Define function to control changes in FineperKm Spinbox widget
95      def getFinePerKm(self, val):
96          v = int(float(val))
97          self.finePerKmSel.config(text = v)
98          self.calculateFine()
99
100     # Define the function to calculate the Fine based on user input
101     def calculateFine(self):
102         currentSpeed, speedLimit, finePerKm = 0, 0.0, 0
103
104         # Make sure the objects are initiated and assigned with values
105         if (self.currentSpeedScale.get()!= ''
106           and self.speedLimitSpinbox.get()!= ''
107           and self.finePerKmScale.get()!= ''):
108             currentSpeed = self.currentSpeedScale.get()
109             speedLimit = float(self.speedLimitSpinbox.get())
110             finePerKm = self.finePerKmScale.get()
111         else:
112             currentSpeed, finePerkKm = 0, 0; speedLimit = 0.0
113
114         # Calculate the fine and display it on the associated label
115         diff = currentSpeed - speedLimit
116         if (diff <= 0):
117             self.fine.config(text = 'No fine')
118         else:
119             self.fine.config(text='Fine in USD: '+str(diff*finePerKm))
```

In the class presented above, the statements that create and run the GUI have been already separated and extracted, ready to be imported to the main application that will eventually create the multithreaded objects. Apart from extracting these particular statements, the class implements the *SpeedControl* application as discussed in the previous section. The class needs to be saved as a text file with the *.py* extension.

Chapter5BubbleSort.py

```
1    # Import modules tk, random and time
2    import tkinter as tk
3    from tkinter import ttk
4    from tkinter import *
5    import random
6    import time
7    import threading
8
9    class BubbleSort(threading.Thread):
10
11       # Initialise the various lists used by the objects of the class
12       unsortedL = []; sortedL = []; statisticsData = [];
13       sizes = [5, 20, 100, 250, 500, 750, 1000, 2000, 5000, 10000, 20000]
14
15
16       # Create and run the main window frame for the application
17       def __init__(self, winFrame):
18           super(BubbleSort, self).__init__()
19           self.winFrame = winFrame
20           self.winFrame.title("Bubble Sort");
21           self.winFrame.config(bg = 'light grey')
22           self.winFrame.resizable(True, True);
23           self.winFrame.geometry('650x300')
24           self.listSize = 0
25           self.createGUI()
26
27       # Define the functions that will create the application GUI
28       def createGUI(self):
29           self.unsortedFrame()
30           self.entryFrame()
31           self.entryButton()
32           self.sortButton()
33           self.sortedFrame()
34           self.clearButton()
35           self.statisticsButton()
36           self.statisticsSelection()
37
38       # Create labelframe; populate with Unsorted Array Listbox widgets
39       def unsortedFrame(self):
40           self.UnsortedFrame=tk.LabelFrame(self.winFrame,
                                               text='Unsorted Array')
41           self.UnsortedFrame.config(bg = 'light grey', fg = 'blue',
42               bd = 2, relief = 'sunken')
```

```
43              # Create a scrollbar widget to attach to UnsortedList
44              self.UnsortedListScrollBar = Scrollbar(self.UnsortedFrame,
45                  orient = VERTICAL)
46              self.UnsortedListScrollBar.pack(side = RIGHT, fill = Y)
47              # Create a listbox in the Unsorted Array frame
48              self.UnsortedList = tk.Listbox(self.UnsortedFrame,
49                  yscrollcommand = self.UnsortedListScrollBar.set,
50                  bg = 'cyan', width = 13, height = 12, bd = 0)
51              self.UnsortedList.pack(side = LEFT, fill = BOTH)
52              # Associate the scrollbar command with its parent widget
53              # (i.e., the UnsortedList yview)
54              self.UnsortedListScrollBar.config(command =
                                                    self.UnsortedList.yview)
55              # Place the Unsorted frame & its components into the interface
56              self.UnsortedFrame.pack()
57              self.UnsortedFrame.place(relx = 0.02, rely = 0.05)
58
59          # Create the labelframe that will contain the Entry widget
60          def entryFrame(self):
61              self.EntryFrame = tk.LabelFrame(self.winFrame, text= 'Actions')
62              self.EntryFrame.config(bg = 'light grey', fg = 'red', bd = 2,
63                  relief = 'sunken')
64              self.EntryFrame.pack(); self.EntryFrame.place(relx=0.25,
                                                            rely=0.05)
65              # Create the label in the Entry frame
66              self.EntryLabel = tk.Label(self.EntryFrame,
67                  text = 'How many integers\nin the list', width = 16)
68              self.EntryLabel.config(bg = 'light grey', fg = 'red', bd = 3,
69                  relief = 'flat', font = 'Arial 14 bold')
70              self.EntryLabel.grid(column = 0, row = 0)
71              # Create combo box to select the number of elements in lists
72              self.ListSizeCombo = ttk.Combobox(self.EntryFrame, width = 10)
73              self.ListSizeCombo['values'] = self.sizes
74              self.ListSizeCombo.current(0)
75              self.ListSizeCombo.grid(column = 1, row = 0)
76
77          # Create the button that will insert new entries into the unsorted
78          # array and list box
79          def entryButton(self):
80              self.EntryButton = tk.Button(self.EntryFrame, relief= 'raised',
81                  text = 'Populate\nUnsorted list', width = 16)
82              self.EntryButton.bind('<Button-1>',
83                  lambda event: self.populateUnsortedList())
84              self.EntryButton.grid(column = 0, row = 2)
85
86          # Populate the unsorted list with random numbers and populate
87          # the unsorted list box
88          def populateUnsortedList(self):
89              self.listSize = int(self.ListSizeCombo.get())
```

```
90
91          # Generate random integers with randint() from the random class
92          for i in range (self.listSize):
93              n = random.randint(-100, 100)
94              # Enter the generated random integer to the relevant place
95              # in the unsorted list
96              self.unsortedL.insert(i, n)
97          # Populate UnsortedList with the unsorted list elements
98          for i in range (0, self.listSize):
99              self.UnsortedList.insert(i, self.unsortedL[i])
100         self.UnsortedListScrollBar.config(command=
                                                self.UnsortedList.yview)
101
102     # Create the button that will sort the numbers and display them
103     # in the sorted array and list box
104     def sortButton(self):
105         self.SortButton = tk.Button(self.EntryFrame, relief = 'raised',
106             text = 'Sort numbers\nwith BubbleSort', width = 16)
107         self.SortButton.bind('<Button-1>',lambda event:
                                                self.sortToSortedList())
108         self.SortButton.grid(column = 1, row = 2)
109
110     # Create the labelframe to include the Sorted Array Listbox widgets
111     def sortedFrame(self):
112         self.SortedFrame=tk.LabelFrame(self.winFrame,
                                                text='Sorted Array')
113         self.SortedFrame.config(bg = 'light grey', fg = 'blue', bd = 2,
114             relief = 'sunken')
115         # Create a scrollbar widget to attach to the SortedList
116         self.SortedListScrollBar = Scrollbar (self.SortedFrame)
117         self.SortedListScrollBar.pack(side = RIGHT, fill = Y)
118         # Create the list box in the Sorted Array frame
119         self.SortedList = tk.Listbox (self.SortedFrame,
120             yscrollcommand = self.SortedListScrollBar.set,
121             bg = 'cyan', width = 13, height = 12, bd = 0)
122         self.SortedList.pack(side = LEFT, fill = BOTH)
123         # Associate the scrollbar command with its parent widget
124         # (i.e., the SortedList yview)
125         self.SortedListScrollBar.config(command =
                                                self.SortedList.yview)
126         # Place the Unsorted frame and its parts into the interface
127         self.SortedFrame.pack(); self.SortedFrame.place(relx = 0.75,
                                                rely = 0.05)
128
129     # Bubble Sort sorts the list & records information for later use
130     def sortToSortedList(self):
131         # Load unsorted list & list box to the sorted list & list box
132         for i in range (0, self.listSize):
133             self.sortedL.insert(i, self.unsortedL[i])
```

```
134
135          # Start timer
136          self.startTime = time.process_time()
137
138          # The Bubble sort algorithm
139          for i in range (self.listSize-1):
140              for j in range (self.listSize-1):
141                  if (self.sortedL[j] > self.sortedL[j+1]):
142                      temp = self.sortedL[j]
143                      self.sortedL[j] = self.sortedL[j+1]
144                      self.sortedL[j+1] = temp
145
146          # End timer
147          self.endTime = time.process_time()
148
149          # Load the sorted list to the relevant list box
150          for i in range (0, self.listSize):
151              self.SortedList.insert(i, self.sortedL[i])
152          self.SortedListScrollBar.config(command=self.SortedList.yview)
153
154      # Create button that will clear the two list boxes & the two lists
155      def clearButton(self):
156          self.ClearButton = tk.Button(self.EntryFrame,
157              text = 'Clear lists', relief = 'raised', width = 16)
158          self.ClearButton.bind('<Button-1>',
                                        lambda event: self.clearLists())
159          self.ClearButton.grid(column = 0, row = 3)
160
161      # Clear all lists, list & combo boxes, & related global variable
162      def clearLists(self):
163          self.sortedL.clear()
164          self.unsortedL.clear()
165          self.UnsortedList.delete('0', 'end')
166          self.SortedList.delete('0', 'end')
167          self.statisticsData.clear()
168          self.StatisticsCombo.delete('0', 'end')
169          self.listSize = 0
170
171      # Create the button that will display sorting information
172      def statisticsButton(self):
173          self.StatisticsButton = tk.Button(self.EntryFrame,
174              text = 'Show statistics', relief = 'raised', width = 16)
175          self.StatisticsButton.bind('<Button-1>',
                                        lambda event: self.statistics())
176          self.StatisticsButton.grid(column = 1, row = 3)
177
178      # Create the option menu that will show the statistical results
179      # from the sorting process
180      def statisticsSelection(self):
181          self.StatisticsSelection = tk.StringVar()
```

```
182          self.statisticsData = ['The statistics will appear here']
183          self.StatisticsSelection.set(self.statisticsData[0])
184          self.StatisticsCombo = ttk.Combobox(self.EntryFrame,
185              textvariable = self.StatisticsSelection, width = 30)
186          self.StatisticsCombo['values'] = self.statisticsData
187          self.StatisticsCombo.grid(column = 0, columnspan = 2, row = 4)
188
189      # Calculate and report the statistics from the sorting process
190      def statistics(self):
191          self.statisticsData.clear()
192          self.statisticsData.insert(1,
193              'The size of the list is ' + str(self.listSize))
194          self.statisticsData.insert(2,
195              'The sum of the list is ' + str(sum(self.sortedL)))
196          self.statisticsData.insert(3, 'The time passed to sort the ' +
197              'list was ' + str(round(self.endTime - self.startTime, 5)))
198          self.statisticsData.insert(4, 'The average of the sorted list '
199              +'is: ' + str(round(sum(self.sortedL) / self.listSize, 2)))
200          self.StatisticsCombo['values'] = self.statisticsData
```

As with the *SpeedControl* class discussed previously, the class presented above is the modified version of the Bubble Sort application. The object-oriented paradigm is adopted by separating and extracting the statements that would create and run the GUI. The remaining code is saved as a *.py* text file in Jupyter, in order to be accessible by the main application.

The following class implements the main application that imports the two classes and runs them as threaded objects. The classes are imported in lines 5–6, and the main GUI object is created in lines 47, 49, and 51. The interface offers a single method: the display of a popup menu when a left-click event takes place. The menu allows for the creation of two threaded objects based on *SpeedControl* and *Bubble Sort* (line 30). The reader should note how the statements separated and extracted from the imported classes were added to the main application in lines 32–37 and 39–44 respectively:

```
1   # Import libraries
2   import tkinter as tk
3   from tkinter import Menu
4   from tkinter import *
5   import Chapter5SpeedControl
6   import Chapter5BubbleSort
7   import threading
8
9   class Application:
10
11      # Create main window frame for the application with the popup menu
12      def __init__(self, winFrame):
13          self.winFrame = winFrame
14          self.winFrame.title("Application with threads")
15          self.winFrame.config(bg = 'light grey')
16          self.winFrame.resizable(False, False)
```

```
17          self.winFrame.geometry('260x220')
18
19          self.popupmenu = tk.Menu(self.winFrame, tearoff = 0)
20          self.popupmenu.add_command(label = "Speed Control",
21              command = self.speedControlThread)
22          self.popupmenu.add_command(label = "Bubble Sort",
23              command = self.bubbleSortThread)
24          self.winFrame.bind('<Button-1>',
                                        lambda event: self.popupMenu(event))
25          self.winFrame.config(menu = self.popupmenu)
26
27          self.winFrame.mainloop()
28
29      def popupMenu(self, event):
30          self.popupmenu.tk_popup(event.x_root, event.y_root)
31
32      def speedControlThread(self):
33          # Prepare the Speed Control GUI
34          speedControlFrame = tk.Tk()
35          speedControl1 =
                          Chapter5SpeedControl.SpeedControl(speedControlFrame)
36          speedControl1.start()
37          speedControlFrame.mainloop()
38
39      def bubbleSortThread(self):
40          # Prepare the Bubble sort GUI
41          bubbleSortFrame = tk.Tk()
42          bubbleSort1 = Chapter5BubbleSort.BubbleSort(bubbleSortFrame)
43          bubbleSort1.start()
44          bubbleSortFrame.mainloop
45
46 # Prepare the application GUI
47 winFrame = tk.Tk()
48
49 application = Application(winFrame)
50
51 winFrame.mainloop()
```

Output 5.4.3:

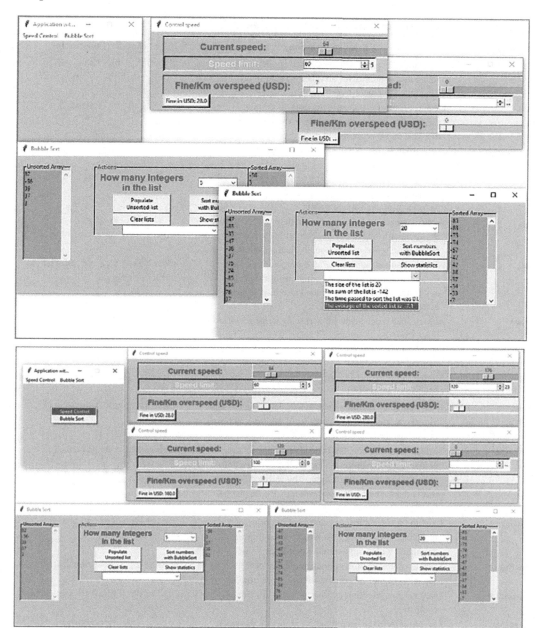

5.5   WRAP UP

Chapters 4 and 5 provided a step-by-step, systematic walkthrough of Graphical User Interface (GUI) programming with Python, and an introduction to GUI objects like menus, tabs, and threads. Key Python widgets were introduced alongside their most common uses and options. This was done through a series of straightforward examples and applications that progressed gradually from simpler to more challenging implementations. Although a detailed coverage of all the available widgets is beyond the scope of this chapter, Table 5.1 provides widget lists with descriptions, and

TABLE 5.1

Frequently Used Widgets and the Module They Belong to

Widget Name	Brief Description	Module/Constructor
Windows frame	The main object of a windows-based application, acting as a container for all other widgets in order to create the Graphical-User-Interface.	`tkinter, tk.Tk()`
Label	Displays a short message to the user. Its content is not expected to change significantly in the program lifecycle and it is not meant to be used for interaction. Nevertheless, it is possible to write code that will enhance its functionality.	`tkinter, tk.Label()`
Button	Used to handle basic interaction between the user and the application. This is usually implemented through movement or click-based events.	`tkinter, tk.Button()`
Entry	A basic widget used to accept a single line of text from the keyboard. As with most other widgets, it can be modified in terms of functionality and appearance.	`ttk, ttk.Entry()`
Scale	A controlled mechanism for accepting numerical user input. Two different implementations of the widget are available, with the one found in tkinter offering more options than that in ttk.	`tkinter/ttk, tk.Scale()/ ttk.Scale()`
Spinbox	A controlled mechanism for accepting numerical user input from the ttk library.	`ttk, ttk.Spinbox()`
Frame	Used for improved control of the GUI. It can contain various other widgets.	`tkinter, tk.Frame()`
Labelframe	Similar to the frame widget, but with the inclusion of a label.	`tkinter, tk.LabelFrame()`
Listbox	Used to display separate lines of text, allowing the user to make a selection. The contents of multiple listboxes can be synchronized.	`tkinter, tk.ListBox()`
Combobox	Similar to the list box, but instead of being permanently expanded it is in a collapsed state and only opens when clicked upon. The selected line of text is displayed on the top level (i.e., the displayed text box when the list is collapsed).	`ttk, ttk.Combobox()`
ScrollBar	Used to improve the appearance and use of associated multiline widgets (e.g., list boxes) when they are populated with a large number of entries.	`tkinter, ScrollBar()`
CheckButton	Used to offer selection options. It allows for the selection of multiple options at any given time.	`tkinter, tk.CheckButton()`
RadioButton	Used to offer selection options. Options are mutually exclusive.	`tkinter, tk.RadioButton()`
Progressbar	Used to inform the user about the state of a particular running method. It can be *determinate*, in which case the widget presents the actual state of the method, or *indeterminate*, where the widget provides a scrolling message indicating that the method is still in progress.	`ttk, ttk.Progressbar()`
Text	Similar to the entry widget, but allowing multiple lines of text.	`tk, tk.Text()`
Canvas	A widget that provides a space to place graphics, text, or other objects.	`tk, tk.Canvas()`
Notebook	Provides the supporting object for tabbed frames.	`ttk, ttk.Notebook()`

the modules/libraries they belong to as a quick reference. This information can be also used as a reference for constructors when creating objects from the respective classes. Additional details on the listed widgets (including `tkinter`) can be found in the official Python documentation.

In addition to the aforementioned widgets, a number of other objects are frequently used to improve the GUI experience. Although many of these are not *standalone* objects, their use in conjunction with other objects is rather common. Table 5.2 lists some of these objects:

The above objects make use of a number of methods that contribute to the creation of the overall user experience. Table 5.3 lists some of the most important of the methods used in the various scripts and applications developed in this chapter:

TABLE 5.2
Notable Objects and Their Modules

Object	Brief Description	Module
`Image`	Used to load and display an image. It supports different file types (e.g., *gif*, *jpg*, *png*). Various different methods are available, depending on the file type.	`PIL`
`tk.StringVar()`, `tk.IntVar()`, `tk.DoubleVar()`, etc.	Used to host text or numbers.	`tkinter`
`askyesno()`, `askokcancel()`, `askretrycancel()`, `askquestion()`	Used to display different types of pre-defined message boxes.	`messagebox`
`showinfo()`, `showerror()`, `showwarning()`	Used to display a simple message box with an info, error, or warning icon.	`messagebox`
`askopenfile()`, `asksaveasfilename()`, `askdirectory()`, `askcolor()`	Used to display the common windows-based dialogs, ranging from *file dialogs* to *color chooser* modules.	`filedialog`, `colorchooser`
`menu`, `popup menu`	Used to display regular windows-based or popup menus.	`tkinter`, `tk.Menu()`
`Thread`	Used to create threaded objects.	`threading`

TABLE 5.3
Frequently Used Methods and Their Respective Widgets (in Alphabetical Order with Constructors First)

Method	Brief Description
`.add_command()`, `.add_checkbutton()`, `.add_radiobutton()`, `.add_cascade()`, `.add_separator()`	Adds the various components of a menu object.
`.after()`	Invokes a method after a set amount of time has elapsed.
`.append()`	Appends a new element to the end of a list.
`.askyesno()`, `askokcancel()`, `.askretrycancel()`, `.askquestion()`	Offers a set of different types of pre-defined message boxes.
`.askopenfile()`, `.asksaveasfilename()`, `.askdirectory()`, `.askcolor()`	Offers a set of different types of pre-defined dialogs.
`.bind()`	Binds the widget with a user interaction event.
`.clear()`	Clears the values from a list.

(Continued)

TABLE 5.3 (*Continued*)

Frequently Used Methods and Their Respective Widgets (in Alphabetical Order with Constructors First)

Method	Brief Description
.config()	Allows the configuration of the widget in terms of its characteristics (e.g., color, font properties).
.current()	Identifies the current selection from a combo box.
.curselection()	Identifies the current selection from a list box.
.delete()	Deletes values from a list box.
.destroy()	Destroys the current frame/interface.
.exit()	Exits the current frame/interface (or the entire application).
.geometry()	Accepts the initial dimensions of the frame in the form of a string (i.e., 'length x width').
.grid()	Places the widget on the grid of the parent widget and at a specific column and row. It can span across multiple columns/rows.
.grid_remove()	Temporarily hides the widget from the grid of the parent without deleting or destroying it.
int(), float(), str()	Converts the specified values to integer, float, or string values respectively.
.insert()	Inserts values to a list box.
.mainloop()	Puts the frame in an idle state, and monitors possible interactions. The latter can take the form of defined events between the user and the GUI.
.maxsize(), .minsize()	Defines the minimum/maximum size of the associated frame.
.open()	Reads an image/picture based on its full path, assigned as an argument.
.pack()	Attaches the widget to the parent, allowing coordinates to be calculated either on a relative or absolute basis.
.PhotoImage()	Creates a memory pointer to a processed image object, by means of the open() method.
.place()	Places the widget at specific coordinates on the parent frame, either on a relative or absolute basis.
.process_time()	Counts the time needed for a particular process to execute.
.randit()	Generates random numbers in the specified range.
.resizable()	Specifies whether the object is resizable based on a Boolean value (True/False) that is provided as a parameter.
.resize()	Specifies the size of the image/picture. It is usually accompanied by the ANTIALIAS expression to ensure the quality of the image is maintained when downsizing.
round(), sum(), len()	Basic mathematical methods.
.selection_set()	Selects a particular indexed element in a list box.
.set (), .get ()	Sets or gets the value of an object.
.showinfo(), .showerror(), .showwarning()	Offer different types of pre-defined message boxes.
.start(), .stop()	Starts or stops a threaded object.
.title()	Provides a title to the windows frame.
.update_idletasks()	Ensures that a widget/object that has been idle for extended periods of time is not destroyed.

For most methods listed on Table 5.3, there exists a number of options/parameters that may be also used for the improvement of the GUI. These are applicable to a variety of widgets/objects. Table 5.4 provides a list of some of the most important ones. The list is not exhaustive, but it is based on cases described in detail in the various examples in this chapter.

TABLE 5.4
Frequently Used Properties and Their Descriptions

Properties/Expressions	Brief Description
activebackground, activeforeground	The background or foreground color when the cursor hovers over the widget.
anchor	Ensures that the particular element it applies to (i.e., text or image) is placed on a position within the parent widget that will remain unchanged.
borderwidth, bd	The width of the border around the widget (e.g., borderwidth=12) as an integer.
command	The method called when the widget is clicked.
compound	Combines two objects in the same position (e.g., an image and a text) in a parent label widget. It can take different values (e.g., left, center, right) that specify the order of the two objects.
expand	Specifies whether the underlying widget is expandable (value is "Y" or non-zero) or not (value is "N" or zero) when the parent widget is resized.
fg (or foreground), bg (or background)	The color of the foreground/background (fg/bg) or the text a particular widget will display (see Table 4.6).
fill	Specifies whether the widget it applies to will expand horizontally (fill=tk.X), vertically (fill=tk.Y) or both (fill=tk.BOTH).
font	Sets/gets the font name and the size of the text to be displayed by the widget (e.g., font = 'Arial 24').
from_ =, to=	Sets the numerical boundaries of the widget.
height, width	The height or width of the widget in characters (for text widgets) or pixels (for image widgets).
highlightcolor	The color of the text of the widget when the widget is in focus.
image	Defines an image to be displayed on the widget instead of text.
justify	Determines how multiple lines of text will be justified in respect to each other. Values are LEFT, CENTER, or RIGHT.
lambda expression	Sets the parameters to be passed on to a method or method when an event is triggered.
onvalue, offvalue	The values assigned to a check button depending on whether it is selected or not.
orient	Specifies the orientation of the widget (horizontal or vertical).
padx, pady	Additional padding left/right (padx) or above/below (pady) in relation to the widget.
relief	Causes the widget to be displayed with a particular visual effect in terms of its border appearance (see Table 4.6 for available values).
resolution	The incremental or decremental step of the scale widget.
relx, rely	The position of the widget relative to the parent object.
show	Replaces the text of the current widget with the specified character(s).
side	Specifies the position of the content of the widget (Left, Center, or Right).
state	The state of responsiveness and/or accessibility of the widget. Values can be NORMAL, ACTIVE, DISABLED.
text	The textual content to be displayed.
textvariable	The textual content of the text-based widget.
troughcolor	The color of the trough of the scale widget.
value	The value assigned to a radio button, depending on the selection/state.
["values"]	Associates/populates a combo box with a particular list of values.
underline	If −1, no character of the button's text will be underlined. If a non-zero value is provided, the corresponding character(s) will be underlined.
wraplength	If non-zero, the text lines of the widget will be wrapped to fit the length of the parent widget.
yscrollcommand, xscrollcommand	Used to activate the scrollbar.
yview, xview	Specifies the orientation of a scrollbar (yview for vertical or xview for horizontal).

TABLE 5.5
Frequently Used Events and Their Descriptions

Event	Brief Description
`<Button-1>`, `<Button-2>`, `<Button-3>`	Triggered when the left, middle, or right button of the mouse is clicked upon the widget.
`<Double-Button-1>`, `<Double-Button-2>`, `<Double-Button-3>`	Triggered when the left, middle, or right mouse button is double clicked upon the widget.
`<Enter>`	Triggered when the mouse is hovering across the widget.
`<Key>`	Triggered when any key on the keyboard is pressed. Use the `event.keycode` option to check the key that was pressed. Note that the values of the keyboard keys vary between operating systems.
`<Leave>`	Triggered when the mouse leaves the parent widget.

It should be evident by the examples provided in this chapter that one of the most important concepts in GUI programming is the user's interaction with the widgets, as this is how events are used to trigger specific tasks. Such interactions usually take the form of mouse clicks or keyboard events. Table 5.5 lists some of the most important methods of interactions as a quick reference.

Finally, some common values of the options mentioned previously are provided on Table 5.6 below.

TABLE 5.6
Possible Values for the Various Different Options

Option	Values Available
Color related	It is possible to set the color of the widget, text, or object, either in the form of a hexadecimal string (e.g., "#000111"), or by using color names (e.g., "white", "black", "red", "green", "blue", "cyan", "yellow", and "magenta").
Font related	The font of a text can be set just after the text is specified, using the following sub-options: • `Family`: The font family names as a string. • `Size`: The font height in points (n) or pixels (–n). • `Weight`: The attributes of the text ("bold" for bold, or "normal" for regular text). • `Slant`: The attributes of the text ("italic" for italic, or "roman" for unslanted). • `Underline`: The attributes of the text (1 for underlined or 0 for normal text). • `Overstrike`: The attributes of the text (1 for overstruck or 0 for normal text).
Anchor related	The possible values for the anchor justification are: NW, N, NE, W, CENTER, E, SW, S, SE.
Relief styles	After specifying the text of a widget, the possible values for the relief option are: `raised`, `sunken`, `flat`, `groove`, `ridge`.
Bitmap styles	Possible bitmap styles include the following: `error`, `gray75`, `gray50`, `gray25`, `gray12`, `hourglass`, `info`, `questhead`, `question`, `warning`. These can be used in combination with, or instead of, text.
Cursor styles	Possible cursor styles include the following: `arrow`, `circle`, `clock`, `cross`, `dotbox`, `exchange`, `fleur`, `heart`, `man`, `mouse`, `pirate`, `plus`, `shuttle`, `sizing`, `spider`, `spraycan`, `star`, `target`, `tcross`, `trek`, `watch`. These can be used after the text is specified.
Pack options	There are 4 options in terms of placing a particular widget in respect to the parent widget through the `pack()` method. Use the side option with values: TOP (default), BOTTOM, LEFT, or RIGHT. There are 3 options to determine whether and how a particular widget should expand when the parent widget expands. Use the fill option with values: NONE (default), X (fill only horizontally), Y (fill only vertically), or BOTH (fill both horizontally and vertically).

(Continued)

TABLE 5.6 (*Continued*)
Possible Values for the Various Different Options

Option	Values Available
Grid options	When placing widgets on the interface using the `grid()` method, the following options are available: • `columnrow`: The column and row the widget will be placed in. The leftmost column (0) and the first row are the defaults. • `columnspan`, `rowspan`: The number of columns or rows a widget will span across. 1 is the default value. • `ipadx`, `ipady`: The number of pixels to pad the widget (horizontally and vertically) within its borders. • `padx`, `pady`: The number of pixels to pad the widget (horizontally and vertically) outside its borders. • `sticky`: Determines how the widget will be aligned if its size is smaller than its cell in the grid. The default value is `centered`. Other possible values are N, E, S, W, NE, NW, SE, and SW.

5.6   CASE STUDY

Complete the integration of the *Basic Widgets* Python script from Chapters 4 with a full menu system in an object-oriented application, using all three types of menus (i.e., regular, toolbar, popup), as described in this chapter. The menu system should include the following options: *Color dialog, Open File dialog, Separator, Basic Widgets, Save As, Open Directory, Separator, About,* and *Exit*.

6 Data Structures and Algorithms with Python

Thaeer Kobbaey
Higher Colleges of Technology

Dimitrios Xanthidis
University College London
Higher Colleges of Technology

Ghazala Bilquise
Higher Colleges of Technology

CONTENTS

6.1 Introduction 208
6.2 Lists, Tuples, Sets, Dictionaries..209
    6.2.1 List...209
    6.2.2 Tuple 214
    6.2.3 Sets..214
    6.2.4 Dictionary 215
6.3 Basic Sorting...217
    6.3.1 Bubble Sort 217
    6.3.2 Insertion Sort 220
    6.3.3 Selection Sort 222
    6.3.4 Shell Sort 225
    6.3.5 Shaker Sort 227
6.4 Recursion, Binary Search, and Efficient Sorting with Lists...............................230
    6.4.1 Recursion 230
    6.4.2 Binary Search 233
    6.4.3 Quicksort 235
    6.4.4 Merge Sort 238
6.5 Complex Data Structures 242
    6.5.1 Stack 242
    6.5.2 Infix, Postfix, Prefix..245
    6.5.3 Queue 248
    6.5.4 Circular Queue..250
6.6 Dynamic Data Structures 253
    6.6.1 Linked Lists...254
    6.6.2 Binary Trees 261
    6.6.3 Binary Search Tree 262
    6.6.4 Graphs...267
    6.6.5 Implementing Graphs and the Eulerian Path in Python.........................269
6.7 Wrap Up..271
6.8 Case Studies..271
6.9 Exercises 272
References...272

DOI: 10.1201/9781003139010-6

207

6.1 INTRODUCTION

Data is defined as a collection of facts. In raw form, data is difficult to process and, thus, in need of further structuring in order to be useful. In computer science, a *data structure* refers to the organization, storage, and management of data in a way that allows its efficient processing and retrieval. In simple terms, a data structure

> **Observation 6.1 – Data Structures:** A way of representing, organizing, storing, and accessing data based on a set of well-defined rules.

represents the associated data on a computer in a specific format, while preserving any underlying logical relationships, and it provides storage and efficient access to the data based on set of performance-enhancing rules.

As an example, one can consider the real-life scenario of searching for a particular name in a phone book. The search is being made easy by organizing the names in the phone book and sorting them in alphabetical order. In this rather primitive example, one is not required to go through the phone book page by page to find the desired name. Other relevant examples include the history of web pages visited through the web browser (implemented as a *linked-list* structure), the undo/redo mechanism available in many applications (implemented as *stack* structure), the *queue* structures used by operating systems for scheduling the various CPU tasks, and the *tree* structure used in many artificial intelligence-based games to track the player's actions.

In a broader context, there are two different types of data structures:

- *Basic* data structures that are usually available in every modern programming language. In Python, these include structures like the *list*, the *dictionary*, the *tuple*, and the *set*. Lists and *tuples* allow the programmer to work with data that is ordered *sequentially*. *Sets* are unordered collections of values with no duplicates.
- *Complex* data structures, like *stacks*, *queues*, and various types of *trees*, that are built on basic data structures. In terms of the way these structures organize data, stacks and queues are classified as *linear* (i.e., the data elements are ordered), whereas trees and graphs as *non-linear* (i.e., the elements do not follow a particular order).

This chapter covers the following topics:

- Basic data structures (i.e., lists, tuples, sets, and dictionaries) and their operations.
- Basic Sorting Algorithms: *bubble sort, insertion sort, selection sort, shell sort, shaker sort.*
- The concept of *recursion* and its application to *binary search*, and the *merge sort* and *quick sort* algorithms.
- Complex data structures (i.e., *stacks* and *queues*).
- Dynamic data structures like *singly* and *doubly linked lists, binary trees/binary search trees*, and *graphs*.

The focus is both on the computational thinking behind these topics, and on a detailed look into the programming concepts used for their implementation. Nevertheless, it must be stated that this chapter aims to provide a thorough introduction of the underlying ideas rather than to cover the aforementioned data structures exhaustively. Fundamental and critically important data structures and the associated algorithms like the heap tree and the heap sort or hashing structures and hashing tables, are not covered here. The reader can find more details on related subjects in the seminal works of Dijkstra et al. (1976), Knuth (1997), and Stroustrup (2013), to whom the modern computer science and information systems and technology community owes much of its existence.

6.2 LISTS, TUPLES, SETS, DICTIONARIES

This section explores the four built-in data structures provided by Python, namely *lists*, *tuples*, *sets* and *dictionaries*. These structures are also briefly discussed in Chapter 2, where they are referred to as *non-primitive* data types. Their main use is to store a collection of values and provide tools for its manipulation.

6.2.1 LIST

A *list* is a data structure that stores a collection of items in specified, and frequently successive, memory locations. Each item in the list has a location number called an *index*. The index starts from zero and follows a *sequential* order. This does not refer to the values of the stored data being ordered in a particular way (e.g., alphabetically), but the *index* values. To access an item at a particular location, the programmer can simply use the index number corresponding to this location. The concept of the list is analogous to a to-do list that contains things that must be accomplished. In terms of functionality, Python provides various operations, such as adding items to, and removing from, a list. Since items in a list can be modified, it is considered to be *mutable*.

At a practical level, lists in Python are denoted by square brackets (i.e., []). The list can be populated by adding items within the brackets, separated by commas. The following script creates a list, and then prints both the list *items* and the *number of items* in the list. It also asks the user to specify the index of an item to print (starting from zero), a range of items to print from the start of the list to a user-specified index, and a range of items to print from a user-specified index to the end of the list:

> **Observation 6.2 – List:** A *list* is a data structure that stores a collection of items in specified, usually successive, memory locations. It is indexed by a *sequential index* that always starts at zero. The items do not have to be in a particular order. A list is a *mutable* object, meaning that each item can be modified.

```
1    # Create the list
2    cars = ["BMW", "Toyota", "Honda", "Mercedes"]
3
4    # Print the list items
5    print("The list of the cars is the following: ", cars)
6
7    # Use the len() function to print the number of items in the list
8    print("The number of items in the list is: ", len(cars))
9
10   # Ask the user for the index number of an item for printing
11   singleIndex = int(input("Enter the index \
12   of the item to print (indexes start from 0): "))
13   print("Your selection for display is: ", cars[singleIndex])
14
15   # Ask the user for the starting index of the print range
16   startingIndex = int(input("Enter the starting index of the range \
17   of items to print (index starts from 0): "))
18   print("Your selected range of items to display is: ",
19       cars[startingIndex:len(cars)-1])
20
21   # Ask the user for the ending index of the print range
22   endingIndex = int(input("Enter the ending index of the range of items \
```

```
23  to print (index starts from 0): "))
24  print("Your selected range of items to display is: ",
    cars[0:endingIndex])
25
26  # Use a negative index to start printing the list from the end
27  print("The last item in the list is: ", cars[-1])
```

Output 6.2.1.a:

```
The list of the cars is the following:  ['BMW', 'Toyota', 'Honda', 'Mercedes']
The number of items in the list is:  4
Enter the index of the item to print (indexes start from 0): 0
Your selection for display is:  BMW
Enter the starting index of the range of items to print (index starts from 0): 1
Your selected range of items to display is:  ['Toyota', 'Honda']
Enter the ending index of the range of items to print (index starts from 0): 2
Your selected range of items to display is:  ['BMW', 'Toyota']
The last item in the list is:  Mercedes
```

In this script, the reader will notice that the syntax for calling a range of items is list[start:end], with start denoting the position of the starting index (inclusive) and end the ending index (not inclusive). It must be stressed that the start and end parameters are optional. For instance, expression cars[0: endingIndex] could be replaced by cars[:endingIndex] and, similarly, expression cars[startingIndex:len(cars)-1] could be replaced by cars[startingIndex:].

The reader should also note that if the user tries to access a list item using an index that does not exist, an IndexError exception will be raised, as illustrated in the example below:

Output 6.2.1.b:

```
The list of the cars is the following:  ['BMW', 'Toyota', 'Honda',
'Mercedes']
The number cf items in the list is:  4
Enter the index of the item to print (indexes start from 0): 4

-----------------------------------------------------------------
IndexError                               Traceback (most recent call last)
<ipython-input-5-695ecl33b0e9> in <module>
    11 singleIndex = int(input("Enter the index \
    12 of the item to print (indexes start from 0): "))
---> 13 print("Your selection for display is: ", cars[singleIndex])
    14
    15 # Ask the user for the starting index of a range of items in the
       list to print

IndexError: list index out of range
```

In addition to the basic functions discussed above, Python also provides a number of additional functions that can be used to manipulate a list (Table 6.1):

TABLE 6.1

Most Important Functions for List Manipulation

Functions	Description
append(item)	Adds an element at the end of the list
clear()	Removes all the elements from the list
copy()	Returns a copy of the list
count()	Returns the number of elements with the specified value
extend(list2)	Adds the elements of a second list (e.g., list2) to the end of the current list
index(item)	Returns the index of the first item with the specified value
insert(pos, item)	Adds an element at the specified position
pop()	Removes and returns the last element of the list
remove(item)	Removes the item with the specified value
reverse()	Reverses the order of the list
sort()	Sorts the list in ascending order

The script below is a modified version of the previously created one, demonstrating the use of append(), insert(), extend(), remove(), and pop() (Table 6.1). The script performs the tasks of adding items at the end of a list (line 9), inserting an item in a particular position specified by an index value (line 11), extending the list by adding items from a second list (lines 16–17), removing a particular item from the list (line 22), and removing the last item of the list (line 26):

```
1   # Create the list
2   cars = ["BMW", "Toyota", "Honda", "Mercedes"]
3
4   # Print the list size and its items
5   print("The list of the cars has ", len(cars),
6         " items which are the following: ", cars)
7
8   # Append/add an item to the end of the list
9   cars.append("Nissan")
10  # Insert an item to position 1 of the list
11  cars.insert(1,"Suzuki")
12  # Print the updated list
13  print("The updated list after the append and insert is: ", cars)
14
15  # Extend the list by adding the items of a second list
16  cars2 = ["Renault", "Audi"]
17  cars.extend(cars2)
18  print("The updated list after extending it with items from "
19        "a second list is: ", cars)
20
21  # Remove a specific item from the list
22  cars.remove("Toyota")
23  print(cars)
24
25  # Remove the last item from the list
26  cars.pop()
27  print(cars)
```

Output 6.2.1.c:

```
The list of the cars has  4  items which are the following:  ['BMW',
'Toyota', 'Honda', 'Mercedes']
The updated list after the append and insert is:  ['BMW', 'Suzuki',
'Toyota', 'Honda', 'Mercedes', 'Nissan']
The updated list after extending it with items from a second list is: ['BMW',
'Suzuki', 'Toyota', 'Honda', 'Mercedes', 'Nissan', 'Renault', 'Audi']
['BMW', 'Suzuki', 'Honda', 'Mercedes', 'Nissan', 'Renault', 'Audi']
['BMW', 'Suzuki', 'Honda', 'Mercedes', 'Nissan', 'Renault']
```

The following variation of the same script showcases the use of reverse(), sort(), sort(reverse=True), and index() in order to reverse the items of the list (line 9), sort them in ascending order (line 13), sort them in descending/reverse order (line 17), and find and return the index of a particular item (line 21). Notice that none of the results of these functions have a permanent effect on the original list:

```
1    # Create the list
2    cars = ["BMW", "Toyota", "Honda", "Mercedes", "Toyota"]
3
4    # Print the list size and its items
5    print("The list of the cars has ", len(cars),
6            " items which are the following: ", cars)
7
8    # Print the items of the list in reverse order
9    cars.reverse()
10   print(cars)
11
12   # Sort the items of the list and print them
13   cars.sort()
14   print(cars)
15
16   # Sort the items of the list in reverse order and print them
17   cars.sort(reverse=True)
18   print(cars)
19
20   # Find and return the index of a specific item in the list
21   print(cars.index("BMW"))
```

Output 6.2.1.d:

```
The list of the cars has  4  items which are the following:
['BMW', 'Toyota', 'Honda', 'Mercedes']
['Mercedes', 'Honda', 'Toyota', 'BMW']
['BMW', 'Honda', 'Mercedes', 'Toyota']
['Toyota', 'Mercedes', 'Honda', 'BMW']
3
```

Finally, with the use of in <list>, copy(), count(), and clear(), the programmer can examine in run-time whether a particular item belongs in a list (lines 8–11 and 13–16), copy the contents of a list (line 23), count the occurrences of an item in the list (line 19), and clear the list (line 27):

```
1    # Create the list
2    cars = ["BMW", "Toyota", "Honda", "Mercedes", "Toyota"]
3
4    # Print the list items
5    print("The list of the cars is the following: ", cars)
6
7    # Print True or False depending on whether an item is included in
     the list
8    if ("Toyota" in cars):
9        print("Toyota is in the list")
10   else:
11       print("Toyota is not in the list")
12
13   if ("Nissan" in cars):
14       print("Nissan is in the list")
15   else:
16       print("Nissan is not in the list")
17
18   # The number of occurrences of an item in the list
19   occurrences = cars.count("Toyota")
20   print("Occurrences of the particular item in the list is: ",
     occurrences)
21
22   # Copy the contents of a list into another
23   newCars = cars.copy()
24   print("The contents of the new list are: ", newCars)
25
26   # Clear the list
27   newCars.clear()
28   print("The newCars list of items is now empty: ", newCars)
```

Output 6.2.1.e:

```
The list of the cars is the following:  ['BMW', 'Toyota', 'Honda',
'Mercedes', 'Toyota']
Toyota is in the list
Nissan is not in the list
Occurences of the particular item in the list is:  2
The contents of the new list are:  ['BMW', 'Toyota', 'Honda',
'Mercedes', 'Toyota']
The newCars list of items is now empty:  []
```

6.2.2 TUPLE

Tuples are a special type of list, with items being organized in a particular order and accessed by referencing index values. The difference between a normal list and a tuple is that the latter is *immutable*, meaning that its items *cannot be modified*. As such, tuples do not offer some of the extended functionality of a list described in the previous section. In terms of syntax, tuples are created using parentheses instead of square brackets. The following script demonstrates the basics of tuple creation and usage:

Observation 6.3 – Tuple: A special type of list that is *immutable* (i.e., its items cannot be modified). Tuples are created using parentheses instead of square brackets.

```
1   # Create a tuple
2   cars = ("BMW", "Toyota", "Honda", "Mercedes")
3
4   # Display all items in the tuple
5   print("The items in the tuple are: ", cars)
6   # Display the first item in the tuple
7   print("The first item in the tuple is: ", cars[0])
8
9   # Raises TypeError exception as the tuple item can't be modified
10  cars[0] = "Tesla"
```

Output 6.2.2:

```
The items in the tuple are: ('BMW', 'Toyota', 'Honda', 'Mercedes')
The first item in the tuple is: BMW
---------------------------------------------------------------------
TypeError                              Traceback (most recent call last)
<ipython-input-1-3c3eee3a45c8> in <module>
      8
      9 # Raises a TypeError exception since the item in the tuple cannot be
        modified
---> 10 cars[0] = "Tesla"

TypeError: 'tuple' object does not support item assignment
```

6.2.3 SETS

A *set* is a collection of unordered and unique items. It is created using curly braces (i.e., {}) (Hoare, 1961). When the print() function is used to display the contents of a set, the duplicates are removed from the output and its contents are not presented in a particular order. In fact, every time the code is executed the order of the elements is different.

There are four particular operators/functions used on sets:

1. **The in Operator:** Examines whether an item is included in the set.

Observation 6.4 – Set: A collection of unordered, unique items. Use the in operator to examine if an item belongs to a set. Use the intersection() function to find the common items between two sets. Use the difference() function to retrieve items from the first set that are not found in the second. The union() function combines the items of two sets, removing any duplicates.

2. **The `intersection()` Function:** Identifies the common items between two sets.

3. **The `difference()` Function:** Retrieves items from a set that do not exist in another set.

4. **The `union()` Function:** Combines the items of two sets and returns a new one after removing any duplicates.

The following script demonstrates the basic use of sets and their main operations:

```
1    # Create the set
2    cars = {"BMW", "Toyota", "Honda", "Mercedes", "Toyota"}
3    # Print the set
4    print("The cars set includes the following items: ", cars)
5
6    # Check whether a particular item exists in the set
7    if ("Honda" in cars):
8        print("Honda is in the cars set")
9    else:
10       print("Honda is not in the cars set")
11
12   # Create and print an additional set
13   german_cars = {"BMW", "Mercedes", "Audi", "Porsche"}
14   print("The german cars set includes the following items: ", german_cars)
15
16   # Find and print the intersection (i.e., common items of the two sets)
17   print("The intersection, i.e., the common items of the two sets, is: ",
18       cars.intersection(german_cars))
19
20   # Find and print the difference of the two sets
21   print("The different items between the two sets are: ",
22       cars.difference(german_cars))
23
24   # Find and print the union of the two sets
25   print("The union of the two sets is: ", cars.union(german_cars))
```

Output 6.2.3:

```
The cars set includes the following items: {'Honda', 'Mercedes',
'BMW', 'Toyota'}
Honda is in the cars set
The german cars set includes the following items: {'Mercedes',
'Porsche', 'BMW', 'Audi'}
The intersection, i.e., the common items of the two sets, is:
{'Mercedes', 'BMW'}
The different items between the two sets are: {'Honda', Toyota'}
The union of the two sets is: {'Audi', 'Porsche', 'Honda',
'Mercedes', 'BMW', 'Toyota'}
```

6.2.4 DICTIONARY

A *dictionary* is a collection of items that stores values in *key-value* pairs. The *key* is a unique identifier and the *value* is the data associated with it. The dictionary is analogous to a phone book that stores the contact name and telephone of a person. The contact name would be the key that is used

TABLE 6.2

Functions of a Dictionary

Function	Description
clear()	Removes all the elements from the dictionary
copy()	Returns a copy of the dictionary
get(key)	Gets an item by the key
has_key(key)	Returns a Boolean value based of whether the key is in the dictionary or not
items()	Returns a list of (key, value) tuples
keys()	Returns a list of keys
values()	Returns a list of values
pop(key)	Removes an item given the key and returns the value
popitem()	Removes the next item, and returns the key/value
update()	Adds or overwrites items from another dictionary

to look up the telephone number (i.e., the value). In a dictionary, keys must be *unique* and of an *immutable* data type, such as strings or integers, while values can be of any type (e.g., strings, integers, lists).

The Python syntax for creating a dictionary is the following:

```
dictionary={key1: value1, key2: value2}
```

Table 6.2 lists the available dictionary functions

Observation 6.5 – Dictionary: A collection of items stored in a *key-value* pair format. The *keys* must use *immutable* data types. The *values* can be of any type and are *mutable*. The syntax is the following:

```
dictionary={key1:   value1,
key2:  value2}
```

The following script presents an example involving a dictionary named employee that holds the employees' names, salaries, and job titles:

```
1    # Create the dictionary
2    employee = {"name": "Maria", "salary": 15000, "job": "Sales Manager"}
3
4    # Print the dictionary
5    print("The employee dictionary is: ", employee)
6    # Access a specific key and print the paired value
7    print("The pair value for the <name> key is: ", employee["name"])
8
9    # Use the get() method to print a pair based on a given key
10   print("The value pair of the <name> key is: ", employee.get("name"))
11   # If the key value does not exist the get() method will return
12   # None (empty)
13   print("The value pair of the <name> key is: ",
14        employee.get("department"))
15
16   # Add a new pair to the dictionary
17   employee["department"] = "Sales"
18   print("The value pair of the new <department> key is: ",
19   employee.get("department"))
20
21   # Modify the value of a given key
22   employee["salary"] = "20000"
23   print("The new employee dictionary includes the following pairs: ",
```

```
24          employee)
25
26  # Use the update() method to modify the dictionary
27  employee.update({"name":"Alex","department":"Sales"})
28  print(employee)
29
30  # Pop/remove a pair based on a given key, assign it to a new
31  # dictionary and print it
32  emp_job = employee.pop("job")
33  print("The original employee dictionary is: ", employee)
34  print("The new emp_job dictionary is: ", emp_job)
```

Output 6.2.4:

```
The employee dictionary is: {'name': 'Maria', 'salary': 15000, 'job':
'Sales Manager'}
The pair value for the <name> key is: Maria
The value pair of the <name> key is: Maria
The value pair of the <name> key is: None
The value pair of the new <department> key is: Sales
The new employee dictionary includes the following pairs: {'name': 'Maria',
'salary': '20000', 'job': 'Sales Manager', 'department': 'Sales'}
{'name': 'Alex', 'salary': '20000', 'job': 'Sales Manager', 'department':
'Sales'}
The original employee dictionary is: {'name': 'Alex', 'salary': '20000',
'department': 'Sales'}
The new empjob dictionary is: Sales Manager
```

The reader should note that it is possible to access the value of a dictionary key either directly (line 7) or through the get() function (line 10). If access to a value of a key that does not exist in the dictionary is requested, get() returns an empty value (line 13 and 14). It is also worth noting that it is possible to add a new pair of values through the update() function (line 27). Finally, line 32 demonstrates how to remove a particular pair from a dictionary through the pop() function and how to create a new dictionary from it.

6.3 BASIC SORTING

Sorting is a major task in computer science and information systems/technology, with as much as 30% of the total computer processing time of everyday business activity allegedly being devoted to it. In a broader context, sorting is the computational process of arranging data in a particular order. As different sorting algorithms can result in differences of minutes, hours, or even days, *efficiency* is an important factor in terms of sorting time. Efficiency is measured by counting the number of *comparisons* and *exchanges/swaps* required to sort a given list of data. A comparison takes place when an element of the list is compared with another, whereas exchanges/swaps happen when two elements of the list switch their positions.

6.3.1 BUBBLE SORT

The *bubble sort is* one of the most well-known sorting algorithms. It is also covered in Chapter 4 of this book, under the topic of *listboxes*. The main idea of the algorithm is to have the element with the highest (or lowest) value in a list moved to the last (or first) place during each iteration. At each iteration, the program repeats this process, moving the next highest (lowest) number in the list to the appropriate place. The number of the main iteration corresponds to the number of the elements of the list. During each main iteration there are as many comparisons (and potentially exchanges/swaps) as the total number of elements in the list. Thus, the *time complexity* of the bubble sort is $O(n^2)$. The detailed explanation of time complexi-

Observation 6.6 – Bubble Sort: Use two nested `for` loops during the inner iterations to successively move the highest/lowest value element to the end of the list until the entire list is sorted.

ties and the *Big O/Theta/Omega* notation is beyond the scope of this book, but the reader can find related information in most of the essential computer science sources and bibliography. For the purposes of this chapter, it should suffice to claim that the bubble sort is not particularly efficient in terms of time. In order to examine the low efficiency of the algorithm, the reader could assume that each comparison takes *1 nanosecond* to complete (*1 nanosecond = 1.0e−9 seconds*). This would translate to the following rough estimates:

- $n = 10$: $n^2 = 81$ comparisons → approximate time 3e−4 seconds.
- $n = 100$: $n^2 = 9.8e3$ comparisons → approximate time 5e−3 seconds.
- $n = 1,000$: $n^2 = 9.98e5$ comparisons → approximate time 0.4 seconds.
- $n = 10,000$: $n^2 = 9.998e7$ comparisons → approximate time 46 seconds.
- $n = 20,000$: $n^2 = 4e7$ comparisons → approximate time 188 seconds

As these calculations are estimates, they are largely dependent on the system at hand, the type of data of the list, and the conditions of the programming platform used. However, the crude assumptions and numbers used here could provide a rough idea of the increasing inefficiency of the bubble sort in line with an increasing size of the list. Indeed, bubble sort works well as long as n is not higher than approximately 10,000. After this point, it becomes heavy and its inefficiency starts to show.

It is possible to slightly improve the efficiency of the algorithm by avoiding unnecessary comparisons. As an example, one could use the following eight-element list: 3, 5, 4, 2, 3, 1, 6, 7. The algorithm will execute $n−1$ times (i.e., seven iterations) during each of the *main* iterations. The *inner* iterations are then responsible to bring each element to the corresponding place successively (Table 6.3).

The reader should note that, firstly, it is not necessary that an exchange/swap of elements will take place in every iteration of the inner loop and, secondly, at the end of the main outer iteration the highest element is pushed to the end of the list. In this case, in the first main outer iteration, element 7 is pushed to the end of the list. The last line is the result of the first main outer iteration, after all seven inner loops are completed. Subsequent iterations will repeat the same process, ensuring that the next highest element moves to the appropriate position, until all elements have taken the correct place in the list.

TABLE 6.3
The Inner Loop inside the First Main Iteration

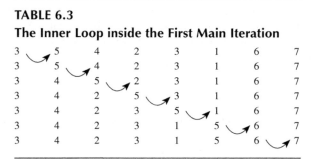

TABLE 6.4
The Results of the Outer Loops

After the 1st pass	3	4	2	3	1	5	6	7
After the 2nd pass	3	2	3	1	4	5	6	7
After the 3rd pass	2	3	1	3	4	5	6	7
After the 4th pass	2	1	3	3	4	5	6	7
After the 5th pass	1	2	3	3	4	5	6	7
After the 6th pass	Comparisons are made with no swaps							
After the 7th pass	Comparisons are made with no swaps							

Table 6.4 presents the results after each of the outer iterations/loops.
A Python implementation of a basic bubble sort and its output is provided below:

```python
1    # Import the random module to generate random numbers
2    import random
3    import time
4
5    comparisons = 0
6    list = []
7
8    # Enter the number of list elements
9    size = int(input("Enter the number of list elements: "))
10   # Use the randint() function to generate random integers
11   for i in range (size):
12       newNum = random.randint(-100, 100)
13       list.append(newNum)
14   print("The unsorted list is: ", list)
15
16   # Bubble sorts the list & records the stats for later use
17   # Start the timer
18   startTime = time.process_time()
19
20   # The bubble sort algorithm
21   for i in range (size-1):
22       for j in range (size-1):
23           comparisons += 1
24           if (list[j] > list[j+1]):
25               temp = list[j]
26               list[j] = list[j+1]
27               list[j+1] = temp
28
29   # End the timer
30   endTime = time.process_time()
31
32   # Display the basic info for the bubble sort
33   print("The sorted list is: ", list)
34   print("The number of comparisons is: ", comparisons)
35   print("The elapsed time in seconds is: ", (endTime - startTime))
```

Output 6.3.1:

```
Enter the number of elements in the list:7
The unsorted list is:  [33, -16, -57, -17, 95, 5, 15]
The sorted list is:  [-57, -17, -16, 5, 15, 33, 95]
The number of comparisons is =  36
The elapsed time in seconds =  0.0
```

6.3.2 INSERTION SORT

Insertion sort is another basic sorting algorithm, similar to bubble sort but somewhat improved. The basic idea is that on the ith *pass* the algorithm inserts the ith *element* into the appropriate place (i.e., *L[i]*) at the end of the *L[1], L[2], ..., L[i-1]* sequence, the elements of which have been previously placed in *sorted order*. As a result, after the insertion, the elements occupying the *L[1], L[2], ..., L[i]* sequence are in sorted order. In simple terms, the algorithm sorts increasingly larger subsets of the original list until the whole list is sorted.

Observation 6.7 – Insertion Sort: Use a `while` loop nested inside a `for` loop to find the highest/lowest value element in the subset of the list in each ith pass. The subset starts with the first two elements (index extends up to $i+1$) and is increased by *1* in each pass.

As an example, assume that the insertion sort is applied to the following seven-element list: *3, 5, 4, 2, 3, 1, 6*, thus executing *n−1* (i.e., *6*) outer iterations/loops. The big difference between this algorithm and bubble sort is that each of the main iterations will not require the same number as the inner iterations, but an increasing iteration number starting from *1* and up to *n−1*. During each inner iteration, the highest element is moved to the last location of the current subset of the list. The following section describes in detail each of the main iterations.

The inner iteration of the first main iteration will put the two elements of the subset in order:

The two-iteration loop of the second main iteration will put the three elements of the subset in order:

The three-iteration loop of the third main iteration will put the four elements of the subset in order:

The four-iteration loop of the fourth main iteration will put the five elements of the subset in order:

```
2    3    4    5    3
2    3    4    3    5
2    3    3    4    5
2    3    3    4    5
2    3    3    4    5
```

The five-iteration loop of the fifth main iteration will put the six elements of the subset in order:

```
2    3    3    4    5    1
2    3    3    4    1    5
2    3    3    1    4    5
2    3    1    3    4    5
2    1    3    3    4    5
1    2    3    3    4    5
```

The six-iteration loop of the sixth main iteration will put the seven elements of the subset in order:

```
1    2    3    3    4    5    6
1    2    3    3    4    5    6
1    2    3    3    4    5    6
1    2    3    3    4    5    6
1    2    3    3    4    5    6
1    2    3    3    4    5    6
1    2    3    3    4    5    6
```

The algorithm relies on the introduction of a temporary element (e.g., temp) and a temporary location (i.e., loc), which are assigned with values *L[1]* and *1* respectively. The following script provides an implementation of the insertion sort algorithm in Python:

```
1    import random
2    import time
3
4    list = []
5    comparisons = 0
6
7    # Enter the number of list elements
8    size = int(input("Enter the number of list elements: "))
9    # Use the randint() function to generate random integers
10   for i in range (size):
11       newNum = random.randint(-100, 100)
12       list.append(newNum)
13   print("The unsorted list is: ", list)
14
15   startTime = time.process_time() # Start the timer
16
17   # The insertion sort algorithm
18   for i in range(1, size):
```

```
19      temp = list[i]
20      loc = i
21      while ((loc > 0) and (list[loc-1] > temp)):
22          comparisons += 1
23          list[loc] = list[loc-1];
24          loc = loc -1
25      list[loc] = temp
26
27  endTime = time.process_time()  # End the timer
28
29  # Display the basic info for the insertion sort
30  print("The sorted list is: ", list)
31  print("The number of comparisons is: ", comparisons)
32  print("The elapsed time in seconds is: ", (endTime - startTime))
```

Output 6.3.2:

```
Enter the number of elements in the list:7
The unsorted list is: [2, -8, 69, 20, -56, -32, -81]
The sorted list is: [-81, -56, -32, -8, 2, 20, 69]
The number of comparisons is = 16
The elapsed time in seconds = 0.0
```

There are a couple of characteristics that make insertion sort significantly more efficient compared to bubble sort. First, since each subset of the list includes fewer elements than the entire list, it performs fewer comparisons. Second, as each pass secures that the subset is in order, fewer swaps are required. However, on average, the algorithm falls under the same time efficiency bracket as bubble sort (i.e., $O(n^2)$), and only shows improvement on the *best case*, where it becomes *linear* and achieves a time complexity of $O(n)$.

An approximation of the time efficiency improvements of the insertion sort over the bubble sort is provided in the list below (assume *1 comparison* takes *1 nanosecond* or *1.0e−9 seconds*; where *Cs* stands for *Comparisons*):

- $n = 10$: ~40 Cs ($n^2 = 81$ in Bubble S.) → approx. 2.0e−4 seconds (3e−4 in Bubble S.)
- $n = 100$: ~4.0e3 Cs ($n^2 = 9.8e3$ in Bubble S.) → approx. 2.5e−3 seconds (4.5e−3 in Bubble S.)
- $n = 1,000$: ~5.0e5 Cs ($n^2 = 9.98e5$ in Bubble S.) → approx. 0.16 seconds (3.7e−1 in Bubble S.)
- $n = 10,000$: ~4.0e7 Cs ($n^2 = 9.998e7$ in Bubble S.) → approx. 15 seconds (46 in Bubble S.)
- $n = 20,000$: ~9.8e7 Cs ($n^2 = 2.0e8$ in Bubble S.) → approx. 57 seconds (188 in Bubble S.)

6.3.3 SELECTION SORT

Selection sort, also considered one of the fundamental sorting algorithms, is similar to insertion sort, but provides some improvements in terms of efficiency as it reduces the number of required swaps. The basic idea is that, on the ith *pass*, the algorithm selects the element with the lowest (or highest) value within a given range (i.e., $A[j], ..., A[n]$), and swaps it with the current position (i.e., $A[j]$). Thus, after the ith *pass*, the ith *lowest* elements will occupy $A[1], A[2], ..., A[i]$ in sorted order.

Observation 6.8 – Selection Sort: Use a `for` loop nested inside another `for` loop to find and replace the highest/lowest value element with the original, i^{th} item in the list. In each successive pass, the subset of the searchable list is reduced by one.

The algorithm utilizes subsets of a list to sort it, moving from the whole list to end up with the smallest divisions of it. In a sense, it is almost the opposite of insertion sort. The algorithm requires one additional variable in order to store the location (index) of the lowest value element within the list.

Using the list from the previous example (i.e., *3, 5, 4, 2, 3, 1, 6*), during the 1st outer iteration of the selection sort, the inner iterations will determine that the lowest value element is in *index 5*. Therefore, the elements in *list[0]* and *list[5]* will be swapped, and the element in *list[0]* will not be involved in any further processing from this point on:

list[0]=3	list[1]=5	list[2]=4	list[3]=2	list[4]=3	list[5]=1	list[6]=6

By the end of the 1st outer iteration, the list has the following structure:

list[0]=1	list[1]=5	list[2]=4	list[3]=2	list[4]=3	list[5]=3	list[6]=6

Given that the 2nd outer loop will move the index to the 2nd element of the list (i.e., $i=1$), the 2nd inner iterations will only deal with the subset of the original list, excluding the sorted part (i.e., *list[0]*). This means that in the unsorted subset of the list, the element with the lowest value will be in *index 3*. Thus, the elements in *list[1]* and *list[3]* will be swapped, while the element in *list[1]* will not be involved in any further processing:

list[0]=1	list[1]=5	list[2]=4	list[3]=2	list[4]=3	list[5]=3	list[6]=6

By the end of the 2nd outer iteration the list will be the following:

list[0]=1	list[1]=2	list[2]=4	list[3]=5	list[4]=3	list[5]=3	list[6]=6

Once again, the 3rd outer loop will move the index to the 3rd element of the list (i.e., $i=2$) and the 3rd inner iterations will only deal with the subset of the original list, excluding the sorted part (i.e., *list[0], list[1]*). As in the previous two iterations, this will result in the element with the lowest value in the unsorted subset of the list being found in *index 4*, and thus the elements in *list[2]* and *list[4]* will be swapped:

list[0]=1	list[1]=2		list[2]=4	list[3]=5	list[4]=3	list[5]=3	list[6]=6

By the end of the 3rd outer iteration the list will be the following:

list[0]=1	list[1]=2	list[2]=3	list[3]=5	list[4]=4	list[5]=3	list[6]=6

Repeating the outer loop for a 4th time will further move the index to the 4th element of the list and the 4th inner iterations will deal with the remaining subset of the list. The inner loop will find the lowest value element to be in *index 5* of that subset, and the elements in *list[3]* and *list[5]* will be swapped:

list[0]=1	list[1]=2	list[2]=3		list[3]=3	list[4]=4	list[5]=5	list[6]=6

The algorithm will continue until there is no subset left unprocessed. By that time, the list will have been sorted. The following script showcases an implementation of selection sort in Python and its output:

```
1    # Import the random module to generate random numbers
2    import random
3    import time
4
5    comparisons = 0
6    list = []
7
8    # Enter the number of list elements
9    size = int(input("Enter the number of list elements: "))
10   # Use the randint() function to generate random integers
11   for i in range (size):
12       newNum = random.randint(-100, 100)
13       list.append(newNum)
14   print("The unsorted list is: ", list)
15
16   # Selection sorts the list & records the stats for later use
17   # Start the timer
18   startTime = time.process_time()
19
20   # The selection sort algorithm
21   for i in range(size):
22       locOfMin = i
23
24       # Find the smallest element in the
25       # remaining subset of the list
26       for j in range(i+1, size):
27           comparisons += 1
28           if (list[locOfMin] > list[j]):
29               locOfMin = j
30
31       # Swap the minimum element with
32       # the first element of the subset
33       list[i], list[locOfMin] = list[locOfMin], list[i]
34
35   # End the timer
36   endTime = time.process_time()
37
38   # Display the basic info for the selection sort
39   print("The sorted list is: ", list)
40   print("The number of comparisons is: ", comparisons)
41   print("The elapsed time in seconds: ", (endTime - startTime))
```

Output 6.3.3:

```
Enter the number of elements in the list:7
The unsorted list is:  [32, 81, -76, -88, 62, -53, -17]
The screed list is:  [-88, -76, -53, -17, 32, 62, 81]
The number of comparisons is =  21
The elapsed time in seconds =  0.0
```

Selection sort is a bit heavier than insertion sort, but it becomes comparatively faster as the list grows larger. Nevertheless, for lists containing between approximately 1,000 and 50,000 elements, both algorithms perform similarly in terms of their efficiency. Their most important difference is that the efficiency of selection sort is quite similar across the *best*, *average*, and *worst* cases, with a time complexity of $O(n^2)$, whereas insertion sort has a complexity that in the best case might even reach $O(n)$. In practice, both algorithms are suitable for relatively small lists.

The following list provides approximate comparative figures highlighting the performance differences between the two algorithms (assume *1 comparison* takes *1 nanosecond* or *1.0e−9 seconds*; Cs stands for *Comparisons*):

- $n = 10$: 45 Cs (up to 40 in Insertion S.) → approx. 6.0e−4 seconds (2.0e−4 in Insertion S.)
- $n = 100$: 4.9e3 Cs (up to 4.0e3 in Insertion S.) → approx. 8.0 e−3 seconds (2.5e−3 in Insertion S.)
- $n = 1,000$: 5.0e5 Cs (up to 5.0e5 in Insertion S.) → approx. 0.18 seconds (0.16 seconds in Insertion S.)
- $n = 10,000$: 5.0e7 Cs (4.0e7 Cs in Insertion S.) → approx. 17 seconds (15 seconds in Insertion S.)
- $n = 20,000$: 2.0e8 Cs (9.8e7 Cs in Insertion S.) → approx. 62 seconds (57 seconds in Insertion S.)
- $n = 30,000$: 4.5e8 Cs (2.2e8 Cs in Insertion S.) → approx. 142 seconds (125 seconds in Insertion S.)

6.3.4 SHELL SORT

In order to improve the performance of sorting larger lists, the reader can use the *shell sort* (also referred to as the *diminishing-increment sort*). The main problem with previously discussed algorithms like insertion, selection and bubble sort, is their time performance of $O(n^2)$, making them extremely slow when sorting big lists. Shell sort, while being based on insertion sort, is using smaller distances between elements. Initially, elements within a specifically defined distance in the list

> **Observation 6.9 – Shell Sort:** An improved variation of the bubble sort, sorting *subsets* of a list based on the distance between the various list elements. The process starts with a defined number that is reduced in each iteration (usually by one).

are sorted. The algorithm then starts working with elements of decreasing distances until all subsequent elements have been processed. The key point in this algorithm is that every pass deals with a relatively small number of elements, or with already sorted elements, and every pass secures an increasing part of the list is ordered. The sequence of the distances can change, provided that *the last distance must be 1*. It is mathematically proven that the algorithm has a time complexity of $O(n^{1.2})$.

As an example, let us consider the following list: *3, 5, 2, 4, 6, 1, 7, 9, 8*. In the *1st pass*, the list is split into three subsets, each of which is processed using the insertion sort. In this particular case, the three subsets have a distance of three between each element:

- **1st Pass/Subset 1:** *3, 4, 7*. Result after insertion sort: *3, 4, 7*
- **1st Pass/Subset 2:** *5, 6, 9*. Result after insertion sort: *5, 6, 9*
- **1st Pass/Subset 3:** *2, 1, 8*. Result after insertion sort: *1, 2, 8*

After the end of the 1st pass the list will be in the following order: *3, 5, 1, 4, 6, 2, 7, 9, 8*.

In the *2nd*, the list is split into two subsets, with each one being processed again using the insertion sort. In this case, the two subsets have a distance of two between each element:

- **2nd Pass/Subset 1:** *3, 1, 6, 7, 8*. Result after insertion sort: *1, 3, 6, 7, 8*
- **2nd Pass/Subset 2:** *5, 4, 2, 9*. Result after insertion sort: *2, 4, 5, 9*

After the end of the 2nd pass, the complete list will be in the following order: *1, 2, 3, 4, 6, 5, 7, 9, 8*.

Finally, in the *3rd pass*, the list is dealt with as a whole, again using the insertion sort. Given that the previous passes ensured that the list is close to being fully sorted, this pass does require multiple swaps but only the necessary comparisons. The following script implements the aforementioned algorithm:

```
1    # Import the random module to generate random numbers
2    import random
3    import time
4
5    comparisons = 0
6    list = []
7
8    # Enter the number of elements for the list
9    size = int(input("Enter the number of list elements: "))
10   # Use the randint() function to generate random integers
11   for i in range (size):
12       newNum = random.randint(-100, 100)
13       list.append(newNum)
14   print("The unsorted list is: ", list)
15
16   # Start the timer
17   startTime = time.process_time()
18
19   # Use shell sort to sort the list and record the statistics for later use
20   # Start with a big distance and reduce it successively
21   distance = int(size/2)
22
23   # Insertion sorts each of the list subsets divided by distance
24   while distance >= 0:
25
26   # The insertion sort algorithm
27       for i in range(size):
28           temp = list[i]
```

```
29              loc = i
30              while ((loc >= distance) and (list[loc-distance] > temp)):
31                   comparisons += 1
32                   list[loc] = list[loc-distance]
33                   loc = loc - distance
34              list[loc] = temp
35
36         distance -= 1
37
38   # End the timer
39   endTime = time.process_time()
40
41   # Display basic info for the shell sort
42   print("The sorted list is: ", list)
43   print("The number of comparisons is: ", comparisons)
44   print("The elapsed time in seconds is: ", (endTime - startTime))
```

Output 6.3.4:

```
Enter the number of elements in the list:10
The unsorted list is:  [-47, 79, -79, 94, -79, -97, -7, -3, 49, 88]
The sorted list is:  [-97, -79, -79, -47, -7, -3, 49, 79, 88, 94]
The number of comparisons is =   10
The elapsed time in seconds =   0.0
```

While the efficiency of the algorithm may not be instantly noticeable, it does make a difference when examined more closely. The following list of approximate results showcases the performance difference between insertion sort and shell sort (assume *1 comparison* takes *1 nanosecond* or *1.0e−9 seconds*; *Cs* stands for *Comparisons*):

* $n = 10$: 8 Cs (up to 40 in Insertion S.) → approx. 3.8e−4 seconds (2.0e−4 in Insertion S.)
* $n = 100$: 4e2 Cs (up to 4.0e3 in Insertion S.) → approx. 3.8e−3 seconds (2.5e−3 in Insertion S.)
* $n = 1,000$: 1.5e4 Cs (up to 5.0e5 in Insertion S.) → approx. 0.27 seconds (0.16 seconds in Insertion S.)
* $n = 10,000$: 1.7e5 Cs (4.0e7 Cs in Insertion S.) → approx. 26 seconds (15 seconds in Insertion S.)
* $n = 20,000$: 3.4e5 Cs (9.8e7 Cs in Insertion S.) → approx. 99 seconds (57 seconds in Insertion S.)
* $n = 30,000$: 5 e5 Cs (2.2e8 Cs in Insertion S.) → approx. 215 seconds (125 seconds in Insertion S.)

6.3.5 SHAKER SORT

The *shaker sort* algorithm is based on the bubble sort, but instead of the list being read always on the same direction, consequent readings occur in opposite directions. This ensures that both the highest and lowest value elements of the list move to the correct positions faster. The main disadvantage of this algorithm is that, since it is based on bubble sort, its time complexity is bound to $O(n^2)$.

Observation 6.10 – Shaker Sort: Use two separates for loops nested inside a while loop to read a list of elements in opposite directions. This ensures that the elements will be positioned to the correct places in the list faster than with bubble sort.

The following list provides approximate comparisons between the shaker and the bubble sort. The examples support the argument that it is not worth using this algorithm unless the size of the list falls within the approximate range of 1,000–50,000 elements. For lists with more elements than the upper threshold of this range (50,000), using the shaker sort is impractical (as in previous examples, *1 comparison* takes *1 nanosecond* to complete and *1 nanosecond = 1.0e−9 seconds*):

- $n = 10$: ~40 Cs ($n^2 = 81$ in Bubble S.) → approx. 7.7e−4 seconds (3e−4 in Bubble S.)
- $n = 100$: ~4.2e3 Cs ($n^2 = 9.8e3$ in Bubble S.) → approx. 3.2e−3 seconds (4.5e−3 in Bubble S.)
- $n = 1,000$: ~3.9e5 Cs ($n^2 = 9.98e5$ in Bubble S.) → approx. 0.28 seconds (0.37 in Bubble S.)
- $n = 10,000$: ~3.8e7 Cs ($n^2 = 9.998e7$ in Bubble S.) → approx. 28 seconds (46 in Bubble S.)
- $n = 20,000$: ~1.5e8 Cs ($n^2 = 2.0e8$ in Bubble S.) → approx. 110 seconds (188 in Bubble S.)

In general, the time complexity of the algorithm for the *average* and *worst* cases are $O(n^2)$, while slight improvements can potentially lead to a running time complexity of $O(n)$ at best.

As an example, let us consider the same list as the one used with bubble sort: 2, 3, 1, 6, 7. During the 1st outer loop, shaker sort will execute two inner iterations successively, with one iteration processing the list to the right and one to the left. Each time an inner loop processes the list to the right, the pointer at the end of the list is reduced by one. Similarly, each time it processes the list to the left, the pointer at the start of the list is increased by one. Starting with the *1st outer iteration*, the inner loop presented in Table 6.5 (processing the list to the right) will take place.

Likewise, in the 1st outer iteration, the inner loop presented in Table 6.6 will process the list to the left.

The reader should note that, at the end of each outer iteration, the highest value element of the current sub-list is pushed to the end of the sub-list and the lowest is pushed to the start. Table 6.7 presents the results of each of the outer iterations. Note that the algorithm will stop at the end the first inner iteration of the *3rd outer pass*, as there are no more swaps to be made:

TABLE 6.5
The First Inner Loop within the First Main Iteration, Reading the List to the Right

3	5	4	2	3	1	6	7
3	5	4	2	3	1	6	7
3	4	5	2	3	1	6	7
3	4	2	5	3	1	6	7
3	4	2	3	5	1	6	7
3	4	2	3	1	5	6	7
3	4	2	3	1	5	6	7

TABLE 6.6
The Second Inner Loop within the First Main Iteration, Reading the List to the Left

3	4	2	3	1	5	6	7
3	4	2	3	1	5	6	7
3	4	2	3	1	5	6	7
3	4	2	1	3	5	6	7
3	4	1	2	3	5	6	7
3	1	4	2	3	5	6	7
1	3	4	2	3	5	6	7

TABLE 6.7

The Results of the Outer Loops

After the 1st pass	1	3	4	2	3	5	6	7
After the 2nd pass	1	2	3	3	4	5	6	7
After the 1st inner of the 3rd outer pass	1	2	3	3	4	5	6	7

The following script demonstrates an implementation of the shaker sort and its output:

```
1    # Import the random module to generate random numbers
2    import random
3    import time
4
5    comparisons = 0
6    list = []
7
8    # Enter the number of list elements
9    size = int(input("Enter the number of list elements: "))
10   # Use the randint() function to generate random integers
11   for i in range (size):
12       newNum = random.randint(-100, 100)
13       list.append(newNum)
14   print("The unsorted list is: ", list)
15
16   # Start the timer
17   startTime = time.process_time()
18
19   # The shaker sort algorithm
20   swapped = True; start = 0; end = size - 1
21
22   # Keep running the shaker sort while swaps are taking place
23   while (swapped == True):
24       # Set swap to false to start the new loop
25       swapped = False;
26
27       # Loop from left to right using bubble sort
28       for i in range(start, end):
29           comparisons += 1
30             if (list[i] > list[i+1]):
31               temp = list[i]; list[i] = list[i+1]; list[i+1] = temp
32               swapped = True;
33       # If there were no swaps, the list is sorted
34       if (swapped == False):
35           break
36       # If at least one swap, then reset swap to false and continue
37       else:
38           swapped = False
39
```

```
40      # Decrease the end of the list to -1, since largest element moved
41      # to the right
42      end -= 1
43
44      # Loop from right to left using bubble sort
45      for i in range (end, start, -1):
46          comparisons += 1
47          if (list[i] < list[i-1]):
48              temp = list[i];  list[i] = list[i-1];  list[i-1] = temp
49              swapped = True
50
51      # Increase the start of the list by 1 since smallest element moved
52      # to the left
53      start += 1
54
55  # End the timer
56  endTime = time.process_time()
57
58  # Display the sorted list
59  print("The sorted list is: ", list)
60  print("The number of comparisons is: ", comparisons)
61  print("The elapsed time in seconds: ", (endTime - startTime))
```

Output 6.3.5:

```
Enter the number of elements in the list:15
The unsorted list is: [98, -23, -29, 17, -11, 2, 77, -20, -53, 66, -2, 33,
63, 33, 68]
The sorted list is: [-53, -29, -23, -20, -11, -2, 2, 17, 33, 33, 63, 66, 68,
77, 98]
The number of comparisons is =  77
The elapsed time in seconds =  0.0
```

6.4   RECURSION, BINARY SEARCH, AND EFFICIENT SORTING WITH LISTS

On a broader context, any attempt to find an algorithm that addresses the problem of sorting a list efficiently is subject to certain restrictions. This is due to the fact algorithms generally fall within the same time complexity of $O(n^2)$, as a result of their *inherent nested loop structures*. As shown in the previous sections, this is true even when improved and optimized versions of the algorithms are used. In order to improve the efficiency of sorting algorithms further, *recursion* must be adopted. This section presents and discusses the concept of recursion, and uses it as a base to implement some common related algorithmic ideas like *binary search* and *factorial*. Subsequently, two notable algorithms that address the problem of sorting large lists in an efficient way are presented: *merge sort* and *quick sort*.

6.4.1   RECURSION

By definition, a *recursive function* is *one that calls itself*. The basic idea is to break a large problem into several smaller parts that are equivalent to the original. These are further broken down successively into even smaller parts, until the problem is small enough for its solution to become evident.

This final point is called a *terminal* or *base case*. The condition that must be met in order to achieve the terminal case is called the *terminal condition*. The associated step followed to break down the problem into smaller parts is called the *basic step*.

In order to contextualize the idea of recursion, one needs to break down what happens on a recursive function call:

Observation 6.11 – Recursion: A *recursive* function is one that calls itself. It takes a large problem and breaks it into smaller ones successively, following a *step*. The step is repeated until the smaller parts are so small that the solution is evident. The final and smallest part is referred to as the *terminal* or *base case*.

- Firstly, the compiler/interpreter passes a parameter to the function.
- The called function and its parameter is pushed to the *program stack* (stacks are discussed in Section 6.5.5), a separate place in memory where the local variables are stored until this particular function call is completed.
- The compiler/interpreter records the *return address*, which will be used as a return to the *calling* function when the current function call is complete.
- When the current function call is complete, the compiler/interpreter records the value to be returned to the calling function (if applicable).

In terms of its results, recursion is similar to the iteration explained in Chapter 2, but differs in terms of the functions used. An iterative algorithm uses a *looping* construct whereas a recursive algorithm uses a *branching* structure. In terms of both time and memory usage, recursive solutions are often less efficient than their iterative counterpart. However, in many occasions they are the only solutions available. Their main advantage is that by simplifying the solution to a single problem they often result in shorter and more readable source code.

The following script presents a basic recursive function that calls itself continuously and indefinitely, printing a particular message:

```
1   def message():
2   print("This is a recursive function")
3   message()
4
5   message()
```

Output 6.4.1.a:

```
This is a recursive function
This is a recursive function
This is a recursive function
This is a recursive function
---------------------------------------------------------------
RecursionError                            Traceback (most recent call last)
<ipython-input-1-e0c7cc045453> in <module>
```

To prevent the function from falling into this *infinite call loop*, the number of repetitions must be controlled. This can be achieved by incorporating the following two steps:

- A *dividing step* must be applied to a subset of the original values in each repetition.
- The *terminal* or *basic case* must be defined and calculated (if applicable).

The following script is a modified version of the message() function presented above. It passes an integer argument that dictates the number of times the function will call itself before the terminal case:

```
1    # The recursive function
2    def message(times):
3        print("Message called with times=", times)
4
5        # Define the dividing step through an if statement
6        if (times > 0):
7            print("\tThis is a recursive function.\n")
8            message(times -1)
9
10   # The terminal or base case stops recursion & "roll back"
11       print("Message returning with times=", times, "\n")
12
13   # Start the recursion by calling the recursive function
14   message(3)
```

Output 6.4.1.b:

```
Message called with times =  3
        This is a recursive function.

Message called with times =  2
        This is a recursive function.

Message called with times =  1
        This is a recursive function.

Message called with times =  0
Message returning with times =  0

Message returning with times =  1

Message returning with times =  2

Message returning with times =  3
```

The application of recursion can be also considered in the context of a purely mathematical function, that of the *factorial*. The complete definition of the factorial is $f(n) = n * f(n-1)$ for $n > 1$, and $f(1) = 1$ for $n = 1$. According to this definition, for $f(4)$ the result would be calculated as follows:

$$f(4) = 4 * f(3) = 4 * 3 * f(2) = 4 * 3 * 2 * f(1) = 4 * 3 * 2 * 1 = 24.$$

Notice that in the case of $f(1)$ there is no further breakdown of the function, as this is considered the *terminal* or *base case* with a result of $f(1) = 1$. The following script implements the solution of the factorial:

```
1    # The factorial function using recursion
2    def factorial(n):
3        # The terminal or base case
4        if (n == 1):
5            return 1
6        # The recursive step
7        else:
8            print(n, "* f(", n-1, ")")
9            return n * factorial(n-1)
10
11   num = int(input("Enter the number to find its factorial: "))
12   print("The factorial for", num, "is ", factorial(num))
```

Output 6.4.1.c:

```
Enter the number to find its factorial: 1
The factorial for 1 is   1

Enter the number to find its factorial: 3
3 * f( 2 )
2 * f( 1 )
The factorial for 3 is   6

Enter the number to find its factorial: 7
7 * f( 6 )
6 * f( 5 )
5 * f( 4 )
4 * f( 3 )
3 * f( 2 )
2 * f( 1 )
The factorial for 7 is   5040
```

6.4.2 BINARY SEARCH

One of the most well-known applications of recursion is the *binary search*. The main idea behind binary search is to find whether a *word exists in a dictionary*. The necessary precondition is to *use it on a sorted list*, regardless of the algorithm used for the sorting. The concept is rather simple:

Observation 6.12 – Binary Search: A recursive algorithm applied to *sorted lists* in order to find the location of a particular element.

- Initially, the algorithm checks whether the word in the middle element of the list exists.
- If it does not and the middle element value is larger than the search value, the list is split into two halves and the middle element of the first half is checked; otherwise, the middle element of the second half is checked.
- The algorithm continues until the desired element is found, in which case the element and its position in the list are reported. If the search element is not found, a relevant message is generated.

An implementation of the binary search algorithm is provided below:

```
# The recursive function for binary search
binarySearch(word, startPage, endPage)
      # if the dictionary consists of one page (base case) search for it in
      # that page
      if startPage = endPage
            search the word in the startPage
      else
            # get to the middle of the dictionary
            middlePage = (endPage + startPage)/2
            # determine which half of the dictionary might contain
            # the chosen word
            # if the word is in the first half
            if the word is located before the middlePage
                  # find the word in the first half of the dictionary
                  binarySearch(word, startPage, middlePage)
            else
                  # find the word in the second half of the dictionary
                  binarySearch(word, middlePage+1, endPage)
```

In this particular algorithm, function *binarySearch* calls itself recursively. At each call, the problem gets smaller as the size is halved. The base case is the *startPage = endPage* statement that dictates that either the word is found or it does not exist in the dictionary.

The following script implements the algorithm:

```
1   # The list of numbers to search in
2   listOfNumbers = [1, 2, 3, 4, 5, 6, 7, 8, 9, 10]
3
4   # The recursive function for binary search
5   def binarySearch(number, startPage, endPage):
6       # If the list consists of one page (base case) search for it
7       # in that page
8       if (startPage == endPage):
9               if (listOfNumbers[startPage] == number):
10                  print("The number was found in the list in "
11                        "position: ", startPage)
12              else:
13                  print("The number was not found in the list")
14      else:
15          # Split the list using the middle point as a reference
16          middlePage = int((endPage + startPage)/2)
17          # Determine which half of the list might contain the number
18          # If the number is in the first half
19          if (number <= listOfNumbers[middlePage]):
20              # Find the number in the first half of the list
21              binarySearch(number, startPage, middlePage)
22          else:
23              # Find the number in the second half of the list
24              binarySearch(number, middlePage + 1, endPage)
25
26  num = int(input("Enter the number to find in the list: "))
```

```
27
28  # Call the binarySearch function
29  binarySearch(num, 0, 9)
```

Output 6.4.2:

```
Enter the number to find in the list: 7
The number was found in the list in position: 6


Enter the number to find in the list: 23
The number was not found in the list
```

6.4.3 QUICKSORT

Quicksort is considered as one of the more advanced sorting algorithms for lists (i.e., static objects), with a better average performance than insertion, selection, and shell sort. It was presented by Hoare in 1962 (Hoare, 1961). Quicksort belongs to a well-known and highly regarded family of algorithms adopting the *divide and conquer* strategy.

The algorithm sorts a list of n elements by picking a key value k in the list as a *pivot* point, around which the list elements are then rearranged. Finding or calculating the ideal pivot point is key, although not absolutely necessary. The pivot point should be either the *median* or close to the median key value, so that the numbers of preceding and succeeding elements in the list are balanced.

Once this pivot key (k) is decided, the elements of the list are rearranged so that those with lower values appear before it and those with higher values after it. Once this process is completed, the list is partitioned into two sub-

> **Observation 6.13 – Quicksort:** Select an element in the list as the *pivot k* element and rearrange the rest so that lower value elements precede it and higher succeed it (or the opposite). Apply the same process to the two resulting sub-lists repeatedly, until there are no more lists to divide. By definition, at the end of this process the list will be sorted.

lists: one containing all values lower than k and one containing k itself (in its original position in the list) plus all values higher than k. This process is applied recursively to the two sub-lists and all subsequent sub-lists created based on them until there are no lists to divide. Once this process is complete, the list is sorted by definition.

As an example, let us consider the following list: *37, 2, 6, 4, 89, 8, 10, 12, 68, 45*. The first element (i.e., *list[0]: 37*) is taken as the pivot element (k). The process will start with the rightmost element of the list, moving in a decremental order from that point on (i.e., *list[9]: 45, list[8]: 68, list[7]: 12*). Each element is compared with k until an element with a lower value is found. In this instance, the process will stop at *list[7]: 12* and this element will be swapped with k (Table 6.8).

TABLE 6.8

The First Round of Comparisons at the Right of the List and Towards the Pivot Element

37	2	6	4	89	8	10	12	68	**45**
37	2	6	4	89	8	10	12	**68**	45
37	2	6	4	89	8	10	**12**	68	45
12	2	6	4	89	8	10	**37**	68	45

TABLE 6.9

The First Round of Comparisons at the Left of the List and Towards the Pivot Element

12	**2**	6	4	89	8	10	**37**	68	45
12	2	**6**	4	89	8	10	**37**	68	45
12	2	6	**4**	89	8	10	**37**	68	45
12	2	6	4	**89**	8	10	**37**	68	45
12	2	6	4	**37**	8	10	**89**	68	45

TABLE 6.10

The First Round of Comparisons Resumes at the Right of the Pivot Element

12	2	6	4	**37**	8	**10**	89	68	45
12	2	6	4	**10**	8	**37**	89	68	45

TABLE 6.11

The First Round of Comparisons Resumes and Finishes at the Left

12	2	6	4	37	**8**	**10**	89	68	45
12	2	6	4	10	8	**37**	89	68	45

Next, the k (37) will be compared with the elements on its left, beginning after 12. The comparisons will continue in an increasing order until an element greater than 37 is found. This will happen for value 89, so 37 and 89 will be swapped (Table 6.9).

After the swap, the process will resume at the left of the previously swapped element (89) and at the right of pivot element k. The first element that will be considered is 10, which is smaller than the pivot element, thus, the two elements will be swapped. The rearranged list is shown in Table 6.10 below.

Finally, the process will start again at the left of the sub-list with 37 as the pivot, and begin with the element after 10. This time, the only remaining element to compare (8) is lower than 37 so no swap will take place between the two elements. This first round of comparisons will end with the 1st pivot element (37) placed in its final place in the list, leaving two unsorted sub-lists on its left and right sides (Table 6.11).

This is the first partitioning of the list into the first two unsorted sub-lists. The exact same comparison process will be next applied to both the left and right sub-lists recursively. When all comparisons and partitions are complete there will be no further sub-lists left to sort and the entire list will be sorted.

The algorithm may seem rather complicated and its efficiency difficult to gauge. Nevertheless, it is indeed much more efficient than all the previously discussed algorithms. A script implementing the quicksort algorithm is provided below:

```
1    # Import the random and time modules
2    # to generate random numbers and keep time
3    import random
4    import time
5    global comparisons
6    list = []
```

```
7
8    # The quicksort algorithm
9    def quickSortReadings(list, start, end):
10       global comparisons
11       pivot = list[start]
12       low = start + 1
13       high = end
14
15       while (True):
16           # Compare elements from the right to find one
17           # that is smaller than the pivot. Stop when one is found
18           while (low <= high and list[high] >= pivot):
19               high -= 1; comparisons += 1
20
21           # Compare elements from the left to find one
22           # that is larger than the pivot. Stop when one is found
23           while (low <= high and list[low] <= pivot):
24               low += 1; comparisons += 1
25
26           # If an element larger or smaller than the pivot is found
27           # swap elements to put things in order & continue the process
28           if (low <= high):
29               list[low], list[high] = list[high], list[low]
30           # Stop and exit if the low index moved beyond the high index
31           else:
32               Break
33
34       list[start], list[high] = list[high], list[start]
35
36       return high
37
38   def quickSortPartition(list, start, end):
39       if start >= end:
40           Return
41
42       p = quickSortReadings(list, start, end)
43       quickSortPartition(list, start, p - 1)
44       quickSortPartition(list, p + 1, end)
45
46   # Enter the number of list elements
47   size = int(input("Enter the number of list elements:"))
48   # Use the randint() function to generate random integers
49   for i in range (size):
50       newNum = random.randint(-100, 100)
51       list.append(newNum)
52   print("The unsorted list is: ", list)
53
54   comparisons = 0
55
56   # Start the timer
57   startTime = time.process_time()
```

```
58
59  quickSortPartition(list, 0, size -1)
60
61  # End the timer
62  endTime = time.process_time()
63
64  # Display the sorted list
65  print("The sorted list is: ", list)
66  print("The number of comparisons is=", comparisons)
67  print("The elapsed time in seconds=", (endTime - startTime))
```

Output 6.4.3:

```
Enter the number of elements in the list:10
The unsorted list is:  [-94, -1, -35, 13, -73, 18, 4, 29, 46, -62]
The sorted list is:  [-94, -73, -62, -35, -1, 4, 13, 18, 29, 46}
The number of comparisons is =  26
The elapsed time in seconds =  0.0
```

The following estimates provide a rough comparison between quicksort and bubble sort, highlighting the fact that the former operates at a completely different efficiency level and, thus, being capable of processing much larger lists. The only possible restrictions in relation to its use have to do with the power of the computer system used and the available memory, as these are determining factors when running recursive calls on lists larger than 100,000 elements (a comparison takes *1 nanosecond* to complete and *1 nanosecond=1.0e−9 seconds*):

- $n = 10$: ~30 Cs ($n^2 = 81$ in Bubble S.) → approx. 1.8e−4 seconds (3e−4 in Bubble S.)
- $n = 100$: ~6.2e2 Cs ($n^2 = 9.8e3$ in Bubble S.) → approx. 4e−4 seconds (4.5e−3 in Bubble S.)
- $n = 1,000$: ~1e4 Cs ($n^2 = 9.98e5$ in Bubble S.) → approx. 9.7e−3 seconds (3.7e−1 in Bubble S.)
- $n = 10,000$: ~3e5 Cs ($n^2 = 9.998e7$ in Bubble S.) → approx. 0.1 seconds (46 in Bubble S.)
- $n = 20,000$: ~1e6 Cs ($n^2 = 2.0e8$ in Bubble S.) → approx. 0.3 seconds (188 in Bubble S.)
- $n = 30,000$: ~3e6 Cs ($n^2 = 2.0e8$ in Bubble S.) → approx. 0.6 seconds (Not practical in Bubble S.)
- $n = 100,000$: ~2e7 Cs ($n^2 = 2.0e8$ in Bubble S.) → approx. 5.6 seconds (Not practical in Bubble S.)
- $n = 300,000$: ~1.8e8 Cs ($n^2 = 2.0e8$ in Bubble S.) → approx. 48 seconds (Not practical in Bubble S.)

In terms of time complexity, while the worst cases run at $O(n^2)$, the average and best cases run at the much more efficient level of $O(n \log(n))$.

6.4.4 MERGE SORT

Merge sort is another advanced algorithm for efficient sorting of large lists, falling into the same divide and conquer approach as quicksort. Merge sort is an excellent choice for sorting data that cannot be kept on the computer memory all at once and are, thus, kept in *secondary storage*.

The essential idea behind merge sort is to split lists into two halves continuously until all sub-lists

Observation 6.14 – Merge Sort: A *divide and conquer* algorithm for sorting *static* lists. The basic idea is to divide the list into two sub-lists repeatedly, until all sub-lists consist of a single element. The divided lists are then merged again following a particular sorting procedure.

consist of a single element and, subsequently, merge the sub-lists while also ordering their elements. Algorithmically, the process is rather straightforward, particularly for the split part. The process the programmer must follow for merging each given set of two sub-lists is summarized below:

- Check if the first sub-list is empty.
- If not, check if the second sub-list is empty.
- If not, compare the first available element in the first sub-list with the first available element in the second sub-list.
- Whichever of the two elements has a lower value must be placed in the first available slot of a new merged list.
- This process should be repeated for all remaining elements of the two sub-lists.
- If all the elements of one of the sub-lists have been used, place the remaining elements of the other sub-list to the new merge list, in the order they appear in the sub-list.
- Recursively repeat this process until all the sub-lists are merged into one ordered merged list.

As an example, let us consider the following list: *25, 13, 9, 32, 17, 5, 33, 25, 43, 21*. Firstly, the list is split into the required set of sub-lists:

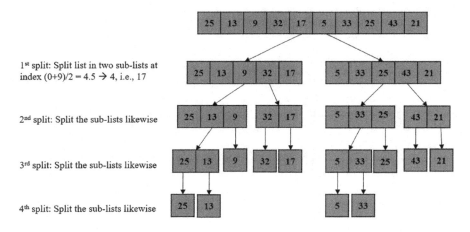

Next, the lists are merged on a bottom-up basis, as shown below:

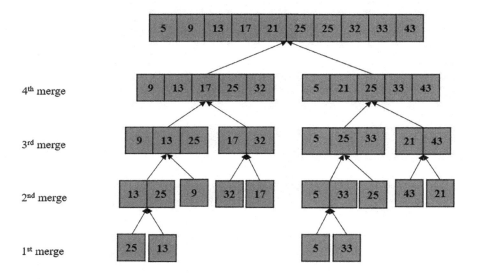

The following script provides an implementation of the merge sort algorithm:

```
1    # Random and time modules generate random numbers & keep time
2    import random
3    import time
4    global comparisons, i, j, k
5    global list
6
7    # Merge two sub-lists, list[first, middle] and list[middle+1, last]
8    def merge(first, middle, last):
9        global list
10       global i, j, k, comparisons
11       size1=middle - first+1; size2=last - middle
12
13       # Create temporary lists
14       leftList=[]; rightList=[]
15
16       # Copy original list to temporary lists leftList & rightList
17       for i in range(0, size1):
18           leftList.append(list[first+i])
19       for j in range(0, size2):
20           rightList.append(list[middle+1+j])
21
22       # Merge temp lists leftList & rightList into original list
23       # until one of the sub-lists is empty
24       i=0; j=0; k=first
25       while (i<size1 and j<size2):
26           if (leftList[i] <= rightList[j]):
27             list[k]=leftList[i]; i += 1; comparisons += 1
28           else:
29             list[k]=rightList[j]; j += 1; comparisons += 1
30           k += 1
31
32       # If list becomes empty, copy remaining elements to original
33       while (i<size1):
34           list[k]=leftList[i]; i += 1; k += 1
35
36       # If list becomes empty, copy remaining elements to original
37       while (j<size2):
38           list[k]=rightList[j]; j += 1; k += 1
39
40   # The merge sort algorithm
41   def mergesort(first, last):
42       global list
43
44       # The recursive step
45       if (first <= last-1):
46           middle=(first+last)//2
```

```
47              mergesort(first, middle)
48              mergesort(middle+1, last)
49              merge(first, middle, last)
50
51   list=[]
52   # Initialize the indices of the sub-lists
53   i, j, k=0, 0, 0
54
55   # Enter the number of list elements
56   size=int(input("Enter the number of list elements: "))
57   # Use the randint() function to generate random integers
58   for i in range (size):
59          newNum=random.randint(-100, 100)
60          list.append(newNum)
61   print("The unsorted list is: ", list)
62
63   comparisons=0
64
65   # Start the timer
66   startTime=time.process_time()
67
68   mergesort(0, size-1)
69
70   # End the timer
71   endTime=time.process_time()
72
73   # Display the sorted list
74   print("The sorted list is: ", list)
75   print("The number of comparisons is=", comparisons)
76   print("The elapsed time in seconds=", (endTime - startTime))
```

Output 6.4.4:

```
Enter the number of elements in the list:15
The unsorted list is: [83, -3, 89, 64, -5, 65, 78, 17, 8, -3, 82, 89, -80, 23, 64]
The sorted list is: [-80, -5, -3, -3, 8, 17, 23, 64, 64, 65, 78, 82, 83, 89, 89]
The number of comparisons is =  42
The elapsed time in seconds =  0.0
```

The efficiency of the algorithm in sorting static lists is comparable to that of quicksort (a comparison takes *1 nanosecond* to complete; *1 nanosecond = 1.0e−9 seconds*):

- $n=10$: ~20 Cs (30 in Quicksort) → approx. 2e−4 seconds (1.8e−4 in Quicksort)
- $n=100$: ~5.4e2 Cs (6.2e2 Cs in Quicksort) → approx. 0.0012 seconds (1.2e−2 in Quicksort)
- $n=1,000$: ~8.6e3 Cs (1e4 Cs in Quicksort) → approx. 0.015 seconds (9.7e−3 seconds in Quicksort)
- $n=10,000$: ~1.2e5 Cs (3e5 Cs in Quicksort) → approx. 0.15 seconds (0.1 seconds in Quicksort)
- $n=30,000$: ~4e5 Cs (3e6 in Quicksort) → approx. 0.44 seconds (0.6 seconds in Quicksort)
- $n=100,000$: ~1.5e6 Cs (2e7 in Quicksort) → approx. 1.6 seconds (5.6 seconds in Quicksort)
- $n=300,000$: ~5e6 Cs (1.8e8 in Quicksort) → approx. 5.5 seconds (48 seconds in Quicksort)

In general, merge sort is more efficient than quicksort as it runs on *O(n logn)* time complexity in all cases (i.e., *best*, *average*, and *worst* case). Most importantly, it becomes significantly better as the size of the list grows larger (e.g., lists consisting of hundreds of thousands of elements or higher) depending on the power, memory, and settings of the system it runs on.

6.5 COMPLEX DATA STRUCTURES

In the previous sections, the focus was on the implementation of sorting by means of relatively simple, *static data structures*, like lists. When it comes to more advanced, real-life applications more complex data structures may be required. This section addresses such data structures, which can take both *linear* and *non-linear* forms (Figure 6.1).

In linear structures, such as *stacks*, *queues*, and *linked lists*, each element occupies a position that is relative to that of previous and succeeding elements within the structure. Consequently, the structure is *traversed* (i.e., read) sequentially. In non-linear structures, such as *trees* and *graphs*, the items are not arranged in a particular, hierarchical order, thus, sequential traverse is not feasible. Non-linear structures are more complex to implement, but they are also more powerful. As such, they are used extensively in real-life applications.

6.5.1 STACK

A *stack* is an ordered list with two ends, the *top* and the *base*. New items are always inserted at the top end in an operation called *push*. Items are also removed from the top end, in what is referred to as *pop*. In a stack, the last item to push is always the first to pop, hence a stack is also called a *last in, first out (LIFO)* list. Besides the item at the top, other items in the stack are not directly accessible. As an analogy, one can think of a stack as a pile of plates stacked upon each other. Each new plate is placed at the top of the pile. In order to be used, a plate is also taken from the top of the pile.

> **Observation 6.15 – Stack:** An ordered, linear list structure with two ends: *top* and *base*. Items are *pushed* to and *popped* from the *top*, and the *last item pushed in the stack is the first to be popped out (LIFO)*. The operations performed on the stack are the following: *initialize, push, pop, isEmpty, top,* and *size*.

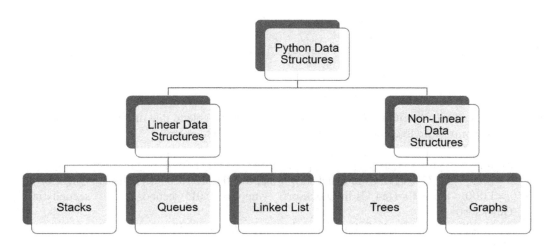

FIGURE 6.1 Classification of data structures.

From a more formal, technical perspective, the *stack ADS (Abstract Data Structure)* consist of the following:

- An index pointing at the top item in the stack, with values ranging from *0* to its *maximum size −1*.
- The body of the stack that stores the values (i.e., the actual data of the list).
- *Initialize* – init(s): A function that initializes the stack (i.e., creating an empty list).
- *Empty* – isEmpty(s): A function that checks whether the stack (s) is empty.
- *Push* – push(x, s): A function that pushes a new item (x) onto the stack (s).
- *Pop* – pop(x, s): A function that deletes the top item (x) from the stack (s).
- *Top* – top(s): A function that returns the item at the top of the stack.
- *Size* – size(s): A function that returns the total number of items in the stack.

The following Python class (filename: *Chapter6Stack.py*) defines the stack structure (*stack ADS*):

```python
class Stack:
    def __init__(self):
        self.items=[]
    def push(self, item):
        self.items.append(item)
    def pop(self):
        return self.items.pop()
    def isEmpty(self):
        return self.items == []
    def top(self):
        if (not self.isEmpty()):
            return self.items[-1]
    def size(self):
        return len(self.items)
    def show(self):
        return self.items
```

Since the class in this form is rather generic, it can be used for a variety of stack-based applications. The following script imports the stack class from *Chapter6Stack.py* in order to implement a simple example of the functionality of the stack:

```python
1    import Chapter6Stack
2
3    fruits=Chapter6Stack.Stack()
4
5    # Confirm that the stack is empty
6    if (fruits.isEmpty() == True):
7        print ("The stack is empty")
8
9    # Push elements to the stack
10   fruits.push('apple')
11   fruits.push('orange')
12   fruits.push('banana')
13
```

```
14  # Confirm that the stack is not empty and print its contents
15  if (fruits.isEmpty()!= True):
16      print("The stack is not empty: It's size is: ", fruits.size())
17      print("The contents of the stack are: ", fruits.show())
18
19  # Return the top item of the stack
20  print("The top item of the stack is: ", fruits.top())
21  # Remove the top item of the stack, print the new top item and the stack
22  print("Remove the top item of the stack: ", fruits.pop())
23  print("The top item of the stack is now: ", fruits.top())
24  print("The contents of the stack now are: ", fruits.show())
```

Output 6.5.1.a:

```
The stack is empty
The stack is not empty: It's size is:  3
The contents of the stack are:  ['apple', 'orange', 'banana']
The top item of the stack is:  banana
Remove the top item of the stack:  banana
The top item of the stack is now:  orange
The contents of the stack now are:  ['apple', 'orange']
```

Stacks are used extensively in computer programs. A rather common example is storing page visits on a web browser. Every page that is visited is added to a stack and when the user clicks on the back button the last page visited is retrieved from the stack. A similar use can be found in the *undo* function included in most computer applications. A stack is used to store all the tasks performed in the application and when the user clicks on the respective button, the last action is retrieved from the stack and its action is reversed. Stacks are also useful in *evaluating expressions*, *backtracking*, and *implementing recursive function calls*.

As an example of a practical use of the stack, let us consider the common utility task of converting a decimal number into binary. The algorithm is quite simple: repeatedly divide the decimal number by 2 until the result is 0, while pushing the remainder of the integer division to the stack. At the end of the process, all the items are popped from the stack to get the binary representation of the decimal number. Assuming that the integer to be converted is number *21*, the above procedure will result in binary number *10101* (Figure 6.2).

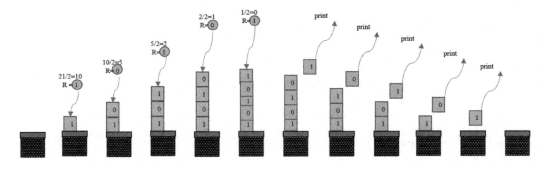

FIGURE 6.2   Decimal to binary number conversion.

The following script implements the stack structure, utilizing *Stack ADS* (*Chapter6Stack.py*) as in the previous example:

```
1    import Chapter6Stack
2
3    # decimal object implements the conversion using the stack
4    decimal = Chapter6Stack.Stack()
5
6    # Accept an integer to convert to binary form
7    userInput = int(input("Enter the integer to convert to binary: "))
8
9    # Repeatedly divide by 2; keep pushing the remainder to the stack
10   while (userInput > 0):
11       decimal.push(userInput % 2)
12       userInput = userInput//2
13
14   # Confirm that the stack is not empty and print its contents
15   if (decimal.isEmpty() != True):
16       print("The stack is not empty: It's size is: ", decimal.size())
17       print("The contents of the stack are: ", decimal.show())
18
19   # Return the number in binary form
20   print("The binary form of the number is: ", end='')
21   for i in range (decimal.size()):
22       print(decimal.pop(), end='')
```

Output 6.5.1.b:

```
Enter the integer to convert to binary: 56
The stack is not empty: It's size is:  6
The contents of the stack are:  [0, 0, 0, 1, 1, 1]
The binary form of the number is: 111000
```

6.5.2 INFIX, POSTFIX, PREFIX

Another application of a stack that is particularly important in computer science is the *evaluation of arithmetic expressions*. In general, the reader should be aware of the fact that there are three kinds of arithmetic notations, namely *infix*, *prefix*, and *postfix*. *Infix* is what humans are mostly used to, as it involves a *binary operator* appearing between two *operands* and determining the type of operation that will take place between them (e.g., *3 + 5*). In a *prefix* notation, the same expression would be converted to *+ 3 5*, where the operator precedes both operands. Likewise, the *postfix* notation would take the form *3 5 +*, with the operator succeeding the two operands. It must be noted that the postfix notation is the one used by compilers when evaluating an arithmetic expression. As such, the conversion of an infix expression that *humans would understand more easily* to a postfix expression that can be *evaluated by compilers* is a rather important task in computer science. The implementation of such a conversion poses three main problems that must be addressed:

Observation 6.16 – Infix, Postfix, Prefix: Three different kinds of notations used to evaluate arithmetic expressions by humans or computers.

- In an infix expression, the operation precedence is forcing multiplication/division to apply before the additions/subtractions, whereas in a postfix expression there is no operator priority.
- When translating an infix to a postfix expression, only the placement of the operators is different. An algorithm that translates from infix to postfix only needs to shift the operators to the right, and possibly reorder them.
- Postfix expressions do not take parentheses.

The following algorithm uses a stack to temporarily store the operators until they can be inserted to the right position into the postfix expression:

- Initialize the stack.
- Scan the infix expression from left to right.
- While the scanned character is valid:
  - If the character is an operand, move it directly to the postfix expression.
  - If the character is an operator, compare it with the operator at the top of the stack.
    - While the operator at the top of the stack is of higher or equal priority than the character just encountered, and is not a left parenthesis character, pop the operator from the stack and move it to the postfix expression. Once all the operators are popped, push the current character/operator to the stack.
  - If the character is a left parenthesis, push the character onto the stack.
  - If the character is a right parenthesis, pop and move the operators off the stack to the postfix expression. Pop the left parenthesis and ignore it.
  - If the operator at the top of the stack is of a lower priority than the character just encountered or if the stack is empty, push the character that was just encountered to the stack.
- After the entire infix expression has been scanned, pop any remaining operators from the stack and move them to the postfix expression.

As an example, Figure 6.3 illustrates the use of a stack to convert infix expression *2 + 3 x 5 + 4* into postfix.

- 2+3=5 → 2 3 +=5
- 2 x 5+3=13 → 2 5 x 3 +=13
- 2+5 x 3=17 → 2 5 3 x 3=17
- 2 x 3+5 x 4=26 → 2 3 x 5 4 x +=26
- 2+3 x 5+4=21 → 2 3 5 x+4 +=21

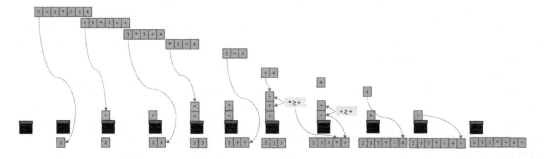

FIGURE 6.3 Infix expression remaining to be evaluated.

Figures 6.4 and 6.5 demonstrate a more complex case of an infix to postfix expression conversion that includes operators in parentheses: $2 \times (7 + 3 \times 4) + 6$.

The evaluation of a postfix expression utilizes the steps described in the algorithm below:

- Scan the postfix expression from left to right.
- If an operand is encountered, push it to the stack.
- If an operator is encountered, apply it to the top two operands of the stack and replace the two operands with the result of the operation.
- After scanning the entire postfix expression, the stack should have one item, which is the value of the expression.

Figure 6.6 illustrates how expression $1\ 6 + 5\ 2 - \times$ is evaluated using a stack.

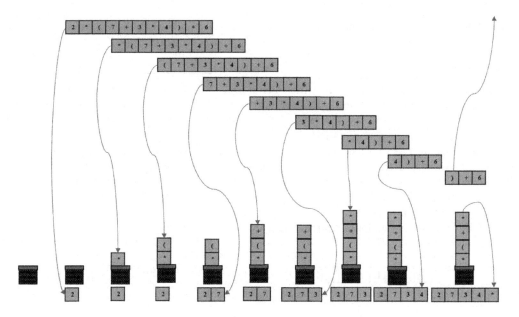

FIGURE 6.4 Infix to postfix with parenthesis – Part A.

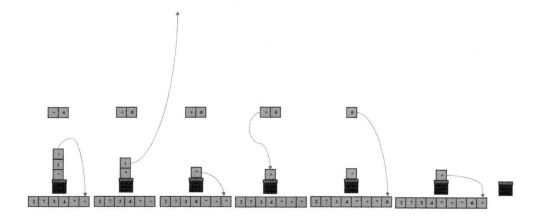

FIGURE 6.5 Infix to postfix with parenthesis – Part B.

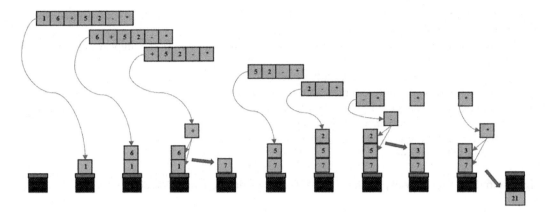

FIGURE 6.6   Evaluating a postfix expression.

6.5.3 QUEUE

A *queue* is also a linear structure in which items are added at one end through a process called *enqueue*, but removed from the other end through what is referred to as *dequeue*. The two ends are called *rear* and *front*. Unlike the stack, in a queue the *items that are added first are also removed first*, hence it is also described as a *first in, first out (FIFO)* structure. A queue is analogous to people waiting in line to purchase a ticket or pay a bill. The person first in line is the first one to be served.

The following is a visual illustration of the queue structure:

Observation 6.17 – Queue: An ordered, linear list structure with two ends: *rear* and *front*. Items are *enqueued* at one end and *dequeued* at the other. The first enqueued item is also the first to be dequeued *(FIFO)*. The operations performed on the queue are the following: *initialize, enqueue, dequeue, isEmpty, peek,* and *size*.

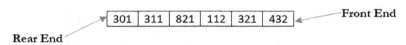

Figure 6.7 below illustrates the execution of a simple queue:

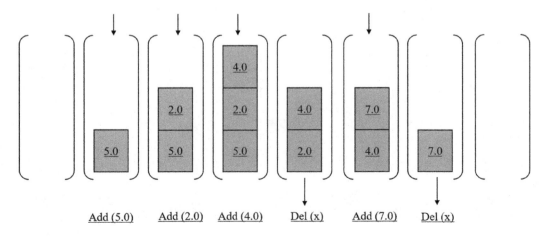

FIGURE 6.7   Execution of a simple queue.

In computer science, queues are used extensively to schedule tasks, such as printing or managing CPU processes. When multiple users submit print jobs, the printer queues all the jobs and prints them in a first-come-first-served basis. Similarly, when multiple processes require to use the CPU, the order of execution is scheduled and performed through a queue structure.

The *queue ADS* consists of the following:

- An index that points to the *front* item of the queue.
- An index that points to the *rear* item of the queue.
- The body of the queue that stores its values (i.e., the actual data in the list).
- *Initialize* – init(q): A function that initializes the queue (i.e., creates the empty list).
- *Empty* – isEmpty(q): A function that checks whether the queue is empty.
- *Enqueue* – enqueue(x, q): A function that adds an item to the rear end of the queue.
- *Dequeue* – dequeue(x, q): A function that returns the item at the front end of the queue and removes it from the queue.
- *Front* – peek(q): A function that returns the item at the front of the queue.
- *Size* – size(q): A function that returns the number of items in the queue.

The Python class provided below (filename: *Chapter6Queue.py*) is an implementation of the queue ADS:

```python
class Queue:
    # Initialize the queue
    def __init__(self):
        self.items=[]

    # Check whether the queue is empty
    def isEmpty(self):
        return self.items == []

    # Add an item to the queue
    def enqueue(self, item):
        self.items.insert(0,item)

    # Delete an item from the queue
    def dequeue(self):
        if not self.isEmpty():
            return self.items.pop()

    def peek(self):
        if not self.isEmpty():
            return self.items[-1]

    def size(self):
        return len(self.items)
    def show(self):
        return self.items
```

The following script (filename: *Chapter6QueueExample*) imports and runs a simple queue ADS:

```
1    import Chapter6Queue
2
3    q = Chapter6Queue.Queue()
4    print(q.isEmpty())
5    q.enqueue('Task A')
```

```
6   print(q.show())
7   q.enqueue('Task B')
8   print(q.show())
9   q.enqueue('Task C')
10  print(q.show())
11  print(q.dequeue()) # removes Task A
12  print(q.show())
13  print(q.dequeue()) # removes Task B
14  print(q.show()) # q has only one task left
15  print(q.size())
```

Output 6.5.3:

```
True
['Task A']
['Task B', 'Task A']
['Task C', 'Task B', 'Task A']
Task A
['Task C', 'Task B']
Task B
['Task C']
1
```

6.5.4 CIRCULAR QUEUE

A *circular queue* is essentially the same as a regular queue, but with two major differences. First, the size of the circular queue does not change. This size restriction can be viewed as the main weakness of the circular queue. Second, its front and rear are continuously moving in a circular form based on the demand for *enqueue* and *dequeue*, provided that there is available empty space and that they do not clash with each other (i.e., the front cannot be in the same list index as the rear). This is an important observation, as it is possible that *the front item is stored before the rear one* on the circular queue. Because of these qualitative differences,

Observation 6.18 – Circular Queue: A structure similar to a *queue* with the difference that its size does not change and the *front* and *rear* are movable. This is based on the demand for *enqueue* and *dequeue* in a circular form, allowing for the *front* item to be stored before the *rear*.

a circular queue ADS needs to check whether the queue is full before enqueuing a new item in it. Figure 6.8 provides an illustration of the circular queue operation.

The following script (filename: *Chapter6CircularQueue*) imports and runs an implementation of the queue ADS:

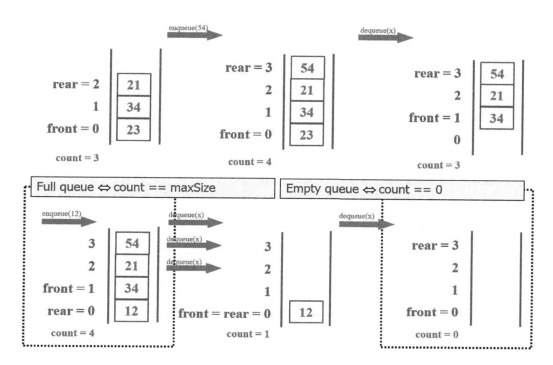

FIGURE 6.8 Example of circular queue.

```
1    class CircularQueue():
2
3        # Initialize the circular queue to the preferred size
4        # with all its items empty and the front and rear starting at -1
5        def __init__(self, maxSize):
6            self.cqSize=maxSize
7            self.queue=[None] * self.cqSize
8            self.front=self.rear=-1
9
10       # Insert an item into the circular queue
11       def enqueue(self, data):
12           # Insert the first item to the queue, start the front and rear
13           if (self.front == -1):
14              self.front=self.rear=0
15              self.queue[self.rear]=self.queue[self.front]=data
16           # Insert items to the queue
17           else:
18              # Only be concerned with the front item; use % and the size
19              # of the queue to move the front in a circular manner
20              self.front=(self.front+1) % self.cqSize
21              self.queue[self.front]=data
22           print("Queue size: ", self.cqSize, "Queue front: ", self.front,
23              "Queue rear: ", self.rear)
24
25       # Delete an item from the circular queue
```

```
26          def dequeue(self):
27              if (self.front == -1):
28                print("The circular queue is empty\n")
29              # If the front item is the same as the rear the queue has only
30              # one item; empty the queue
31              elif (self.front == self.rear):
32                self.front = self.rear = -1
33              else:
34                # Only be concerned with the rear item; use % and the size
35                # of the queue to move the rear in a circular form
36                self.queue[self.rear] = [None]
37                self.rear = (self.rear + 1) % self.cqSize
38              print("Queue size: ", self.cqSize, "Queue front: ", self.front,
39                "Queue rear: ", self.rear)
40
41          # The printCQueue will display the contents of the circular queue
42          def printCQueue(self):
43              # If the front value is -1 the circular queue is still empty
44              if(self.rear == -1):
45                print("No element in the circular queue")
46              # If front index is larger than rear then queue is still valid
47               elif (self.front >= self.rear):
48                for i in range(self.rear, self.front + 1):
49                  print(self.queue[i], end=" ")
50              # If front less than rear, queue has completed a circle
51              else:
52                for i in range(self.front + 1):
53                  print(self.queue[i], end=" ")
54                for i in range(self.rear, self.cqSize):
55                  print(self.queue[i], end=" ")
56              print()
57
58          # Check whether the circular queue is full
59          def isFull(self):
60              if ((self.front + 1) % self.cqSize == self.rear):
61                return True
62              else:
63                return False
64
65   # Ask the user for the preferred size for the circular queue
66   maxSize = int(input("Enter the size of the circular queue:"))
67   cq = CircularQueue(maxSize)
68
69   # Keep working on the circular queue until input is not E or D
70   while (True):
71       # Ask the user for the next move, enqueue or dequeue
72       choice = input("(E)nqueue or (D)equeue or (Q)uit?")
73       if (choice == "E"):
74           if (cq.isFull() != True):
75               newItem = int(input("Enter the next item of the circular
                                                             queue:"))
76           cq.enqueue(newItem)
```

```
77                    else:
78                        print("The queue is full. Cannot insert a new item")
79              elif (choice == "D"):
80                    cq.dequeue()
81              else:
82                    break
83              print("The updated Queue is: ", end=" ")
84              cq.printCQueue()
```

Output 6.5.4:

```
Enter the size of the circular queue:3
(E)nqueue or (D)equeue or (Q)uit?E
Enter the next item of the circular queue:10
Queue size:   3 Queue front:   0 Queue rear:   0
The updated Queue is:   10
(E)nqueue or (D)equeue or (Q)uit?E
Enter the next item of the circular queue:20
Queue size:   3 Queue front:   1 Queue rear:   0
The updated Queue is:   10 20
(E)nqueue or (D)equeue or (Q)uit?E
Enter the next item of the circular queue:30
Queue size:   3 Queue front:   2 Queue rear:   0
The updated Queue is:   10 20 30
(E)nqueue or (D)equeue or (Q)uit?E
The queue is full. Cannot insert a new item
The updated Queue is:   10 20 30

(E)nqueue or (D)equeue or (Q)uit?
```
```
D
```
```

(E)nqueue or (D)equeue or (Q)uit?D
Queue size:   3 Queue front:   2 Queue rear:   1
The updated Queue is:   20 30
(E)nqueue or (D)equeue or (Q)uit?D
Queue size:   3 Queue front:   2 Queue rear:   2
The updated Queue is:   30
(E)nqueue or (D)equeue or (Q)uit?E
Enter the next item of the circular queue:40
Queue size:   3 Queue front:   0 Queue rear:   2
The updated Queue is:   40 30

(E)nqueue or (D)equeue or (Q)uit?
```
```

```

6.6   DYNAMIC DATA STRUCTURES

The data structures described in the previous sections are characterized as *static*, since they all use *inherently static list structures*. To some extent, issues like restrictions associated with the requirement for large amounts of memory, generally weak performance due to the heavy nature of the tasks, and a certain inflexibility, can be traced in all of these structures. The previously discussed cases have demonstrated that the execution of even the most advanced algorithms tends to become impractical as the size of the structures increases. In order to address this issue, there is a need for more effective data structures that allocate the available computer memory only as and when necessary, and in the most efficient way possible. Structures that fall under this category are collectively known as *dynamic data structures*. Some of the most important of these structures are introduced and briefly discussed in the following sections.

6.6.1 LINKED LISTS

A *linked list* is a collection of *nodes* linked to each other through *pointers*. The structure is recursive by definition. Each node includes a data value and a pointer to the first node of a subsequent linked list, or to *null* if the latter is empty. In order to navigate a linked list, it is necessary to create a separate object, called *head*, that always points to the first node of the list. Subsequent nodes are accessed via the associated pointers, stored in each node. If the list is empty, the head will simply point to a *null* value. In a similar fashion, the link pointer of

Observation 6.19 – Linked List: A structure of connected *nodes*. Each node contains a data value and a pointer to the first node of the subsequent list. A *head* pointer is always pointing to the first node. The last node points to *null*. The rest of the nodes are defined as *intermediate*.

the last node is set to *null* to mark the end of the list. There is only one head, and it is always pointing to the first node of the linked list. Similarly, there is only one *tail* (i.e., the last node), pointing to *null*. All other nodes are called *intermediate* nodes and have both a predecessor and a successor. *Traversing* (i.e., moving through) intermediate nodes towards the tail starts at the first node of the list, pointed to by the head. For this purpose, it is best to create another object, usually called *current*, that is used to move between the intermediate nodes in the list.

The strength of the linked list is that its data are stored *dynamically*, with new nodes created only if and when necessary, and unwanted nodes deleted if they are not in use. Separately from the data, the pointer of every newly created node is set to point to *null*. Nodes can store any data type, but all nodes of a linked list need to store the same data type.

Figure 6.9 illustrates the structure of a linked list. Notice how the head points to the first node and that the last node points to *null*:

FIGURE 6.9 Linked list.

The implementation of a linked list requires two classes. The first is the *node* class that contains a data and a pointer to the *next* item. For any new node that is created, *next* will point to *null*. The second, is the linked list itself that contains the *head* pointer to the first item in the list and the *current_node* that is used to move through the list. Both the *head* and the *current_node* will initially point to *null* since there are no items in the list.

The *linked list ADS (Abstract Data Structure)* includes the following operations:

- **Instantiating & initializing the list:** This function is used to create the head and the current object that initially point to *null* (i.e., the empty list; Figure 6.10). The Python code for this function is the following:

```
def __init__(self):
    self.head=self.current_node=None
```

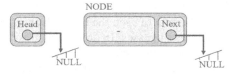

FIGURE 6.10 New linked list.

- **Checking if the list is empty:** This function checks whether the linked list is empty, in which case no more nodes can be deleted and any newly inserted node must be the first in the list. The Python code is the following:

```
def isEmpty(self):
    current_node=self.head
    if (current_node == None):
        return True
```

- **Reading and printing the list:** It is often useful to print the nodes of the list and provide information about its size (i.e., the number of nodes it contains). In order to do this, it is necessary to *traverse* (i.e., read through) the list starting at the first node. While the *current_node value is not null*, current node values are read/printed successively as the list is traversed. Figure 6.11 illustrates this process diagrammatically. The related Python code is presented below:

```
def readList(self):
    count=0
    current_node=self.head
    print("The current list is: ", end=" ")
    while (current_node):
        count += 1
        print(current_node.data, " ", end="")
        current_node=current_node.next
    print("\nThe size of the linked list is: ", count)
```

- **Inserting a new node in the list:** A new node can be either inserted as a first element when the list is empty or as the last element *appended* to the list. In the former case, a new node is created (including the associated data) and its *next* element is set to point to *null*. Finally, the *head* is set to point to the new node (Figure 6.12). In the case of appending a new element to the list, after the new node is created, the list is traversed until the last node is reached. Once this is done, the *next* element of the last node is set to point to the newly created node (Figure 6.13). The related Python code is presented below:

```
def append(self, data):
    # Create the newNode to append the linked list
    newNode = Node(data)

    # Case 1: List is empty
    if (self.head == None):
        self.head = newNode
        Return

    # Case 2: If the list is not empty start the
    # current node at the head of the list
    current_node = self.head

    # Loop through the linked list untill the current node
    # has Next pointing to None
    while (current_node.next):
        current_node = current_node.next

    # Add new node to the end of the list
    current_node.next = newNode
```

- **Deleting a node:** This operation starts by checking if the linked list is empty. If not, it searches for the data that must be deleted. If the data are not found, the list remains as is. If the data are found, the node they belong to is deleted and the list is updated accordingly. There are two cases to consider in relation to this process. The first case is that the node to be deleted is the first one in the list. In this case, the process simply involves the allocation of the *head* to the *next* node, and the assignment of the pointer that points to the deleted node to *null*. The second case is that the node to be deleted is not the first one in the list. In this case, it is necessary to also find the nodes before and after the deleted, and keep references to them. With this information at hand, the *next* pointer of the node *preceding* the deleted one is made to point to the node *succeeding* it. Finally, the pointers of the deleted node are removed. Figure 6.14 illustrates this process diagrammatically.

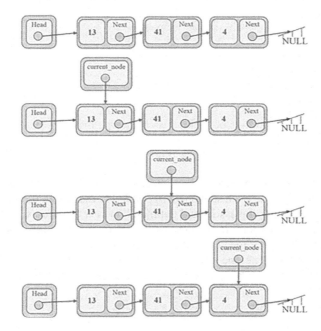

FIGURE 6.11 Traversing the linked list.

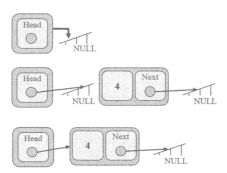

FIGURE 6.12 Inserting the first node.

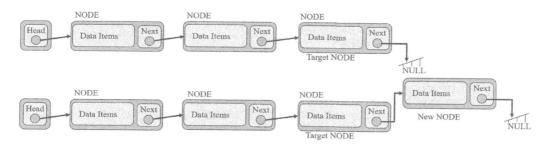

FIGURE 6.13 Appending a node to the list.

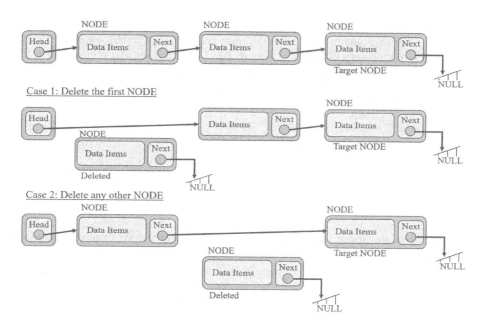

FIGURE 6.14 Deleting a node from a linked list.

The following Python script demonstrates the deletion process:

```python
def delete(self, data):
    if (self.isEmpty()):
        print("There is no node available to delete. "
            "The linked list is empty.")
    else:
        current_node = self.head
        # Case 1: If the node to be deleted is the first node
        if (current_node and current_node.data == data):
            # Set the head of the list of the next item
            self.head = current_node.next
            # Set the current item's pointer to null
            current_node.next = None
            Return
```

```
    # Keep track of the previous node while searching
    # for the node to be deleted
    previous_node = None

    while (current_node and current_node.data != data):
        previous_node = current_node
        current_node = current_node.next

    # Check if the node was found
    if (current_node is None):
        return
    previous_node.next = current_node.next
    current_node = None
```

- **Destroying the list:** Since building a linked list involves the dynamic allocation of memory in the form of pointers, it is advisable that before the underlying application stops, any pointers and memory allocated during its lifecycle are freed and released back to the system. The following Python code demonstrates a possible implementation of this task:

```
def destroyList(self):
    temp=self.head
    if (temp is None):
        print("\n The linked list is deleted")
    while (temp):
        self.head=temp.next
        temp=None
        temp=self.head
        self.readList()
```

The reader can merge the above functions and commands as in the code example provided below (the code is arranged into two classes, stored in file *Chapter6LinkedList.py*):

```
class Node:
    def __init__(self, data):
        self.data=data
        self.next=None
class LinkedList:
    def __init__(self):
        ...
    def append(self,data):
        ...
    def delete(self, data):
        ...
    def destroyList(self):
        ...
    def readList(self):
        ...
    def isEmpty(self):
        ...
```

The following script (filename: *Chapter6LinkedListExample*) implements the class, as discussed above:

```
1    import Chapter6LinkedList
2
3    ll = Chapter6LinkedList.LinkedList()
4
5    while (True):
6        print("[A]: Append a new node")
7        print("[D]: Delete a particular node")
8        print("[Q]: Clear all list and exit")
9        print("[P]: Print the current list")
10       choice = input("Enter your choice: ")
11       if (choice == "A"):
12           newNode = int(input("Enter the new node value to append the
   list: "))
13           ll.append(newNode)
14       elif (choice == "D"):
15           deleteNode = int(input("Enter the node to delete: "))
16           ll.delete(deleteNode)
17       elif (choice == "P"):
18           ll.readList()
19       else:
20           ll.destroyList()
21           break
```

Output 6.6.1:

```
[A]: Append a new node
[D]: Delete a particular node
[Q]: Clear all list and exit
[P]: Print the current list
Enter your choice: A
Enter the new node value to append the list: 5


[A]: Append a new node
[D]: Delete a particular node
[Q]: Clear all list and exit
[P]: Print the current list
Enter your choice: A
Enter the new node value to append the list: 3


[A]: Append a new node
[D]: Delete a particular node
[Q]: Clear all list and exit
[P]: Print the current list
Enter your choice: A
Enter the new node value to append the list: 7
```

```
[A]: Append a new node
[D]: Delete a particular node
[Q]: Clear all list and exit
[P]: Print the current list
Enter your choice: P
The current list is:  5  3  7
The size of the linked list is:  3


[A]: Append a new node
[D]: Delete a particular node
[Q]: Clear all list and exit
[P]: Print the current list
Enter your choice: D
Enter the node to delete: 3


[A]: Append a new node
[D]: Delete a particular node
[Q]: Clear all list and exit
[P]: Print the current list
Enter your choice: P
The current list is:  5  7
The size of the linked list is:  2
```

In addition to the operations discussed above, the effectiveness of the linked list could be also improved by:

- Inserting a new node before/after an existing node based on its data.
- Searching for a node using key data, and retrieving the data and the positional index of the node.
- Modifying the data of a particular node within the list.
- Sorting the linked list.

Some key points when implementing linked lists or related structures are summarized in the list below:

- To access the nth node of a linked list, it is necessary to pass through the first $n-1$ nodes.
- If nodes are added at a particular position instead of just being appended, the insertion will result in a node index change.
- Deletion of nodes will result in a node index change.
- Trying to store the node indices in a linked list is of no use, since they are constantly changing (indeed, there are no actual indices in such a list).
- To append a node, one has to traverse the whole list and reach the last node.
- In addition to the *head* and *current_node* pointers, adding a *tail* pointer to the last node of the list makes appending easier and more efficient.
- To delete the last node, one has to traverse the whole list and find the two last positions.
- If for any reason the *head* pointer is lost, the linked list cannot be read and retrieved.

A particular variation of the linked list is the *circular linked list*, in which the last node is linked to the first. It is used when the node next to the last corresponds to the first one, such as in the cases of the weekdays or the *ring network topology*. The advantage of the circular linked list is that it can be traversed starting at any node and is able to reach the node it has started with again in a circular manner. Figure 6.15 provides an illustration of a simple circular linked list.

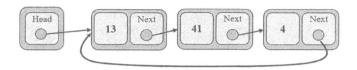

FIGURE 6.15   A circular linked list.

6.6.2   BINARY TREES

The previous section focused in the *singly linked list*, in which the pointer of each node points to the next node. The main problem with this type of linked list is that it *does not offer direct access to the previous node*. This can make the process of deleting nodes from the list rather complicated. *Doubly linked lists* can address this problem. As the name implies, the main difference between singly and doubly linked lists is that the latter consist of two pointers instead of one, with the additional pointer pointing to the previous node. Despite

> **Observation 6.20 – Doubly Linked List:** A structure similar to a *singly linked list*, but containing two pointers pointing to both the *next* and *previous* nodes instead of just one (*next*).

the obvious functional advantage of this additional pointer, it tends to make operations more complicated and causes additional overhead, as an extra pointer is added to every node. Figure 6.16 provides an illustration of the inner structure of a doubly linked list node and an example of a three-node doubly linked list connections:

Among the most important types of doubly linked lists is the *binary tree* (Figure 6.17), a *rooted tree* in which every node has at most two *children* (i.e., *degree 2*). Its recursive definition declares that *a binary tree is either an external node (leaf) or an internal node (root/parent) and up to two sub-trees (a left subtree and a right subtree).* In simple terms, if a node is a root, it has one or two children nodes but no parent, if it is a leaf, it has a parent node but no children, and every node is an element that contains data. The number of levels in the tree is defined as its *depth*.

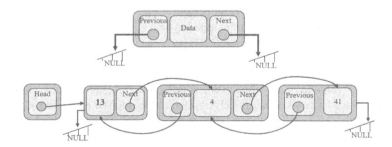

FIGURE 6.16   A NODE of a double linked list.

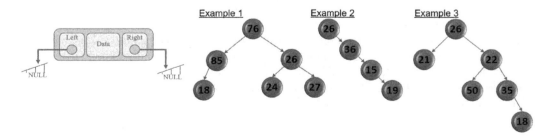

FIGURE 6.17   Binary trees.

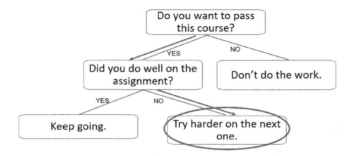

FIGURE 6.18   Decision trees.

Example 1 in Figure 6.17 shows an *unfinished* binary tree with *degree 2* and a *depth* of three levels. The tree has 76 as its root, 26 and 85 as children nodes, and 27, 24, and 18 as leaf nodes. Example 2 shows a completely unbalanced binary tree and Example 3 a mixed case.

Binary trees are commonly used in *decision tree* structures (Figure 6.18), although this may often go unnoticed.

Observation 6.21 – Binary Tree: A *rooted tree* in which every node is either an *external* node (leaf) or an *internal* node (root/parent), with up to two sub-trees (a left subtree and a right subtree).

6.6.3   Binary Search Tree

A particular type of a binary tree is the *binary search tree*. Its definition is the same as that of the regular binary tree, but with the following additional properties:

- All elements rooted at the right child of a node have higher values than that of the parent node.
- All elements rooted at the left child of a node have lower values than that of the parent node.

Observation 6.22 – Binary Search Tree: A structure based on a binary tree with the difference that all elements rooted at the right child of a node are greater and those rooted at its left child lower than the value of the parent node.

In the example provided in Figure 6.19 the reader would notice that every node on the left subtree of the root has a lower value than 43, while every node on the right subtree has a higher value. The reader should also notice that this is recursively applied to the internal nodes too (e.g., as in the case of node with value 56). This could be potentially reversed by having the smaller values on the right and the larger on the left subtrees respectively, but the logic of the binary tree structure remains the same.

There are three systematic ways to visit all the nodes of a binary search tree: *preorder*, *inorder*, and *postorder*. If the left subtree contains values that are lower than the root node, all three of these will traverse the left subtree before the right subtree. Their only difference lies on when the root node is visited and read (Table 6.12).

The implementation of a linked list requires two classes. The first is the *node* class, containing the data and a pointer to the *next* item. For any new node that is created, *next* will point to *null*. The second is the *linked list* itself, and contains the *head* pointer (pointing to the first item in the list) and the *current_node* that is used to move through the list. Both the *head* and the *current_node* will initially point to *null* since there are no items in the list.

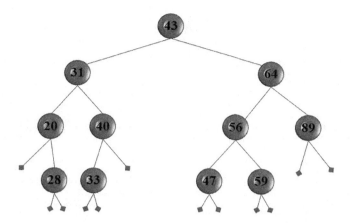

FIGURE 6.19   Binary search tree.

TABLE 6.12
Searching a Node in a Binary Search Tree

Inorder Traversal	Preorder Traversal	Postorder Traversal
Traverse the left subtree.	Visit/read the root node.	Traverse the left subtree.
Visit/read the root node.	Traverse the left subtree.	Traverse the right subtree.
Traverse the right subtree.	Traverse the right subtree.	Visit/read the root node.
Resulting list: 20, 28, 31, 33, 40, 43, 47, 56, 59, 64, 89	Resulting list: 43, 31, 20, 28, 40, 33, 64, 56, 47, 59, 89	Resulting list: 28, 20, 33, 40, 31, 47, 59, 56, 89, 64, 43

In its most basic form, the *binary search tree ADS* includes the following operations:

- **Instantiating & initializing the Binary Search Tree (BST):** This function is used to create each new node in the BST, allocating the necessary memory and initializing its pointers to both the left and right subtrees to *null*. Figure 6.20 provides a visual representation of the new node and the following code excerpt illustrates its implementation:

```
def __init__(self, key):
    self.left=None
    self.right=None
    self.data=key
```

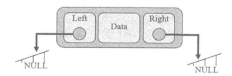

FIGURE 6.20   New node for the BST.

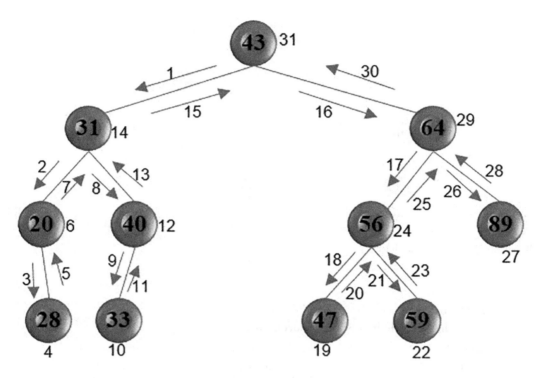

FIGURE 6.21   Traversing the BST inorder.

- **Inorder traversal of the BST:** The *inorder* function, one of the most well-known functions associated with dynamic data structures, happens to be also among the easiest ones. The following Python code and Figure 6.21 illustrate its operation:

```
def traverseInorderBST(root):
    # If the BST current node is not a leaf traverse
    # the left subtree. If it is, print its data and
    # then traverse the right subtree
    if (root):
        traverseInorderBST(root.left)
    print(root, root.data)
    traverseInorderBST(root.right)
```

- **Inserting a new node to the list:** The goal of this function is to place the newly imported data to the desired place in the BST. When the BST is empty, the new node simply initializes it. In all other cases, the function recursively checks whether the data value in the new node is lower, equal to, or higher than the data in the current node, and keeps on moving to the respective subtree accordingly until the current node is empty. At that point, it finally assigns the new node. Figure 6.22 illustrates this process by inserting nodes from the following list to a BST: *43, 31, 64, 56, 20, 40, 59, 28, 33, 47, 89*. The Python code for this function is the following:

```
def insert(root, key):
    # If there is no BST create its first node
    if (root is None):
        return BinarySearchTree(key)
    else:
```

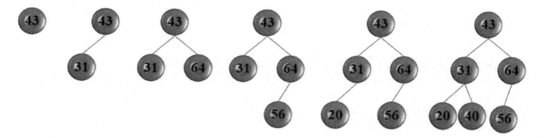

FIGURE 6.22   Inserting nodes to the BST.

```
# If the current node's data is less than or equal
# to the new key, move into the right subtree;
# otherwise, move to the right subtree recursively
if (root.data <= key):
    root.right=insert(root.right, key)
else:
    root.left=insert(root.left, key)
return root
```

- **Searching for a key value in the BST:** This function searches the BST for a key value provided by the user. As with the previous functions, it recursively calls itself on either the left or right subtree in an effort to find a match for the key value. If the key value is not found after all the BST has been searched, an empty BST is returned. This raises an error and crashes the application unless it is handled by the calling function. Figure 6.23 illustrates both a case where the key is being found and one where it is not. The following Python code provides an implementation of this function:

```
def search(root, key):
    # Recursively visit the left and right subtrees to find
    # the node that matches the key searched for
    if (root.data == key):
        return root
    if (root.data<key):
        return search(root.right,key)
    else:
        return search(root.left,key)
```

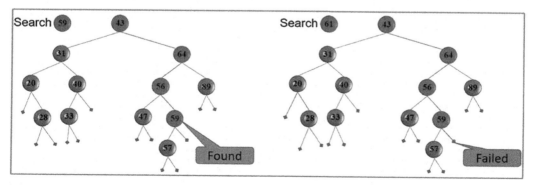

FIGURE 6.23   Data search in a BST.

```
    # If the key is not found, return the empty BST
    if (root is None):
        return None
```

- **Deleting a node from the BST:** Arguably, this is the most complex function in the BST ADS. If the current root is empty, which may be because the key was not found, there is nothing to be done and the current BST is returned as is. In any other case, the key is found in the current node, or its left or right subtree. If the key is found in the current node and the right subtree is empty, the function replaces the current node with its left subtree. Accordingly, if the left subtree is empty it is replaced with the right subtree. If none of these are empty, the function finds the minimum data in the right subtree, replaces the data in the current node, and the current node with the right subtree, while also deleting the node of the subtree with the lowest value data. If the key is not found in the current node, the function is called recursively on the left and the right subtrees, depending on whether the key value is lower or higher than the current node data. Figure 6.24 illustrates this process and the related Python script is provided below:

```
def delete_Node(root, key):
    """ If the root is empty, return it; if not, if the key is
    larger than the current root, find it in the right subtree;
    Otherwise, if it is smaller, find it in the left subtree
    If the key is matched, delete the current root """
    if (root == None):
        return root
    elif (root.data > key):
        root.left = delete_Node(root.left, key)
    elif (root.data < key):
        root.right= delete_Node(root.right, key)
    """ If the key is matched, then, if there is no right
    subtree just replace the current node with the left
    subtree; similarly in this case, if there is no left
    subtree just replace the current node with the right
    subtree."""
    elif (root.data == key):
        if (root.right == None):
            return root.left
        if (root.left == None):
            return root.right
        """ If none of the left or right subtrees is empty
        replace the data in the current node with the minimum
        data in the right subtree and delete the node with
        that minimum data from the right subtree"""
        temp = root.right
```

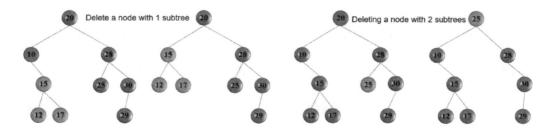

FIGURE 6.24 Deleting a node from a BST.

```
              mini_data = temp.data
              while (temp.left):
                  temp = temp.left
                  mini_data = temp.data
              root.data = mini_data
              root.right = delete_Node(root.right,root.data)
        return root
```

- **Destroying the BST:** As with most structures occupying computer memory space, it is advisable that the BST is deleted (i.e., *destroyed*) when exiting the application. The following Python code excerpt provides a possible implementation of this task:

```
def destroyBST(root):
    if (root):
        destroyBST(root.left)
        destroyBST(root.right)
        print("Node destroyed before exiting: ", root, root.data)
        root=None
```

Finally, it must be noted that the performance of the BST in terms of searching, inserting, or deleting depends on how *balanced* it is. In the case of well-balanced BSTs, the performance is always *O(logn)*, while in extremely unbalanced cases the performance can be improved to *O(n)*.

6.6.4 GRAPHS

A *graph* is a non-linear data structure consisting of nodes, also called *vertices*, which may or may not be connected to other nodes. The line or path connecting two nodes is called an *edge*. If edges have particular flow directions, the graph is said to be *directed*. Graphs with no directional edges are referred to as *undirected* graphs (Figure 6.25).

A directed graph consists of a set of *vertices* and a set of *arcs*. The vertices are also called *nodes* or *points*.

> **Observation 6.23 – Graph:** A non-linear structure of *nodes/vertices* interconnected through *edges*. Edges may have a particular direction (*directed graphs*) or not (*undirected graphs*). Graphs can be presented as *static adjacency matrices* or as *dynamic adjacency lists*.

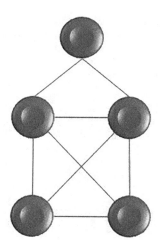

FIGURE 6.25   An undirected graph.

FIGURE 6.26   Arc (*V, W*).

An *arc* is an ordered pair of vertices (*V, W*); *V* is called the *tail* and *W* is called the *head* of the arc. Function *arc (V, W)* is often expressed as $V \rightarrow W$ (Figure 6.26).

A path in a directed graph can be described as a sequence of vertices $V_1, V_2, ...V_n$, thus $V_1 \rightarrow V_2$, $V_2 \rightarrow V_3, ..., V_{n-1} \rightarrow V_n$ can be viewed as arcs. In this occasion, the path from vertex V_1 to vertex V_n, passes through vertices $V_2, V_3, ..., V_{n-1}$, and ends at vertex V_n. The *length* of the path is the number of arcs on the path, in this particular case $n-1$. A path is *simple* if all vertices, except possibly the first and last, are *distinct*. A simple *cycle* is a simple path of a length of *at least one* that begins and ends at the same vertex. A *labeled* graph is one in which each arc and/or vertex can have an associated label that carries some kind of information (e.g., a name, cost, or other values associated with the arc/vertex).

There are two ways to represent a directed graph: as a *static adjacency matrix* or as a *dynamic adjacency list*. The prefix *static* refers to the use of a static structure (i.e., a list), whereas the prefix *dynamic* refers to the use of a dynamic structure in the form of a linked list. In the case of the former, assuming that $V = \{1, 2, ..., N\}$, the *adjacency matrix* of G is an *NxN* matrix A of *booleans*, where *A[i, j]* is *true* if and only if there is an arc from vertex *i* to *j*. An extension of this scheme is what is called a *labelled adjacency matrix*, where *A[i, j]* is the *label* of the arc going from vertex *i* to vertex *j*; if there is no arc from *i* to *j*, it is not possible to have an associated value referring to it. The main disadvantage of the adjacency matrix is that it requires storage in the region of $O(n^2)$. In contrast, in the case of the adjacency list, which is essentially a list of pointers representing every vertex of the graph that is adjacent to vertex *i*, the whole structure is dynamic and, therefore, can have its memory size increased or decreased on demand.

Figure 6.27 presents examples of an adjacency matrix and an adjacency list.

An undirected graph consists of a set of vertices and a set of arcs. As in the case of the directed graph, the vertices are also called *nodes* or *points*. Its main difference from a directed graph is that edges are unordered, implying that *(V, W) = (W, V)*.

The applications of graphs, both directed and undirected, are numerous. Examples include, but are not limited to, the airlines industry, the logistics and freight industries, or the various GPS and navigation systems. In all these cases, the solution to most of their operational problems is a form of the famous *shortest path* algorithm. The idea behind this algorithm is pretty simple.

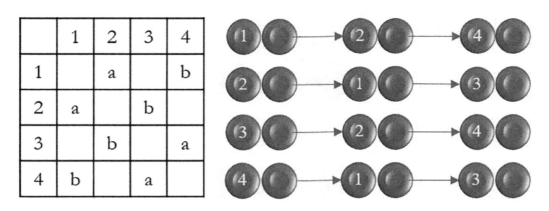

FIGURE 6.27   Adjacency matrix vs. adjacency list.

- A directed graph $G = (V, E)$ is drawn, in which each arc has a *non-negative label* and a vertex is specified as the source.
- The cost of the *shortest path* from the source back to itself is calculated through *every other vertex in V* (i.e., the length of the path).

Dijkstra's famous *greedy* algorithm, also called the *Eulerian* path, provides the solution to this problem. The algorithm can be summarized in the following steps:

- **Step 1:** Determine if the solution is feasible, which is true only if every vertex is connected to an *even* number of other vertices.
- **Step 2:** Start with the *source* vertex and move to the first next available vertex in the adjacency matrix (or adjacency list).
- **Step 3:** Print/store the identified vertex and delete it from the adjacency matrix (or adjacency list).
- **Step 4:** Repeat Steps 2 and 3 until there are no more connections to use.

6.6.5 IMPLEMENTING GRAPHS AND THE EULERIAN PATH IN PYTHON

Implementing an undirected graph implies the implementation of either an adjacency matrix or an adjacency list. Although the implementations may differ, the algorithm is basically the same in both cases: the Eulerian path (Dijkstra's algorithm) is used to find and display the shortest path between the vertices.

Based on the undirected graph provided in Figure 6.28, the following script offers three different scenarios (i.e., scenarios can be selected by enabling/disabling the associated commented statements). The scenario firstly prompts the user to enter the number of vertices in the graph. Next, it accepts the connections in the form of an adjacency matrix as *0s* or *1s* (`fillAdjacencyMatrix()`), checks whether the Eulerian path algorithm can be applied to this particular matrix, and traverses the graph and displays the shortest path. Note that this process may result in one path being inside another. In this case, in the second round, the vertex that opens the path must also close it. The reader should also notice that, in order to merge two paths, the vertex that opens and closes

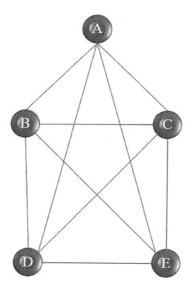

FIGURE 6.28 An undirected graph.

the second path is the one that associates the two separate cases, in the form of a *zoom-in* path residing inside another. The second and third scenarios involve two different, pre-defined matrices that represent graphs and are addressed accordingly:

```
1   def fillAdjacencyMatrix(matrix, vertices):
2       for i in range(vertices):
3           col = []
4           for j in range (vertices):
5               print("Enter 1 if there is a connection between ", i, \
6                   " and ", j, " or 0 if not: ", end = " ")
7               connectionExists = int(input())
8               col.append(connectionExists)
9           matrix.append(col)
10      return matrix
11
12  def displayAdjacencyMatrix(matrix, vertices):
13      for i in range(vertices):
14          print(matrix[i])
15
16  def checkEulerian(matrix, vertices):
17      newStartVertex = -1
18      for i in range(vertices-1, -1, -1):
19          sumPerCol = 0
20          for j in range (vertices):
21              sumPerCol = sumPerCol + matrix[i][j]
22          if (sumPerCol != 0):
23              newStartVertex = i
24      return newStartVertex
25
26  # Ask the user for the number of graph vertices
27  numVertices = int(input("Number of graph vertices: "))
28  #graph = []
29  graph =[[0,1,1,1,1], [1,0,1,1,1], [1,1,0,1,1], [1,1,1,0,1], [1,1,1,1,0]]
30  #graph = [[0,1,0,0,0,1], [1,0,1,0,1,1], [0,1,0,1,1,1], [0,0,1,0,1,0],
31  [0,1,1,1,0,1], [1,1,1,0,1,0]]
32
33  # Fill the adjacency matrix
34  # graph = fillAdjacencyMatrix(graph, numVertices)
35  # Display the adjacency matrix before running the Eulerian Path
36  displayAdjacencyMatrix(graph, numVertices)
37  # Check if the Eulerian Path algorithm can be applied in this case
38  startVertex = checkEulerian(graph, numVertices)
39  endVertex = vertex = startVertex
40  col = 0
41
42  if (startVertex == -1):
43      print("Eulerian Path cannot be applied in this case")
44  else:
45      print("The first round: ", graph[vertex][0], end = "")
46      while (vertex < numVertices and col < numVertices):
```

```
47              if (graph[vertex][col] == 0):
48                 col += 1
49                 if (col == numVertices or vertex == numVertices):
50                     startVertex = checkEulerian(graph, numVertices)
51                     if (startVertex == -1):
52                         print("\nPath closed")
53                     else:
54                         endVertex = startVertex
55                         vertex = startVertex; col = 0
56                         print("\nZoom into", startVertex,
57                             "for the round: ", startVertex, end = " ")
58              elif (graph[vertex][col] == 1):
59                 print("->", col, end = "")
60                 graph[vertex][col] = graph[col][vertex] = 0
61                 vertex = col; col = 0
```

Output 6.6.5:

```
How many vertices in the graph? 5
[0, 1, 1, 1, 1]
[1, 0, 1, 1, 1]
[1, 1, 0, 1, 1]
[1, 1, 1, 0, 1]
[1, 1, 1, 1, 0]
The first round:   0-> 1-> 2-> 0-> 3-> 1-> 4-> 0
Zoom into 2 for the round:  2 -> 3-> 4-> 2
Path closed
```

6.7   WRAP UP

In this chapter an effort was made to briefly explain some of the most important data structures in programming and the algorithms to support those. The various scripts were showcasing how Python can be utilized to implement those." Apparently, there are several other data structures available and, perhaps, more efficient algorithms to implement those which was beyond the scope of this chapter.

6.8   CASE STUDIES

1. Create an application that implements the algorithms and tasks specified below. The application should use a *GUI interface* in the form of a *tabbed notebook*, using *one tab for each algorithm*. The application requirements are the following:
   a. Implement the following static sorting algorithms: *bubble sort, insertion sort, shaker sort, merge sort.*
   b. Ask the user to enter a regular arithmetic expression in a form of a phrase, with each of the operators limited to *single-digit integer numbers.* Convert the infix expression to postfix.
   c. Ask the user to enter a sequence of integers, insert them into a binary search tree and implement the BST ADS algorithm with both inorder and postorder traversals.

6.9   EXERCISES

1. Use a notebook GUI to implement the selection sort, the shell sort and the quicksort (one on each tab).
2. Use a stack to implement the following tasks:
   a. Reversing a string.
   b. Calculating the sum of integers *1...N*.
   c. Calculating the sum of squares *1 ^ 2 +...+ N ^ 2*.
   d. Checking if a number or word is a *palindrome*.
   e. Evaluating a postfix expression by using a stack.
3. Implement a *deque* structure with an example to test it. A deque is a linear structure of items similar to a queue in the sense that it has two ends (i.e., front and rear). However, it can enqueue and dequeue *from both ends of the structure*. Deque supports the following operations:
   a. `add _ front(item)`: Adds an item to the front of the deque.
   b. `add _ rear(item)`: Adds an item to the rear of the deque.
   c. `remove _ front(item)`: Removes an item from the front of the deque.
   d. `remove _ rear(item)`: Removes an item from the rear of the deque.
   e. `isEmpty()`: Returns a Boolean value indicating whether the deque is empty or not.
   f. `peek _ front()`: Returns the item at the front of the deque without removing it.
   g. `peek _ rear()`: Returns the item at the rear of the deque without removing it.
   h. `size()`: Returns the number of items in the deque.
4. Using a graph do the following:
   a. Ask the user to enter the number of vertices in the undirected graph.
   b. Ask the user to enter the name of each of the vertices in the undirected graph.
   c. Ask the user to enter the connected vertices to each of the edges in the undirected graph.
   d. Determine whether the Eulerian Path solution (Dijkstra's algorithm) is feasible.
   e. In case it is not, ask the user to add new connections to the missing ones.
   f. Create the adjacency matrix for the graph and display it.
   g. Create the adjacency list for the graph and display it.
   h. Run the Dijkstra's algorithm to find the shortest path, starting from a source entered by the user.
   i. Display the solution of the shortest path.

REFERENCES

Dijkstra, E. W., Dijkstra, E. W., Dijkstra, E. W., & Dijkstra, E. W. (1976). *A Discipline of Programming* (Vol. 613924118). Prentice-Hall: Englewood Cliffs.
Hoare, C. A. R. (1961). Algorithm 64: Quicksort. *Communications of the ACM, 4*(7), 321.
Knuth, D. E. (1997). *The Art of Computer Programming* (Vol. 3). Pearson Education.
Stroustrup, B. (2013). *The C++ Programming Language*. India: Pearson Education.

7 Database Programming with Python

Dimitrios Xanthidis
University College London
Higher Colleges of Technology

Christos Manolas
The University of York
Ravensbourne University London

Tareq Alhousary
University of Salford
Dhofar University

CONTENTS

7.1 Introduction 273
7.2 Scripting for Data Definition Language 274
    7.2.1 Creating a New Database in MySQL 276
    7.2.2 Connecting to a Database 279
    7.2.3 Creating Tables 280
    7.2.4 Altering Tables...289
    7.2.5 Dropping Tables...294
    7.2.6 The DESC Statement..296
7.3 Scripting for Data Manipulation Language..296
    7.3.1 Inserting Records..296
    7.3.2 Updating Records 301
    7.3.3 Deleting Records 303
7.4 Querying a Database and Using a GUI 305
    7.4.1 The SELECT Statement..306
    7.4.2 The SELECT Statement with a Simple Condition..307
    7.4.3 The SELECT Statement Using GUI 310
7.5 Case Study 316
7.6 Exercises 316
References..317

7.1 INTRODUCTION

Most IT professionals and scholars may agree on what makes computers special and useful: they can perform operations at lightning speed and on large volumes of data. Stemming from these two fundamental computational thinking elements are the notions of *algorithms* and *programs* as a means to process and manipulate data. In the scope of computer science, information systems, and information technology, the logical and physical organization of data falls under the broader context of *databases*. A thorough analysis of the various concepts related to databases and their structural

DOI: 10.1201/9781003139010-7

design is outside the scope of this book. The reader can find relevant information on Elmasri & Navathe (2017). The focus of this chapter is on the crossroads between computer programming with Python and a common type of database structure: the *relational database*.

In relational databases, there are three main types of scripting techniques and/or languages that are used to perform the various associated tasks, namely *Data Definition Language (DDL)*, *Data Manipulation Language (DML)*, and *Queries*. DDL is used to create, display, modify, or delete the database and its structures and tables, and it is associated with the *database schema* or *metadata*. DML is used to insert data into the various tables, and modify or delete this data as required. It relates to the *database instance* or *state*. Queries are used to display the data in various different ways. Most commercially available *Database Management Systems*

> **Observation 7.1 – Types of Scripting in Relational Databases:** There are three types of scripts addressing relational databases: *Data Definition Language (DDL)*, *Data Manipulation Language (DML)*, and *Queries*.

> **Observation 7.2 – Database Schema, Database Instance:** The structure of a database, including table *metadata*, is also referred to as the *database schema*. The data stored on the tables at any given time are called the *database instance* or *state*.

(DBMS) incorporate facilities and tools that utilize these three mechanisms.

The DBMS of choice for this chapter is *MySQL* (2021). This is part of a package that includes both the DBMS and a local server solution called *Apache* (2021). The package supports both Windows and Mac OS systems, and the two associated versions come under the name *MAMP*. The packages are free for download from MAMP (2021) and Oracle (2021b) and the installation is pretty intuitive and straightforward. While it is always beneficial for one to study and understand the tools and technologies of any given system to a good extent, it must be noted that no prior knowledge or practical experience with MAMP is needed in order to practice and execute the examples presented in this chapter. While the examples make use of the MySQL DBMS and the Apache Server, this is just a matter of simply logging in and activating them, and accessing the created databases. The scripts provided in this chapter will do all the necessary work, while the results will appear in the relevant MySQL database.

This chapter will cover the following topics:

- **DDL (Data Definition Language):** Creating a database and connecting to it. Modifying, deleting, or displaying DB tables, structures, and attributes.
- **DML (Data Manipulation Language):** Inserting, modifying, and deleting records in a table.
- **Queries:** Displaying the records of one or more tables in various different ways.
- Using **GUI programming**, and in particular the *Grid* widget, to create presentable database applications with Python.

It should be noted that while expertise in databases is not essential, a good understanding of the concepts and techniques introduced in *Chapter 4: Graphical User Interface Programming with Python* and *Chapter 5: Application Development with Python* may be required. Ideally, the reader should be comfortable with the major concepts introduced in all the previous introductory chapters, as many of these concepts will be utilized or integrated in the examples presented here.

7.2 SCRIPTING FOR DATA DEFINITION LANGUAGE

As mentioned, MAMP will provide some of the tools that are necessary for the examples presented in this chapter. The MAMP packages must be downloaded and installed, as required. Once installation is complete, the MAMP application must be launched. This will start the Apache local server and the MySQL DBMS, both of which are required in order to run a *client-server* application. Figures 7.1 and 7.2 illustrate the MAMP server and the MySQL DBMS interfaces, respectively:

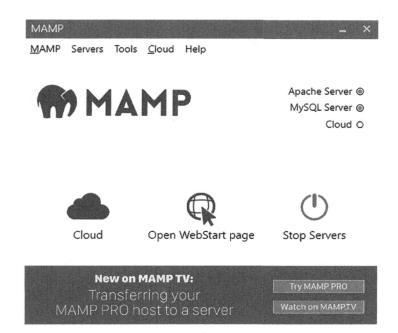

FIGURE 7.1   MAMP server.

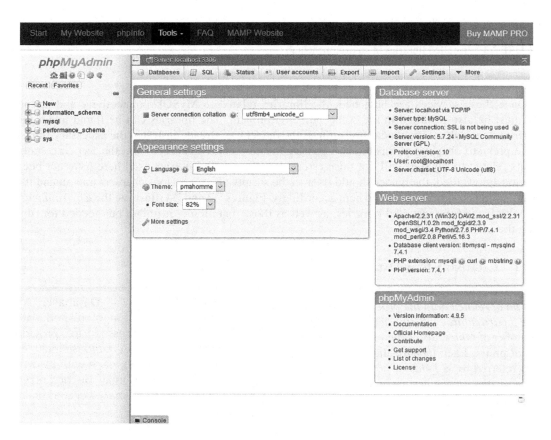

FIGURE 7.2   MySQL phpMyAdmin.

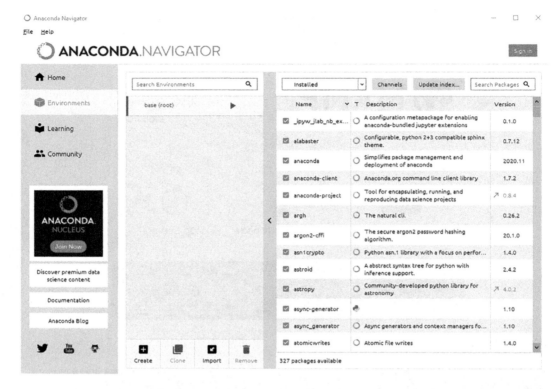

FIGURE 7.3   Installed libraries in environments tab.

Once these services are launched, the libraries related to MySQL connectivity and script-ing must be also installed in the Anaconda environment. The libraries can be found under the *Environments* tab in Anaconda Navigator. If the reader has already installed the necessary librar-ies in previous chapters of this book, installing the new libraries ensures that the `import` state-ments related to MySQL will not raise errors. If some of the libraries used here have not been previously installed, the reader should refer to the scripts of the previous chapters and amend the installation and scripts presented here accordingly. Figures 7.3 and 7.4 illustrate the *Environments* tab with lists of the installed libraries, as well as those that are not installed but needed for run-ning the examples.

7.2.1   CREATING A NEW DATABASE IN MYSQL

A database can be formally defined as *an organized col-lection of related data the processing of which can pro-vide a particular, explicit meaning.* A database includes a number of *tables,* also called *relations,* hence the *rela-tional* prefix. Each table/relation consists of *attributes,* also referred to as *fields* or *columns.* Typically, one or more of these attributes serve as unique record identi-fiers called *primary keys* and are often organized using *indices.* These structural elements of the database are collectively referred to as the database *metadata.* As

Observation 7.3 – Database: An organized collection of related data which are processed to provide explicit meaning. A *database* includes a number of *tables,* each with its own *attributes.* Tables may be organized using a unique *primary key* and make use of *indices.*

mentioned, the creation and control of metadata can be handled using the DDL.

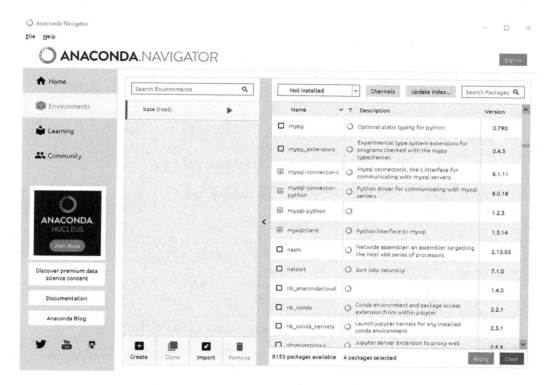

FIGURE 7.4 Not installed but necessary libraries.

It goes without saying that the database itself needs to be created prior to the creation of the metadata. In MySQL, the creation of a new database is as simple as clicking on the *New* option on the left panel of *phpMyAdmin* (Figure 7.2). When creating a new database, the user must specify a name, the database format (usually *GuiDB*) and the default character set (usually *utf8*). In Python, the creation process involves a number of steps:

- Obtaining the log-in credentials for the MySQL environment. These can be found in the *Welcome* page in the *Example* area in MySQL.
- Using the `config` object (list) to set the credentials in the *dictionary* form: `config = {'user': 'root', 'password': 'root', 'host': 'localhost'}`.
- Writing the statements to connect to the database, setting the SQL statement, and executing the commands.

Writing a Python script to create a database may be as simple as writing the basic statements in a *command-prompt* mode or as sophisticated as offering a full GUI environment. The following Python script is an example of the latter. Notice that, upon execution, the application should not produce an output, which simply means that no problems were encountered while connecting to MySQL. Instead of an output, the program should display the newly created database as an *available* database. It must be also stressed that SQL statements are simply treated as strings that *are not case sensitive*. As such, they can be written with capital or lower-case letters, or a combination of both. In this chapter, it was decided to use capital letters for the keywords of the statements, in line with the style adopted in the official MySQL documentation (Oracle, 2021a). This decision had to do mainly with distinguishing the SQL keywords from the SQL database table and attribute names and from the Python code, thus improving clarity and readability:

```
1   import tkinter as tk
2   from tkinter import ttk
3   import mysql.connector
4
5   config = {'user': 'root', 'password': 'root', 'host': 'localhost'}
6
7   def createDB(dbName):
8       GUIDB = 'GuiDB'
9
10      connect = mysql.connector.connect(**config)
11      cursor = connect.cursor()
12      sqlString = "CREATE DATABASE " + dbName.get() + \
13          "DEFAULT CHARACTER SET utf8"
14      cursor.execute(sqlString.format(GUIDB))
15
16  # Create the basic window frame and give it a title
17  winFrame = tk.Tk()
18  winFrame.title("Create a new database")
19  # Create the interface
20  winLabel = tk.Label(winFrame,
21          text = "Enter the name of the new database", bg = "grey")
22  winLabel.grid(column = 0, row = 0)
23  # Create the StringVar object that will accept user input from the
24  # keyboard,and initialize it
25  textVar = tk.StringVar()
26  textVar.set("Enter the name here")
27  winText = ttk.Entry(winFrame, textvariable = textVar, width = 30)
28  winText.grid(column = 0, row = 1)
29  winButton = tk.Button(winFrame, font = "Arial 16",
30  text = "Click to create the new DB\nin the localhost")
31  winButton.bind("<Button-1>", lambda event, a = textVar: createDB(a))
32  winButton.grid(column = 0, row = 2)
33
34  winFrame.mainloop()
```

Output 7.2.1:

The part of the script specifically relating to the database is in lines 3–13. In line 3, the `mysql. connector` function that handles the connection with MySQL is imported. A standard connection configuration is implemented in line 5. Once the GUI is built, a click button event calls the `creat-eDB()` function that assigns the most frequently used database format (GuiDB) to the relevant variable (line 8). Next, it connects to MySQL using the `mysql.connector.connect(**config)` adaptor (line 10), prepares the pending execution statement in the form of a `sqlString` (line 12), and executes the statement (line 13).

7.2.2 Connecting to a Database

As in the previous example, once the database is created a *connection* must be established. Connecting to a database involves the creation of a link to it inside the relevant DMS (e.g., MySQL) through a server, such as Internet Information Server (ISS) or Apache. Once the connection is established, the database must be opened and a link must be created and attached to it. This usually requires some credentials, including login username, password, the host address (i.e., the network address of the server that hosts the database), and the name of the database itself. In the case of databases stored and used from within a local computer system and a local server (e.g., MySQL through Apache), the host address is usually "localhost".

The following Python script connects to the newly created database. It sets the configuration string (`config`) that holds the credentials for the connection to the database (lines 2–3). Next, it links the execution statement with the MySQL database through `mysql. connector` (line 5). Once the connection is successfully established, the results are loaded to the `cursor` object, which always receives the results of all executed SQL statements (line 6). Lastly, the database tables are displayed by executing the `cursor.execute("SHOW TABLES")` (line 7) and `cursor.fetchall()` (line 8) commands.

In this example the reader should note the use of the `try...except` statement (lines 4 and 10) to display the appropriate messages in the cases of both successes and failures. This ensures that statements execution that may return incorrect or unexpected values will not cause the application to crash. As an example, running this script with `newDB` as the database name will display the tables as expected. However, if the data-

> **Observation 7.4 – Connecting to a Database:**
> 1. Import the `mysql.connector` library.
> 2. Use the `cursor` object and the `mysql.connector. connect(**config)` function to connect to the database.
> 3. Prepare the SQL statement.
> 4. Execute the SQL statement using the `cursor.execute()` function.

> **Observation 7.5 – The SHOW TABLES Statement:** Use the SHOW TABLES statement to locate tables in the database. If successful, use the `cursor. fetchall()` function to load the results to the `cursor` object for later use.

> **Observation 7.6 – Exception Handling:** It is highly advisable that the `try...except` exception handling structure is used for each statement related to SQL scripts, as it is likely that the execution of such statements will frequently cause errors that can lead to the abnormal termination (crash) of the application.

base name were to be changed to a non-existing one (e.g., `newDB1`), the *exception handling* code in lines 9 and 10 would be executed, launching an error message. It is worth mentioning that the execution of the `except` segment of the script will be triggered for any reason that might cause a failure in connecting to the database. Nevertheless, if the database is empty, an empty set of tables will be displayed:

```
1   import mysql.connector
2   config = {'user': 'root', 'password': 'root',
3           'host': 'localhost', 'database': 'newDB'}
4   try:
5           link = mysql.connector.connect(**config)
6           cursor = link.cursor()
7           cursor.execute("SHOW TABLES")
8           print(cursor.fetchall())
9   except:
10          print("There is an error with the connection")
```

Output 7.2.2.a:

```
[('STUDENT',), ('Table1',)]
```

Output 7.2.2.a shows the results for a database including tables Student and Table 1.

Output 7.2.2.b:

```
There is an error with the connection
```

Output 7.2.2.b shows the results for an empty database. In this case, the exception handling mechanism is activated and the corresponding error message is displayed. Returning an empty cursor after the execution of the SHOW TABLES statement is considered an internal error, and it is thus raising an exception.

7.2.3 Creating Tables

The first action needed once a new database is created is the creation of its table(s). This is accomplished by the execution of the CREATE TABLE statement in SQL. The CREATE TABLE statement is very similar or identical across different DBMS. A detail description of the small syntax variations between different DBMS systems is beyond the scope of this chapter, but the basic structure remains the same.

Assuming the commonly used *relational model*, seven particular elements need to be specified when creating a table:

1. The *table name* (i.e., the name of each structure that will store data in its columns or fields, also called *attributes*).
2. The *number of attributes* of the table.
3. *The name* of each attribute, preferably as a single, descriptive word.
4. The *data type* for each of the attributes (e.g., CHAR, INT, or DATE).
5. The *length/size* of the data for each attribute in bytes.
6. Whether any of the attributes is the *primary key,* or part of a *combined primary key* of the table.
7. Whether any of the attributes is a *foreign key,* referencing a corresponding attribute in another table.

Observation 7.7 – The CREATE TABLE Statement: Use the CREATE TABLE statement to create a table, define its attributes, data types, and sizes, and set possible *primary* and *foreign keys*.

Observation 7.8 – Create Tables with No Primary or Foreign Key: Use the following statement to create a table with no *primary* or *foreign keys:*

```
CREATE TABLE (<attribute1>
<DATA TYPE>(<size>),...,
<attributeN> <DATA
TYPE>(<size>))
```

Provided that these seven elements are specified, there are three possible cases when creating a table:

1. The table *does not have a primary key* and does not have any of its attributes *referencing the attributes of another table*. In this case, the table is part of a single-table database or it is a *parent* table for other tables to refer to.
2. The table has *one or more of its attributes designated as a primary key*, ensuring that each of its records is *unique*.
3. There are *more than one tables in the database* and they are somehow *related to each other*. This occurs when one or more of the attributes reference an identical column in another table within the same database.

Python provides support for all three cases. Starting with the first case, one could create a table with a number of attributes, but no primary or foreign keys. This can be done either *statically* or *dynamically*. A static approach entails pre-defined statements and pre-determined results. A dynamic approach allows the programmer to determine the table structure at run-time. The following script and output is an example of the latter:

```
1    import mysql.connector
2
3    # The database config details
4    config = {'user': 'root', 'password': 'root',
5      'host': 'localhost', 'database': 'newDB'}
6
7    # The name of the table and its attributes
8    tableName = input("Enter the name of the table to create: ")
9    sqlString = "CREATE TABLE " + tableName + "("
10   numOfAt = int(input("Enter the number of attributes in the table"))
11   atName = [""]*numOfAt
12   atType = [""]*numOfAt
13   atSize = [0]*numOfAt
14
15
16   # Define the table structure (i.e., attribute details)
17   for i in range(numOfAt):
18       atName[i] = input("Enter the attribute " + str(i) + ": ")
19       atType[i]=str(input("Enter 'char' for char, 'int' for int type: "))
20       atSize[i] = int(input("Enter the size of the attribute: "))
21       sqlString += atName[i]+ " " + atType[i]+"("+str(atSize[i])+")"
22
23       if (i < numOfAt-1):
24           sqlString += ","
25       else:
26           sqlString += ")"
27
28   # The SQL statement and exception handling mechanism
29   print("The SQL statement to run is: ", sqlString)
```

```
30
31  try:
32      link = mysql.connector.connect(**config)
33      cursor = link.cursor()
34      cursor.execute(sqlString)
35      sqlString = "DESC " + tableName
36      cursor.execute(sqlString)
37      attributes = cursor.fetchall()
38      # Desc/show the metadata of the new table
39      print("The metadata for the new table "+str(tableName)+" are: ")
40      for row in attributes:
41          print(row)
42  except:
43      print("There is an error with the connection")
```

Output 7.2.3.a:

```
Enter the name of the table to create: Student
Enter the number of attributes in the table: 3
Enter the attribute 0: Name
Enter 'char' for char type, 'int' for int type: char
Enter the size of the attribute: 10
Enter the attribute 1: Address
Enter 'char' for char type, 'int' for int type: char
Enter the size of the attribute: 15
Enter the attribute 2: Year
Enter 'char' for char type, 'int' for int type: int
Enter the size of the attribute: 4
The SQL statement to run is:  Create Table Student(Name char(10),
Address char(15),Year int(4))
The metadata for the new table Student are:
('Name', 'char(10)', 'YES', '', None, '')
('Address', 'char(15)', 'YES', '', None, '')
('Year', 'int(4)', 'YES', '', None, '')
```

The script consists of three distinct parts. In the first part (lines 7–13), the user is prompted to enter a name for the new table and the number of its attributes. The SQL string that is subsequently used for the creation of the table is also constructed. In the second part (lines 15–25), the user is prompted to enter the required details for each attribute (e.g., name, data type, size), and the SQL string is updated accordingly. The third part involves code that connects to the database and executes the SQL string. As mentioned, this is wrapped in an exception handling block in order to prevent a possible uncontrolled termination of the program due to failures of database-related activities (lines 30–42). This is one the most straightforward cases of creating tables using Python scripts. Indeed, this implementation simply involves the incorporation and execution of SQL statements through the Python script wrapper, similarly to what one would do with any other modern programming language.

In the output of this particular example, the user enters the rather trivial and common example of a Student table with three basic attributes: Name, Address, and Year (of birth). After execution, the

> **Observation 7.9 – Primary Key:** An attribute or a combination of attributes with values that *uniquely identify* each particular record in the table.

reader should be able to verify that the table has been created with the desired structure (e.g., with no primary or foreign keys) by checking database *newDB* in MySQL.

The second case involves the addition of *primary keys* to the table. As a reminder, a formal definition of the primary key is that of *an attribute of a table the value of which identifies records uniquely*. Simply put, the primary key designation ensures that there are no duplicate values for the related attribute(s). It must be stressed again that two distinct possibilities exist in relation to primary keys. The first is that it consists of a single attribute. In this case the syntax is the following:

```
CREATE TABLE <table name> (<attribute1>
<DATA TYPE>(<size>) PRIMARY KEY,...,
<attributeN> <DATA TYPE>(<size>))
```

The second is that the primary key consists of a combination of two or more attributes. In this case the syntax is slightly different:

```
CREATE TABLE <table name> (<attribute1>
<DATA TYPE>(<size>),..., <attributeN>
<DATA TYPE>(<size>), PRIMARY KEY
(<attributeX>,... <attributeY>))
```

The following script is another version of the one presented previously, modified in order to addresses the creation of a table with a single primary key (lines 15–31):

Observation 7.10 – Foreign Key: An attribute that *references* the values of a corresponding attribute on another table of the same database that is also the primary key for the *referenced* table.

Observation 7.11 – Create a Table with a Single Primary Key but No Foreign Key:

```
CREATE TABLE <table name>
(<attribute1> <DATA
TYPE>(<size>) PRIMARY KEY,
..., <attributeN> <DATA
TYPE>(<size>))
```

Observation 7.12 – Create a Table with Combined Primary Key but No Foreign Key:

```
CREATE TABLE <table name>
(<attribute1> <DATA
TYPE>(<size>),...,
<attributeN> <DATA
TYPE>(<size>), PRIMARY KEY
(<attributeX>,...
<attributeY>))
```

```python
1   import mysql.connector
2
3   # The database config details
4   config = {'user': 'root', 'password': 'root',
5     'host': 'localhost', 'database': 'newDB'}
6
7   # The name of the table and its attributes
8   tableName = input("Enter the name of the table to create: ")
9   sqlString = "CREATE TABLE " + tableName + "("
10  numOfAt = int(input("Enter the number of attributes in the table: "))
11  atName = [""]*numOfAt
12  atType = [""]*numOfAt
13  atSize = [0]*numOfAt
14
15  key = 0
16  # Define the structure of the table (i.e., attribute details)
17  for i in range(numOfAt):
```

```
18       atName[i] = input("Enter the attribute " + str(i) + ": ")
19       atType[i]=str(input("Enter 'CHAR' for char, 'INT' for int type: "))
20       atSize[i] = int(input("Enter the size of the attribute: "))
21       sqlString += atName[i] + " " + atType[i] + \
22           "(" + str(atSize[i]) + ")"
23       if (key == 0):
24           primaryKey = str(input("Is this a primary key (Y/N)? "))
25           if (primaryKey == "Y"):
26               sqlString += " PRIMARY KEY"
27               key = 1
28       if (i < numOfAt-1):
29           sqlString += ", "
30       else:
31           sqlString += ")"
32
33   # The SQL statement to run using exception handling
34   print("The SQL statement to run is: \n", sqlString)
35   try:
36       link = mysql.connector.connect(**config)
37       cursor = link.cursor()
38       cursor.execute(sqlString)
39       sqlString = "DESC " + tableName
40       cursor.execute(sqlString)
41       columns = cursor.fetchall()
42       print("The structure/metadata of the table ",str(tableName),"is:")
43       for row in columns:
44           print(row)
45   except:
46       print("There is an error with the connection")
```

Output 7.2.3.b:

```
Enter the name of the table to create: Customers
Enter the number of attributes in the table: 3
Enter the attribute 0: CustomerID
Enter 'char' for char, 'int' for int type: int
Enter the size of the attribute: 3
Is this a primary key (Y/N)? Y
Enter the attribute 1: CustLastName
Enter 'char' for char, 'int' for int type: char
Enter the size of the attribute: 15
Enter the attribute 2: CustFirstName
Enter 'char' for char, 'int' for int type: char
Enter the size of the attribute: 10
The SQL statement to run is:
 Create Table Customers(CustomerID int(3) Primary key, CustLastName
 char(15), CustFirstName char(10))
There is an error with the connection
```

Output 7.2.3.c:

```
Enter the name of the table to create: Items
Enter the number of attributes in the table: 3
Enter the attribute 0: ItemID
Enter 'char' for char, 'int' for int type: char
Enter the size of the attribute: 6
Is this a primary key (Y/N)? Y
Enter the attribute 1: ItemDesc
Enter 'char' for char, 'int' for int type: char
Enter the size of the attribute: 25
Enter the attribute 2: ItemPrice
Enter 'char' for char, 'int' for int type: int
Enter the size of the attribute: 5
The SQL statement to run is:
 Create Table Items(ItemID char(6) Primary key, ItemDesc char(25),
 ItemPrice int(5))
The structure/metadata of the table  Items is:
('ItemID', 'char(6)', 'NO', 'PRI', None, '')
('ItemDesc', 'char(25)', 'YES', '', None, '')
('ItemPrice', 'int(5)', 'YES', '', None, '')
```

The output demonstrates the creation of two of the three tables (i.e., Customers and Items) from Table 7.1.

The third case involves the connection of more than one tables connecting to each other through a common attribute. In this case, this common attribute is usually designated as a primary key in one of the tables and a foreign key in the others, although this is not the only possible arrangement. This practice is often termed as *referencing*, as the foreign key of the child table *references* the primary key of the parent table. The syntax for the creation of the table and the key designation is the following:

Observation 7.13 – Create a Table with One or More Foreign Keys:

```
CREATE TABLE <table name>
(<attribute1> <DATA
TYPE>(<size>), FOREIGN KEY
(<attribute name>) REFERENCES
<table name> (<attribute
name>),..., <attributeN>
<DATA TYPE>(<size>), FOREIGN
KEY (<attribute name>)
REFERENCES <table name>
(<attribute name>))
```

```
CREATE TABLE <table name> (
<attribute1> <DATA TYPE>(<size>), FOREIGN KEY (<attribute name>)
REFERENCES <table name> (<attribute name>),...
<attributeN> <DATA TYPE>(<size>) FOREIGN KEY (<attribute name>) REFERENCES
<table name> (<attribute name>))
```

TABLE 7.1
Customers – Items – Orders

Customers		Items		Orders	
Attribute	**Type**	**Attribute**	**Type**	**Attribute**	**Type**
CustomerID	INT(3) *PK*	ItemID	CHAR(6) *PK*	OrderID	INT(3) *PK*
CustLastName	CHAR(15)	ItemDesc	CHAR(25)	CustID	INT(3) *FK*
CustFirstName	CHAR(10)	ItemPrice	INT(5)	ItemID	INT(6) *FK*
				OrderYear	INT(4)
				OrderQuantity	INT(3)

The following Python script is another amendment to the previously developed script, allowing for the specification of a foreign key attribute, and the corresponding tables and reference attributes. It is beyond the scope of this chapter to discuss the numerous possibilities of such tasks in detail, and to provide safety measures against the multitude of cases of incorrect entries that could cause abnormal termination of the program. The goal of this example is to demonstrate how to use Python to facilitate the creation of such relationships in their simplest form using database table `Orders` from Table 7.1:

```
1    import mysql.connector
2
3    # The database config details
4    config = {'user': 'root', 'password': 'root',
5      'host': 'localhost', 'database': 'newDB'}
6
7    # The name of the table and its attributes
8    tableName = input("Enter the name of the table to create: ")
9    sqlString = "CREATE TABLE " + tableName + "("
10   numOfAt = int(input("Enter the number of attributes in the table: "))
11   atName = [""]*numOfAt
12   atType = [""]*numOfAt
13   atSize = [0]*numOfAt
14
15   pkey = 0
16   # Define the structure of the table (i.e., attribute details)
17   for i in range(numOfAt):
18       atName[i] = input("\nEnter the attribute " + str(i) + ": ")
19       atType[i]=str(input("Enter 'CHAR' for char, 'INT' for int type: "))
20       atSize[i] = int(input("Enter the size of the attribute: "))
21       sqlString += atName[i] + " " + atType[i] + \
22           "(" + str(atSize[i]) + ")"
23       if (pkey == 0):
24           primaryKey = input("Is this a primary key (Y/N)? ")
25           if (primaryKey == "Y"):
26               sqlString += " PRIMARY KEY"
27               pkey = 1
28       foreignKey = input("Is this a foreign key (Y/N)? ")
29       if (foreignKey == "Y"):
30           availableTables = "SHOW TABLES"
31           link = mysql.connector.connect(**config)
32           cursor = link.cursor()
33           cursor.execute(availableTables)
34           tables = cursor.fetchall()
35           print(tables)
36           refTable = input("Select the table to reference: ")
37           availableAttributes = "DESC " + str(refTable)
38           link = mysql.connector.connect(**config)
39           cursor = link.cursor()
40           cursor.execute(availableAttributes)
```

```
41              columns = cursor.fetchall()
42              print(columns)
43              refAt = input("Select the attribute to reference: ")
44              sqlString += ", FOREIGN KEY (" + atName[i]
45              sqlString += ") REFERENCES " + str(refTable) + "(" + \
46                  str(refAt) + ")"
47          if (i < numOfAt-1):
48              sqlString += ", "
49          else:
50              sqlString += ")"
51
52      # The SQL statement and the exception handling mechanism
53      print("\nThe SQL statement to run is: \n", sqlString)
54      try:
55          link = mysql.connector.connect(**config)
56          cursor = link.cursor()
57          cursor.execute(sqlString)
58          sqlString = "DESC " + tableName
59          cursor.execute(sqlString)
60          columns = cursor.fetchall()
61          print("\nThe structure/metadata of the table ",
62              str(tableName), "is:")
63          for row in columns:
64              print(row)
65      except:
66          print("There is an error with the connection")
```

Output 7.2.3.d:

```
Enter the name of the cable to create: Orders
Enter the number of attributes in the table: 5

Enter the attribute 0: OrderiD
Enter 'char' for char type, 'int. for int type: int
Enter the size of the attribute: 3
Is this a primary key (Y/N)? Y
Is this a foreign key (Y/N)? n

Enter the attribute 1: CustID
Enter 'char' for char type, 'int. for int type: int
Enter the size of the attribute: 3
Is this a foreign key (Y/N)? Y
[('customers',), ('items',), ('student',), ('table1',)]
Select the table to reference: Customers
[('CustomerID', 'int(3)', 'NO', 'PRI', None, ''), ('CustLastName',
'char(15)', 'YES', '', None, ''), ('CustFirstName', 'char(10)', 'YES', '',
None, '')]
Select the attribute to reference: CustomerID
```

```
Enter the attribute 2: ItemID
Enter 'char' for char type, 'int' for int type: char
Enter the size of the attribute: 6
Is this a foreign key (Y/N)? Y
[('customers',), ('items',), ('student',), ('table1',)]
Select the table to reference: Items
[('ItemID', 'char(6)', 'NO', 'PRI', None, ''), ('ItemDesc', 'char(25)',
'YES', '', None, ''), ('ItemPrice', 'int(5)', 'YES', '', None, '')]
Select the attribute to reference: ItemID

Enter the attribute 3: OrderYear
Enter 'char' for char type, 'int. for int type: int
Enter the size of the attribute: 4
Is this a foreign key (Y/N)? N

Enter the attribute 4: OrderQty
Enter 'char' for char type, 'int' for int type: int
Enter the size of the attribute: 3
Is this a foreign key (Y/N)? N

The SQL statement to run is:
 Create Table Orders(OrderID int(3) Primary key, CustID int(3), Foreign
 Key (CustID) References Customers(CustomerID), ItemID char(6), Foreign
 Key (ItemID) References Items(ItemID), OrderYear int(4), OrderQty int(3))

The structure/metadata of the table Orders is:
('OrderID', 'int(3)', 'NO', 'PRI', None, '')
('CustID', 'int(3)', 'YES', 'MUL', None, '')
('ItemID', 'char(6)', 'YES', 'MUL', None, '')
('OrderYear', 'int(4)', 'YES', '', None, '')
('OrderQty', 'int(3)', 'YES', '', None, '')
```

Once the table is created and references to tables Customers and Items are established, the following Entity Relationship Diagram (ERD) should appear in MySQL Designer (Figure 7.5):

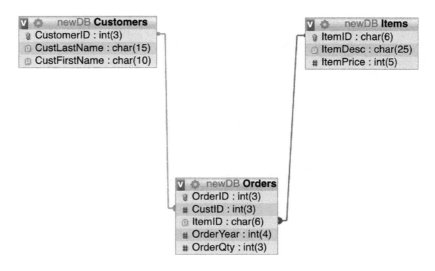

FIGURE 7.5   Entity relationship diagram for the customers-items-orders database.

7.2.4 ALTERING TABLES

As discussed, the CREATE TABLE statement creates new tables and defines their attributes and characteristics. In other words, it is used to create and specify the *metadata* of the table. This metadata is not expected to change frequently; indeed, the better the design of the database the lower the possibility of metadata modification being required. Nevertheless, when necessary, the most drastic way to do so is to destroy and re-create the entire table. This is also the easiest solution provided that the table contains no data. However, the feasibility of using this function is inversely related to the amount of existing data, as destroying the table would also lead to permanent data loss.

This is where the ALTER TABLE statement comes into play. The statement has numerous variations, but they all serve the purpose of altering the structure and metadata of an *existing* table. The most important and frequently used of these variations cover the following:

Observation 7.14 – The ALTER TABLE Statement:

```
ALTER TABLE <name> ADD <new
attribute> <DATA TYPE>(<size>)
ALTER TABLE <name> DROP
<attribute name>
ALTER TABLE <name> CHANGE
<attribute name><attribute
new name> <attribute new DATA
TYPE>(<new size>)
ALTER TABLE <name> ADD (new
attribute) <DATA TYPE>(<size>)
PRIMARY KEY
ALTER TABLE <name> DROP
PRIMARY KEY
```

1. Adding/deleting/modifying an attribute in an existing table.
2. Adding/deleting a primary key constraint.

The first set of statements relates to the manipulation of simple attributes. For instance, if a new attribute is to be added to an existing table, the ALTER TABLE syntax would be the following:

```
ALTER TABLE <table name> ADD <new attribute> <DATA TYPE>(<size>)
```

Accordingly, to delete an existing attribute from a table the statement can be used with following syntax:

```
ALTER TABLE <table name> DROP <attribute name>
```

Modifications of the data type and/or size of an attribute would take the following form:

```
ALTER TABLE <table name> CHANGE <attribute name> <attribute new name>
<attribute new DATA TYPE>(<new size>)
```

The second set of statements involves the addition of a new attribute that also serves as a (composite) primary key or the deletion of the primary key function of an attribute. In the first case, the following syntax should be used:

```
ALTER TABLE <table name> ADD <new attribute> <DATA TYPE>(<size>) PRIMARY KEY
```

In the case of the latter, the syntax would be the following:

```
ALTER TABLE <table name> DROP PRIMARY KEY
```

The following Python script demonstrates the use of all the aforementioned cases in a single application:

```
1    import mysql.connector
2
3    # The database config details
4    config = {'user': 'root', 'password': 'root',
5      'host': 'localhost', 'database': 'newDB'}
6
7    # Show the available tables
8    availableTables = "SHOW TABLES"
9    link = mysql.connector.connect(**config)
10   cursor = link.cursor()
11   cursor.execute(availableTables)
12   tables = cursor.fetchall()
13   print(tables)
14
15   # Select the table to alter and show its attributes
16   selectedTable = input("Select the table to alter: ")
17   availableAttributes = "DESC " + str(selectedTable)
18   link = mysql.connector.connect(**config)
19   cursor = link.cursor()
20   cursor.execute(availableAttributes)
21   columns = cursor.fetchall()
22   for row in columns:
23       print(row)
24
25   # Decide to add a column in the selected table, modify it, or drop it
26   alterType = input("(A)dd a new column\n(M)odify its size\n(D)rop one?\
27         \n(APK)Add Primary Key\n(DPK)Drop Primary Key?\
28         n\Select preferred task: ")
29   if (alterType == "A"):
30       atName = input("\nEnter the attribute name: ")
31       atType = input("Enter 'char' for char type, 'int' for int type: ")
32       atSize = int(input("Enter the size of the attribute: "))
33   if (alterType == "D"):
34       atName = input("\nEnter the name of the attribute to drop: ")
35   if (alterType == "M"):
36       atName = input("\nEnter the name of the attribute to change: ")
37       atNewName = input("\nEnter the new name of the attribute: ")
38       atNewType=input("Enter 'char' for char type, 'int' for int type: ")
```

```
39        atNewSize = int(input("Enter the size of the attribute: "))
40    if (alterType == "APK"):
41        atName = input("\nEnter the name of the attribute to \
42            convert to Primary Key: ")
43        atNewType=input("Enter 'char' for char type, 'int' for int type: ")
44        atNewSize = int(input("Enter the size of the attribute: "))
45
46    # Prepare and execute the alter statement
47    if (alterType == "A"):
48        sqlString = "ALTER TABLE " + str(selectedTable) + " ADD " + \
49            atName + " " + str(atType) + "(" + str(atSize) + ")"
50    elif (alterType == "D"):
51        sqlString = "ALTER TABLE " + str(selectedTable) + \
52            " DROP COLUMN " + str(atName)
53    elif (alterType == "M"):
54        sqlString = "ALTER TABLE " + str(selectedTable) + " CHANGE " + \
55            atName + " " + atNewName + " " + atNewType + \
56                "(" + str(atNewSize) + ");"
57    elif (alterType == "APK"):
58        sqlString="ALTER TABLE "+str(selectedTable)+" ADD "+atName + \
59            " " + atNewType + "(" + str(ateNewSize) + ") PRIMARY KEY"
60
61    elif (alterType == "DPK"):
62        sqlString="ALTER TABLE "+str(selectedTable)+" DROP PRIMARY KEY"
63
64    print(sqlString)
65    try:
66        link = mysql.connector.connect(**config)
67        cursor = link.cursor()
68        cursor.execute(sqlString)
69        print(cursor)
70        sqlString = "DESC " + selectedTable
71        cursor.execute(sqlString)
72        columns = cursor.fetchall()
73        print("\nThe structure/metadata of the table ",
74            str(selectedTable), "is:")
75        for row in columns:
76            print(row)
77    except:
78        print("There is an error with the connection")
```

Output 7.2.4.a: Adding a new attribute

```
[('customers',), ('items',), ('orders',), ('student',), ('table1',)]
Select the table to alter: Student
('Name', 'char(10)', 'YES', '', None, '')
('Address', 'char(15)', 'YES', '', None, '')
('Year', 'int(4)', 'YES', '', None, '')
(A)dd a new column
(M)odify its size
(D)rop one?
(APK)Add Primary Key
(DPK)Drop Primary Key?
Select preferred task: A

Enter the attribute name: MobileNumber
Enter 'char' for char type, 'int' for int type: char
Enter the size of the attribute: 15
Alter table Student add MobileNumber char(15)
MySQLCursor: Alter table Student add MobileNumber cha..

The structure/metadata of the table Student is:
('Name', 'char(13)', 'YES', '', None, '')
('Address', 'char(15)', 'YES', '', None, '')
('Year', 'int(4)', 'YES', '', None, '')
('MobileNumber', 'char(15)', 'YES', '', None, '')
```

Output 7.2.4.b: Modifying an attribute

```
[('customers',), ('items',), ('orders',), ('student',), ('table1',)]
Select the table to alter: Student
('Name', 'char(10)', 'YES', '', None, '')
('Address', 'char(15)', 'YES', '', None, '')
('Year', 'int(4)', 'YES', '', None, '')
('MobileNumber', 'char(15)', 'YES', '', None, '')
(A)dd a new column
(M)odify its size
(D)rop one?
(APK)Add Primary Key
(DPK)Drop Primary Key?
Select preferred task: M

Enter the name of the attribute to change: MobileNumber

Enter the new name of the attribute: PhoneNumber
Enter 'char' for char type, 'int' for int type: char
Enter the size of the attribute: 20
Alter table Student change MobileNumber PhoneNumber char(20);
MySQLCursor: Alter table Student change MobileNumber ..

The structure/metadata of the table  Student is:
('Name', 'char(10)', 'YES', '', None, '')
('Address', 'char(15)', 'YES', '', None, '')
('Year', 'int(4)', 'YES', '', None, '')
('PhoneNumber', 'char(20)', 'YES', '', None, '')
```

Output 7.2.4.c: Deleting/Dropping an attribute

```
[('customers',), ('items',), ('orders',), ('student',), ('table1',)]
Select the table to alter: Student
('Name', 'char(10)', 'YES', '', None, '')
('Address', 'char(15)', 'YES', '', None, '')
('Year', 'int(4)', 'YES', '', None, '')
('PhoneNumber', 'char(20)', 'YES', '', None, '')
(A)dd a new column
(M)odify its size
(D)rop one?
(APK)Add Primary Key
(DPK)Drop Primary Key?
Select preferred task: D

Enter the name of the attribute to drop: PhoneNumber
Alter table Student drop column PhoneNumber
MySQLCursor: Alter table Student drop column PhoneNum..

The structure/metadata of the table  Student is:
('Name', 'char(10)', 'YES', '', None, '')
('Address', 'char(15)', 'YES', '', None, '')
('Year', 'int(4)', 'YES', '', None, '')
```

Output 7.2.4.d: Adding a primary key

```
[('customers',), ('items',), ('orders',), ('student',), ('table1',)]
Select the table to alter: student
('Name', 'char(10)', 'YES', '', None, '')
('Address', 'char(15)', 'YES', '', None, '')
('Year', 'int(4)', 'YES', '', None, '')
(A)dd a new column
(M)odify its size
(D)rop one?
(APK)Add Primary Key
(DPK)Drop Primary Key?
Select preferred task: APK

Enter the name of the attribute to     convert to Primary Key: StudentID
Enter 'char' for char type, 'int' for int type: char
Enter the size of the attribute: 10
Alter table student add StudentID char(10) Primary key
MySQLCursor: Alter table student add StudentID char(1..

The structure/metadata of the table  student is:
('Name', 'char(10)', 'YES', '', None, '')
('Address', 'char(15)', 'YES', '', None, '')
('Year', 'int(4)', 'YES', '', None, '')
('StudentID', 'char(10)', 'NO', 'PRI', None, '')
```

Output 7.2.4.e: Dropping a primary key

```
[('customers',), ('items',), ('orders',), ('student',), ('table1',)]
Select the table to alter: student
('Name', 'char(10)', 'YES', '', None, '')
('Address', 'char(15)', 'YES', '', None, '')
('Year', 'int(4)', 'YES', '', None, '')
('StudentID', 'char(10)', 'NO', 'PRI', None, '')
(A)dd a new column
(M)odify its size
(D)rop one?
(APK)Add Primary Key
(DPK)Drop Primary Key?
Select preferred task: DPK
Alter table student Drop Primary Key
MySQLCursor: Alter table student Drop Primary Key

The structure/metadata of the table student is:
('Name', 'char(10)', 'YES', '', None, '')
('Address', 'char(15)', 'YES', '', None, '')
('Year', 'int(4)', 'YES', '', None, '')
('StudentID', 'char(10)', 'NO', '', None, '')
```

The script allows the user to select the table the metadata of which must be altered. The user is presented with a simple menu that can be used for choosing the type of the execution statement. Upon execution the result is displayed on screen, but can be also verified in MySQL. As the concepts related to the programming aspects of the script have been covered in previous sections, they are not discussed here. The outputs showcase some testing cases based on the developed script.

7.2.5 DROPPING TABLES

The deletion of an entire table, and especially of one that contains data, is not something that one should resort to frequently. Nevertheless, there are occasions that this may be necessary. Assuming that there are no referential integrity relationships between the table in question and any other tables, the deletion can be implemented with the DROP TABLE statement and a simple reference to the name of the table:

> **Observation 7.15 – The DROP TABLE Statement:** Destroys (deletes) a table and all the data contained in it, as in the example below.
> ```
> DROP TABLE <table name>
> ```

```
DROP TABLE <table name>
```

The following Python script demonstrates this by displaying the available tables to the user and offering a mechanism for table selection and deletion to the user:

```
1    import mysql.connector
2
3    # The database config details
4    config = {'user': 'root', 'password': 'root',
5      'host': 'localhost', 'database': 'newDB'}
6
7    # Show the available tables
```

```
8    def showTables():
9        availableTables = "SHOW TABLES"
10       link = mysql.connector.connect(**config)
11       cursor = link.cursor()
12       cursor.execute(availableTables)
13       tables = cursor.fetchall()
14       print(tables)
15
16   # Show the available tables
17   showTables()
18
19   # Select the table to drop and show its attributes
20   selectedTable = input("Select the table to drop: ")
21   availableAttributes = "DESC " + str(selectedTable)
22   link = mysql.connector.connect(**config)
23   cursor = link.cursor()
24   cursor.execute(availableAttributes)
25   columns = cursor.fetchall()
26   for row in columns:
27       print(row)
28
29   # Confirm the decision to drop the table
30   dropConfirmation = input("Are you sure you want to drop \
31           the table (Y/N)? ")
32   if (dropConfirmation == "Y"):
33       sqlString = "DROP TABLE " + str(selectedTable)
34
35       print(sqlString)
36       try:
37           link = mysql.connector.connect(**config)
38           cursor = link.cursor()
39           cursor.execute(sqlString)
40           # Show the available tables
41           showTables()
42       except:
43           print("There is an error with the connection")
```

Output 7.2.5:

```
[('customers',), ('items',), ('orders',), ('student',), ('table1',), ('test',)]
Select the table to drop: test
('test1', 'char(10)', 'NO', 'PRI', None, '')
('test2', 'char(10)', 'YES', '', None, '')
Are you sure you want to drop the table (Y/N)? Y
Drop table test
[('customers',), ('items',), ('orders',), ('student',), ('table1',)]
```

The output shows how to use the DROP TABLE statement to delete/destroy a table and its data. Note that before trying to drop a table (in this instance table Test), one has to ensure that the table has been created and is in existence.

7.2.6 THE DESC STATEMENT

In previous sections, there were instances where the structure or *metadata* of a table had to be displayed. The statement used in such cases was the following:

```
DESC <table name>
```

Observation 7.16 – The DESC Statement: Returns the metadata of a table as in the example below.

```
DESC <table name>
```

This statement returns a list of tuples with the attributes of the table and the associated details, such as its name, size, and primary key designation. The reader can refer to the scripts provided in previous sections as practical examples of its functionality and use.

7.3   SCRIPTING FOR DATA MANIPULATION LANGUAGE

The previous sections introduced the various DDL statements used to create, alter, and drop the metadata of the tables in a database. This is often called the *database schema*. As mentioned, it is not expected nor desired that this schema changes frequently. Once the schema is finalized, one can start working on its *state* or *instance*. A *database instance* contains all the data stored in the database at any particular moment in time. The statements used for working with the database instance are usually referred to as the Data Manipulation Language (DML). As in DDL and the database schema, DML statements are used to create or insert new records to a table, modify and amend data, or delete existing records from a table. The following sections introduce the most basic and common uses of these statements.

7.3.1   INSERTING RECORDS

The INSERT statement is used to insert a single *record (row)* to a table. The general syntax of the statement is the following:

```
INSERT INTO <table name>
VALUES (<attribute1 value>... <attributeN value>)
```

If the user is allowed to insert data to a table in a different order than the one specified in the corresponding table metadata or to enter data selectively to a subset of the table attributes, the following syntax could be used:

```
INSERT INTO <table name>
(<attributeX name>... <attributeZ name>)
VALUES (<attributeX value>... <attributeZ value>)
```

Observation 7.17 – Insert Records:

```
INSERT INTO <table name>
VALUES (<attribute1 value>...
<attributeN value>)
```

If the data order is different than that of the table attributes, or if some attributes are not supposed to receive data, the following syntax can be used:

```
INSERT INTO <table name>
(<attributeX name>...
<attributeZ name>)
VALUES (<attributeX value>...
<attributeZ value>)
```

The following Python script demonstrates the use of the INSERT statement, using a case where the user is also allowed to select the table to which the statement applies first:

```
1    import mysql.connector
2
3    # Provide the established database config
4    GUIDB = 'GuiDB'
```

```
5   config = {'user': "root", 'password': "root",
6       'host': "localhost", 'database': "newDB"}
7
8   # Connect to the newDB database
9   connect = mysql.connector.connect(**config)
10  cursor = connect.cursor()
11
12  try:
13      # Attempt to show the tables of the newDB database
14      cursor.execute("SHOW TABLES")
15      tables = cursor.fetchall()
16      print("DB tables are: " + str(tables))
17  except:
18      print("There was a problem showing tables")
19
20  tableName = input("Enter the table selected: ")
21  try:
22      # Show the table metadata
23      cursor.execute("DESC " + tableName)
24      columns = cursor.fetchall()
25      print("Selected table is: ", tableName)
26      print("Its attributes are: ")
27      for row in columns:
28          print(row)
29
30      # Show the current instance of the table
31      cursor.execute("SELECT * FROM " + str(tableName))
32      records = cursor.fetchall()
33      print("The records in the table are: ")
34      for row in records:
35          print(row)
36  except:
37      print("There was a problem showing the table attributes")
38
39  # Prepare the insert statement
40  numColumns = len(columns)
41  attributes = [""]*numColumns
42  sqlString = "INSERT INTO " + tableName + " VALUES ("
43
44  # Invite user's input for each attribute
45  for i in range(numColumns):
46      attributes[i] = input("Enter data for attribute " + str(i) + ": ")
47      if (columns[i][1][0] == "c"):
48          sqlString += "\"" + attributes[i] + "\""
49      elif (columns[i][1][0] == "i"):
50          sqlString += attributes[i]
51      if (i < numColumns-1):
52          sqlString += ", "
53  sqlString += ")"
54
55  # Execute the prepared insert statement
```

```
56  print("SQL statement to execute is: ")
57  print(sqlString)
58  cursor.execute(sqlString)
59  # Commit the results to ensure they are permanently stored
60  connect.commit()
61
62  # Show the new instance of the table
63  print("The records in the " + str(tableName) + " table are: ")
64  sqlString = "SELECT * FROM " + tableName
65  cursor.execute(sqlString)
66  records = cursor.fetchall()
67  for row in records:
68      print(row)
```

Output 7.3.1.a: Inserting a new record to Student

```
DB tables are: [('customers',), ('items',), ('orders',), ('student',)]
Enter the table selected: student
Selected table is:  student
Its attributes are:
('Name', 'char(10)', 'YES', '', None, '')
('Address', 'char(15)', 'YES', '', None, '')
('Year', 'int(4)', 'YES', '', None, '')
('StudentlD', 'char(10)', 'NO', '', None, '')
The records in the table are:
Enter data for attribute 0: Alex
Enter data for attribute 1: Westwood 7
Enter data for attribute 2: 2002
Enter data for attribute 3: 001
SQL statement to execute is:
Insert into student values ("Alex", "Westwood 7", 2002, "001")
The records in the student table are:
('Alex', 'Westwood 7', 2002, '001')
```

Upon execution, the script displays the tables in the current database and prompts the user to select one of them. Once a selection is made, the user is provided with both the metadata and the instance of the table. Next, the user is invited to enter values for each of the attributes of the table, one at a time. In this case, the more generic, basic syntax is adopted, so the user must enter values for all the attributes of the table in the order dictated when the table was created. After all values are collected, the related INSERT statement is prepared and executed, and its result is *committed*. Finally, the script provides the new instance of the table.

The following observations are also noteworthy in relation to the script and its output. Firstly, any text value that is inserted to a table *always takes single quotes,* while numbers do not. Dates also have a particular, unique format. Secondly, in this particular example, the user attempts to insert a record to the Student table, which has no primary key attribute, and is neither referencing nor being referenced by another table. As this is a rather straightforward case, should any issues arise with the statement these should be likely related to technical connectivity issues between the database, the server, and the connections in the script. Thirdly, when committing the results of the INSERT statement, it is important that the newly inserted data are indeed stored in the table.

One could use the Customers, Items, and Orders tables as a working example. Firstly, the user would enter a new record to the Customers table (note that the table has an attribute that serves as a primary key). The following output illustrates this with the following data: 001, "John", and "Good":

Output 7.3.1.b: Inserting a new record to `Customers`

```
DB tables are: [('customers',), ('items',), ('orders',), ('student',)]
Enter the table selected: customers
Selected table is:  customers
Its attributes are:
('CustomerID', 'int(3)', 'NO', 'PRI', None, '')
('CustLastName', 'char(15)', 'YES', '', None, '')
('CustFirstName', 'char(10)', 'YES', '', None, '')
The records in the table are:
Enter data for attribute 0: 001
Enter data for attribute 1: John
Enter data for attribute 2: Good
SQL statement to execute is:
Insert into customers values (001, "John", "Good")
The records in the customers table are:
(1, 'John', 'Good')
```

Next, let us assume that the user attempts to enter a new record with the following data: 001, "Maria", and "Green". The problem in this case is that the user is attempting to insert a new record with the same value for the primary key (i.e., 001). This will raise an internal error, since MySQL does not allow duplicate values for this attribute. The output shows the error that would be raised in such a case:

Output 7.3.1.c: Attempting to insert a new record to `Customers` with duplicate primary key

```
DB tables are: [('customers',), ('items',), ('orders',), ('student',)]
Enter the table selected: customers
Selected table is:  customers
Its attributes are:
('CustomerID', 'int(3)', 'NO', 'PRI', None, '')
('CustLastName', 'char(15)', 'YES', '', None, '')
('CustFirstName', 'char(10)', 'YES', '', None, '')
The records in the table are:
(1, 'John', 'Good')
Enter data for attribute 0: 001
Enter data for attribute 1: Maria
Enter data for attribute 2: Green
SQL statement to execute is:
Insert into customers values (001, "Maria", "Green")

---------------------------------------------------------------------
~\anaconda3\lib\site-packages\mysql\connector\connection.py in handle_
result(self, packet)
    571             return self._handle_eof(packet)
    572         elif packet[4] == 255:
--> 573             raise errors.get_exception(packet)
    574
    575         # We have a text result set

IntegrityError: 1062 (23000): Duplicate entry '1' for key 'PRIMARY'
```

Following up on the same example, let us assume that the user attempts to insert a record in the Items table, as displayed on the output below:

Output 7.3.1.d: Inserting a record to Items

```
DB tables are: [('customers',), ('items',), ('orders',), ('student',)]
Enter the table selected: items
Selected table is:  items
Its attributes are:
('ItemID', 'char(6)', 'NO', 'PRI', None, '')
('ItemDesc', 'char(25)', 'YES', '', None, '')
('ItemPrice', 'int(5)', 'YES', '', None, '')
The records in the table are:
Enter data for attribute 0: 100
Enter data for attribute 1: Refrigerator
Enter data for attribute 2: 600
SQL statement to execute is:
Insert into items values ("100", "Refrigerator", 600)
The records in the items table are:
('100', 'Refrigerator', 600)
```

The user may also attempt to insert a record in the Orders table. Firstly, let us assume that the user correctly inputs data that correspond to the other two tables (i.e., Customers and Items). The following output illustrates a successful attempt:

Output 7.3.1.e: Inserting a record to Orders

```
DB tables are: [('customers',), ('items',), ('orders',), ('student',)]
Enter the table selected: orders
Selected table is:  orders
Its attributes are:
('OrderID', 'int(3)', 'NO', 'PRI', None, '')
('CustID', 'int(3)', 'YES', 'MUL', None, '')
('ItemID', 'char(6)', 'YES', 'MUL', None, '')
('OrderYear', 'int(4)', 'YES', '', None, '')
('OrderQty', 'int(3)', 'YES', '', None, '')
The records in the table are:
Enter data for attribute 0: 1
Enter data for attribute 1: 1
Enter data for attribute 2: 100
Enter data for attribute 3: 2021
Enter data for attribute 4: 15
SQL statement to execute is:
Insert into orders values (1, 1, "100", 2021, 15)
The records in the orders table are:
(1, 1, '100', 2021, 15)
```

In contrast, if we assume that the user attempts to insert another record to Orders with no consideration towards the corresponding Customers table, an error will be raised:

Output 7.3.1.f: Violating a referential integrity constraint in an INSERT statement

```
DB tables are: (('custorers',), ('items',), ('orders',), ('student',)]
Enter the table selected: orders
Selected table is:  orders
Its attributes are:
('OrderID', 'int(3)', 'NO', 'PRI', None, '')
('CustID', 'int(3)', 'YES', 'MUL', None, '')
('ItemID', 'char(6)', 'YES', 'MUL', None, '')
('OrderYear', 'int(4)', 'YES', '', None, '')
('OrderQty', 'int(3)', 'YES', '', None, '')
The records in the table are:
(1, 1, '100', 2021, 15)
Enter data for attribute 0: 2
Enter data for attribute 1: 2
Enter data for attribute 2: 100
Enter data for attribute 3: 2021
Enter data for attribute 4: 10
SQL statement to execute is:
Insert into orders values (2, 2, "100", 2021, 10)
```

- -

```
IntegrityError                        Traceback (most recent call last)
~\anaconda3\lib\site-packages\mysql\connector\connection.py in _handle_
result(self, packet)
    571               return self._handle_eof(packet)
    572           elif packet[4] == 255:
--> 573               raise errors.get_exception(packet)
    574
    575           # We have a text result set

IntegrityError: 1452 (23000): Cannot add or update a child row: a foreign
            key constraint fails ('newdb'.'orders', CONSTRAINT
            'orders_ibfk_1' FOREIGN KEY ('CustID') REFERENCES
            'customers' ('CustomerID'))
```

These examples provide a basic demonstration of various cases of data insertion to tables, and of potential violations of important constraints like primary and foreign keys. Of course, this is not an exhaustive collection of all possible cases, but it should provide some clarity in terms of working with INSERT statements in Python. Ideally, exception handling should be employed to control as many violation scenarios as possible.

7.3.2 UPDATING RECORDS

Contrary to *data definition* statements, where the case of changing the metadata of a table after its creation is generally undesirable and quite rare, when it comes to *data manipulation* it is necessary to be able to change the data of particular records rather frequently. This is accomplished with the use of the UPDATE statement:

Observation 7.18 – The UPDATE Statement:

```
UPDATE <table name>
SET <attribute1> = <value1>,...,
<attributeN> = <valueN>
WHERE <condition that involves
one or more attributes>
```

```
UPDATE <table name>
SET <attribute1> = <value1>,..., <attributeN> = <valueN>
WHERE <condition that involves one or more attributes>
```

The following Python script is based on the examples developed in the previous sections, and adopts the same user prompts and table selection functions in order to showcase the use of the UPDATE statement, using the Customers table:

```
1    import mysql.connector
2
3    # Provide the established database config
4    GUIDB = 'GuiDB'
5    config = {'user': "root", 'password': "root",
6            'host': "localhost", 'database': "newDB"}
7
8    # Connect to the newDB database
9    connect = mysql.connector.connect(**config)
10   cursor = connect.cursor()
11
12   try:
13           # Attempt to show the tables of the newDB database
14           cursor.execute("SHOW TABLES")
15           tables = cursor.fetchall()
16           print("DB tables are: " + str(tables))
17   except:
18           print("There was a problem showing tables")
19
20   tableName = input("Enter the table selected: ")
21   try:
22           # Show the table metadata
23           cursor.execute("DESC " + tableName)
24           columns = cursor.fetchall()
25           print("Selected table is: ", tableName)
26           print("Its attributes are: ")
27           for row in columns:
28                   print(row)
29
30           # Show the current instance of the table
31           cursor.execute("SELECT * FROM " + str(tableName))
32           records = cursor.fetchall()
33           print("The records in the table are: ")
34           for row in records:
35                   print(row)
36   except:
37           print("There was a problem showing the table attributes")
38
39   # Prepare the update statement
40   attributeSelected = input("Select the attribute to change its values: ")
41   newValue = input("Enter the new value")
42   oldValue = input("Enter the old value")
43   sqlString = "UPDATE " + tableName + " SET " + attributeSelected + \
```

```
44          " = " + "\'" + newValue + "\'" + " WHERE " +
      attributeSelected + \
45          " = " + "\'" + oldValue + "\'"
46
47    # Execute the prepared Update statement
48    print("SQL statement to execute is: ")
49    print(sqlString)
50    cursor.execute(sqlString)
51    # Commit the results to ensure they are permanently stored
52    connect.commit()
53
54    # Show the new instance of the table
55    print("The records in the " + str(tableName) + " table are: ")
56    sqlString = "SELECT * FROM " + tableName
57    cursor.execute(sqlString)
58    records = cursor.fetchall()
59    for row in records:
60          print(row)
```

Output 7.3.2: Updating a record in `Customers`

```
DB tables are: [('customers',), ('items',), ('orders',), ('student',)]
Enter the table selected: customers
Selected table is:  customers
Its attributes are:
('CustomerID', 'int(3)', 'NO', 'PRI', None, '')
('CustLastName', 'char(15)', 'YES', '', None, '')
('CustFirstName', 'char(10)', 'YES', '', None, '')
The records in the table are:
(1, 'John', 'Good')
Select the attribute to change its values: CustLastName
Enter the new valueJames
Enter the old valueJohn
SQL statement to execute is:
Update customers set CustLastName = 'James' where CustLastName = 'John'
```

In addition to the UPDATE statement and its execution, the reader should pay close attention to the requirement to *commit* the results of the execution. The commit() function ensures that the results are permanently stored in the table. It must be also noted that there are several variations of the UPDATE statement, the detailed coverage of which is out of the scope of this chapter. For more detailed information on this topic, the reader is advised to refer to the official MySQL documentation.

7.3.3   DELETING RECORDS

In DML, the deletion of one or more records from a table is handled through the DELETE statement. The general syntax of the statement is the following:

DELETE <table name> WHERE <condition>

Observation 7.19 – The `DELETE` Statement:

DELETE <table name> WHERE <condition>

If the WHERE clause is omitted, all the records of the table are deleted. Nevertheless, the empty table will be still in existence, as the table deletion is a task achieved only through the DROP statement. It must be also noted that the <condition> part is quite flexible and can include various expressions and parameters, such as one or more attributes of the same table, queries related to the same table, or queries from different tables. Finally, it is important to remember that the DELETE statement cannot be executed if the result is *violating referential integrity constraints*.

Using the same example as in previous sections, the following Python script demonstrates a simple use of the DELETE statement:

```
1   import mysql.connector
2
3   # Provide the established database config
4   GUIDB = 'GuiDB'
5   config = {'user': "root", 'password': "root",
6            'host': "localhost", 'database': "newDB"}
7
8   # Connect to the newDB database
9   connect = mysql.connector.connect(**config)
10  cursor = connect.cursor()
11
12  try:
13      # Attempt to show the tables of the newDB database
14      cursor.execute("SHOW TABLES")
15      tables = cursor.fetchall()
16      print("DB tables are: " + str(tables))
17  except:
18      print("There was a problem showing tables")
19
20  tableName = input("Enter the table selected: ")
21  try:
22      # Show the table metadata
23      cursor.execute("DESC " + tableName)
24      columns = cursor.fetchall()
25      print("Selected table is: ", tableName)
26      print("Its attributes are: ")
27      for row in columns:
28          print(row)
29
30      # Show the current instance of the table
31      cursor.execute("SELECT * FROM " + str(tableName))
32      records = cursor.fetchall()
33      print("The records in the table are: ")
34      for row in records:
35          print(row)
36  except:
37      print("There was a problem showing the table attributes")
38
39  # Prepare the Delete statement
40  attributeSelected = input("Select the attribute based on \
41          which to delete a record(s): ")
42  deleteValue = input("Enter the data to delete: ")
```

```
43   sqlString = "DELETE FROM " + tableName + " WHERE " + \
44          attributeSelected + " = " + "\'" + deleteValue + "\'"
45
46   # Execute the prepared Update statement
47   print("SQL statement to execute is: ")
48   print(sqlString)
49   cursor.execute(sqlString)
50   # Commit the results to ensure they are permanently stored
51   connect.commit()
52
53   # Show the new instance of the table
54   print("The records in the " + str(tableName) + " table are: ")
55   sqlString = "SELECT * FROM " + tableName
56   cursor.execute(sqlString)
57   records = cursor.fetchall()
58   for row in records:
59        print(row)
```

Output 7.3.3: Updating a record in Customers

```
DB tables are: [('customers',), ('items',), ('orders',), ('student',)]
Enter the table selected: orders
Selected table is:   orders
Its attributes are:
('OrderID', 'int(3)', 'NO', 'PRI', None, '')
('CustID', 'int(3)', 'YES', 'MUL', None, '')
('ItemID', 'char(6)', 'YES', 'MUL', None, '')
('OrderYear', 'int(4)', 'YES', '', None, '')
('OrderQty', 'int(3)', 'YES', '', None, '')
The records in the table are:
(1, 1, '100', 2021, 15)
Select the attribute based on        which to delete a record(s): 100
Enter the data to delete: 100
SQL statement to execute is:
Delete from orders where 100 = '100'
The records in the orders table are:
```

In the example illustrated in the output, the user selects the only record that has a value of 100 for attribute ItemID in the Orders table. The reader should note how DELETE is prepared based on the user's selections, and how the result is committed using the commit() function.

7.4   QUERYING A DATABASE AND USING A GUI

Querying and reporting data from database tables is arguably the most useful part of database management from the perspective of the user. Thus, it should come as no surprise that the remaining SQL statements are specifically used for these purposes. The available clauses are numerous, and the possibilities for nested queries and for *conditional query execution* render the potential combinations virtually limitless. As such, an exhaustive coverage of every possible case of querying and reporting is not only outside the scope of this chapter, but also a rather futile attempt in general. The focus of this section is to showcase some basic ways to execute querying and reporting tasks, and to demonstrate how GUIs could be utilized for presentation purposes.

7.4.1   THE SELECT STATEMENT

The SELECT statement is used to query and report data from tables. Its most basic and generic syntax does not involve any clauses that dictate additional functionality or selection criteria:

```
SELECT * FROM <table name> WHERE *
```

Such a statement will return all the attributes of the specified table, as the asterisk (*) character is used to

---

Observation 7.20 – The SELECT Statement:

```
SELECT <list of attributes
from one or more tables> OR *
FROM <list of tables>
WHERE <conditions>
```

---

include *all attributes and all conditions.* Selections based on more specific criteria can be built by adding the required clauses:

```
SELECT <list of attributes from one or more tables> OR *
FROM <list of tables>
WHERE <conditions>
```

The <conditions> part specifies the particular requirements that the data must meet in order to be reported, ranging from no conditions to very complicated multi-attribute and multi-table ones. Similarly, the <list of tables> part specifies the tables that must be included in the report. The reader can refer to the rich and readily available collection of related textbooks and resources, providing thorough descriptions of the numerous forms of the detailed syntax clauses and possible refinements (Oracle, 2021a).

The following Python script builds on the previous examples to demonstrate querying and reporting on data from a table (i.e., Customers, Products, Orders), as specified by the user:

```
1    import mysql.connector
2
3    # Provide the established database config
4    GUIDB = 'GuiDB'
5    config = {'user': "root", 'password': "root",
6            'host': "localhost", 'database': "newDB"}
7
8    # Connect to the newDB database
9    connect = mysql.connector.connect(**config)
10   cursor = connect.cursor()
11
12   try:
13           # Attempt to show the tables of the newDB database
14           cursor.execute("SHOW TABLES")
15           tables = cursor.fetchall()
16           print("DB tables are: " + str(tables))
17   except:
18           print("There was a problem showing tables")
19
20   tableName = input("Enter the table selected: ")
21   try:
22           # Show the table metadata
23           cursor.execute("DESC " + tableName)
24           columns = cursor.fetchall()
```

```
25              print("====================")
26              print("Selected table is: ", tableName)
27              print("====================")
28              print("Its attributes are:")
29              for row in columns:
30                      print(row)
31
32              # Show the current instance of the table
33              cursor.execute("SELECT * FROM " + str(tableName))
34              records = cursor.fetchall()
35              print("==============================")
36              print("The records in the table are: ")
37              print("==============================")
38              for row in records:
39                      print(row)
40     except:
41              print("There was a problem showing the table attributes")
```

Output 7.4.1: Reporting data from a table based on user selection

```
DB tables are: [('customers',), ('items',), ('orders',), ('student',)]
Enter the table selected: customers
====================
Selected table is:  customers
====================
Its attributes are:
('CustomerID', 'int(3)', 'NO', 'PRI', None, '')
('CustLastName', 'char(15)', 'YES', '', None, '')
('CustFirstName', 'char(10)', 'YES', '', None, '')
==============================
The records in the table are:
==============================
(1,  'John',  'Good')
(2,  'Norman',  'Chris')
(3,  'Flora',  'Alex')
```

In the case presented here, the output reports all the records from the Customers table.

7.4.2 THE SELECT STATEMENT WITH A SIMPLE CONDITION

The previous section demonstrated the use of simple SELECT statements to report on data of a MySQL table. The complexity of the queries is limited only by the imagination and capabilities of the programmer and the task at hand, since Python provides the facilities and support for highly complex querying and reporting tasks. As a starting point for building more complex tasks, the following Python script invites the user to select a table from an example database and build a query based on the selection. Next, it prompts the user for a particular attribute to base the condition on, and for setting particular preferences for the condition depending on whether the attribute is numerical or text-based:

```
1    import mysql.connector
2
3    # Provide the established database config
```

```
4    GUIDB = 'GuiDB'
5    config = {'user': "root", 'password': "root",
6        'host': "localhost", 'database': "newDB"}
7
8    # Connect to the newDB database
9    connect = mysql.connector.connect(**config)
10   cursor = connect.cursor()
11
12   try:
13       # Attempt to show the tables of the newDB database
14       cursor.execute("SHOW TABLES")
15       tables = cursor.fetchall()
16       print("DB tables are: " + str(tables))
17   except:
18       print("There was a problem showing tables")
19
20   tableName = input("Enter the table selected: ")
21
22   # Show the table metadata
23   cursor.execute("DESC " + tableName)
24   columns = cursor.fetchall()
25   print("=================================================")
26   print("Selected table is: ", tableName)
27   print("=================================================")
28   print("Its attributes are:")
29   for row in columns:
30       print(row)
31
32   # Select the attribute to build the condition
33   print("=================================================")
34   condAttribute = input("Enter the attribute to build the condition: ")
35   typeAttribute = input("Is it a numeric attribute or a text (Num/Text):")
36   if (typeAttribute == "Num"):
37       minCond = int(input("Enter the min value for the attribute"))
38       maxCond = int(input("Enter the max value for the attribute"))
39       sqlStatementCondition = " WHERE "+str(condAttribute)+" >= "+ \
40           str(minCond)+" AND "+str(condAttribute)+" <= "+str(maxCond)
41   if (typeAttribute == "Text"):
42       startingText = input("Enter the starting text of the value to \
43           search for: ")
44       sqlStatementCondition = " WHERE "+str(condAttribute)+" LIKE \'"+ \
45           str(startingText) + "%\'"
46
47   # Show the current instance of the table
48   sqlStatement = "SELECT * FROM " + str(tableName) + sqlStatementCondition
49   print(sqlStatement)
50   cursor.execute(sqlStatement)
51   records = cursor.fetchall()
52   print("===================================")
53   print("The records in the table are: ")
```

```
54  print("=====================================")
55  for row in records:
56      print(row)
```

Output 7.4.2.a – Example 1: Conditionally reporting data based on user selection

```
DB tables are: [('customers',), ('items',), ('orders',), ('student',)]
Enter the table selected: items
=================================================
Selected table is:   items
=================================================
Its attributes are:
('ItemID', 'char(6)', 'NO', 'PRI', None, '')
('ItemDesc', 'char(25)', 'YES', '', None, '')
('ItemPrice', 'int(5)', 'YES', '', None, '')
=================================================
Enter the attribute to build the condition: ItemPrice
Is it a numeric attribute or a text (Num/Text):Num
Enter the min value for the attribute300
Enter the max value for the attribute450
Select * from items where ItemPrice >= 300 and ItemPrice <= 450
=================================
The records in the table are:
=================================
('100', 'RF-100', 300)
('200', 'TV-LG100', 400)
('303', 'PC-3', 400)
```

Output 7.4.2.b – Example 2: Conditionally reporting data based on user selection

```
DB tables are: [('customers',), ('items',), ('orders',), ('student',)]
Enter the table selected: items
=================================================
Selected table is:   items
=================================================
Its attributes are:
('ItemID', 'char(6)', 'NO', 'PRI', None, '')
('ItemDesc', 'char(25)', 'YES', '', None, '')
('ItemPrice', 'int(5)', 'YES', '', None, '')
=================================================
Enter the attribute to build the condition: ItemDesc
Is it a numeric attribute or a text (Num/Text):Text
Enter the starting text of the value to search for: TV
Select * from items where ItemDesc like 'TV%'
=================================
The records in the table are:
=================================
('200', 'TV-LG100', 400)
('201', 'TV-Samsung 100', 550)
('202', 'TV-BenQ', 600)
```

In the output of Example 1 above, the user firstly selects table Items. Next, a list of all the available attributes is presented to the user as a choice for the condition of the SELECT statement. The user selects ItemPrice and is prompted to choose whether it is a numerical or text attribute. As the attribute is numerical, the script offers the option to enter the min and max values. On the contrary, in the output of Example 2, the user selects an attribute that is text-based. Hence, the script offers a different set of prompts and statements, appropriate for the use of the SELECT statement with text-based conditions.

The reader should note that the SELECT statements in both cases are the same as those used in MySQL. The only challenge in this instance is that the programmer has to prepare the final SQL script with the dynamic elements in place. Expectedly, if no dynamic elements are involved in the query (e.g., if the table and the condition are predefined), the preparation of the SELECT statement is less complicated.

7.4.3 THE SELECT STATEMENT USING GUI

Arguably, if one aims to develop a user-oriented application, it is necessary to wrap the application with a user-friendly GUI. An extensive introduction to the most important GUI widgets (e.g., *labels, entry boxes, radio buttons, buttons*) and their application is provided in earlier chapters of this book. In the current context, it is assumed that the focus is on the creation of a grid-based layout that will be used to host the results of the SQL queries. In such a case, a *grid layout manager* could be used. The following Python script showcases the development and execution of a condition-based MySQL SELECT query using a fully deployed GUI:

```
1    import mysql.connector
2    import tkinter as tk
3    from tkinter import ttk
4
5    global tableName, attributeName, radioButton, textVar
6    global minLabel, maxLabel, textualLabel; global textualEntry
7    global selectionsFrame, resultsFrame; global columnName, columnType
8    global minCondScale, maxCondScale; global tablesCombo, columnsCombo
9    global connect, cursor, config; global tables, columns
10   global minCond, maxCond; global minValue, maxValue, numCols
11
12   # Create the frame to select the table for the query and its attributes
13   def selectionGUI():
14       global tables, columns; global tablesCombo, columnsCombo
15       global tableName, radioButton, textVar
16       global selectionsFrame, resultsFrame
17       global minLabel, maxLabel, textualLabel
18       global minCondScale, maxCondScale; global textualEntry
19
20       # The frame for the query selections of the user
21       selectionsFrame=tk.LabelFrame(winFrame, text='Query selections')
22       selectionsFrame.config(bg = 'light grey', fg = 'red', bd = 2,
23               relief = 'sunken')
24       selectionsFrame.grid(column = 0, row = 0)
25
26       # Create the combobox to hold the tables available in the db
27       tablesLabel = tk.Label(selectionsFrame,
```

```
28                 text = "Tables available:", bg = "light grey")
29          tablesLabel.grid(column = 0, row = 0)
30          tablesCombo = ttk.Combobox(selectionsFrame,
31                  textvariable = tableName, width = 15)
32          tablesCombo['values'] = tables; tablesCombo.current(0)
33          tablesCombo.grid(column = 1, row = 0)
34
35          # Button updates the attributes combo based on the table selection
36          updateAttributesButton = tk.Button(selectionsFrame,
37              text = 'Update Attributes', relief = 'raised', width = 15)
38          updateAttributesButton.bind('<Button-1>',
39              lambda event: updateAttributes())
40          updateAttributesButton.grid(column = 2, row = 0)
41
42          # Create the button to run the query
43          runButton = tk.Button(selectionsFrame, text = 'Run Query',
44              relief = 'raised', width = 15)
45          runButton.bind('<Button-1>', lambda event: runQuery())
46          runButton.grid(column = 3, row = 0)
47
48          # Update the columns combo based on the table selection
49          columnsLabel = tk.Label(selectionsFrame,
50              text = "Select attribute:", bg = "light grey")
51          columnsLabel.grid(column = 0, row = 1)
52          columnsCombo = ttk.Combobox(selectionsFrame,
53              textvariable = attributeName, width = 15)
54          columnsCombo.grid(column = 1, row = 1)
55
56          # Check whether selected attribute is numeric or text
57          numericalAttribute = tk.Radiobutton (selectionsFrame,
58              text = 'Numerical\nattribute', width = 10, height = 2,
59              bg = 'light green', variable = radioButton, value = 1,
60              command = radioClicked).grid(column = 2, row = 1)
61          textAttribute = tk.Radiobutton (selectionsFrame,
62              text = 'Text\nattribute', width = 10, height = 2,
63              bg = 'light green', variable = radioButton, value = 2,
64              command = radioClicked).grid(column = 3, row = 1)
65          radioButton.set(1)
66
67          # Create the GUI for the numerical conditional parameters
68          minLabel=tk.Label(selectionsFrame,text="Min value:",bg="light grey")
69          minLabel.grid(column = 0, row = 4); minLabel.grid_remove()
70          minCond = tk.IntVar()
71          minCondScale = tk.Scale (selectionsFrame, length = 200,
72              from_ = 0, to = 10000)
73          minCondScale.config(resolution = 10,
74              activebackground = 'dark blue', orient = 'horizontal')
75          minCondScale.config(bg = 'light blue', fg = 'red',
76              troughcolor = 'cyan', command = onScaleMin)
77          minCondScale.grid(column = 1, row = 4); minCondScale.grid_remove()
```

```
78          maxLabel = tk.Label(selectionsFrame, text = "Max value:",
79              bg = "light grey")
80          maxLabel.grid(column = 2, row = 4); maxLabel.grid_remove()
81          maxCond = tk.IntVar()
82          maxCondScale = tk.Scale (selectionsFrame, length = 200,
83              from_ = 0, to = 10000)
84          maxCondScale.config(resolution = 10, activebackground = 'dark blue',
85              orient = 'horizontal')
86          maxCondScale.config(bg = 'light blue', fg = 'red',
87              troughcolor = 'cyan', command = onScaleMax)
88          maxCondScale.grid(column = 3, row = 4); maxCondScale.grid_remove()
89
90          # Create the GUI for the textual parameters
91          textualLabel = tk.Label(selectionsFrame,
92              text = "Enter text to find:", bg = "light grey")
93          textualLabel.grid(column = 0, row = 5); textualLabel.grid_remove()
94          textVar = tk.StringVar()
95          textualEntry = ttk.Entry(selectionsFrame,
96              textvariable = textVar, width = 20)
97          textualEntry.grid(column = 1, row = 5); textualEntry.grid_remove()
98
99  # Update the attributes table based on the table selection
100 def updateAttributes():
101     global cursor; global tableName, textVar; global tables, columns
102     global tablesCombo, columnsCombo; global numCols
103     global columnName, columnType; global mindCondScale, maxCondScale
104
105     try:
106         # Show the selected table metadata
107         if (str(tableName.get()) != ""):
108             sqlString = "DESC " + str(tableName.get())
109             cursor.execute(sqlString)
110             columns = cursor.fetchall()
111
112             # Reformat the columns list to new useful ones
113             numCols = len(columns)
114             columnName = []; columnType = []
115             for i in range (numCols):
116                 columnName.append(columns[i][0])
117                 columnType.append(columns[i][1])
118                 columns[i] = str(columns[i][0]) + " " + \
119         str(columns[i][1])
120             columnsCombo['values'] = columns
121             columnsCombo.current(0)
122     except:
123         print("There was a problem showing the attributes")
124
125 # Update the attributes table based on the table selection
126 def runQuery():
127     global cursor; global tableName; global tables, columns
```

```
128        global columnsCombo; global numCols, numRows
129        global selectedAttribute; global columnName, columnType
130        global minValue, maxValue; global resultsFrame
131
132        # Empty the results list and the results frame
133        records = []
134        if (resultsFrame != None):
135            resultsFrame.destroy()
136
137        # Prepare the query to run
138        selectedIndex = columnsCombo.current()
139        if (radioButton.get() == 1):
140            sqlStatementCondition = " WHERE " + \
141  str(columnName[selectedIndex]) + \
142            " >= " + str(minValue) + " AND " + \
143  str(columnName[selectedIndex]) + \
144            " <= " + str(maxValue)
145        elif (radioButton.get() == 2):
146            startingText = str(textVar.get())
147            sqlStatementCondition = " WHERE " + \
148  str(columnName[selectedIndex]) + \
149        " LIKE \'" + str(startingText) + "%\'"
150
151        # The frame for the query selections of the user
152        resultsFrame = tk.LabelFrame(winFrame, text = "Query data")
153        resultsFrame.config(bg = 'light grey', fg = 'red', bd = 2,
154            relief = 'sunken')
155        resultsFrame.grid(column = 0, row = 1)
156
157        # Show the current instance of the table
158        sqlStatement = "SELECT * FROM " + str(tableName.get()) + \
159            sqlStatementCondition
160        cursor.execute(sqlStatement)
161        records = cursor.fetchall()
162
163        numRows = len(records)
164
165        for i in range(numRows):
166            for j in range(numCols):
167                # Create the labels to display the columns of results
168                newLabel = tk.Label(resultsFrame, width = 24)
169                if (i%2 == 0):
170                    newLabel.config(text = records[i][j],
171                            bg = "light grey", relief = "sunken")
172                else:
173                    newLabel.config(text = records[i][j],
174                            bg = "light cyan", relief = "sunken")
175                newLabel.grid(column = j, row = i)
176
```

```
177  # Display/hide the relevant conditional parameters depending on
178  # the type of the attribute
179  def radioClicked():
180      global minLabel, maxLabel; global minCondScale, maxCondScale
181      global textualLabel, textualEntry
182
183      if (radioButton.get() == 1):
184          minLabel.grid(); minCondScale.grid(); maxLabel.grid()
185          maxCondScale.grid(); textualLabel.grid_remove();
186          textualEntry.grid_remove()
187
188      if (radioButton.get() == 2):
189          minLabel.grid_remove(); minCondScale.grid_remove()
190          maxLabel.grid_remove()
191          maxCondScale.grid_remove(); textualLabel.grid()
192          textualEntry.grid()
193
194  # Define the method to control the min condition value
195  def onScaleMin(val):
196      global minValue
197      minValue = int(val)
198
199  # Define the method to control the max condition value
200  def onScaleMax(val):
201      global maxValue
202      maxValue = int(val)
203  #=====================================================================
204  # Provide the established database config
205  GUIDB = 'GuiDB'
206  config = {'user': "root", 'password': "root", 'host': "localhost",
207            'database': "newDB"}
208
209  # Connect to the newDB database
210  connect = mysql.connector.connect(**config)
211  cursor = connect.cursor()
212
213  # Basic window frame with the title through tk.Tk() constructor
214  winFrame = tk.Tk()
215  winFrame.config(bg = "grey")
216  winFrame.title("Queries through GUIs")
217
218  try:
219      # Attempt to show the tables of the newDB database
220      cursor.execute("SHOW TABLES")
221      tables = cursor.fetchall()
222  except:
223      print("There was a problem with reporting the tables")
224
225  tableName = tk.StringVar()
```

```
226 attributeName = tk.StringVar()
227 radioButton = tk.IntVar()
228 resultsFrame = None
229 updateAttributes()
230 selectionGUI()
231
232 winFrame.mainloop()
```

Output 7.4.3.a: Using the grid layout manager with a numerical condition query

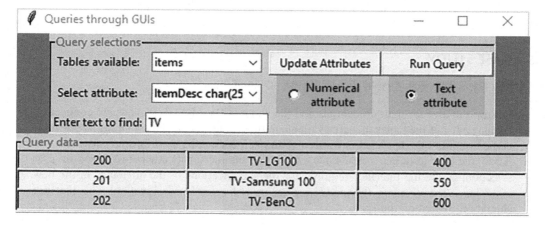

Output 7.4.3.b: Using the grid layout manager with a text-based condition query

Conceptually, this script is divided into four parts. The first part (lines 12–92) provides the GUI element using the `selectionGUI()` function. This covers the main body of the GUI but excludes the grid where the query data will be reported on. When running the application, the user must perform the following actions:

1. Select a table from the connected database through the relevant combo box.
2. Update the combo box using the attributes of the selected table.
3. Select the attribute upon which the condition for the query will be based.
4. Identify whether the attribute is numerical (`int`) or text-based (`char`).

The second part (lines 165–180) provides the necessary functionality for the user to be able to decide the type of the attribute, through the selection of the relevant radio button. This provides the appropriate partial interface that will enable the creation of the condition. The reader should note how the selection causes the partial interfaces to appear/disappear and be replaced by the most appropriate option based on the selection. This can be further enhanced and automated to include as many conditions as needed.

In the third part, function `updateAttributes()` (lines 93–116) is used to update the attributes combo box based on the selected table. Functions `onScaleMin()` and `onScaleMax()` (lines 182–190) are also part of this process, as they allow the user to determine the limits of the condition when a numerical attribute is selected.

Arguably, the most important part of the application is the `runQuery()` function (lines 118–163). The function firstly prepares the query based on the user's preferences, and subsequently runs it based on the prepared condition. Upon execution, the data grid is displayed as required, with a number of columns dictated by the results of the query. The grid is merely an arrangement of a sequence of columns (i.e., per line of the grid layout manager) that is created on-the-spot and loaded with the results of the previously executed query.

In relation to appearance and aesthetics, the reader should also note how the variation of the background color of each new line creates a specific color theme for the grid. It must be stressed that, in this particular application, *the grid consists of labels and it is, thus, not possible to work on it directly*. If a different widget were to be used instead (e.g., *entry boxes*), the contents would be editable and processing (e.g., updating the value of a particular attribute on selected table records) could be applied to the data directly through the grid.

The simple application presented here is just a sample of the use and functionality of the SQL and GUI features provided by Python. As mentioned, SQL provides numerous options and possibilities, and this is reflected on the virtually limitless potential when designing and implementing database applications in Python or other compatible programming languages.

7.5 CASE STUDY

Create an application that provides the following functionality:

a. Prompt the user for their credentials and the name of the MySQL database to connect to. Display a list of the tables that are available in the connected database in a *status bar* form at the bottom of the application window (Hint: A *label* can be used for this purpose).
b. Allow the user to define a new table and set the number of its attributes. Based on user selection, create the interface required for the specifications of the attributes in the new table (i.e., *attribute name, type* and *size, primary* or *foreign key* designation). The interface should be created on-the-spot.

The application must use a GUI interface and the MySQL facilities for the database element.

7.6 EXERCISES

Based on the `Employee` example, write Python scripts to perform the following tasks using MySQL:

1. Create table DEPT to host departmental data for a company, with the following attributes:
   a. Code → DeptNo, Number (2), not null, primary key.
   b. Department name → Dname, 20 characters.
2. Create table EMP to host employee data, with the following attributes:
   c. Code → Empno, Number (4), not null, primary key.
   d. Name (Last and First) → Ename, 40 characters.
   e. Job → Job, 10 characters.
   f. Manager Code → Mgr, Number (4), internal foreign key to Emp → Empno.
   g. Date Hired → Hiredate, date.
   h. Monthly salary → Sal, Number (7, 2), between 100 and 10,000.
   i. Department code → DeptNo, Number (2), foreign key to Dept → DeptNo.
3. Alter table DEPT to include the following attribute: Location → DLocation, 20 characters.
4. Alter table EMP to include the following attribute: Sales Commission → Comm, Number (7, 2), no more than Sal.
5. Insert five records into DEPT.
6. Insert ten records into EMP, two for each department.
7. Delete the record of the department entered last.

REFERENCES

APACHE. (2021). APACHE Software Foundation. https://apache.org.

Elmasri, R., & Navathe, S. (2017). *Fundamentals of Database Systems* (Vol. 7). Pearson, Hoboken, NJ.

MAMP. (2021). Download MAMP & MAMP PRO. https://www.mamp.info/en/downloads.

MySQL. (2021). Oracle Corporation. https://www.mysql.com.

Oracle. (2021a). MySQL Documentation. https://dev.mysql.com/doc.

Oracle. (2021b). Oracle.com.

8 Data Analytics and Data Visualization with Python

Dimitrios Xanthidis
University College London
Higher Colleges of Technology

Han-I Wang
The University of York

Christos Manolas
The University of York
Ravensbourne University London

CONTENTS

8.1 Introduction 320
8.2 Importing and Cleaning Data 322
    8.2.1 Data Acquisition: Importing and Viewing Datasets 322
    8.2.2 Data Cleaning: Delete Empty or NaN Values 324
    8.2.3 Data Cleaning: Fill Empty or NaN Values 326
    8.2.4 Data Cleaning: Rename Columns 327
    8.2.5 Data Cleaning: Changing and Resetting the Index 329
8.3 Data Exploration 329
    8.3.1 Data Exploration: Counting and Selecting Columns 329
    8.3.2 Data Exploration: Limiting/Slicing Dataset Views 331
    8.3.3 Data Exploration: Conditioning/Filtering 332
    8.3.4 Data Exploration: Creating New Data 333
    8.3.5 Data Exploration: Grouping and Sorting Data 336
8.4 Descriptive Statistics 339
    8.4.1 Measures of Central Tendency 340
    8.4.2 Measures of Spread 343
    8.4.3 Skewness and Kurtosis 347
    8.4.4 The describe() and count() Methods 350
8.5 Data Visualization 352
    8.5.1 Continuous Data: Histograms 352
    8.5.2 Continuous Data: Box and Whisker Plot 354
    8.5.3 Continuous Data: Line Chart 356
    8.5.4 Categorical Data: Bar Chart 357
    8.5.5 Categorical Data: Pie Chart 363
    8.5.6 Paired Data: Scatter Plot 364
8.6 Wrapping Up 366
8.7 Case Study 371
References 371

DOI: 10.1201/9781003139010-8

8.1 INTRODUCTION

Python is one of the most popular modern programming languages for *data analytics, data visualization,* and *data science* tasks in general. Indeed, its reputation as a programming language comes from its efficiency in such tasks and the wealth of related facilities and tools it provides. Its power in addressing data analytics problems comes from its numerous built-in libraries, including *Pandas, Numpy, Matplotlib, Scipy,* and *Seaborn.* These libraries provide functionality to read data from a variety of sources, clean data, and perform descriptive and inferential statistics operations. In addition, the libraries provide data visualization facilities, supporting the generation of all types of charts based on the data at hand. Finally, the platform is capable of performing the aforementioned tasks on large collections of data, a task commonly referred to as *big data analytics.*

Observation 8.1 – Data Analytics: Analysis of data from various sources to produce meaningful results that aid the process of decision-making.

Observation 8.2 – Data Visualization: The process of illustrating the results of data analytics through visual means.

Observation 8.3 – Big Data: Data obtained from a large variety of sources, at great velocity, in large amounts of volumes, and in a variety of formats.

A formal definition of the term data analytics may be difficult to come up with, as it is a relatively new and rather broad concept in the contemporary business and academic context. However, a possible description could be that the term refers to *the efficient analysis of data from various sources to produce meaningful results that aid the process of decision-making.* If this was to be extended in order to also capture big data analytics, the associated data would be expected to come from *a large variety of sources, at great velocity (i.e., speed), in vast amounts of volume, and in a serious variety of formats,* as pointed in relevant, contemporary literature. The term *data visualization,* another relatively new concept, refers to common mechanisms of illustrating the results of data analytics in the form of various *charts,* available as visual tools or through built-in methods in programming libraries.

A quick look into any book or resource related to data analytics would unveil that the process is more or less the same, with any minor variations most likely having to do with the terminology rather than the functionality and structure. The latter includes the following seven steps:

- **Research Objectives/Research Question(s):** The first part of the data analytics process is frequently omitted, as it can be deemed as an obvious step. However, it is the most essential part of the process and requires effort to develop. To complicate things further, it is a task of purely investigative nature, so limited support is available in terms of specific and automated tools. It is basically a process seeking to establish the objectives and questions the process is aiming to address for the task at hand at any given instance. It is beyond the scope of this chapter to address these concepts in more detail. For more information, the reader is encouraged to refer to literature related to research methods and methodologies.
- **Data Acquisition:** The process of reading data stored in a variety of formats and sources, including *spreadsheets, comma separated files, web pages,* and *databases.* Once the data is read, it is stored in a specific type of variable called *data frame* for further processing.
- **Cleaning Data:** While the collection of complete and error-free data during the acquisition process is highly desirable, this is seldom the case. Given that the data are entered by users who are often not familiar with the data entry process, it is highly probable and expected to encounter such problems. The process of *data cleaning* focuses on the removal of these types of errors.
- **Exploratory Analysis:** This is a process that comes after data cleaning, with the aim of identifying and summarizing the main characteristics of the data. It often involves the application of *descriptive statistics* methods and analysis.

- **Modeling and Validation:** This process involves the deployment of advanced tools and techniques, such as machine learning, for building models relating to the data. This task covers broad and deep areas of study and expertise that is beyond the scope of this chapter.
- **Visualizing Results:** This task relates to the use of various facilities and programming libraries to create charts that help in visualizing the data and assisting in the process of decision-making.
- **Reporting:** Writing-up of the final reports relating to the data, including any conclusions and recommendations.

It is apparent that the process involves various fields of expertise, including *databases* and *data mining, artificial intelligence/machine learning, statistics, social science*, and others. It is this interdisciplinary nature of the overall process that results in the widely used *data science* term.

As mentioned, the main Python packages and libraries used for data processing and visualization are *Pandas, Numpy, Matplotlib, Scipy,* and *Seaborn.* More specifically, the main characteristics of these libraries are the following:

Observation 8.4 – Data Science: An interdisciplinary field that involves *databases* and *data mining, AI/ machine learning, statistics, social sciences* and other relevant means to analyze and interpret data.

- **NumPy:** A library optimized for working with single and multi-dimensional arrays. A tool suitable for machine learning and statistical analysis tasks.
- **Pandas:** An easy-to-use, open-source library that is based on NumPy. It works particularly well with one and two-dimensional data (*Series* and *DataFrame* respectively). It is a good choice for statistical analysis tasks.
- **SciPy:** Another library based on NumPy. It offers additional functionality compared to NumPy, making it a solid choice for both machine learning and statistical analysis tasks.
- **Matplotlib:** A low level *plotting* library suitable for creating basic graphs. While it provides a lot of freedom to the programmer, it may be rather demanding in terms of coding requirements. One must be also aware of the fact that Matplotlib cannot deal directly with analysis. As such, this needs to be addressed prior to plotting.
- **Python's Statistics:** A built-in Python library for descriptive statistics. It works rather well when datasets are not too large (Statistics — Mathematical Statistics Functions, 2021).

In this chapter, the reader will have the opportunity to acquire basic skills required for cleaning and describing data, and performing data visualization, while familiarising with some of the most popular libraries associated with these tasks. This chapter is divided into four main sections:

- **Data Acquisition and Cleaning:** Import, re-arrange, and clean data from various types of sources.
- **Data Exploration:** Report data by selecting, sorting, filtering, grouping, and/or re-calculating rows/columns, as necessary.
- **Data Processing/Descriptive Statistics:** Apply simple descriptive statistics on the data frame.
- **Data Visualization:** Use the available methods from the various Python packages for data visualization.

Excel files *Grades.xlsx* and *Grades2.xlsx* are used for the various examples presented throughout this chapter.

8.2 IMPORTING AND CLEANING DATA

Before discussing the process of importing data for analysis, there are two key terms that need to be presented: *arrays/lists* and *data frames*. Unlike other common programming languages like C++ or Java, in Python there is no distinct array object. Instead, this functionality is provided by the list object, as discussed in Chapter 2. As a quick reminder, a list is a sequence of variables that hold data of the same data type, sharing the same name, and being distinguished only by their index.

> **Observation 8.5 – Data Frame:** Typically, a two-dimensional data structure with rows representing the data records. Records are divided into columns, and indices are used to speed up the searching process within the data frame.

A *data frame* is a data structure that resembles a relational database table, or an Excel spreadsheet consisting of rows and columns. The *rows* correspond to the actual records of the data frame and are accessed by their index number. The *columns* correspond to the attributes/columns/fields in a database table and are accessed by their names. The *index* is the first column of a data frame (i.e., starting at zero).

8.2.1 DATA ACQUISITION: IMPORTING AND VIEWING DATASETS

The Pandas library is required in order to create the object used to both read the data from the source and create the data frame to which data analysis will be applied. Various sources and formats of data are supported, including *Excel* and *Comma Separated Values (CSV)* files, *tables*, *plain text*, *databases*, or *web-based* sources. In all cases, the basic process of reading from the source remains the same. However, the method and the parameters used may vary slightly, depending on the source.

In the case of reading data from Excel files, the general syntax is the following:

> **Observation 8.6 – The Pandas Library:** The Pandas library provides support for the creation of objects that can be used for various data analytics tasks.

> **Observation 8.7 – Reading from data sets:** Use the read _ excel(), read _ csv(), or read _ html() methods to import (read) data from Excel, CSV, or html files into the data frame.

```
<name of data frame> = <name of Pandas
object>.read_excel("<Filename>", sheet_
name = "<Sheet name>")
```

The following example demonstrates the process of reading data from a particular spreadsheet (*Grades 2020*) within an Excel file (*Grades.xlsx*):

```
1   import pandas as pd
2   dataset = pd.read_excel("Grades.xlsx", sheet_name = "Grades 2020")
3   print(dataset)
```

Output 8.2.1.a:

	Final Grade	Final Exam	Quiz 1	Quiz 2	Midterm Exam	Project
0	58.57	50.5	76.0	70.7	60.0	55
1	65.90	49.0	89.0	63.0	54.0	90
2	69.32	63.5	73.0	54.7	70.0	80

3	72.02	60.5	99.0	74.7	76.0	70
4	73.68	74.0	84.0	53.3	64.0	87
5	61.32	45.5	94.0	42.7	66.0	70
6	67.87	66.5	73.0	53.7	54.0	87
7	75.57	66.0	94.0	58.7	92.0	70
8	61.28	50.5	84.0	37.3	58.0	78
9	0.00	NaN	NaN	NaN	NaN	69
10	62.35	48.0	78.0	49.0	70.0	71
11	66.13	61.0	83.0	45.3	70.0	70
12	69.43	50.0	80.0	49.3	90.0	76
13	82.60	74.0	94.0	65.0	86.0	92
14	0.00	NaN	NaN	NaN	NaN	75
15	62.62	45.5	78.0	56.7	72.0	70
16	0.00	NaN	NaN	NaN	NaN	0

The above script reports 16 rows/records across 6 columns. A few key things are noteworthy in the script output. Firstly, the name of the `read_excel()` method is *case sensitive*. This is in line with the general Python syntax rule for methods and statements used in data analytics tasks. Secondly, as mentioned, it is highly unlikely to deal with perfect, clean data during data analysis. More often than not, one has to deal with erroneous, corrupt, or missing data. The latter applies to both designated *NaN* entries or empty cells. Fortunately, there are easy ways to tackle such problems, some of which are described in the following sections. Finally, it is worth mentioning that in order to report a given dataset the `print()` method can be used. The method comes handy in several situations related to reporting data from datasets and it is further discussed latter in this chapter.

In the case of reading data from a flat CSV file, the general syntax is the following:

```
<name of data frame> = <name of Pandas object>.read_csv("<Filename.csv",
delimiter = ', ')
```

The following script reads and reports the data included in file *Grades2.csv*:

```
1   import pandas as pd
2   dataset = pd.read_csv('Grades2.csv', delimiter = ', ')
3   Dataset
```

Output 8.2.1.b:

	Final Grade	Final Exam	Quiz 1	Quiz 2	Midterm Exam	Project
0	67.47	59.0	70	72.7	70	72
1	75.13	61.5	76	68.3	82	87
2	66.85	77.5	84	52.0	40	80
3	54.45	34.5	62	44.0	44	90
4	76.95	66.5	68	67.0	82	92
5	45.13	26.0	52	26.3	50	68

6	73.23	63.5	96	68.3	62	89
7	81.87	83.0	97	82.7	84	72
8	62.63	54.5	54	31.3	64	87
9	58.75	46.5	54	39.0	52	90
10	49.75	27.5	48	37.0	62	70
11	44.25	21.5	55	18.0	42	80
12	62.52	31.0	85	54.7	68	89
13	47.33	16.5	38	33.3	52	89
14	68.97	55.0	65	49.7	70	94

In the case of reading data from a web page, the general syntax is the following:

```
<name of data frame> = <name of Pandas object>.read_html("<url>")
```

8.2.2 DATA CLEANING: DELETE EMPTY OR NaN VALUES

There are two main techniques to clean a dataset. One has to do with correcting erroneous data and the other with dealing with missing values. The cleaning process may include the partial or complete deletion of the related rows or the replacement of cells that contain missing data with specific calculated or predefined values.

In the case of the former, there are two possible scenarios. Rows may contain missing or designated NaN

Observation 8.8 – Drop NaN or Empty Values: Use the dropna() method to delete rows with NaN or empty values from a data frame. The method must be used with the how parameter ("all" or "any" values).

values, in some or all of its columns. If it is decided to delete all the rows that contain missing data, the following syntax should be used:

```
<name of new Data Frame> = <name of original Data Frame>.dropna()
```

The following script demonstrates the application of the dropna() method that deletes all rows with cells that include NaN values:

```
1   import pandas as pd
2   dataset = pd.read_excel('Grades.xlsx', sheet_name = "Grades 2020")
3   dframe_no_missing_data = dataset.dropna()
4   dframe_no_missing_data
```

Using the dropna(how = "any") method form instead of the simple dropna() form will produce the same result, similarly to deleting any row that contains either NaN or empty values. The full syntax in this case is very similar to the previous one:

```
<name of new Data Frame> = <name of original Data Frame>.dropna(how = "any")
```

The following Python script provides an example of this method applied to the same data frame:

```
1   import pandas as pd
2   dataset = pd.read_excel('Grades.xlsx', sheet_name = "Grades 2020")
3   dframe_delete_rows_with_any_na_values = dataset.dropna(how = "any")
4   dframe_delete_rows_with_any_na_values
```

Output 8.2.2.a:

	Final Grade	Final Exam	Quiz 1	Quiz 2	Midterm Exam	Project
0	58.57	50.5	76.0	70.7	60.0	55
1	65.90	49.0	89.0	63.0	54.0	90
2	69.32	63.5	73.0	54.7	70.0	80
3	72.02	60.5	99.0	74.7	76.0	70
4	73.68	74.0	84.0	53.3	64.0	87
5	61.32	45.5	94.0	42.7	66.0	70
6	67.87	66.5	73.0	53.7	54.0	87
7	75.57	66.0	94.0	58.7	92.0	70
8	61.28	50.5	84.0	37.3	58.0	78
10	62.35	48.0	78.0	49.0	70.0	71
11	66.13	61.0	83.0	45.3	70.0	70
12	69.43	50.0	80.0	49.3	90.0	76
13	82.60	74.0	94.0	65.0	86.0	92
15	62.62	45.5	78.0	56.7	72.0	70

The reader should note that 2 of the 16 original rows are deleted from the data frame as a result of running the two versions of the script, irrespectively of whether the dropna() or dropna(how "any") method form is used.

If it is decided to delete only the rows with *all columns* containing NaN or empty values, the following syntax of the dropna() method should be used:

```
<name of new Data Frame> = <name of original Data Frame>.dropna(how = "all")
```

The following script and its output demonstrate the use of the dropna() method, with parameters that result in *the deletion of rows consisting exclusively of cells with NaN values*. Note that none of the 16 original rows are deleted from the data frame as a result of the method call.

```
1   import pandas as pd
2   dataset = pd.read_excel('Grades.xlsx', sheet_name = "Grades 2020")
3   dframe_delete_rows_with_all_na_values = dataset.dropna(how = "all")
4   dframe_delete_rows_with_all_na_values
```

Output 8.2.2.b:

	Final Grade	Final Exam	Quiz 1	Quiz 2	Midterm Exam	Project
0	58.57	50.5	76.0	70.7	60.0	55
1	65.90	49.0	89.0	63.0	54.0	90
2	69.32	63.5	73.0	54.7	70.0	80
3	72.02	60.5	99.0	74.7	76.0	70
4	73.68	74.0	84.0	53.3	64.0	87
5	61.32	45.5	94.0	42.7	66.0	70
6	67.87	66.5	73.0	53.7	54.0	87
7	75.57	66.0	94.0	58.7	92.0	70
8	61.28	50.5	84.0	37.3	58.0	78
9	0.00	NaN	NaN	NaN	NaN	69
10	62.35	48.0	78.0	49.0	70.0	71
11	66.13	61.0	83.0	45.3	70.0	70
12	69.43	50.0	80.0	49.3	90.0	76
13	82.60	74.0	94.0	65.0	86.0	92
14	0.00	NaN	NaN	NaN	NaN	75
15	62.62	45.5	78.0	56.7	72.0	70
16	0.00	NaN	NaN	NaN	NaN	0

8.2.3　Data Cleaning: Fill Empty or NaN Values

It is often the case that empty cells or cells with NaN values are filled with either predefined values or values calculated based on the rest of the data. In such cases, instead of the dropna() method (in any of its forms), one can use the fillna(<value>, [inplace = true]) method. The general syntax of the method is the following:

> **Observation 8.9 – Fill NaN or Empty Values:** Use the fillna() method to define replacement values for any NaN or empty values encountered.

```
<name of new Data Frame> = <name of original Data Frame>.fillna(value[,
how = 'all'] [, inplace = True])
<name of new Data Frame> = <name of original Data Frame>.fillna(value[,
how = 'any'] [, inplace = True])
```

The value can be defined before running the script, based on existing dataset values and/or other calculations (e.g., using the *mean* of the existing data in the same column). The inplace parameter enables the permanent change of the data in the dataset, if set to true. While the false value can be also used, this would not make much sense, since it is the default value when inplace *is not used*.

The following script and its output demonstrate the use of the `fillna()` method, while also applying the `inplace` parameter to enable the permanent change of the data. The default value used for the modification of empty or missing values is zero. The reader should note that the `inplace` parameter *affects only the dataset resulting from the execution of the script, and not the data source*:

```
1   import pandas as pd
2   dataset = pd.read_excel('Grades.xlsx', sheet_name = "Grades 2020")
3   dataset.fillna(0, inplace = True)
4   dataset
```

Output 8.2.3:

	Final Grade	Final Exam	Quiz 1	Quiz 2	Midterm Exam	Project
0	58.57	50.5	76.0	70.7	60.0	55
1	65.90	49.0	89.0	63.0	54.0	90
2	69.32	63.5	73.0	54.7	70.0	80
3	72.02	60.5	99.0	74.7	76.0	70
4	73.68	74.0	84.0	53.3	64.0	87
5	61.32	45.5	94.0	42.7	66.0	70
6	67.87	66.5	73.0	53.7	54.0	87
7	75.57	66.0	94.0	58.7	92.0	70
8	61.28	50.5	84.0	37.3	58.0	78
9	0.00	0.0	0.0	0.0	0.0	69
10	62.35	48.0	78.0	49.0	70.0	71
11	66.13	61.0	83.0	45.3	70.0	70
12	69.43	50.0	80.0	49.3	90.0	76
13	82.60	74.0	94.0	65.0	86.0	92
14	0.00	0.0	0.0	0.0	0.0	75
15	62.62	45.5	78.0	56.7	72.0	70
16	0.00	0.0	0.0	0.0	0.0	0

8.2.4 DATA CLEANING: RENAME COLUMNS

It is sometimes required to change the column headings in a dataset. This is especially true in the case of formal reports, where clarity and appearance are key. In such cases, the `rename()` method is used. The method allows for the temporary change of the column heading without affecting the original dataset at the source.

Observation 8.10 – rename(): Use the `rename()` method to change the column heading appearance. Use the `set` notation to dictate the old and new (temporary) column names.

The general syntax is the following:

```
df.rename(columns = {"oldname": "newname", } [, inplace=True])
```

As in the previous case, if the `inplace` parameter is used, the column names will be changed for the resulting dataset, but the source data will not be affected. The most crucial aspect of the syntax is that the programmer can change any number of column names just by separating them using commas:

```
1  import pandas as pd
2  dataset = pd.read_excel('Grades.xlsx', sheet_name = "Grades 2020")
3  dataset_new = dataset.rename(columns = {"Final Grade": "Total Grade",
4      "Quiz 1": "Test 1", "Quiz 2": "Test 2", "Midterm Exam": "Midterm"})
5  dataset_new
```

Output 8.2.4:

	Total Grade	Final Exam	Test 1	Test 2	Midterm	Project
0	58.57	50.5	76.0	70.7	60.0	55
1	65.90	49.0	89.0	63.0	54.0	90
2	69.32	63.5	73.0	54.7	70.0	80
3	72.02	60.5	99.0	74.7	76.0	70
4	73.68	74.0	84.0	53.3	64.0	87
5	61.32	45.5	94.0	42.7	66.0	70
6	67.87	66.5	73.0	53.7	54.0	87
7	75.57	66.0	94.0	58.7	92.0	70
8	61.28	50.5	84.0	37.3	58.0	78
9	0.00	NaN	NaN	NaN	NaN	69
10	62.35	48.0	78.0	49.0	70.0	71
11	66.13	61.0	83.0	45.3	70.0	70
12	69.43	50.0	80.0	49.3	90.0	76
13	82.60	74.0	94.0	65.0	86.0	92
14	0.00	NaN	NaN	NaN	NaN	75
15	62.62	45.5	78.0	56.7	72.0	70
16	0.00	NaN	NaN	NaN	NaN	0

The reader should note the use of the `set` notation to declare the pairs of column names (i.e., old and new) when changing them. It must be also noted that, in order for the change to apply, the result of the `rename()` method must be assigned to a new dataset before it is reported.

8.2.5   DATA CLEANING: CHANGING AND RESETTING THE INDEX

The *index* of a dataset is important, as it can speed up the process of data searching. This is particularly relevant when searching for or sorting data on a column of the dataset different than the one the focus is on. In such a case, it is convenient to temporarily change the indexed column to perform the task at hand, and return back to the original state by resetting the index to its original column once this is completed. The general syntax for changing and resetting the index in a dataset is the following:

Observation 8.11 – set_index(), reset_index(): Use the set_index() and reset_index() methods to set the index of the dataset to another column and restore it back to the original one.

```
<name of dataset>.set_index("<column name>" [, inplace=True])
<name of dataset>.reset_index([inplace=True])
```

8.3   DATA EXPLORATION

Data exploration is an umbrella term, encompassing processes used to report data in various different ways. For example, it may refer to the process of row/column selection for inclusion in the report, or to facilities used to sort and/or filter data based on certain, defined conditions. If necessary, it offers options to group the data in one or more columns and the functionality to create new columns based on calculations on existing ones. This section will explore some of the most important concepts and methods related to data exploration.

8.3.1   DATA EXPLORATION: COUNTING AND SELECTING COLUMNS

Three of the basic methods and parameters used in order to view the data of a dataset are len(), columns, and shape. The len() method reports the number of records in the dataset. The general syntax is the following:

Observation 8.12 – len(): Use the len() method and the columns and shape attributes of a dataset to report the number of its records, the names of its attributes, and the number of its records and columns, respectively.

```
len(<name of dataset>)
```

The columns attribute can be used to get a list of the available columns in the dataset, with the following syntax:

```
<name of dataset>.columns
```

Finally, the *shape* attribute can be used to report the number of records and columns in a dataset:

```
<name of dataset>.shape
```

The following script uses all three of the above, while also including a basic statement to display all the data in the dataset:

```
1   import pandas as pd
2   dataset = pd.read_excel('Grades.xlsx', sheet_name = "Grades 2020")
3   dataset[["Final Grade", "Final Exam", "Quiz 1", "Quiz 2",
4        "Midterm Exam", "Project"]]
5   dataset
6   len(dataset)
7   dataset.columns
8   dataset.shape
```

Output 8.3.1.a: Basic exploration methods without print

```
(17, 6)
```

It should be noted that the script fails to display all the requested output. Instead, it displays only the result of the application of shape: the number of records and columns. If it is necessary to display all the requested information, the print() method should be used, as in the amended version of the script below:

```
1   import pandas as pd
2   dataset = pd.read_excel('Grades.xlsx', sheet_name = "Grades 2020")
3   dataset[["Final Grade", "Final Exam", "Quiz 1", "Quiz 2",
4       "Midterm Exam", "Project"]]
5   print(dataset)
6   print("The dataset has", len(dataset), "records")
7   print("The columns in the dataset are:", dataset.columns)
8   print("The number of records is:", dataset.shape[0])
9   print("The number of columns is:", dataset.shape[1])
```

Output 8.3.1.b: Basic exploration methods using print

```
    Final Grade  Final Exam  Quiz 1  Quiz 2  Midterm Exam  Project
0         58.57        50.5    76.0    70.7          60.0       55
1         65.90        49.0    89.0    63.0          54.0       90
2         69.32        63.5    73.0    54.7          70.0       80
3         72.02        60.5    99.0    74.7          76.0       70
4         73.68        74.0    84.0    53.3          64.0       87
5         61.32        45.5    94.0    42.7          66.0       70
6         67.87        66.5    73.0    53.7          54.0       87
7         75.57        66.0    94.0    58.7          92.0       70
8         61.28        50.5    84.0    37.3          58.0       78
9          0.00         NaN     NaN     NaN           NaN       69
10        62.35        48.0    78.0    49.0          70.0       71
11        66.13        61.0    83.0    45.3          70.0       70
12        69.43        50.0    80.0    49.3          90.0       76
13        82.60        74.0    94.0    65.0          86.0       92
14         0.00         NaN     NaN     NaN           NaN       75
15        62.62        45.5    78.0    56.7          72.0       70
16         0.00         NaN     NaN     NaN           NaN        0
The dataset has 17 records
The columns in the dataset are: Index(['Final Grade', 'Final Exam',
'Quiz 1', 'Quiz 2', 'Midterm Exam', 'Project'], dtype='object')
The number of records is: 17
The number of columns is: 6
```

As shown above, it is possible to improve the output appearance by adding appropriate text through the print() method. Obviously, the presentation of the results could be further improved with the use of more elaborate presentation techniques and tools, such as an appropriate GUI.

8.3.2 DATA EXPLORATION: LIMITING/SLICING DATASET VIEWS

It is often the case that it is impractical to display all the data in a single report. This is especially true when working with very large datasets. In such cases, it is preferable to display just a sample of the dataset, by limiting the number of records and/or columns. There are a number of methods that can be used for this task. Methods head(n) and tail(n) restrict the number of the displayed records, either at the top or the bottom of the dataset. The general syntax is the following:

Observation 8.13 – head(), tail(): Use the head(n) and tail(n) methods to restrict the number of displayed records from the top and bottom of the dataset. Use the loc[] or iloc[] attributes to restrict the report to the specified rows and columns using labels or indices.

```
<name of dataset>.head(number of rows from the top)
<name of dataset>.tail(number of rows from the bottom)
```

Methods loc[] and iloc[] can be used to restrict the displayed results based on specific rows and/or columns:

```
<name of dataset>[start record number: end record number [: step]
<name of dataset>.loc[start record number: end record number [: step],
"<start column name>": "<end column name>"]
<name of dataset>.iloc[[start record number: end record number, start
column index: end column index]
```

The practical application of these methods and attributes is demonstrated in the following script:

```
1    import pandas as pd
2    dataset = pd.read_excel('Grades.xlsx', sheet_name = "Grades 2020")
3    print(dataset.head(5))
4    print(dataset.tail(5))
5    print(dataset[0:37:5])
6    print(dataset.loc[0:5,"Final Grade": "Final Exam"])
7    print(dataset.iloc[0:5,0:3])
```

Output 8.3.2:

	Final Grade	Final Exam	Quiz 1	Quiz 2	Midterm Exam	Project
0	58.57	50.5	76.0	70.7	60.0	55
1	65.90	49.0	89.0	63.0	54.0	90
2	69.32	63.5	73.0	54.7	70.0	80
3	72.02	60.5	99.0	74.7	76.0	70
4	73.68	74.0	84.0	53.3	64.0	87
	Final Grade	Final Exam	Quiz 1	Quiz 2	Midterm Exam	Project
12	69.43	50.0	80.0	49.3	90.0	76
13	82.60	74.0	94.0	65.0	86.0	92
14	0.00	NaN	NaN	NaN	NaN	75
15	62.62	45.5	78.0	56.7	72.0	70
16	0.00	NaN	NaN	NaN	NaN	0
	Final Grade	Final Exam	Quiz 1	Quiz 2	Midterm Exam	Project
0	58.57	50.5	76.0	70.7	60.0	55
5	61.32	45.5	94.0	42.7	66.0	70
10	62.35	48.0	78.0	49.0	70.0	71
15	62.62	45.5	78.0	56.7	72.0	70

```
      Final Grade    Final Exam
0            58.57         50.5
1            65.90         49.0
2            69.32         63.5
3            72.02         60.5
4            73.68         74.0
5            61.32         45.5
      Final Grade    Final Exam   Quiz 1
0            58.57         50.5     76.0
1            65.90         49.0     89.0
2            69.32         63.5     73.0
3            72.02         60.5     99.0
4            73.68         74.0     84.0
```

In the output, the reader will notice that with the application of head(5) and tail(5), only the five first and last records of the dataset are displayed (with all their columns). Next, records are displayed in intervals of five, starting from zero and ending with the last records of the dataset. The next section displays six records of the dataset using only the first three columns (inclusive of the index of the dataset). In a similar way, the last section shows the first five records using only the first four columns (inclusive of the index of the dataset), but the columns are specified by their index and not their names. If it is required to report on non-sequential columns, these columns must be included in square brackets ([]) and separated by commas.

8.3.3 DATA EXPLORATION: CONDITIONING/FILTERING

Expectedly, Pandas also offers a set of methods that allow for the filtering of the displayed data through *conditioning*. For instance, the unique() method displays only the first occurrence of *recurring* data values from the specified column:

Observation 8.14 – unique(): Use the unique() method and the square bracket ([]) list notation to report unique data in a dataset based on a specified column and to set the conditions for the reported records.

```
<name of dataset>["<name of column>"].
unique()
```

It is also possible to define a particular condition that limits the displayed results like in the case of an if statement. The condition can be simple (single) or complex. The general syntax is the following:

```
<name of dataset>[<condition>]
<name of dataset> [<condition>[&/|] <condition>]]
```

The following script uses the data from the *Grades.xlsx* file to identify unique grades for the project, and report all final grades with a percentage higher than 80% and between 1% and 59%:

```
1   import pandas as pd
2   dataset = pd.read_excel('Grades.xlsx')
3   print("Unique grades for project:", dataset["Project"].unique())
4   print("Final grades more than 80%:\n",
5         dataset[dataset["Final Grade"] > 80])
6   print("Final grades 1% to 60%:\n", dataset[(dataset["Final Grade"] > 0)
7         & (dataset["Final Grade"] < 60)])
```

Output 8.3.3:

```
Unique grades for project: [55 90 80 70 87 78 69 71 76 92 75  0]
Final grades more than 80%:
     Final Grade  Final Exam  Quiz 1  Quiz 2  Midterm Exam  Project
13          82.6        74.0    94.0    65.0          86.0       92
Final grades 1% to 60%:
     Final Grade  Final Exam  Quiz 1  Quiz 2  Midterm Exam  Project
0           58.57       50.5    76.0    70.7          60.0       55
```

The reader should note that it is possible to limit the displayed columns if the `loc[]` parameter is also used, although this is not shown in the current script and its output. It is also worth mentioning that, in a compound condition like the second one in the example, instead of using the `and` or `or` keywords one can use `&` and `|` operators respectively.

8.3.4 DATA EXPLORATION: CREATING NEW DATA

As part of the data exploration process, it is sometimes necessary to create new data. This can take four different forms:

- Merging two or more datasets into one.
- Creating a new column with data *derived* from other available data sources, in the same or other datasets.
- Creating a new column with data *calculated* from other available data sources, in the same or other datasets.
- Creating a new file of a certain file type (e.g., Excel, CSV).

The `append()` method is used to merge two or more datasets. The basic syntax is the following:

```
<name of new dataset> = <name of first
old dataset>.append(<name of second old
dataset>)
```

To create a new column with values calculated based on data of other columns one can use the following command:

```
<name of dataset>["<name of new column>"]
= expression with other columns
```

If the newly created column is based on certain conditions applied to data from other columns the following commands could be used instead:

```
<name of dataset>["<name of new column>"]
= np.where(condition, value if True,
value if False)
```

or

```
<name of dataset>["<name of new column>"] =
np.select(<condition set>, <set of values>)
```

Observation 8.15 – Create New Column: Use the following expression and syntax to create a new column based on the values of other columns from the same or other datasets:

```
<name of dataset>["<name of
new column>"] = expression
with other columns
```

Observation 8.16 – Create a New Column Using `np.where()` or `np.select()`: Use Numpy's `np.where()` or `np.select()` methods and the following syntax to create a new column based on a simple or complex condition. This can include other columns from the same or other datasets:

```
<name of dataset>["<name of
new column>"] = np.where
(condition, value if True,
value if False)
<name of dataset>["<name of
new column>"] = np.select
(<condition set>, <set of
values>)
```

Finally, to create a new dataset and store it in a file, one of the following command structures could be used. The examples provided here cover Excel and CSV files, but the same logic also applies to other data file formats.

```
Excel files:
<name of new Excel file object> =
pd.ExcelWriter("<name of new Excel file>")
<name of dataset>.to_excel(<name of new
Excel file object>, "sheet name")
<name of new Excel file object>.save()

CSV files:
<name of dataset>.to_csv("<name of new
CSV file>")
```

Observation 8.17 – Create a New Excel File: Use the following syntax to create a new Excel file from a given dataset:

```
<name of new Excel file
object> = pd.ExcelWriter
("<name of new Excel file>")
<name of dataset>.to_excel
(<name of new Excel file
object>, "sheet name")
<name of new Excel file
object>.save()
```

Using the *Grades.xlsx* dataset as an example, student grades are stored in a particular section of a course and in a particular semester. If another dataset for the same course but a different section exists in another file (e.g., *Grades2.csv*), it may be useful to merge the two and perform the necessary processes in the newly created dataset. The following script reads two different files (i.e., Excel and CSV), reports their data, appends the second dataset at the end of the first, defines the condition, and creates a new column with values calculated from the data of other columns. Finally, it saves the new dataset in both Excel and CSV formats:

Observation 8.18 – Create a New CSV File: Use the following syntax to create a new CSV file from a given dataset:

```
<name of dataset>.to_csv
("<name of new CSV file>")
```

```
1    import pandas as pd
2    import numpy as np
3
4    dataset1 = pd.read_excel("Grades.xlsx")
5    print("The data in Grades file are:"); print(dataset1.head(3))
6    dataset2 = pd.read_csv('Grades2.csv')
7    print("The data in Grades2 file are:"); print(dataset2.tail(3))
8    dataset = dataset1.append(dataset2)
9    print("The new merge dataset is:"); print(dataset.head(3))
10   print(dataset.tail(3))
11
12   # The conditions for the Letter Grades
13   conditions = [(dataset["Final Grade"] > 90.0),
14       (dataset["Final Grade"] > 80.0) & (dataset["Final Grade"] <= 89.9),
15       (dataset["Final Grade"] > 70.0) & (dataset["Final Grade"] <= 79.9),
16       (dataset["Final Grade"] > 60.0) & (dataset["Final Grade"] <= 69.9),
17       (dataset["Final Grade"] < 59.9)
18       ]
19
20   # The list of Grade Letters based on the conditions
21   gradeLetters = ["A", "B", "C", "D", "F"]
22   # Create a new Letter Grades column in the new dataset using numpy
23   dataset["Letter Grade"] = np.select(conditions, gradeLetters)
```

```
24  dataset["Course Work"] = dataset["Quiz 1"]*0.1+dataset["Quiz 2"]*0.1+ \
25      dataset["Midterm Exam"]*0.25 + dataset["Project"]*0.25
26  print("A partial view of the new dataset:")
27
28  # Find the number of records in the dataset
29  rowNum = len(dataset)
30  # Select the columns to be displayed in the report
31  cols = [7, 1, 0, 6]
32  print(dataset.iloc[:rowNum:5, cols])
33
34  # Save the new dataset as an Excel file
35  newExcel = pd.ExcelWriter("NewGrades.xlsx")
36  dataset.to_excel(newExcel, "New Data")
37  newExcel.save()
38
39  # Save the new dataset as a CSV file
40  dataset.to_csv("newGrades.csv")
```

Output 8.3.4:

```
The data in Grades file are:
     Final Grade   Final Exam   Quiz 1   Quiz 2   Midterm Exam   Project
0          58.57         50.5     76.0     70.7           60.0        55
1          65.90         49.0     89.0     63.0           54.0        90
2          69.32         63.5     73.0     54.7           70.0        80
The data in Grades2 file are:
     Final Grade   Final Exam   Quiz 1   Quiz 2   Midterm Exam   Project
12         62.52         31.0       85     54.7             68        89
13         47.33         16.5       38     33.3             52        89
14         68.97         55.0       65     49.7             70        94
The new merge dataset is:
     Final Grade   Final Exam   Quiz 1   Quiz 2   Midterm Exam   Project
0          58.57         50.5     76.0     70.7           60.0        55
1          65.90         49.0     89.0     63.0           54.0        90
2          69.32         63.5     73.0     54.7           70.0        80
     Final Grade   Final Exam   Quiz 1   Quiz 2   Midterm Exam   Project
12         62.52         31.0     85.0     54.7           68.0        89
13         47.33         16.5     38.0     33.3           52.0        89
14         68.97         55.0     65.0     49.7           70.0        94
A partial view of the new dataset:
     Course Work   Final Exam   Final Grade   Letter Grade
0          43.42         50.5         58.57              F
5          47.67         45.5         61.32              D
10         47.95         48.0         62.35              D
15         48.97         45.5         62.62              D
3          44.10         34.5         54.45              F
8          46.28         54.5         62.63              D
13         42.38         16.5         47.33              F
```

Some key observations can be made based on this script. Firstly, it is possible, and indeed common, for the programmer to require the merging of datasets from files of different file types. In this instance, the script merges a dataset stored in an Excel file with one in a CSV file. Secondly, although it is possible to use multiple lines of code to define the values of a new column based on different conditions, a more efficient option is to use the np.where() method to define the conditions and their paired values in advance, and subsequently use the np.select() method from the *Numpy* library. Thirdly, it is possible to create a new column based on simple or complex expressions that include other columns. Fourthly, it may be more convenient to define the displayed records and columns as variables and use them in a statement, rather than directly adding the associated constraints to the statement. Finally, the reader should note that the sequence of statements used to create a new Excel file is different than that for a CSV file. Such differences also exist for files of other formats.

8.3.5 DATA EXPLORATION: GROUPING AND SORTING DATA

Data grouping is one of the most important data processing tasks, and is usually carried out before other tasks commence. This is commonly coupled with *data sorting*, and the two tasks together constitute a key building block for the production of professional reports. Unsurprisingly, Python provides facilities for both of these tasks.

In order to group data within a dataset, the groupby() method can be used. The general syntax is the following:

Observation 8.19 – Grouping Data: Use the groupby() method to group a dataset based on one or more columns. The method must be used with either an *aggregate* method (e.g., mean()) or with the apply(lambda x: x[...]) statement for non-aggregate groupings.

```
<name of dataset>.groupby(["<name of column>" [, "<name of column>",
...]]).<aggregate function>
```

It must be noted that the method requires the application of an *aggregation* (e.g., mean) to the grouped data, a concept covered in the following section. Alternatively, if the goal is to simply display the report grouped by a specific column, the apply() method can be used with the following syntax:

```
<name of dataset>.groupby(["<name of column>" [, "<name of column>",
...]]).apply(lambda x: x[<rows>, <cols>])
```

The apply() method replaces the aggregation with the lambda x: x[...] expression in order to specify the records and columns that should be displayed in the report.

The reader should also note that if more than one column is used for the grouping, the data will be *initially grouped based on the firstly selected column*. After that point, data will be grouped in each separate group based on the second column.

For the purposes of data sorting, the sort_values() method is used. The general syntax is the following:

```
<name of dataset>.sort_values(["<name of
column>" [, "<name of column>", ...]] [,
ascending = False])
```

Observation 8.20 – Sorting Data: Use the sort_values() method to sort a dataset based on one or more specified columns.

As with data grouping, the reader should note that if more than one column is specified, the data with the same value are sorted based on the first column.

Finally, it is possible to combine the functionality of groupby() and sort_values() by firstly applying the former and assigning the result to the lambda expression, and then applying the sort_values() method to the lambda expression.

The following script reads a CSV file and groups and reports its data based on the *Letter Grade* column, displaying only columns *Letter Grade* and *Final Grade*. Next, it creates a second dataset and sorts the values based on the *Final Grade* in ascending order. Finally, it utilizes the `apply()` method to group the data based on *Letter Grade* and sort them based on *Final Grade*:

```
1    import pandas as pd
2
3    dataset = pd.read_csv('newGrades.csv')
4
5    # Report the number of records in the dataset
6    rows = len(dataset)
7
8    # Report the records grouped by Letter Grade
9    dataset1 = dataset[["Letter Grade", "Final Grade"]]
10   print(dataset1.groupby(["Letter Grade"]).apply(lambda x: x[0:rows]))
11
12   # Report the records sorted by Final Grade
13   dataset2 = dataset[["Letter Grade", "Final Grade"]]
14   print(dataset2.sort_values(["Final Grade"], ascending = False))
15
16   # Report the records firstly grouped by Letter Grade and
17   # then sorted by Final Grade (within groups)
18   dataset3 = dataset[["Letter Grade", "Final Grade"]]
19   print(dataset3.groupby(["Letter Grade"]).
20           apply(lambda x: x.sort_values(["Final Grade"], ascending=False)))
```

Output 8.3.5.a–8.3.5.c:

		Letter Grade	Final Grade
Letter Grade			
B	13	B	82.60
	24	B	81.87
C	3	C	72.02
	4	C	73.68
	7	C	75.57
	18	C	75.13
	21	C	76.95
	23	C	73.23
D	1	D	65.90
	2	D	69.32
	5	D	61.32
	6	D	67.87
	8	D	61.28
	10	D	62.35
	11	D	66.13
	12	D	69.43
	15	D	62.62
	17	D	67.47
	19	D	66.85

	25	D	62.63
	29	D	62.52
	31	D	68.97
F	0	F	58.57
	9	F	0.00
	14	F	0.00
	16	F	0.00
	20	F	54.45
	22	F	45.13
	26	F	58.75
	27	F	49.75
	28	F	44.25
	30	F	47.33

	Letter Grade	Final Grade
13	B	82.60
24	B	81.87
21	C	76.95
7	C	75.57
18	C	75.13
4	C	73.68
23	C	73.23
3	C	72.02
12	D	69.43
2	D	69.32
31	D	68.97
6	D	67.87
17	D	67.47
19	D	66.85
11	D	66.13
1	D	65.90
25	D	62.63
15	D	62.62
29	D	62.52
10	D	62.35
5	D	61.32
8	D	61.28
26	F	58.75
0	F	58.57
20	F	54.45
27	F	49.75
30	F	47.33
22	F	45.13
28	F	44.25
14	F	0.00
9	F	0.00
16	F	0.00

		Letter Grade	Final Grade
Letter Grade			
B	13	B	82.60
	24	B	81.87
C	21	C	76.95
	7	C	75.57

	18	C	75.13
	4	C	73.68
	23	C	73.23
	3	C	72.02
D	12	D	69.43
	2	D	69.32
	31	D	68.97
	6	D	67.87
	17	D	67.47
	19	D	66.85
	11	D	66.13
	1	D	65.90
	25	D	62.63
	15	D	62.62
	29	D	62.52
	10	D	62.35
	5	D	61.32
	8	D	61.28
F	26	F	58.75
	0	F	58.57
	20	F	54.45
	27	F	49.75
	30	F	47.33
	22	F	45.13
	28	F	44.25
	9	F	0.00
	14	F	0.00
	16	F	0.00

The output shows the results of the reports for the three datasets. From left to right, the output shows the results of `groupby()` based on *Letter Grade*, the results of `sort_values()` based on *Final Grade*, and the dataset grouped by *Letter Grade* and sorted by *Final Grade*. The reader should note that, in this instance, the outputs are presented side-by-side for demonstration purposes, but in a more realistic scenario they should be presented in succession, as dictated by the actual output.

8.4 DESCRIPTIVE STATISTICS

Descriptive statistics are defined as the analysis of data that describe, show, or summarize information in a meaningful manner. They are simply a way of describing the data and they *do not draw conclusions, make predictions, or test hypotheses based on the data*, all of which form a specific branch of statistical analysis referred to as *inferential statistics* (covered in Chapter 9). This section provides introductions to basic concepts relating to descriptive statistics and how Python is used to carry out various descriptive analysis tasks.

Before performing any statistical task, it is useful to distinguish and identify the type(s) of data that will be analysed, as this largely dictates the most appropriate descriptive statistics and data visualisation techniques for the task at hand.

Observation 8.21 – Descriptive Statistics: A branch of data analysis that describes, displays, or summarizes information without drawing conclusions, making predictions, or testing hypotheses.

Observation 8.22 – Categorical and Continuous Data: *Categorical data* are data that can be divided into groups or classes but with no numerical relationship. *Continuous data* are numerical data that can be used for counting or measurements.

In a broad context, data can be simply categorized into two types: *categorical* and *continuous*. Categorical data are data that can be divided into groups or classes that *do not have a numerical or hierarchical relationship* (e.g., gender). Continuous data are numerical, and can include *counting* (i.e., integers) *or measurements* (i.e., any numerical values). The reader should become familiar with these two terms, as they are used extensively throughout this section.

8.4.1 MEASURES OF CENTRAL TENDENCY

There are two main ways to explore and describe continuous data: (a) measuring their *central tendency* and, (b) measuring their *spread*. The following sections introduce and briefly discuss these two concepts.

The measures of central tendency show the central or middle values of datasets. Hence, this is also frequently referred to as *measures of central location*. There are three different measures that can be considered as the centre of a dataset, namely *mean, median,* and *mode*.

The *mean*, also called the *arithmetic mean*, is a popular measure of central tendency. It is the *average* of the data in a dataset, and is calculated as the sum of all the data values divided by the number of cases in the dataset. The mean can fail to describe the central location of the data if there are outliers present or if the data are *skewed*.

The *median* is the middle point of a dataset that has been sorted in either ascending or descending order. The main difference between the mean and the median is that the former is heavily affected by outliers or skewed data, while the latter is affected only slightly or not at all.

The following Python script reads the data frame from the *newGrades.csv* file introduced in previous script samples, and calculates the means, medians, and modes of each of the columns:

> **Observation 8.23 – Measures of Central Tendency:** Measures that describe the central or middle values of a dataset. The three different measures are the *mean*, the *median*, and the *mode*.

> **Observation 8.24 – (Arithmetic) Mean:** The average of the data in a dataset, calculated as the sum of all the data values divided by the number of cases.

> **Observation 8.25 – Median:** The middle point of a *sorted* dataset.

> **Observation 8.26 – Mode:** The most frequently occurring value in the dataset. If more than one such values exist, the dataset is characterized as *multimodal*.

```
1    import pandas as pd
2
3    # Define the format of float numbers
4    pd.options.display.float_format = '${:,.2f}'.format
5
6    dataset = pd.read_csv('newGrades.csv')
7
8    # Define the number of rows and columns in the data frame
9    rows = len(dataset)
10   cols = ["Final Grade", "Final Exam", "Quiz 1", "Quiz 2",
11          "Midterm Exam", "Project"]
12
13   # Calculate the mean of all columns and append the dataset
14   mean1 = dataset["Final Grade"].mean()
15   mean2 = dataset["Final Exam"].mean()
16   mean3 = dataset["Quiz 1"].mean()
```

```
17  mean4 = dataset["Quiz 2"].mean()
18  mean5 = dataset["Midterm Exam"].mean()
19  mean6 = dataset["Project"].mean()
20  means = {"Final Grade": mean1, "Final Exam": mean2, "Quiz 1": mean3,
21        "Quiz 2": mean4, "Midterm Exam": mean5, "Project": mean6}
22  dataset = dataset.append(means, ignore_index = True)
23
24  # Calculate the median of all columns and append the dataset
25  median1 = dataset["Final Grade"].median()
26  median2 = dataset["Final Exam"].median()
27  median3 = dataset["Quiz 1"].median()
28  median4 = dataset["Quiz 2"].median()
29  median5 = dataset["Midterm Exam"].median()
30  median6 = dataset["Project"].median()
31
32  medians = {"Final Grade": median1, "Final Exam": median2,
33        "Quiz 1": median3, "Quiz 2": median4, "Midterm Exam": median5,
34        "Project": median6}
35  dataset = dataset.append(medians, ignore_index = True)
36
37  # Find the mode in all columns and append the dataset
38  mode1 = dataset["Final Grade"].mode(dropna = True).values
39  if (len(mode1) > 1):
40        mode1 = "Multimode"
41  mode2 = dataset["Final Exam"].mode(dropna = True).values
42  if (len(mode2) > 1):
43        mode2 = "Multimode"
44  mode3 = dataset["Quiz 1"].mode(dropna = True).values
45  if (len(mode3) > 1):
46        mode3 = "Multimode"
47  mode4 = dataset["Quiz 2"].mode(dropna = True).values
48  if (len(mode4) > 1):
49        mode4 = "Multimode"
50  mode5 = dataset["Midterm Exam"].mode(dropna = True).values
51  if (len(mode5) > 1):
52        mode5 = "Multimode"
53  mode6 = dataset["Project"].mode(dropna = True).values
54  if (len(mode6) > 1):
55        mode6 = "Multimode"
56  modes = {"Final Grade": mode1, "Final Exam": mode2, "Quiz 1": mode3,
57        "Quiz 2": mode4, "Midterm Exam": mode5, "Project": mode6}
58  dataset = dataset.append(modes, ignore_index = True)
59
60  # Report the dataset
61  dataset1 = dataset[["Final Grade", "Final Exam", "Quiz 1", "Quiz 2",
62        "Midterm Exam", "Project"]]
63  print(dataset1.iloc[0:rows:1])
64
65  #Report the rows with the means, medians, modes
66  print("Means"); print(dataset1.iloc[32:33])
67  print("Medians"); print(dataset1.iloc[33:34])
68  print("Modes"); print(dataset1.iloc[34:35])
```

Output 8.4.1:

	Final Grade	Final Exam	Quiz 1	Quiz 2	Midterm Exam	Project
0	$58.57	$50.50	$76.00	$70.70	$60.00	$55.00
1	$65.90	$49.00	$89.00	$63.00	$54.00	$90.00
2	$69.32	$63.50	$73.00	$54.70	$70.00	$80.00
3	$72.02	$60.50	$99.00	$74.70	$76.00	$70.00
4	$73.68	$74.00	$84.00	$53.30	$64.00	$87.00
5	$61.32	$45.50	$94.00	$42.70	$66.00	$70.00
6	$67.87	$66.50	$73.00	$53.70	$54.00	$87.00
7	$75.57	$66.00	$94.00	$58.70	$92.00	$70.00
8	$61.28	$50.50	$84.00	$37.30	$58.00	$78.00
9	$0.00	NaN	NaN	NaN	NaN	$69.00
10	$62.35	$48.00	$78.00	$49.00	$70.00	$71.00
11	$66.13	$61.00	$83.00	$45.30	$70.00	$70.00
12	$69.43	$50.00	$80.00	$49.30	$90.00	$76.00
13	$82.60	$74.00	$94.00	$65.00	$86.00	$92.00
14	$0.00	NaN	NaN	NaN	NaN	$75.00
15	$62.62	$45.50	$78.00	$56.70	$72.00	$70.00
16	$0.00	NaN	NaN	NaN	NaN	$0.00
17	$67.47	$59.00	$70.00	$72.70	$70.00	$72.00
18	$75.13	$61.50	$76.00	$68.30	$82.00	$87.00
19	$66.85	$77.50	$84.00	$52.00	$40.00	$80.00
20	$54.45	$34.50	$62.00	$44.00	$44.00	$90.00
21	$76.95	$66.50	$68.00	$67.00	$82.00	$92.00
22	$45.13	$26.00	$52.00	$26.30	$50.00	$68.00
23	$73.23	$63.50	$96.00	$68.30	$62.00	$89.00
24	$81.87	$83.00	$97.00	$82.70	$84.00	$72.00
25	$62.63	$54.50	$54.00	$31.30	$64.00	$87.00
26	$58.75	$46.50	$54.00	$39.00	$52.00	$90.00
27	$49.75	$27.50	$48.00	$37.00	$62.00	$70.00
28	$44.25	$21.50	$55.00	$18.00	$42.00	$80.00
29	$62.52	$31.00	$85.00	$54.70	$68.00	$89.00
30	$47.33	$16.50	$38.00	$33.30	$52.00	$89.00
31	$68.97	$55.00	$65.00	$49.70	$70.00	$94.00

Means

	Final Grade	Final Exam	Quiz 1	Quiz 2	Midterm Exam	Project
32	$58.87	$52.71	$75.28	$52.36	$65.72	$76.84

Medians

	Final Grade	Final Exam	Quiz 1	Quiz 2	Midterm Exam	Project
33	$62.63	$53.60	$77.00	$52.83	$65.86	$78.00

Modes

	Final Grade	Final Exam	Quiz 1	Quiz 2	Midterm Exam	Project
34	[0.0]	Multimode	Multimode	Multimode	[70.0]	[70.0]

The script and its output demonstrate a few important points:

- Given that the various calculations occasionally produce *floating point* numbers with several decimal digits, it may be desirable to limit the latter to a more manageable scale (i.e., two digits). The statement in line four formats the output accordingly.
- The statements in lines 13–15 calculate the mean of each of the columns of the dataset. Next, these values are appended at the end of the dataset as a new row.

- In a similar fashion, the statements in lines 21–26 calculate the median of each of the columns of the dataset and append them as a new row at the end of the dataset. It should be noted that, since it is necessary to have the data sorted in order to make such a calculation, this particular method performs this task too.
- The statements in lines 33–50 calculate the mode for each of the columns. Since it is undesirable in this particular example to have more than one such value reported, the code includes appropriate `if` statements to ensure that *the mode is a single value per column* or report that the output is *multimodal*, (i.e., it includes more than one values).
- Finally, the reader should note the use of the `dropna = True` parameter in the statements that ensure empty or NaN values are not considered in the mode calculation. The `.values` parameter also discards the information related to the resulting series and its object type, leaving only the pure value.

8.4.2 Measures of Spread

Another way to describe and summarize continuous data is through *measures of spread*. Such measures *quantify the variability of data points;* hence they are also called *measures of dispersion*. Measures of spread are frequently used in conjunction with measures of central tendency to provide a clearer and more rounded overview of the data at hand. The importance of measures of spread lies in the fact that they can describe how well the mean represents the data. If the data spread is large (i.e., if there are large differences between the data points), the mean may not be as good a representation of the data as the median or the mode.

The *data range* is the difference between the minimum and maximum data points in the dataset. It is calculated as *range = max−min*.

Quartiles describe the data spread by breaking the data into four parts (i.e., quarters), using three quartiles. The 1st quartile (Q1) is the 25th percentile of the sample, dividing roughly the lowest 25% from the rest of the data, while the 2nd quartile (Q2) is the 50th percentile or the median, and the third (Q3) the 75th percentile. Quartiles are a useful measure of spread, as they are much less affected by outliers or skewed datasets than other measures like *variance* or *standard deviation*.

Variance shows numerically how far the data points are from the mean. Variance is useful as, unlike quartiles, it takes into account all data points in the dataset and provides a better representation of the data spread. The variance of dataset x with n data points is expressed as $s^2 = \Sigma_i(x_i - mean(x))^2/(n-1)$, where $i = 1, 2, ..., n$ and $mean(x)$ is the mean of x. In order to get a better understanding of why the sum has to be divided with $n-1$ instead of n, the reader can refer to *Bessel's correction*.

Standard deviation also demonstrates how the data points spread out from the mean. It is the *positive square root of the variance*. A small standard deviation

Observation 8.27 – Measures of Spread: Measures that quantify the variability of data points in a dataset. If the spread is large, the measures of tendency are not good representations of the data.

Observation 8.28 – `min()`, `max()`: Use the `min()` and `max()` methods to find the minimum and maximum values in a dataset. Calculate their difference to find the *range* of these values.

Observation 8.29 – Quartiles: Use the `quantile()` method to specify and report the relevant quartile of data in a dataset. For instance, `quantile(0.1)` will report the lowest 10% of the data values in the dataset.

Observation 8.30 – `variance()`: Use the `variance()` method to find the variance of a dataset and show the distance of the data points from the mean.

Observation 8.31 – Standard Deviation (SD): Standard deviation shows the distance of the data points from the mean. The larger its values the larger the spread of the data points from the mean. It is frequently preferable to the *measure of variance*.

indicates that the data are close to the mean, while a large one shows a high outwards data spread from the mean. Standard deviation is often the preferred choice in order to present the data spread, and it is more convenient compared to variance, as it utilizes *the same unit as the data points.*

The following script uses the Pandas and Statistics Python packages to read the *newGrades. csv* file, find the *max* and *min* values for each column in the dataset, find the 25% (1ˢᵗ) quartile and calculate the variance and the standard deviation using both the regular `std()` and the `stdev()` methods from the *statistics* package. Finally, it creates a new dataset with all the related values, and reports the dataset:

```
1   import pandas as pd
2   import statistics
3
4   # Define the format of float numbers
5   pd.options.display.float_format = '${:,.2f}'.format
6
7   dataset = pd.read_csv('newGrades.csv')
8
9   rows = len(dataset)
10  cols = ["Final Grade", "Final Exam", "Quiz 1", "Quiz 2",
11         "Midterm Exam", "Project"]
12
13  # Find the max values in each column
14  max1 = dataset["Final Grade"].max(); max2 = dataset["Final Exam"].max()
15  max3 = dataset["Quiz 1"].max(); max4 = dataset["Quiz 2"].max()
16  max5 = dataset["Midterm Exam"].max(); max6 = dataset["Project"].max()
17
18  # Find the min values in each column
19  min1 = dataset["Final Grade"].min(); min2 = dataset["Final Exam"].min()
20  min3 = dataset["Quiz 1"].min(); min4 = dataset["Quiz 2"].min()
21  min5 = dataset["Midterm Exam"].min(); min6 = dataset["Project"].min()
22
23  # Find the lower 25% quartile in all columns
24  quartile25a = dataset["Final Grade"].quantile(0.25);
25  quartile25b = dataset["Final Exam"].quantile(0.25)
26  quartile25c = dataset["Quiz 1"].quantile(0.25);
27  quartile25d = dataset["Quiz 2"].quantile(0.25)
28  quartile25e = dataset["Midterm Exam"].quantile(0.25)
29  quartile25f = dataset["Project"].quantile(0.25)
30
31  # Calculate the variance in all columns
32  variance1 = statistics.variance(dataset["Final Grade"].dropna())
33  variance2 = statistics.variance(dataset["Final Exam"].dropna())
34  variance3 = statistics.variance(dataset["Quiz 1"].dropna())
35  variance4 = statistics.variance(dataset["Quiz 2"].dropna())
36  variance5 = statistics.variance(dataset["Midterm Exam"].dropna())
37  variance6 = statistics.variance(dataset["Project"].dropna())
38
39  # Calculate the standard deviation of all columns using std()
40  std1 = dataset["Final Grade"].std(); std2 = dataset["Final Exam"].std()
41  std3 = dataset["Quiz 1"].std(); std4 = dataset["Quiz 2"].std()
```

```
42  std5 = dataset["Midterm Exam"].std(); std6 = dataset["Project"].std()
43
44  # Calculate the standard deviation in all columns using stdev()
45  stdev1 = statistics.stdev(dataset["Final Grade"].dropna())
46  stdev2 = statistics.stdev(dataset["Final Exam"].dropna())
47  stdev3 = statistics.stdev(dataset["Quiz 1"].dropna())
48  stdev4 = statistics.stdev(dataset["Quiz 2"].dropna())
49  stdev5 = statistics.stdev(dataset["Midterm Exam"].dropna())
50  stdev6 = statistics.stdev(dataset["Project"].dropna())
51
52  # Report the dataset
53  dataset1 = dataset[["Final Grade", "Final Exam", "Quiz 1", "Quiz 2",
54         "Midterm Exam", "Project"]]
55  print(dataset1.iloc[0:rows:1])
56
57  # Append the dataset with the max values
58  maxs = {"Final Grade": max1, "Final Exam": max2, "Quiz 1": max3,
59         "Quiz 2": max4, "Midterm Exam": max5, "Project": max6}
60  dataset1 = dataset1.append(maxs, ignore_index = True)
61
62  mins = {"Final Grade": min1, "Final Exam": min2, "Quiz 1": min3,
63          "Quiz 2": min4, "Midterm Exam": min5, "Project": min6}
64  dataset1 = dataset1.append(mins, ignore_index = True)
65
66  quartiles = {"Final Grade": quartile25a, "Final Exam": quartile25b,
67               "Quiz 1": quartile25c, "Quiz 2": quartile25d,
68               "Midterm Exam": quartile25e, "Project": quartile25f}
69  dataset1 = dataset1.append(quartiles, ignore_index = True)
70
71  variances = {"Final Grade": variance1, "Final Exam": variance2,
72               "Quiz 1": variance3, "Quiz 2": variance4,
73               "Midterm Exam": variance5, "Project": variance6}
74  dataset1 = dataset1.append(variances, ignore_index = True)
75
76  stds = {"Final Grade": std1, "Final Exam": std2, "Quiz 1": std3,
77          "Quiz 2": std4, "Midterm Exam": std5, "Project": std6}
78  dataset1 = dataset1.append(stds, ignore_index = True)
79
80  stdevs = {"Final Grade": stdev1, "Final Exam": stdev2, "Quiz 1": stdev3,
81          "Quiz 2": stdev4, "Midterm Exam": stdev5, "Project": stdev6}
82  dataset1 = dataset1.append(stdevs, ignore_index = True)
83
84  # Report the rows with the max, min, quartile, variance, and std values
85  print("Max"); print(dataset1.iloc[32:33])
86  print("Min"); print(dataset1.iloc[33:34])
87  print("25% Quartile"); print(dataset1.iloc[34:35])
88  print("Variance"); print(dataset1.iloc[35:36])
89  print("Standard Deviation (using: std())"); print(dataset1.iloc[36:37])
90  print("Standard Deviation (using: stdev())")
91  print(dataset1.iloc[37:38])
```

Output 8.4.2:

	Final Grade	Final Exam	Quiz 1	Quiz 2	Midterm Exam	Project
0	$58.57	$50.50	$76.00	$70.70	$60.00	55
1	$65.90	$49.00	$89.00	$63.00	$54.00	90
2	$69.32	$63.50	$73.00	$54.70	$70.00	80
3	$72.02	$60.50	$99.00	$74.70	$76.00	70
4	$73.68	$74.00	$84.00	$53.30	$64.00	87
5	$61.32	$45.50	$94.00	$42.70	$66.00	70
6	$67.87	$66.50	$73.00	$53.70	$54.00	87
7	$75.57	$66.00	$94.00	$58.70	$92.00	70
8	$61.28	$50.50	$84.00	$37.30	$58.00	78
9	$0.00	NaN	NaN	NaN	NaN	69
10	$62.35	$48.00	$78.00	$49.00	$70.00	71
11	$66.13	$61.00	$83.00	$45.30	$70.00	70
12	$69.43	$50.00	$80.00	$49.30	$90.00	76
13	$82.60	$74.00	$94.00	$65.00	$86.00	92
14	$0.00	NaN	NaN	NaN	NaN	75
15	$62.62	$45.50	$78.00	$56.70	$72.00	70
16	$0.00	NaN	NaN	NaN	NaN	0
17	$67.47	$59.00	$70.00	$72.70	$70.00	72
18	$75.13	$61.50	$76.00	$68.30	$82.00	87
19	$66.85	$77.50	$84.00	$52.00	$40.00	80
20	$54.45	$34.50	$62.00	$44.00	$44.00	90
21	$76.95	$66.50	$68.00	$67.00	$82.00	92
22	$45.13	$26.00	$52.00	$26.30	$50.00	68
23	$73.23	$63.50	$96.00	$68.30	$62.00	89
24	$81.87	$83.00	$97.00	$82.70	$84.00	72
25	$62.63	$54.50	$54.00	$31.30	$64.00	87
26	$58.75	$46.50	$54.00	$39.00	$52.00	90
27	$49.75	$27.50	$48.00	$37.00	$62.00	70
28	$44.25	$21.50	$55.00	$18.00	$42.00	80
29	$62.52	$31.00	$85.00	$54.70	$68.00	89
30	$47.33	$16.50	$38.00	$33.30	$52.00	89
31	$68.97	$55.00	$65.00	$49.70	$70.00	94

Max

	Final Grade	Final Exam	Quiz 1	Quiz 2	Midterm Exam	Project
32	$82.60	$83.00	$99.00	$82.70	$92.00	$94.00

Min

	Final Grade	Final Exam	Quiz 1	Quiz 2	Midterm Exam	Project
33	$0.00	$16.50	$38.00	$18.00	$40.00	$0.00

25% Quartile

	Final Grade	Final Exam	Quiz 1	Quiz2	Midterm Exam	Project
34	$57.54	$45.50	$65.00	$42.70	$54.00	$70.00

Variance

	Final Grade	Final Exam	Quiz 1	Quiz 2	Midterm Exam	Project
35	$461.46	$289.85	$267.49	$242.37	$197.06	$291.88

Standard Deviation (using: std())

	Final Grade	Final Exam	Quiz 1	Quiz 2	Midterm Exam	Project
36	$21.48	$17.02	$16.36	$15.57	$14.04	$17.08

Standard Deviation (using: stdev())

	Final Grade	Final Exam	Quiz 1	Quiz 2	Midterm Exam	Project
37	$21.48	$17.02	$16.36	$15.57	$14.04	$17.08

8.4.3 Skewness and Kurtosis

Skewness measures the *asymmetry* of the data and describes the amount by which the distribution differs from a normal distribution. There are several mathematical definitions of skewness. A commonly used one is *Pearson's skewness coefficient*, which can be derived using the size of a dataset, the mean, and the standard deviation of the data. *Negative skewness* values indicate a dominant tail on the left side, while *positive* values correspond to a long tail on the right side. If the skewness is close to 0 (i.e., between −0.5 and 0.5), the data are considered to be *symmetric* (Figure 8.1). When the skewness is between −1 and −0.5 or between 0.5 and 1, the data are considered to be *moderately skewed*. If skewness is less than −1 or more then 1, the data are considered to be *highly skewed*.

Kurtosis shows whether the data is heavy-tailed or light-tailed compared to a normal distribution. In other words, kurtosis identifies whether the data contains extreme values. A high kurtosis indicates a heavy tail and more outliers in the data, while a low kurtosis shows a light tail and fewer outliers. An alternative and effective way to show kurtosis and skewness is the *histogram*, as it visually demonstrates the shape of the data distribution.

There are three main types of kurtosis: *mesokurtic*, *leptokurtic*, and *platykurtic* (Figure 8.2).

> **Observation 8.32 – Skewness:** Use the `skew()` method to calculate the *skewness* of a dataset. Based on *Pearson's skewness coefficient*, skewness between −0.5 and 0.5 is considered to be *symmetric*, while values between −1 and −0.5 or 0.5 and 1 indicate that skewness is *moderate* and values less than −1 or more than 1 that it is *high*.

> **Observation 8.33 – Kurtosis:** Use the `kurtosis()` method to calculate the *kurtosis* of a dataset. Data can be characterized as *mesokurtic* (normal distribution with value of 3), *leptokyrtic* (data heavily-tailed with profusion of outliers and value higher than 3), or *platykurtic* (data light-tailed with less extreme values than normal distribution and value lower than 3).

- **Mesokurtic (Kurtosis = 3):** Data are normally distributed.
- **Leptokurtic (Kurtosis > 3):** Data are heavy-tailed with profusion of outliers.
- **Platykurtic (Kurtosis < 3):** Data are light-tailed and/or contain less extreme values than normal distribution.

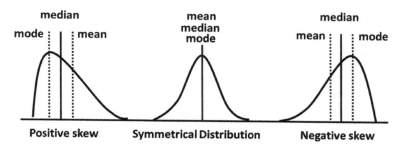

FIGURE 8.1 Symmetric, positive, and negative skewness.

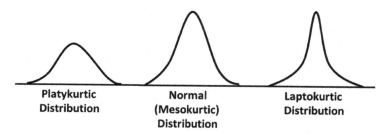

FIGURE 8.2 Main types of kurtosis.

The following script reads the *newGrades.csv* file, calculates the *skewness, kurtosis*, and *sum* values of all columns, and reports them alongside the rest of the dataset:

```
1    import pandas as pd
2
3    # Define the format of float numbers
4    pd.options.display.float_format = '${:,.2f}'.format
5
6    dataset = pd.read_csv('newGrades.csv')
7
8    rows = len(dataset)
9    cols = ["Final Grade", "Final Exam", "Quiz 1", "Quiz 2",
10          "Midterm Exam", "Project"]
11
12   # Find the skewness (Pearson's coefficient) values for each column
13   skew1 = dataset["Final Grade"].skew()
14   skew2 = dataset["Final Exam"].skew()
15   skew3 = dataset["Quiz 1"].skew()
16   skew4 = dataset["Quiz 2"].skew()
17   skew5 = dataset["Midterm Exam"].skew()
18   skew6 = dataset["Project"].skew()
19
20   # Find the kurtosis values for each column
21   kurtosis1 = dataset["Final Grade"].kurtosis()
22   kurtosis2 = dataset["Final Exam"].kurtosis()
23   kurtosis3 = dataset["Quiz 1"].kurtosis()
24   kurtosis4 = dataset["Quiz 2"].kurtosis()
25   kurtosis5 = dataset["Midterm Exam"].kurtosis()
26   kurtosis6 = dataset["Project"].kurtosis()
27
28   # Find the sum of all values for each column
29   sum1 = dataset["Final Grade"].sum()
30   sum2 = dataset["Final Exam"].sum()
31   sum3 = dataset["Quiz 1"].sum()
32   sum4 = dataset["Quiz 2"].sum()
33   sum5 = dataset["Midterm Exam"].sum();
34   sum6 = dataset["Project"].sum()
35
36   # Report the dataset
37   dataset1 = dataset[["Final Grade", "Final Exam", "Quiz 1", "Quiz 2",
38          "Midterm Exam", "Project"]]
39   print(dataset1.iloc[0:rows:1])
40
41   # Append the dataset with the max values
42   skewness = {"Final Grade": skew1, "Final Exam": skew2, "Quiz 1": skew3,
43          "Quiz 2": skew4, "Midterm Exam": skew5, "Project": skew6}
44   dataset1 = dataset1.append(skewness, ignore_index = True)
45
46   kurtosis = {"Final Grade": kurtosis1, "Final Exam": kurtosis2,
47          "Quiz 1": kurtosis3, "Quiz 2": kurtosis4,
48          "Midterm Exam": kurtosis5, "Project": kurtosis6}
49   dataset1 = dataset1.append(kurtosis, ignore_index = True)
50
```

```
51  sums = {"Final Grade": sum1, "Final Exam": sum2, "Quiz 1": sum3,
52        "Quiz 2": sum4, "Midterm Exam": sum5, "Project": sum6}
53  dataset1 = dataset1.append(sums, ignore_index = True)
54
55  # Report the rows with the skewness, kurtosis, and sums
56  print("Skewness"); print(dataset1.iloc[32:33])
57  print("Kurtosis"); print(dataset1.iloc[33:34])
58  print("Sum values"); print(dataset1.iloc[34:35])
```

Output 8.4.3:

	Final Grade	Final Exam	Quiz 1	Quiz 2	Midterm Exam	Project
0	$58.57	$50.50	$76.00	$70.70	$60.00	55
1	$65.90	$49.00	$89.00	$63.00	$54.00	90
2	$69.32	$63.50	$73.00	$54.70	$70.00	80
3	$72.02	$60.50	$99.00	$74.70	$76.00	70
4	$73.68	$74.00	$84.00	$53.30	$64.00	87
5	$61.32	$45.50	$94.00	$42.70	$66.00	70
6	$67.87	$66.50	$73.00	$53.70	$54.00	87
7	$75.57	$66.00	$94.00	$58.70	$92.00	70
8	$61.28	$50.50	$84.00	$37.30	$58.00	78
9	$0.00	NaN	NaN	NaN	NaN	69
10	$62.35	$48.00	$78.00	$49.00	$70.00	71
11	$66.13	$61.00	$83.00	$45.30	$70.00	70
12	$69.43	$50.00	$80.00	$49.30	$90.00	76
13	$82.60	$74.00	$94.00	$65.00	$86.00	92
14	$0.00	NaN	NaN	NaN	NaN	75
15	$62.62	$45.50	$78.00	$56.70	$72.00	70
16	$0.00	NaN	NaN	NaN	NaN	0
17	$67.47	$59.00	$70.00	$72.70	$70.00	72
18	$75.13	$61.50	$76.00	$68.30	$82.00	87
19	$66.85	$77.50	$84.00	$52.00	$40.00	80
20	$54.45	$34.50	$62.00	$44.00	$44.00	90
21	$76.95	$66.50	$68.00	$67.00	$82.00	92
22	$45.13	$26.00	$52.00	526.30	$50.00	68
23	$73.23	$63.50	$96.00	$68.30	$62.00	89
24	$81.87	$83.00	$97.00	5E2.70	$84.00	72
25	$62.63	$54.50	$54.00	$31.30	$64.00	87
26	$58.75	$46.50	$54.00	$39.00	$52.00	90
27	$49.75	$27.50	$48.00	$37.00	$62.00	70
28	$44.25	$21.50	$55.00	$18.00	$42.00	80
29	$62.52	$31.00	$85.00	$54.70	$68.00	89
30	$47.33	$16.50	$38.00	$33.30	$52.00	89
31	$68.97	$55.00	$65.00	$49.70	$70.00	94

Skewness

	Final Grade	Final Exam	Quiz 1	Quiz 2	Midterm Exam	Project
32	$-1.96	$-0.43	$-0.51	$-0.18	$0.05	$-3.03

Kurtcsis

	Final Grade	Final Exam	Quiz 1	Quiz 2	Midterm Exam	Project
33	$3.52	$-0.35	$-0.53	$-0.39	$-0.60	$13.01

Sum values

	Final Grade	Final Exam	Quiz 1	Quiz 2	Midterm Exam	Project
34	$1,883.94	$1,528.50	$2,183.00	$1,518.40	$1,906.00	$2,459.00

8.4.4 THE DESCRIBE() AND COUNT() METHODS

Two more methods that are worth mentioning are describe() and count(). These methods come rather handy when describing *categorical* data, but can be also used with *continuous* data. The describe() method provides a simple way to describe data, reporting the *max, min, variance, quartiles, mean,* and *standard deviation* without having to deal with each of them separately. The count() method reports the number of occurrences of each case of *categorical* data in the dataset (i.e., it denotes *frequency of occurrence*). It can be also calculated on a percentage basis in order to obtain a representation of the *part-to-whole* relationship.

Observation 8.34 – describe(): Use the describe() method to automatically report a set of basic descriptive statistics.

Observation 8.35 – count(): Use the count() method to report the frequency of occurrence of categorical data.

The following script uses *newGrades.csv* to report basic descriptive statistics for *Final Grade*, while also counting the *As, Bs, Cs, Ds,* and *Fs* in the report:

```
1   import pandas as pd
2
3   # Define the format of float numbers
4   pd.options.display.float_format = '${:,.2f}'.format
5
6   dataset c pd.read_csv('newGrades.csv')
7
8   rows = len(dataset)
9   cols = ["Final Grade", "Letter Grade"]
10
11  # Report the basic descriptive statistics for Final Grade
12  print("Basic descriptive statistics on Final Grade")
13  print(dataset["Final Grade"].describe(), "\n")
14
15  # Create a new dataset with Letter Grade only
16  dataset1 = dataset[["Letter Grade"]]
17
18  # Find the number of occurrences of Letter Grades
19  countAll = dataset1.count()
20  print("Total students:", countAll.values)
21
22  dataset2 = dataset1[dataset1["Letter Grade"] == "A"]
23  if (not dataset2.empty):
24          countA = dataset2.count()
25  else:
26          countA = 0
27  print("Students awarded an A:", countA)
28
```

```
29  dataset2 = dataset1[dataset1["Letter Grade"] == "B"]
30  if (not dataset2.empty):
31        countB = dataset2.count().values
32  else:
33        countB = 0
34  print("Students awarded an B:", countB)
35
36  dataset2 = dataset1[dataset1["Letter Grade"] == "C"]
37  if (not dataset2.empty):
38        countC = dataset2.count().values
39  else:
40        countC = 0
41  print("Students awarded an C:", countC)
42
43  dataset2 = dataset1[dataset1["Letter Grade"] == "D"]
44  if (not dataset2.empty):
45        countD = dataset2.count().values
46  else:
47        countD = 0
48  print("Students awarded an D:", countD)
49
50  dataset2 = dataset1[dataset1["Letter Grade"] == "F"]
51  if (not dataset2.empty):
52        countF = dataset2.count().values
53  else:
54        countF = 0
55  print("Students awarded an F:", countF)
```

Output 8.4.4:

```
Basic descriptive statistics on Final Grade
count    $32.00
mean     $58.87
std      $21.48
min       $0.00
25%      $57.54
50%      $64.27
75%      $70.08
max      $82.60
Name: Final Grade, dtype: float64

Total students: [32]
Students awarded an A: 0
Students awarded an 3: [2]
Students awarded an C: [6]
Students awarded an D: [14]
Students awarded an F: [10]
```

8.5   DATA VISUALIZATION

We are all familiar with the expression *a picture is worth a thousand words*. *Data visualisation* refers to the use of graphical means to represent and summarize data. It can help the analyst identify and conceptualize patterns, trends, and correlations present in the data that may be otherwise difficult to spot. It is also an efficient

> **Observation 8.36 – Data Visualization:** The use of visual means, such as various types of charts, to represent and summarize data.

way to convey insights or summaries to wider audiences and, thus, it is widely used for data presentation (particularly when working with *big data*). Data visualisation is also an essential step before undertaking *inferential statistics analysis* (Chapter 9) and *machine learning* (Chapter 10) tasks, as it provides an overview of some of the structures and techniques used in these fields. In general, data visualisation is useful for the following tasks:

- Recognizing the structure and patterns of the data.
- Detecting errors or outliers.
- Exploring relationships between variables.
- Discovering new trends.
- Suggesting appropriate inferential statistical analysis and machine learning methods.
- Identifying the need for data correction (e.g., transforming data to log-scale).
- Communicating data to wider audiences.

Python is a popular data visualization choice for data scientists, as it provides various packages and libraries suitable for visualization tasks. Some popular plotting libraries are the following:

- **Matplotlib:** As mentioned in earlier sections, Matplotlib is a low-level plotting library, suitable for creating basic graphs and providing a lot of options relating to this task to the programmer.
- **Pandas:** Pandas is based on Matplotlib and, in addition to plotting, it also provides extra analysis functionality.
- **Seaborn:** Seaborn is a high-level plotting library with a solid collection of usable, default styles. It also allows for graph plotting with minimal coding, and it provides advanced visuals, making it the tool of choice for many data scientists.

The above libraries and packages provide a wealth of available methods to produce any type of visualization. In this section, only Pandas and Matplotlib are used. This is mainly for simplicity and clarity reasons.

8.5.1   Continuous Data: Histograms

A *histogram* is a type of graph that can depict the *distribution of continuous numerical data* by displaying the data frequency using bars of different heights. Due to the use of bars, prior to plotting histograms, one first has to *bin* the range of data values. The term *bin* is used to describe the process of dividing the entire range of data values into a *series of intervals*. Subsequently, data falling into each interval are counted and the resulting frequencies are plotted in the form of bars. Bins are usually specified as *consecutive, non-overlapping intervals* and often have equal or comparable sizes, although this is not a strict requirement (Freedman et al., 1998).

> **Observation 8.37 – Histograms:** Use the `plot.hist()` method (*Pandas* library) to visualize continuous data, dividing the entire range of values into a series of intervals referred to as *bins*. Parameters such as `subplots`, `layout`, `grid`, `xlabelsize`, `ylabelsize`, `xrot`, `yrot`, `figsize`, and `legend` allow for the detailed configuration of the histogram.

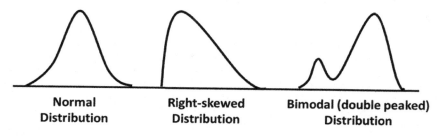

Normal Distribution   **Right-skewed Distribution**   **Bimodal (double peaked) Distribution**

FIGURE 8.3 Types of histograms.

Histograms can be used when investigating and demonstrating the shape of the data distribution (i.e., its center, spread, and skewness), as well as its various modes and the presence of outliers. They help the analysis by visually determining whether two or more data distributions are different, like in the example above (Figure 8.3).

At first, histograms may look like bar charts, but these two graph formats are notably different. Histograms are used for *summarising and grouping continuous data into ranges*, while bar charts are used for displaying the *frequency of categorical data*. Another difference is that the proportion of the data in a histogram is represented as a unified area of the graph, while in a bar chart through the length of individual bars. Bar charts are discussed in more detail in later parts of this chapter.

To plot a histogram in Python, one can use the `plot.hist()` method from the Pandas library. For basic plotting, no further arguments are needed. However, the method accepts additional arguments in order to optionally control specific plotting details, such as the bin size (the default value is 10). It is also possible to have multiple histograms generated and illustrated in one single plot. The `subplots` parameter allows the programmer to plot each feature in the dataset separately, and the `layout` parameter specifies the number of plots per row and column of a given diagram. By default, the histogram appears inside a grid, but it is possible to avoid this by setting the `grid` parameter to `False`. The letter size of the x or y axis can be controlled by setting the `xlabelsize` or `ylabelsize` parameters, respectively. The histogram can be rotated by a specified number of degrees on the x or y axis, by setting the `xrot` or `yrot` parameters. The size of the figures can be specified (in inches) using the `figsize` parameter.

The following script uses the *newGrades.csv* dataset used in previous examples to display six histograms in one plot (i.e., two lines and three columns):

```
1   import pandas as pd
2
3   dataset = pd.read_csv('newGrades.csv')
4   dataset1 = dataset[["Final Grade", "Final Exam",  "Quiz 1", "Quiz 2",
5        "Midterm Exam", "Project"]]
6
7   # Prepare a histogram with 2 lines of subplots, visible grid & legend
8   # in 2 rows & 3 columns, with figures of size 10x10 inches, & 10 bins
9   plt = dataset1.plot.hist(subplots = 2, grid = True, legend = True,
10       layout = (2, 3), figsize = (10, 10), bins = 10)
```

Output 8.5.1:

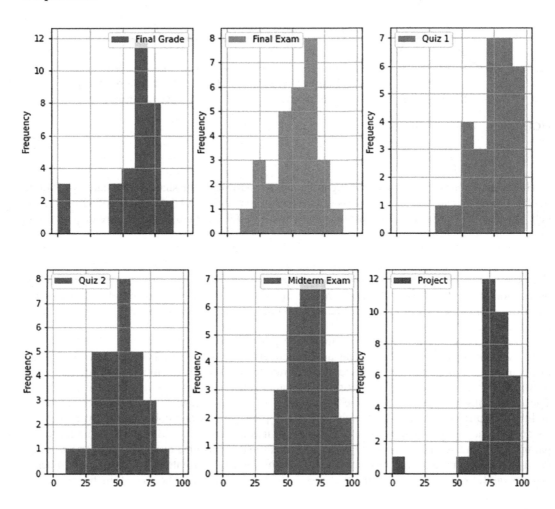

8.5.2 Continuous Data: Box and Whisker Plot

A *box and whisker* plot, also called *box plot*, is a graphical representation of the *spread of continuous data,* based on a five number summary: the *minimum*, the *maximum*, the *sample median*, the *first quartile* (Q1), and *the third quartile* (Q3). As the name suggests, the plot contains two parts: a box and a set of whiskers. The two ends of the whiskers show the minimum and the maximum values of the dataset, while the top and the bottom of the box represent Q3 and Q1, respectively. The horizontal

Observation 8.38 – Box and Whisker Plot: Use the boxplot() method (Pandas library) to draw a *box and whisker* plot. Plot aspects like the grid, the figure size, and the labels can be configured using the grid, figsize, and labels parameters, respectively.

line in the middle of the box denotes the median. The data point that is located outside the whiskers of the box plot is defined as an *outlier,* which is *the value that is more than one and a half times the length of the box.* It is worth noting that box plots work better with data that only contain a limited number of categories (Figure 8.4).

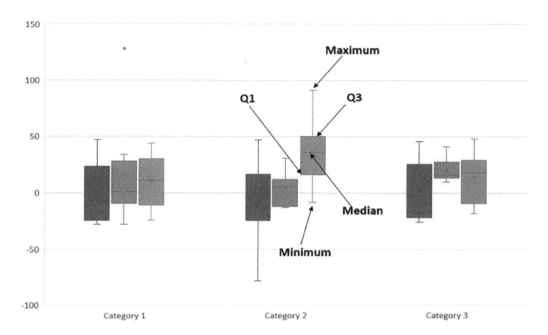

FIGURE 8.4   Box and whisker plot.

Box plots can be used when:

- Working with numerical data.
- Presenting the spread of the data and the central value.
- Comparing data distribution across different categories.
- Identifying outliers.

Box plots can be created using the boxplot() method from the Pandas library. The x and y axis values can be modified using the by and column parameters, respectively (Pandas, 2021a). For an improved visual effect, one can alternatively use the sns.boxplot() method from the Seaborn library.

The following script draws a box and whisker plot for the *newGrades.csv* dataset:

```
1    import pandas as pd
2
3    dataset = pd.read_csv('NewGrades.csv')
4
5    # The names of the columns on the x-axis
6    cols = ["Final Grade", "Final Exam", "Quiz 1", "Quiz 2",
7           "Midterm Exam", "Project"]
8
9    dataset1 = dataset[["Final Grade", "Final Exam", \
10          "Quiz 1", "Quiz 2", "Midterm Exam", "Project"]]
11
12   # Prepare a box and whisker diagram with all the 6 columns represented
13   # in a single plot of size 10x10 inches
14   dataset1.boxplot(grid = True, figsize = (10, 10), showcaps = True, \
15          showbox = True, showfliers = True, labels = cols)
```

Output 8.5.2:

```
<AxesSubplot:>
```

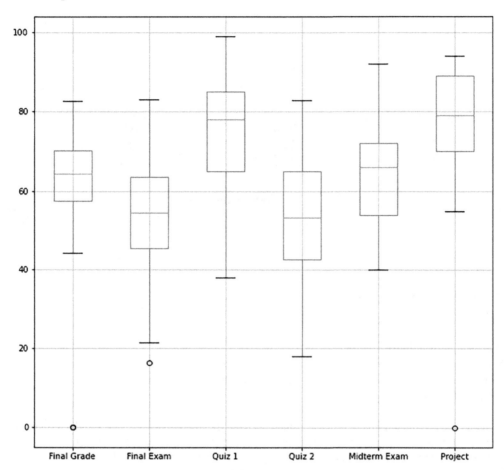

8.5.3   CONTINUOUS DATA: LINE CHART

A *line chart* is a graphical method to represent *trend* data as a continuous line. It connects a series of historical data points by line segments in order to depict the variations of the data continuously over time. The x-axis corresponds to time or continuous progression, while the y-axis represents the corresponding values.

Line charts can be used when:

> **Observation 8.39 – Line Chart:** Use the `plot.line()` method (Pandas library) to draw a *line* chart. There are several parameters available for the detailed configuration of the chart.

- Working with numerical data (y-axis) that follow a continuous progression (x-axis).
- Emphasizing changes in values over time or as a continuous progression.
- Comparing between different series of trends.

To create a line chart, one can call the `plot.line()` method from the Pandas library. If multiple lines are plotted in a single line chart, Pandas automatically creates a legend. This is a rather useful feature when comparing data trends.

The following script uses the *newGrades.csv* dataset to draw a line chart plotting all six columns of the dataset:

```
1    import pandas as pd
2
3    dataset = pd.read_csv('newGrades.csv')
4
5    # The names of the columns on the x-axis
6    cols = ["Final Grade", "Final Exam", "Quiz 1", "Quiz 2",
7            "Midterm Exam", "Project"]
8
9    dataset1 = dataset[["Final Grade", "Final Exam", \
10           "Quiz 1", "Quiz 2", "Midterm Exam", "Project"]]
11
12   # Prepare a line chart with all the 6 columns represented
13   # in a single plot of size 7x7 inches
14   dataset1.plot.line(grid = True, figsize = (7, 7),
15           title = "Grades Line Chart")
```

Output 8.5.3:

```
<AxesSubplot:title={'center':'Grades Line Chart'}>
```

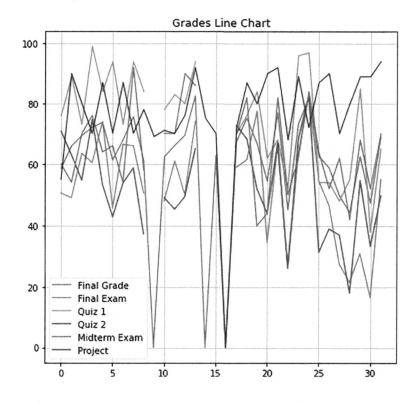

8.5.4 CATEGORICAL DATA: BAR CHART

A *bar chart* is a graph that displays *counts of categorical data* or *data associated with categorical data* in the form of vertical or horizontal rectangular bars. The x-axis (vertical bar chart) represents the data by category, while the y-axis can take any value depending on the dataset used. Bar charts are useful for describing

Observation 8.40 – Bar Chart: Use the `plot.bar()` method (Pandas library) to draw a bar chart. There are several parameters available for the detailed configuration of the chart.

categorical data that have less than approximately 30 categories, as anything close to or above this rough threshold tends to make them rather unreadable. In such cases, a more efficient grouping or re-grouping approach should be considered.

Bar charts can be used when:

- Working with categorical data.
- Investigating the frequency of the data.

To plot a bar chart for categorical data one can use the plot.bar() method (Pandas library). The reader must note that before this method is called, the frequency for each category must be counted using the value_count() method. Methods plt.xlabel(), plt.ylabel(), and plt.title() can be used to add appropriate descriptions to the bar chart.

The following script uses plot.bar() to draw and configure a vertical bar chart (default) based on the *Letter Grade* column of *newGrades2.xlsx* (*New Data* sheet):

```
1    import pandas as pd
2
3    dataset = pd.read_excel('newGrades2.xlsx', sheet_name = "New Data")
4
5    barChart = dataset["Letter Grade"].value_counts().plot.bar(grid = True,
6            legend = True, figsize = (7, 7), rot = 0)
7    barChart.set_title("Final Letter Grades")
8    barChart.set_ylabel("Frequencies")
9    barChart.set_xlabel("Letter Grades")
```

Output 8.5.4.a:

```
Text(0.5, 0, 'Letter Grades')
```

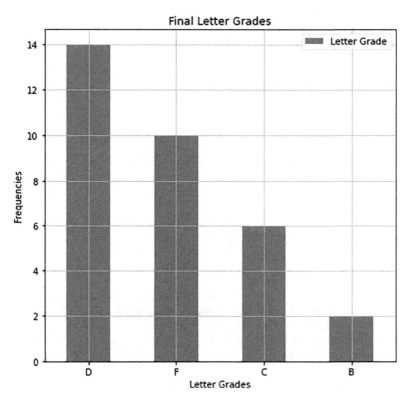

The reader should note the use of the `grid`, `legend`, `figsize`, and `rot` parameters to configure the basic appearance of the chart (i.e., show the grid and the legend, define the size of the figure in inches, and ensure the correct orientation of the x-axis labels, respectively). It must be also noted how methods `set_title()`, `set_ylabel()`, and `set_xlabel()` are used to set the title of the chart and define the headings for the x and y axes.

When horizontal bars are needed instead of vertical ones the `plot.barh()` method should be used instead of the `plot.bar()`. The following script demonstrates this option, while its output illustrates how slight parameter variations can help with the new horizontal orientation:

```
1    import pandas as pd
2
3    dataset = pd.read_excel('newGrades2.xlsx', sheet_name = "New Data")
4
5    barChart = dataset["Final Exam Letter"].value_counts().plot.barh(
6            grid = True, legend = True, figsize = (7, 7), rot = 0)
7    barChart.set_title("Final Exam Letter Grades")
8    barChart.set_ylabel("Letter Grades")
9    barChart.set_xlabel("Frequencies")
```

Output 8.5.4.b:

```
Text(0.5, 0, 'Frequencies')
```

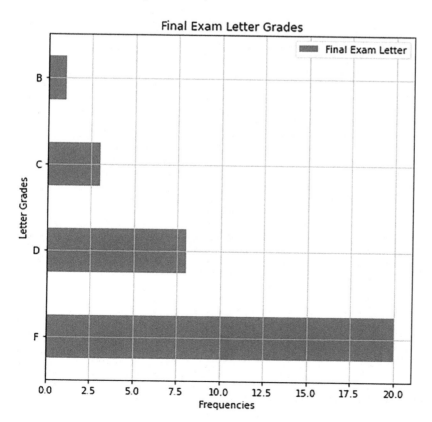

It is also possible to have two or more different bar charts within the same figure. This can take three different forms. The first is to have a single plot with two separate charts as in the script below. The script uses the `subplots()` method from the `plt` object of the `matplotlib.pyplot` package to create two different plots:

```
1    import pandas as pd
2    import matplotlib.pyplot as plt
3
4    dataset = pd.read_excel('newGrades2.xlsx', sheet_name = "New Data")
5
6    # Draw first subplot
7    plt.subplot(1, 2, 1)
8    plot1 = dataset["Letter Grade"].value_counts().plot.bar(grid = True,
9          figsize = (10, 7), legend = True, sharey = True, rot = 0)
10   plot1.set_title("Final Letter Grades")
11   plot1.set_ylabel("Frequencies")
12   plot1.set_xlabel("Letter Grades")
13
14   # Draw second subplot
15   plt.subplot(1, 2, 2)
16   plot2 = dataset["Final Exam Letter"].value_counts().plot.bar(grid=True,
17          figsize = (10, 7), legend = True, sharey = True, rot = 0)
18   plot2.set_title("Final Exam Letter Grades")
19   plot2.set_ylabel("Frequencies")
20   plot2.set_xlabel("Letter Grades")
```

Output 8.5.4.c:

```
Text(0.5, 0, 'Letter Grades')
```

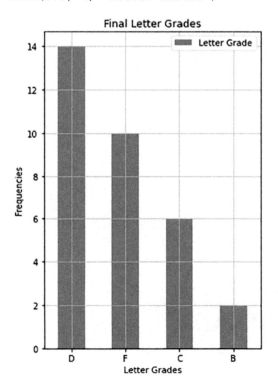
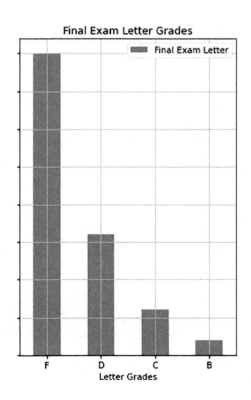

The second form is to create a *compound* or *nested* bar chart, allowing two or more sets of data associated with the same categorical data to be plotted in a single diagram. This is useful in situations requiring visual comparison. The following script is a variation of previously used examples, demonstrating this form of bar chart:

```
1    import pandas as pd
2    import matplotlib.pyplot as plt
3
4    # Read the Excel dataset
5    dataset = pd.read_excel('newGrades2.xlsx', sheet_name = "New Data")
6
7    # Count the frequencies of Letter Grade and Final Exam Letter
8    dataset1 = dataset["Letter Grade"].value_counts()
9    dataset2 = dataset["Final Exam Letter"].value_counts()
10
11   barChart = pd.DataFrame({"Final Letter Grade": dataset1,
12          "Final Exam Letter Grade": dataset2})
13   barChart.plot.bar(grid = True,
14          title = "Final Exam and Final Grade Letter Grades",
15          rot = 0, figsize = (8, 8), color = ["lightblue", "lightgrey"])
16
17   # Use the plt object to set the labels of the x and y axis
18   plt.xlabel("Letter Grades")
19   plt.ylabel("Frequencies")
```

Output 8.5.4.d:

```
Text(0, 0.5, 'Frequencies')
```

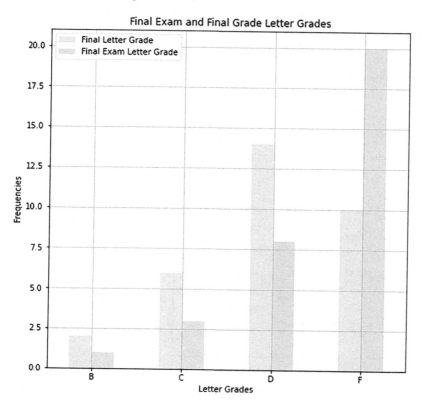

The third form is the *stacked* bar chart. In this case, the various components are stacked upon each other to create a single, unified bar. The following script presents columns *Letter Grade* and *Final Exam Letter* from the *newGrades2.xlsx* dataset (*New Data* sheet). The reader should note that, in addition to the previously mentioned parameters of the regular `plot.bar()` method, the script also uses the `stacked = True` parameter that is responsible for stacking the two datasets:

```
1    import pandas as pd
2    import matplotlib.pyplot as plt
3
4    # Read the Excel dataset
5    dataset = pd.read_excel('newGrades2.xlsx', sheet_name = "New Data")
6
7    # Count the frequencies of the "Letter Grade" & the "Final Exam Letter"
8    dataset1 = dataset["Letter Grade"].value_counts()
9    dataset2 = dataset["Final Exam Letter"].value_counts()
10
11   barChart = pd.DataFrame({"Final Letter Grade": dataset1,
12                            "Final Exam Letter Grade": dataset2})
13   barChart.plot.bar(stacked = True, grid = True,
14         title = "Final Exam and Final Grade Letter Grades",
15         rot = 0, figsize = (8, 8), color = ["lightblue", "lightgrey"])
16
17   # Use the plt object to set the labels of the x-axis and the y-axis
18   plt.xlabel("Letter Grades")
19   plt.ylabel("Frequencies")
```

Output 8.5.4.e:

```
Text(0, 0.5 'Frequencies')
```

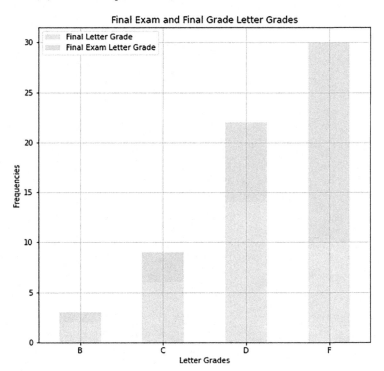

8.5.5 CATEGORICAL DATA: PIE CHART

A *pie chart* is a circular graph that uses the size of *pie slices* to illustrate *proportion*. It displays a part-to-whole relationship of categorical data. Like in the case of the bar chart, the pie chart should be avoided for data with a significant number of categories (i.e., slices), as this would compromise readability. Ideally, data with five or less categories are preferable. If the pie chart is to be used for data with more than five categories, re-categorising or aggregating the data should be considered.

Pie charts can be used when the presentation of the part-to-whole relationship of the data is more important than the precise size of each category, and when it is required to visually compare the size of categories in relation to the whole. However, unlike bar charts, they cannot explicitly demonstrate *absolute numbers or values for each category*. To plot a pie chart, one can use the plot.pie() method from the Pandas library (Pandas, 2021b), while its appearance can be further configured using the plt object from the matplotlib.pyplot package.

Observation 8.41 – Pie Chart: Use the pie() method (Pandas library) to create a pie chart based on a dataset. Use the plt object from matplotlib.pyplot to configure and improve the appearance of the chart.

The following script reads the *New Data* dataset from *newGrades2.xlsx* and creates a pie chart based on the *Letter Grade* column. Next, it demonstrates the use of the labels, autopct, shadow, and startangle parameters to define and format the labels (in percentages), to display shadows, and to dictate the orientation and angle of the slices. Finally, it uses the axis, legend, and title methods to adjust the size of the slices, and to add titles to the chart and the legend:

```
1   import pandas as pd
2   import matplotlib.pyplot as plt
3
4   # Read the Excel dataset
5   dataset = pd.read_excel('newGrades2.xlsx', sheet_name = "New Data")
6
7   labels1 = dataset["Letter Grade"].unique()
8
9   # Count the frequencies of Letter Grade
10  dataset1 = dataset["Letter Grade"].value_counts()
11  plt.pie(dataset1, labels = labels1, autopct = "%1.1f%%", shadow = True,
12          startangle = 90)
13  plt.axis("equal")
14  plt.legend(title = "Final Letter Grades")
15  plt.title("Final Letter Grades")
```

Output 8.5.5:

```
Text(0.5, 1.0, 'Final Letter Grades')
```

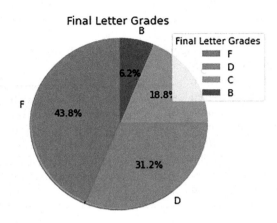

8.5.6   Paired Data: Scatter Plot

A *scatter plot* is a visual representation of *the relationship between two sets of data* using *dots* or *circles*. The dots/circles can report the values of individual data points, but also patterns of the data as a whole. Relationships between variables can be described in the following ways: *positive* or *negative*, *strong* or *weak*, *linear* or *nonlinear* (Figure 8.5).

Scatter plots can be used when:

Observation 8.42 – Scatter Plot: Use `plot.scatter()` (Pandas library) to create a scatter plot. Scatter plots illustrate the relationship between two sets of data using dots or circles.

- Working with paired numerical data.
- Identifying whether the data are correlated.
- Investigating data patterns (e.g., cluster, data gap, outliers) (Figure 8.6).

To create a scatter plot, one can call the `plot.scatter()` method from the Pandas library, and use the x and y arguments to define the paired data. The following script draws a scatter plot chart using the *Final Exam Grades* and *Final Grades* columns from *newGrades2.xlsx*:

```
1    import pandas as pd
2
3    # Read the Excel dataset
4    dataset = pd.read_excel('newGrades2.xlsx', sheet_name = "New Data")
5
6    dataFrame = pd.DataFrame(data = dataset, columns = ["Final Exam",
7         "Final Grade"])
8    dataFrame.plot.scatter(x = "Final Exam", y = "Final Grade",
9         title = "Scatter chart between final exams and final grades ",
10        figsize = (7, 7))
```

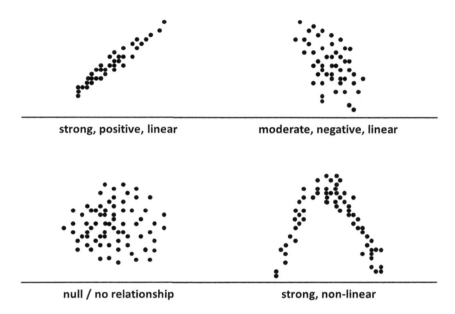

FIGURE 8.5   Types of scatter plots.

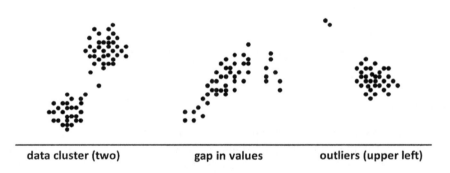

FIGURE 8.6   Investigating data patterns.

Output 8.5.6:

```
<AxesSubplot:title={'center':'Scatter chart between final exams
and final grades '}, xlabel='Final Exam', ylabel='Final Grade'>
```

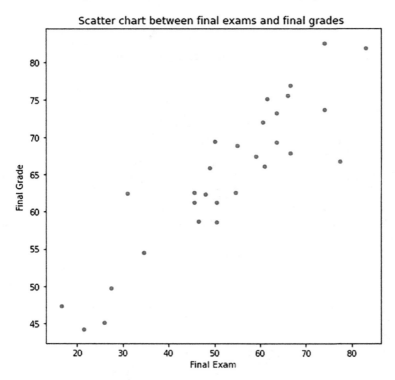

8.6   WRAPPING UP

This chapter covered some of the basic concepts and tasks used in data analysis. Considering the large number of possibilities and analysis combinations that may be utilized in order to provide thorough data analytics results, this chapter was not meant to provide exhaustive analysis of all options, but introductions to some of the main ones that highlight the general approaches and perspectives. For instance, topics like *heatmaps, word clouds, bubble charts, area charts,* and *geospatials* were not covered, although they are rather popular and common data visualization tools. The reader can find more detailed information on such topics in the rather extensive body of work that is readily available in related publications or web sources. At the level of detail and abstraction used in this chapter, Table 8.1 can be used as a quick guide for some of the methods covered, and their use in the context of data analytics.

TABLE 8.1

Quick Guide of Methods and Their Functionality and Syntax

Functionality	Syntax/Example
Data Acquisition	

Import the Pandas library.	`import pandas as <pandas object>`
	`Example:`
	`import pandas as pd`

(Continued)

TABLE 8.1 (*Continued*)

Quick Guide of Methods and Their Functionality and Syntax

Functionality	Syntax/Example

Create a data frame through data read.	`<name of data frame> = <name of pandas object>.read_csv("<Filename.csv", delimiter = ',')`
	Example:
	`dataset=pd.read_csv('WPP2019_TotalPopulationBySex.csv', delimiter = ',')`
	
	`<name of data frame> = <name of pandas object>.read_excel("<Filename.xlsx>", sheet_name = "<Sheet name>")`
	Example:
	`dataset=pd.read_excel(WPP2019_Total_Population.xlsx', sheet_name = "ESTIMATES")`
	
	`<name of data frame> = <name of pandas object>.read_html("<url>")`

Data Cleaning

Functionality	Syntax/Example

Delete all rows containing missing data.	`<name of new Data Frame> = <name of original Data Frame>.dropna()`
	Example:
	`dframe_no_missing_data = dataset.dropna()`

Delete all rows containing any missing data.	`<name of new Data Frame> = <name of original Data Frame>.dropna(how = "any")`
	Example:
	`dframe_delete_rows_with_any_na_values = dataset.dropna(how = "any")`

Delete all rows with missing data in all columns.	`<name of new Data Frame> = <name of original Data Frame>.dropna(how = "all")`
	Example:
	`dframe_delete_rows_with_all_na_values = dataset.dropna(how = "all")`

Replace missing values with a predefined or calculated value.	`<name of new Data Frame> = <name of original Data Frame>.fillna(value[, how = 'all'] [, inplace = True])`
	`<name of new Data Frame> = <name of original Data Frame>.fillna(value[, how = 'any'] [, inplace = True])`
	Example:
	`dataset.fillna(0, inplace = True)`

Change the names of columns with new ones.	`<name of Data Frame>.rename(columns = {"oldname": "newname", } [, inplace=True])`
	Example:
	`dataset_new = dataset.rename (columns = {"Final Grade": "Total Grade", "Quiz 1": "Test 1", "Quiz 2": "Test 2", "Midterm Exam": "Midterm"})`

Change the index of a dataset and reset it back to the original column.	`<name of dataset>.set_index("<column name>"[, inplace=True])` `<name of dataset>.reset_index([inplace=True])`

Data Exploration

Functionality	Syntax/Example

Find the number of records in the dataset.	`len(<name of dataset>`
	Example:
	`len(dataset)`

(Continued)

TABLE 8.1 (*Continued*)
Quick Guide of Methods and Their Functionality and Syntax

Functionality	Syntax/Example

Report the columns of the dataset.	`<name of dataset>.(columns)`
	`Example:`
	`dataset.columns`

Report the number of records and columns in the dataset.	`<name of dataset>.shape`
	`Example:`
	`dataset.shape`

Report the first *n* records of the dataset.	`<name of dataset>.head(n)`
	`Example:`
	`dataset.head(5)`

Report the last *n* records of the dataset.	`<name of dataset>.tail(n)`
	`Example:`
	`dataset.tail(5)`

Report a number of records and columns from the dataset, based on their name and/or index value.	`<name of dataset>[start row: end row: step]`
	`<name of dataset>.loc[start row: end row, "<name of starting column>": "<name of ending column>"]`
	`<name of dataset>.iloc[start row: end row, start column (index): end column (index)`
	`Example:`
	`print(dataset[0:37:5])`
	`print(dataset.loc[0:5," Final Grade" : "Final Exam"])`
	`print(dataset.iloc[0:5,0:3])`

Report only the unique values from a selected column in the dataset.	`<name of dataset>["<name of column>".unique()]`
	`Example:`
	`dataset["Project"].unique())`

Report data based on simple or compound condition.	`<name of dataset>[<condition>]`
	`<name of dataset> [<condition>[&/│] <condition>]]`
	`Examples:`
	`dataset["Final Grade"] > 80`
	`dataset["Final Grade"] > 0) & (dataset["Final Grade"] < 60)`

Merge two datasets into a new one.	`<name of new dataset> = <name of first old dataset>.append(<name of second old dataset>)`
	`Example:`
	`dataset = dataset1.append(dataset2)`

Create a new column based on an expression using data from other columns.	`<name of dataset>["<name of new column>"] = expression with other columns`
	`Example:`
	`dataset["Course Work"] = dataset ["Quiz"]*0.2 + dataset ["Midterm Exam"] *0.25 + dataset["Project"]*0.25`

Create a new column based on a condition.	`<name of dataset>["<name of new column>"] = np.where (condition, value if True, value if False)`
	`Example:`
	`dataset["Letter Grade"] = np.where (dataset["Final Grade"] > 89, "A")`

Create a new column based on a set of conditions and paired values.	`<name of dataset>["<name of new column>"] = np.select (conditions, paired values)`
	`Example:`
	`dataset["Letter Grade"] = np.select (conditions, gradeLetters)`

(*Continued*)

TABLE 8.1 (*Continued*)

Quick Guide of Methods and Their Functionality and Syntax

Functionality	Syntax/Example

Group a dataset based on one or more columns, and apply any aggregate method necessary (e.g., sum(), mean()).	`<name of dataset>.groupby(["<name of column>" [, "<name of column>",...]]).<aggregate function>`
	Example:
	`dataset1.groupby(["Letter Grade"]).mean()`

Group a dataset based on one or more columns. Use apply() to organize the records and columns in the dataset.	`<name of dataset>.groupby(["<name of column>" [, "<name of column>",...]]).apply(lambda x: x[<rows>, <cols>])`
	Example:
	`dataset1.groupby(["Letter Grade"]).apply(lambda x: x[0:rows])`

Sort the data in a dataset.	`<name of dataset>.sort_values(["<name of column>" [, "<name of column>",...]] [, ascending = False])`
	Example:
	`dataset3.groupby(["Letter Grade"]).apply (lambda x: x.sort_values (["Final Grade"], ascending=False))`

Descriptive Statistics

Use mean() to find the mean/average in a dataset.	`<name of dataset>["<name of column>"].mean()`
	Example:
	`dataset["Final Grade"].mean()`

Use median() to find the median in a dataset.	`<name of dataset>["<name of column>"].median()`
	Example:
	`dataset["Final Grade"].median()`

Use mode() to find the most frequent value in a dataset.	`<name of dataset>["<name of column>"].mode()`
	Example:
	`dataset["Final Grade"].mode(dropna = True).values`

Use .values to discard all output from the mode() report except its value.	`<name of dataset>["<name of column>"].mode().values`
	Example:
	`dataset["Final Grade"].mode(dropna = True).values`

Use max() to find the max value in a dataset.	`<name of dataset>["<name of columna>"].max()`
	Example:
	`dataset["Final Grade"].max()`

Use min() to find the min value in a dataset.	`<name of dataset>["<name of columna>"].min()`
	Example:
	`dataset["Final Grade"].min()`

Use quantile(x) to find the xth quantile in a dataset.	`<name of dataset>["<name of columna>"].quantile(0.0-1.0)`
	Example:
	`dataset["Final Grade"].quantile(0.25)`

Use variance() (*Statistics* package) to calculate data *variance*.	`statistics.variance(<name of dataset>["<name of column>"].dropna()`
	Example:
	`statistics.variance(dataset["Final Grade"].dropna())`

Use std() or stdev() (*Statistics* package) to calculate *standard deviation*.	`<name of dataset>["<name of column>"].dropna statistics.stdev(<name of dataset>["<name of column>"].dropna()`
	Example:
	`dataset["Final Grade"].std()`
	`statistics.stdev(dataset["Final Grade"].dropna())`

(Continued)

TABLE 8.1 (*Continued*)

Quick Guide of Methods and Their Functionality and Syntax

Functionality	Syntax/Example

Use skew() to calculate data *skewness*.	`<name of dataset>["<name of column>"].skew()`
	`Example:`
	`dataset["Final Grade"].skew()`

Use kurtosis() to calculate data *kurtosis*.	`<name of dataset>["<name of column>"].kurtosis()`
	`Example:`
	`dataset["Final Grade"].kurtosis()`

Use count() to calculate the *frequency of occurrence* of a value.	`<name of dataset>["<name of column>"].count()`
	`Example:`
	`dataset["Final Grade"].count()`

Use describe() to automatically report a set of basic descriptive statistics.	`<name of dataset>["<name of column>"].describe()`
	`Example:`
	`dataset["Final Grade"].describe()`

	Data Visualization

Use the hist() function (*Pandas* library) to draw *histograms*.	`plt = <name of dataset>.plot.hist(subplots =`
	`<integer>, grid = True/False, legend = True/False,`
	`layout = (<number of rows>, <number of columns>,`
	`figsize = (<size on x axis in inches>, <size on y`
	`axis in inches>), bins = <number of bins>)`
	`Example:`
	`plt = dataset1.plot.hist(subplots = 2, grid = True,`
	`legend = True, layout = (2, 3), figsize = (10, 10),`
	`bins = 10)`

Use the boxplot() function (*Pandas* library) to draw *box and whiskers* plots.	`<name of dataset>.boxplot ([grid = True/False],`
	`[figsize = (<integer>, <integer>), [showcaps = True/`
	`False], [showbox = True/False], [showfliers = True/`
	`False], [labels = <names of columns>)`
	`Example:`
	`dataset1.boxplot(grid = True, figsize = (10, 10),`
	`showcaps = True, showbox = True, showfliers = True,`
	`labels = cols)`

Use the line() function (*Pandas* library) to draw a *line* chart.	`<name of dataset>.plot.line ([grid = True/False],`
	`[figsize = (<integer>, <integer>], [title =`
	`"<title>"])`
	`Example:`
	`dataset1.plot.line(grid = True, figsize = (7, 7),`
	`title = "Grades Line Chart")`

Use the bar() function (*Pandas* library) to draw a *bar* chart. Use the subplots(), and stacked() functions with appropriate code to create different types of bar charts.	`<name of dataset>.plot.bar()`
	`Example:`
	`see relevant script in the text`

Use the pie() function (*Pandas* library) to draw a *pie* chart. Use the plt object of the *matplotlib.pyplot* package to configure and improve the appearance of the chart.	`<name of dataset>.pie()`
	`Example:`
	`see relevant script in the text`

(Continued)

TABLE 8.1 (*Continued*)
Quick Guide of Methods and Their Functionality and Syntax

Functionality	Syntax/Example

Use the `scatter()` function (*Pandas* library) to draw a *scatter plot* based on two datasets.	`<dataFrame>.plot.scatter(x = "<column 1>", y = "<column 2>", [title = "<title>",...])`
	`Example:`
	`dataFrame.plot.scatter(x = "Final Exam", y = "Final Grade", title = "Final exams and final grades ", figsize = (7, 7))`

8.7 CASE STUDY

Readmission is considered a quality measure of hospital performance and a driver of healthcare costs. Studies have shown that patients with diabetes are more likely to have higher early readmissions (readmitted within 30 days of discharge), compared to those without diabetes (American Diabetes Association, 2018; McEwen & Herman, 2018). To reduce early readmission, one solution is to provide additional assistance to patients with high risk of readmission. For this purpose, the US Department of Health would like to know how to identify the patients with high risk of readmission using the collected clinical records of diabetes patients from 130 US hospitals between 1999 and 2008.

As an attempt to assist the US Department of Health in understanding the data, you are asked to explore, analyse (descriptively), and visualize the data of readmission (**readmitted**) and the potential risk factors, such as time in hospital (**time_in_hospital**) and hemoglobin A1c results (**HA1Cresult**), using techniques covered in this chapter.

More specifically, your work should cover the following:

1. **Data Acquisition:** Import the related data file (i.e., Diabetes.csv).
2. **Data Exploration:** Report the number of records/samples and the number of columns/variables in the dataset.
3. **Descriptive Statistics:** Use suitable techniques to summarize or describe the three variables we are interested in: **readmitted, time_in_hospital**, and **HA1Cresult**.
4. **Data Visualisation:** Use appropriate techniques to visualize the three variables and the relationships between **readmitted** and **time_in_hospital**, and **readmission** and **HA1Cresult**.

REFERENCES

American Diabetes Association. (2018). Economic costs of diabetes in the US in 2017. *Diabetes Care, 41*(5), 917–928. https://doi.org/https://doi.org/10.2337/dci18-0007.

Freedman, D., Pisani, R., & Purves, R. (1998). *Statistics* (3rd ed.). New York: WW Norton & Company.

McEwen, L. N., & Herman, W. H. (2018). Health care utilization and costs of diabetes. *Diabetes in America* (3rd ed.), 40-1–40-78. NIDDK.

Pandas. (2021a). *pandas.DataFrame.boxplot.* Version: 1.2.5. https://pandas.pydata.org/docs/reference/api/pandas.DataFrame.boxplot.html.

Pandas. (2021b). *pandas.DataFrame.plot.pie.* https://pandas.pydata.org/docs/reference/api/pandas.DataFrame.plot.pie.html.

Statistics — Mathematical statistics functions. (2021). Python. https://docs.python.org/3/library/statistics.html.

9 Statistical Analysis with Python

Han-I Wang
The University of York

Christos Manolas
The University of York
Ravensbourne University London

Dimitrios Xanthidis
University College London
Higher Colleges of Technology

CONTENTS

9.1 Introduction 374
   9.1.1 What is Statistics?...374
   9.1.2 Why Use Python for Statistical Analysis?..375
   9.1.3 Overview of Available Libraries...375
9.2 Basic Statistics Concepts 376
   9.2.1 Population vs. Sample: From Description to Inferential Statistics 376
   9.2.2 Hypotheses and Statistical Significance 377
   9.2.3 Confidence Intervals 378
9.3 Key Considerations Prior to Conducting Statistical Analysis 379
   9.3.1 Level of Measures: Categorical and Numerical Variables 379
   9.3.2 Types of Variables: Dependent and Independent Variables 380
   9.3.3 Statistical Analysis Types and Hypothesis Tests..................................381
      9.3.3.1 Statistical Analysis for Summary Investigative Questions.......381
      9.3.3.2 Statistical Analysis for Comparison Investigative Questions....381
      9.3.3.3 Statistical Analysis for Relationship Investigative Questions....383
   9.3.4 Choosing the Right Type of Statistical Analysis 385
9.4 Setting Up the Python Environment...386
   9.4.1 Installing Anaconda and Launching the Jupyter Notebook 387
   9.4.2 Installing and Running the Pandas Library 387
   9.4.3 Review of Basic Data Analytics 387
9.5 Statistical Analysis Tasks 388
   9.5.1 Descriptive Statistics 388
   9.5.2 Comparison: The Mann-Whitney U Test 391
   9.5.3 Comparison: The Wilcoxon Signed-Rank Test 391
   9.5.4 Comparison: The Kruskal-Wallis Test 392
   9.5.5 Comparison: Paired t-test 393
   9.5.6 Comparison: Independent or Student t-Test...395
   9.5.7 Comparison: ANOVA...396
   9.5.8 Comparison: Chi-Square 397
   9.5.9 Relationship: Pearson's Correlation 398
   9.5.10 Relationship: The Chi-Square Test 399

DOI: 10.1201/9781003139010-9

      9.5.11 Relationship: Linear Regression..400
      9.5.12 Relationship: Logistic Regression 402
9.6   Wrap Up...404
9.7   Exercises 405
References...407

9.1 INTRODUCTION

When working with data, one of the main questions one seeks to answer is whether the observed value fluctuations and differences *are random or not*. If not by chance, what are the key factors that cause such changes, and what are their relationships with the data? *Statistical analysis*, and in particular *inferential statistics*, is the key tool for answering these questions.

In this chapter, some commonly used statistical functions and the relationship between different types of measurements and statistical tests are introduced, accompanied by demonstrations of how to conduct relevant statistical analysis tasks in Python. The analysis functions follow a linear and incremental order, and build on concepts introduced previously, in order to assist readers with little or no prior experience in this area. For those familiar with the various concepts and functions discussed, this chapter can be used as a refresher or as a practical guide to implementing and executing common statistical functions using the Python platform.

The reader should note that before embarking on any substantial task involving statistical analysis, it is important to consult statistics experts in order to determine the appropriate data collection functions and measurement units, as well as the types of statistical tests required and the best approaches for interpreting and reporting the results.

9.1.1 WHAT IS STATISTICS?

Statistics is a branch of applied mathematics involving the tasks of data collection, manipulation, interpretation, and prediction. Two broad categories can be identified in the field of statistics: *descriptive* and *inferential*. *Descriptive statistics* (covered in part in Chapter 8 on Data Analytics and Data Visualization) focus on *identifying and describing patterns in the data*, by utilizing straightforward functions like *frequencies* and *mean calculations*. In descriptive statistics, there is no uncertainty or unknown factors. The goal is to summarize large volumes of data, making it easier to visualize and understand. On the other hand, *inferential statistics* focus on *putting forward hypotheses* (or *inferences*) related to a *sample* taken from a wider *population*. The hypotheses can be then generalized and applied to the entire population. Hence, as the sample does not contain the entirety of the population, analytical tasks utilizing inferential statistics *are bound to contain an element of uncertainty*.

The reader must note that the term statistics is commonly used to refer to inferential statistics, while the term descriptive statistics is used when analytical tasks are conducted solely for describing existing data. In line with this convention, in this chapter the term statistics will be most frequently used to refer to inferential statistics, unless stated otherwise.

Observation 9.1 – Statistics: A branch of applied mathematics that involves the tasks of data collection, manipulation, interpretation, and prediction. Two broad categories can be identified: *descriptive* and *inferential statistics*.

Observation 9.2 – Descriptive Statistics: The focus is on *identifying and describing patterns in the data* through *frequencies* and *mean calculations*.

Observation 9.3 – Inferential Statistics: The focus is on *putting forward hypotheses (inferences)* related to a *sample* from a wider *population*. If the hypotheses are proven correct, they are generalized and applied to the entire population.

9.1.2 WHY USE PYTHON FOR STATISTICAL ANALYSIS?

A large number of specialized statistical software tools are available, such as *SAS*, *Stata*, *R*, and *SPSS*, and are widely used for both academic and commercial purposes. However, as each of these software packages come from different developers, they use customized features and specialized commands and syntax that cannot be directly translated and exchanged across different platforms. On the contrary, Python is a general-purpose programming language with extensive cross-platform capabilities. This characteristic gives Python an advantage when it comes to complex statistical analysis tasks that mix statistics with other data science fields, such as *image analysis*, *text mining*, or *artificial intelligence* and *machine learning*. In such cases, the richness and flexibility of Python, provided by its ability to adapt its functionality by means of appropriate *modules*, make it a better choice compared to other specialized statistical software packages. Furthermore, the Python language is relatively easy to learn compared to those found in the more specialized statistical software tools. Its syntax is reminiscent of the English language, making it easy to learn and use, and thus accessible to users from diverse backgrounds and programming expertise levels. Finally, Python is an open-source and free-to-use language, unlike most of the specialized statistical packages that frequently come at a considerable cost.

> **Observation 9.4:** Python, as a general-purpose programming language, allows the user to integrate statistics with other data science fields and tasks like *image analysis, text mining, artificial intelligence,* or *machine learning.*

9.1.3 OVERVIEW OF AVAILABLE LIBRARIES

A number of Python libraries, such as *NumPy, SciPy, Scikit-learn*, and *Pandas,* provide functions and tools that allow the user to conduct specific statistical analysis tasks. As the names suggest, NumPy and SciPy focus on numeric and scientific computations, as they support basic operations on *multidimensional arrays*. Accordingly, Scikit-learn is mostly used for *machine learning* and *data mining*, as it offers simple and efficient tools for common data analysis tasks. Pandas is derived from the term *panel data*, and is designed for data manipulation and analysis (McKinney & Team, 2020). For pure statistical analysis purposes, the Pandas library is one of the most suitable options, as it provides high-performance data analysis tools (Anaconda Inc., 2020).

> **Observation 9.5:** The *NumPy* and *SciPy* libraries focus on numeric and scientific computations, *Scikit-learn* is used for *machine learning* and *data mining*, and *Pandas* for data manipulation and analysis.

The reader will notice that the library of choice for a large part of the work covered in this chapter is Pandas. This is due to three main reasons. Firstly, the library is highly suitable for the types of statistical analysis tasks covered in this chapter. Secondly, it supports different data formats like *comma-separated values (.csv), plain text, Microsoft Excel (.xls)*, and *SQL*, allowing the user to import, export, and manipulate databases easily. Thirdly, it is built on top of the SciPy library, so the results can be easily fed into functions of associated libraries like Matplotlib for plotting and Scikit-learn for machine learning tasks (McIntire et al., 2019). This highlights another concept that is central to the structure and rationale of this chapter: *the selective use of different libraries and functions for different analytical tasks*. For instance, functions from the SciPy library may be used for a specific analytical task alongside functions from the Matplotlib library for plotting the output data. This approach aims at promoting the idea that, as long as the fundamental principles and logic for the various different analytical tasks remain the same, *the reader should feel confident to explore different toolkits and solutions.*

9.2 BASIC STATISTICS CONCEPTS

Readers unfamiliar with the intricacies of statistical analysis who come across the notions of *significant difference*, *p-value*, or *confidence intervals* may wonder what exactly these terms mean, and why they are so central in statistics. In this section, key statistics concepts, and the frequently intimidating jargon that accompanies them, are discussed and contextualized using simple examples. This aims at assisting the reader establishing an understanding of the connections and differences between *descriptive* and *inferential* statistics, and how and why scientists frequently make the transition from the former to the latter.

9.2.1 POPULATION VS. SAMPLE: FROM DESCRIPTION TO INFERENTIAL STATISTICS

Population can be defined as the whole set of individuals or subjects for which generalized observations or assumptions are needed, whereas *sample* is the actual part of this population from which data are actually collected. As such, the sample is bound to be *a small part of the entire population*.

> **Observation 9.6 – Population, Sample:** *Population* is the whole set of individuals or subjects for which generalized observations or assumptions are needed. The *sample* is the part of the population from which data are actually collected. The sample is always a small part of the entire population.

In an ideal scenario, individual information from the entirety of the population would be retrieved. In this case, descriptive statistical functions could be utilized to describe the patterns observed in the data. However, this scenario is extremely rare. In most cases, budget and time constraints related to the data collection and analysis tasks at hand impose significant limitations. This is especially true when the study population is substantial, a rather common situation indeed. For example, if a national survey about the quality of life of all patients with diabetes in the UK is to be carried out, researchers would have to interview a population of approximately 4.7 million people (Diabetes UK, 2019). Arguably, it would be much more efficient to survey a group of diabetes patients rather than the entire population. In such cases, since researchers would get access to the information of a sample, statistical functions that allow one to make *inferences* to the population based on the sample are required. Measuring the national *Body Mass Index (BMI)* scores can be used as an example to demonstrate the underlying rationale. Assume that one wants to measure the BMI scores of all smokers in the UK. Since it is not plausible to get information from the entire UK smoker population, a sample will be drawn, which will be then used to draw conclusions. Ultimately, findings will be generalized to the entire UK smoker population using inferential statistics.

In order to determine the required sample size, various different *sampling functions* are available. These include, but are not limited to, *random*, *cluster*, and *stratified sampling*. Depending on the research question behind the study and on the characteristics of the study population, a particular sampling function may be preferable to others. A detailed analysis of sampling functions and how to choose one is outside the scope of this chapter. However, a large number of related resources, like specialized statistics books and online materials are available for those interested in learning more about the topic.

In terms of generalizing findings and observations from the sample to the entire population, one may wonder how such a generalization can be possible and trustworthy. In its simplest form, this is achieved by *conforming to a strict set of minimum requirements*, summarized below:

1. The sample must be *representative of the population* to which the results will be generalized. Representative means that the sample should reflect specific characteristics of the population, such as age, gender, or ethnic background, as closely as possible.

> **Observation 9.7 – Sample Characteristics:** A sample must be *representative of the population*, suitable for answering the research question *quantitatively*, and allowing for *hypothesis testing*.

2. It must be suitable for answering the research question *quantitatively*.
3. It must allow for *hypothesis testing*, as implied by the research question.
4. The *data analysis must match the type of the data being analyzed*. In other words, one needs to use the right statistical function for the data at hand.

These concepts are further discussed in the following sections.

9.2.2 HYPOTHESES AND STATISTICAL SIGNIFICANCE

Once a representative sample is drawn from the study population, *hypotheses* are drawn based on the underlying *research questions*. These hypotheses are, subsequently, systematically tested in order to measure the strength of the evidence and to draw conclusions about the entire population. This is commonly known as *hypothesis testing*. Hypothesis testing is, therefore, the process of *making a claim about the study population and using the sample data to check whether the claim is valid*. A common and long-established convention within the scientific community is that this claim is based on the assumption that the hypothesis *will not be true*, or in other words, that the analysis will show that the intervention or condition under investigation will have *no difference* or *no effect* in the context of the population. This is a specific and standardized type of assumption that is essential in statistical testing, and is commonly referred to as the *null hypothesis (H₀)*. For those unfamiliar with scientific methodologies, the fact that the expectation is that the analysis will unveil *no difference* as opposed to *some difference* may seem counterintuitive. However, the reader should note that the idea behind this is that the analyst seeks to *reject* the null hypothesis rather than *confirm* it. In other words, the assumption is that if one can *disprove* the null hypothesis (i.e., no difference), a difference or effect *must exist within the population*.

> **Observation 9.8 – Null Hypothesis:** The hypothesis that the intervention or condition under investigation associated with the research question will have *no effect in the population*.

To check the validity of the null hypothesis, one needs to conduct a detailed and strictly-defined type of testing, commonly referred to as *statistical significance* testing. There are numerous statistical significance tests to choose from, depending on the research questions and the data at hand (see Section 9.3 for more details on test selection and on how to conduct such tests in Python). A common attribute of all these tests is that they calculate the *probability* of the results observed in the sample being consistent with the results one would likely get from the entire population. This is known as the p-value, which describes how likely it is that the data would have occured by random chance if the null hypothesis is true. Hence, if the p-value is high, the observed sample data will *confirm* the null hypothesis, and thus there *must be no difference in the population*. If the p-value is low, it is a sign that the observed sample data are inconsistent with the null hypothesis (H₀), which is, therefore, *rejected*. In this case, one can conclude that there *must be a difference* present *in the population* and the difference is *statistically significant* or that a *significant difference* has been detected.

> **Observation 9.9 – Hypothesis or Statistical Significance Testing:** Tests that calculate the *probability* of the results being consistent with those from the entire population. If probability is high, the null hypothesis is confirmed and there is no difference in the population; if it is low, the observed sample data are inconsistent with the null hypothesis, which is therefore *rejected*.

As a working example of the above, the reader can assume a study of the effectiveness of a new hypertension drug, by comparing the blood pressure levels of those using it with the levels of those using conventional hypertension drugs. A hypothesis test can be carried out to detect whether the

---

TABLE 9.1

p–value and Significance

p-value	Significance
>0.1	Little or no evidence of a difference or relationship.
0.05–0.1	Weak evidence of a difference or relationship.
0.01–0.05	Evidence of a difference or relationship.
0.001–0.01	Strong evidence of a difference or relationship.
<0.001	Very strong evidence of a difference or relationship.

---

new drug intervention has any effects on the sample or not. The null hypothesis will be based on the claim that there will be no difference of blood pressure levels between the users of the two different drugs in the sample. Hypothesis testing will be conducted and a p-value will be generated. If the p-value is low and the null hypothesis is rejected, there is evidence that there must be a difference in terms of the effectiveness of the two drugs in the general population.

At this point, the reader may start wondering how low the p-value should be in order to be considered low. The answer to this is that it depends on the *significance level* one chooses for the research question. In other words, for each research question, one needs to determine how high or low the probability (i.e., the p-value) must be in order to conclude whether the sample data is consistent with the null hypothesis or not. Conventionally, differences are considered to be significant if the p-value is less than 0.05 (5%). Essentially, the p-value can be regarded as an indicator of the *strength* of the evidence. The reader can use the classification of p-values as a rough guide for determining whether statistical significance requirements are met for a specific analysis task (Table 9.1).

Using the same hypertension drug example, if the p-value of the hypothesis test is found to be 0.03, it indicates that there is a 3% chance that the same treatment effect would occur in the randomly sampled data. Since the 3% chance is lower than the 5% statistical significance threshold, the null hypothesis can be rejected, leading to the conclusion that a significant difference between the two drugs exists in terms of the treatment effects within the general population. It is worth mentioning that the p-value here only indicates a *statistical relationship and not causation*. For identifying causation, more sophisticated inferential statistical analysis methods, such as regression, are needed (see Sections 9.5.11, 9.5.12).

9.2.3 CONFIDENCE INTERVALS

Another key concept used frequently in statistics is that of *confidence intervals*. The term is used to describe the use of *a range of values within which the actual value of the tests may fall* instead of a single estimated value. More specifically, in inferential statistics, one of the primary goals is to estimate population parameters. However, such parameters like *population mean* and *standard deviation* are always unknown, as it is very difficult, or even impossible, to be measured accurately across the entire population. Instead, estimates are made based on the samples. In order to avoid *selection bias* when the sample is selected and to achieve an accurate and objective representation of the population, methods like *random sampling* are commonly used. However, even when such methods are used, uncertainty about the population estimates still exists to a certain degree, due to the possibility of *sampling errors*. It must be noted that, despite the term used, sampling errors do not refer to actual errors. They appear due to the inevitable variability occurring by chance, as random samples are used rather than an

Observation 9.10 – Confidence Intervals: A range of values within which the actual value of the tests may fall. They act as mediators that take into account potential sampling errors and, therefore, provide a higher level of confidence during the statistical analysis process.

entire population. Nevertheless, they are treated as errors for the purposes of statistical testing, as they may lead to inaccurate conclusions.

Although sampling errors cannot be completely eliminated, *confidence intervals* act as a mediator by taking these potential errors into account and providing a range of values the actual population parameter value is likely to fall within. As an example of this, one can assume that researchers want to know the average height of all secondary school students in the UK. Since it is impossible to measure the height of every single student, a random sample of 1,000 secondary school students could be used. If the analysis of the sample measurements results in an average height of 165 cm, it is unlikely that the population mean will also have this exact value, despite the fact that random sampling was used for sample selection. However, if the average height of the sample is expressed as a value within a confidence interval between 160 and 170, researchers can be confident that the true average height of all UK secondary school students among the entire population is captured within this range.

9.3 KEY CONSIDERATIONS PRIOR TO CONDUCTING STATISTICAL ANALYSIS

Before conducting statistical analysis in Python, key aspects of the data collection process, as well as the tools and methods that will be used for the analysis of the collected data, must be considered. At a basic level, such considerations include:

- the *measurement scales* and the *types of variables* that will be used for data collection,
- the *hypothesis* being tested, and
- the *statistical tests* that will be used for data analysis.

> **Observation 9.11 – Variable:** A characteristic, factor, or quantity that can be measured. As the name suggests, it varies between subjects and/or changes over time. It is directly related to the type of statistical analysis adopted for a given task.

A *variable* is a characteristic, factor, or quantity that can be measured, and which may vary between subjects or change over time (or both). For example, *age* is a variable that varies between individuals and changes over time, while *income* also varies between individuals but may, or may not, change over time. The reason the type of the variable is important is that it is directly related to the type of statistical analysis adopted for a given task. This is true for both descriptive and inferential statistics. Certain statistical analysis tests can be used only with certain types of data. For instance, if statistical methods suitable for *categorical* data are used with *continuous* data, the results are bound to be inconsistent and inaccurate. Hence, knowing the type of data that will be collected in advance enables one to choose the appropriate analysis method.

Variables are generally categorized according to the type of measurement they are used for and the level of detail of this measurement. The following sections briefly introduce the different types of variables, the associated types of statistical tests, and how to choose the right statistical test based on the type of variable at hand.

9.3.1 Level of Measures: Categorical and Numerical Variables

Categorical variables, also known as *qualitative* variables, describe *categories* or *factors* of objects, events, or individuals. An example is *gender*, which contains a finite number of categories (e.g., female, male). Categorical variables can also take numerical values (e.g., 1 for female, 2 for male). However, these values are only used for coding and

> **Observation 9.12 – Categorical Variables (Nominal, Ordinal):** Categorical (or qualitative) variables describe categories or factors of objects, events, or characteristics of individuals *with no mathematical meaning. Nominal* variables take discrete values that have no particular order, while *ordinal* variables take discrete, ordered values.

indexing purposes and *do not have any mathematical meaning*. There are two types of categorical variables: *nominal* and *ordinal*. A brief description of each type is provided below.

- *Nominal* variables can have two or more discrete states, but there is no implied order for these states. For example, *gender* (i.e., female, male) is a nominal variable. *Marital status* (i.e., unmarried, married, divorced) and *ethnic background* (e.g., African, Asian, Caucasian) are also examples of nominal variables. Similarly, in medical research, patients that are either *in treatment* or *not in treatment* can be also described by a nominal variable.
- *Ordinal* variables can have also two or more discrete states, but contrary to nominal variables, they can be *ordered* or *ranked*. For example, a *satisfaction scale* that lets respondents choose a value between 1 (strongly disagree) and 5 (strongly agree) is an example of an ordinal variable. *Age group* (e.g., 20–29, 30–39 and so on) and *income* can be also expressed as ordinal variables.

Continuous variables, also known as *quantitative* variables, are variables that can increase or decrease steadily, or by a quantifiable degree or amount. There are two types of continuous variables, namely *interval* and *ratio*. A brief description of each type is provided below.

- *Interval* variables can be *measurable* and *ordered*, and the intervals between the different values are equally spaced. For example, *temperature* measured in degrees (e.g., Celsius) is an interval variable, as the difference between 40°C and 30°C, and 30°C and 20°C is an equidistant interval of 10°C. Other examples of interval variables include *age* (when measured in years, months or days instead of the ordinal *age groups* of the previous example), or *pH*. Another characteristic of interval variables is that they do not have a *true zero*. For

> **Observation 9.13 – Continuous Variables (Interval, Ratio):** Continuous (or quantitative) variables take continuous numerical values describing measured objects, events, or characteristics of individuals. They can take the form of *intervals* with no *true zero* values, or *ratios* where a *true zero* value has a logical meaning.

instance, there is no such thing as *no temperature*, as a temperature of 0°C is still a measurable temperature. Hence, interval variable values can be also *added* or *subtracted* (but not multiplied or divided).
- *Ratio* variables are similar to interval variables, with one important difference: *they do have a true zero point*. When a ratio variable equals to zero, this means there is *none* of this variable. Examples of ratio variables include *height*, *weight*, and *length*. Also, due to the existence of a true zero point, the ratio between two measurements takes a new meaning. For instance, an object weighing 10 kg is twice as heavy as an object weighing 5 kg. However, a temperature of 30°C (interval variable) cannot be considered twice as hot as 15°C. One can only claim that the 30°C temperature is higher than 15°C.

9.3.2 Types of Variables: Dependent and Independent Variables

Variables are typically classified as either *independent* or *dependent*. *Independent* variables, also called *predictor*, *explanatory*, *controlled*, *input*, or *exposure* variables, have an influence on the dependent variables, but are not affected by any other variables themselves, hence their name. Accordingly, *dependent* variables, also known as *observed*, *outcome*, *output*, or *response* variables, are variables that are changing based on changes in the

> **Observation 9.14 – Dependent and Independent Variables:** *Independent* variables are changed/controlled in an experiment that tests their effect on the *dependent* variables. Both independent and dependent variables can be either *categorical* or *continuous*.

associated independent variables. Ultimately, in a scientific experiment, one seeks to change or control the independent variables in order to test the effects of these changes on the dependent variables.

As an example, one can consider the following research question:

Does the length of treatment result in improved health outcomes?

In this case, the *length of treatment* is the independent variable, while *health outcomes* are the dependent variables. Similarly, if one poses the question:

How aspirin dosage affects the frequency of second heart attacks?

The *aspirin dosage* would be the independent variable, while the *heart attack frequency* would be the dependent variable.

It is worth mentioning that any type of categorical or continuous variables can be either independent or dependent, based on the context. A summary of the various different types of variables is provided in Figure 9.1 below.

9.3.3 STATISTICAL ANALYSIS TYPES AND HYPOTHESIS TESTS

There are various different statistical analysis types and hypothesis tests. In general, statistical analysis can solve three main types of investigative questions: *summary*, *comparison*, and *relationship*. A more detailed list of common statistical analysis types, and the categories of problems they are used to address, are presented on Table 9.2 below.

> **Observation 9.15 – Types of Statistical Analysis:** There are three statistical analysis types: *summary* analysis using descriptive statistics, and *comparison* and *relationship* analysis both using inferential statistics.

9.3.3.1 Statistical Analysis for Summary Investigative Questions

Statistical analysis of this type is mainly used for summarizing and describing a single variable *at a given time*. The most common statistical methods associated with this type of analysis are those calculating the *mean* and *median* for continuous variables and the *frequency* for categorical variables.

9.3.3.2 Statistical Analysis for Comparison Investigative Questions

This type of statistical analysis is related to the comparison of the *means* of a single variable between two or more groups. For example, it can be used if one needs to know whether the *Body Mass Index (BMI)* numbers of men and women are significantly different to each other, or whether a new drug can reduce blood pressure (i.e., measuring blood pressure before and after treatment). In this type of analysis, *p-value* is used to determine whether the difference is *statistically significant*.

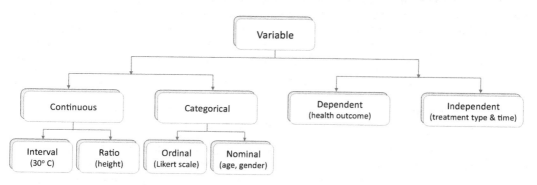

FIGURE 9.1 Types of variables.

TABLE 9.2
Common Types of Statistical Tests

Statistics	Investigative Question	Common Statistical Tests
Descriptive	Summary	Continuous variable: Mean, Median, Mode Categorical variable: Frequency
Inferential	Comparison	Continuous variable: Nonparametric Mann-Whitney U test Wilcoxon Signed-Rank test Kruskal-Wallis, Mood's median test Parametric Student's t-test Paired Student's t-test Analysis of Variance test (ANOVA) Categorical variable: Chi-Square test
Inferential	Relationship	Association strength without causal relationship Pearson's correlation coefficient Chi-Squared test Association strength with causal relationship Linear regression Logistic regression

Overall, there are six common types of tests that can be used for comparative hypothesis cases. The choice of the appropriate test for a particular task depends on a number of factors, such as the sample size, the data characteristics, and the comparison groups. Tests of this type can be further divided into two main categories: *parametric* and *non-parametric* (Table 9.3).

The main difference between parametric and non-parametric analysis is that the former tests the group *means*, while the latter tests the group *medians*. When the sample size of each group is large enough and the comparison data are *continuous* and *normally distributed*, parametric statistical tests are preferable. Parametric tests have more statistical power than their non-parametric counterparts, and can thus detect an existing, underlying effect more efficiently. However, in cases where the sample size is small, or the comparison data are *skewed* or *non-continuous* (e.g., five-point Likert scales) (De Winter & Dodou, 2010), non-parametric statistical methods are more appropriate. Table 9.4 provides a simple indicative list of sample size thresholds for choosing whether parametric and non-parametric tests should be used. The reader can find more on this topic in the various available sources assisting users with statistical test selection, such as Minitab (2015).

Irrespectively of the sample size, when one compares two different means or medians, statistical analysis can be further divided into two types, depending on whether the mean or median comes

TABLE 9.3
Common Types of Comparison Statistical Tests

Parametric Tests (Means)	Non-Parametric Tests (Median)
Independent Student t-test	Mann-Whitney U test
Dependent (Paired) Student t-test	Wilcoxon Signed-Rank test
Analysis of Variance Test (ANOVA)	Kruskal-Wallis, Mood's median test

TABLE 9.4

Simple Guide for Choosing between Parametric and Non-Parametric Tests

Non-Parametric Tests	Sample Size	Parametric Tests
Mann-Whitney U test	$N=15$ in each group	Independent Student t-test
Wilcoxon Signed-Rank test	$N=30$	Dependent (Paired) Student t-test
Kruskal-Wallis, Mood's median test	Compare 2–9 groups, $n=15$ in each group	Analysis of Variance test (ANOVA)
	Compare 10–12 groups, $n=20$ in each group	

from *independent* groups or from *repeated measurements* within the same group. If it comes from independent groups, *independent t-tests* should be used for parametric analysis and *Mann-Whitney U tests* for non-parametric analysis. Examples of such cases are analysis based on measurements of BMI for men and women, or the height of UK and US population. If the mean or median comes from repeated measurements within the same group, *dependent t-tests* should be used for parametric analysis and *Wilcoxon Signed-Rank tests* for non-parametric analysis. An example of this is the measurement of blood pressure before and after using a new drug.

One can also compare three or more different means or medians. An example of this is the comparison of height across different ethnic groups. In this case, *Analysis of Variance (ANOVA) tests* should be used. In simple terms, ANOVA can be viewed as different implementations of *t*-tests that allow one to compare means or medians of more than two groups.

9.3.3.3 Statistical Analysis for Relationship Investigative Questions

This type of statistical analysis is used to investigate the *relationship* between two or more variables. Depending on the type of variable and the purpose of the analysis, it can be further divided into four sub-categories, as outlined in Table 9.5.

In general terms, relationship statistical analysis is suitable for:

- *hypothesis testing*,
- measuring the *association strength*, and
- investigating *causal relationships*.

Hypothesis testing is an attempt to check whether two variables are associated with each other. For example, one may wish to know whether an increase in daily sodium intake results in blood pressure changes Figure 9.2. If the test results in a p-value of 0.05, a significant relationship is assumed to exist between salt intake and blood pressure.

Association strength is a measurement of how closely the two variables are correlated (Table 9.6). This is usually expressed in terms of the R or R^2 value, ranging from −1.0 to 1.0 or 0 to 1.0 respectively. Positive numbers indicate a *positive correlation* (e.g., if one variable increases the other increases too) and negative numbers an *inverse correlation* (e.g., if one variable increases the other

TABLE 9.5

Common Types of Relationship Statistical Tests

Type of Variable	Statistical Test	Association Strength	Causal Relationship
Continuous Variable	Correlation (Linear Regression)	Correlation	Linear Regression
Categorical Variable	Chi-Square (Logistic Regression)	–	Logistic Regression

TABLE 9.6

***R* value and Strength of Correlation**

R value	Strength of Correlation
1.0	Perfect positive correlation
0.7	Strong positive correlation
0.5	Moderate positive correlation
0.3	Weak positive correlation
0	No correlation
−0.3	Weak negative correlation
−0.5	Moderate negative correlation
−0.7	Strong negative correlation
−1.0	Perfect negative correlation

decreases). In this context, a value of 1.0 indicates a perfect correlation, and 0 no correlation. A rule of thumb is that when *R* is higher than 0.7 or lower than −0.7 the two variables are considered to be *highly correlated*. When *R* is between −0.3 and 0.3, the correlation between the two variables is regarded as *weak*. In the example presented in Figure 9.2, *R* is 0.82. Thus, there is a positive relationship between sodium intake and blood pressure. In other words, increasing the daily sodium intake is highly correlated with high blood pressure.

The investigation of *causal relationships* is an attempt to relate the two variables via the equation of a line that stretches across a *cloud of points*. The equation is usually expressed as $Y = a + bX$, and it can be used for prediction. In the example presented in Figure 9.2, the causal relationship results show that blood pressure equals to *114.5 + 3.5 * daily sodium intake*. This indicates that if the daily sodium intake of individuals is known it is possible to predict their approximate blood pressure. For instance, when the daily salt intake is 3 g the blood pressure would be 125 mmHg, and would go up by 3.5 mmHg for every 1 g increase of the daily sodium intake. This example provides a rather simplified, but informative description of the causal relationship concept.

When the two variables are *continuous*, two common types of statistical analysis can be used to test their relationship: *correlation* and *linear regression* (McDonald, 2014). In simple terms, correlation measures the p-value in order to test the hypothesis, and can quantify the direction and strength of the relationship between two continuous variables by summarizing the result with an *R* value. However, correlation cannot infer a *cause-and-effect* relationship. On the other

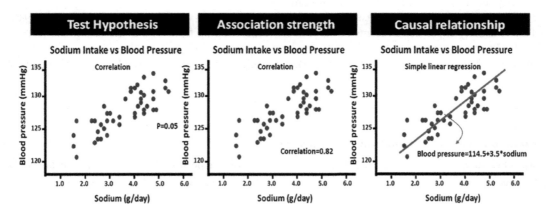

FIGURE 9.2 Relationships between daily salt intake and blood pressure.

TABLE 9.7
Cheat Sheet for Choosing the Right Statistical Test

No. of Variables	Question Type	Dependent Variable	Independent Variable	Statistical Test
1	Summary	Continuous	–	Mean, Mode
1	Summary	Categorical	–	Frequency
1	Comparison	Continuous	2 groups	t-Test
1	Comparison	Continuous	3+ groups	ANOVA
1	Comparison	Categorical	2+ groups	Chi-Square
2	Relationship	Continuous	1 continuous	Correlation
2	Relationship	Categorical	1 categorical	Chi-Square
2+	Relationship	Continuous	1+ variables	Linear Regression
2+	Relationship	Categorical	1+ variables	Logistic Regression

hand, linear regression provides a p-value for hypothesis testing similarly to correlation, but can also summarize the causal relationship with an equation that describes the relationship between variables.

When the variables are *categorical* (i.e., nominal and ordinal), their relationship can be tested using two additional types of statistical analysis: *chi-square test* and *logistic regression*. The chi-square test is used to test the association by providing a p-value. For example, if one is interested in the relationship between *gender* and *smoking status*, the chi-square test can be used. If the result is a p-value of 0.015, a strong association between gender and smoking status can be assumed. As in correlation, the chi-square test cannot infer a cause-and-effect relationship. To do so, logistic regression is required. The latter works like linear regression in the sense that it can summarize the causal relationship with an equation and use the equation for prediction. The only difference between the two is that logistic regression is used for categorical data, while linear regression is used for continuous data.

The reader can find a list and a brief description of a number of common statistical analysis tests discussed in this section on Table 9.7.

9.3.4 Choosing the Right Type of Statistical Analysis

Selecting the right type of statistical analysis is one of the most important considerations when conducting analytical work. This decision is generally based on the type and number of variables, and it can be a challenging process for those with less experience in this field of study. Table 9.7 presents a *cheat sheet* that can be used to determine when to choose the statistical tests mentioned in Section 9.3.3, Table 9.2. The first column contains the *number of variables* under investigation and the second the type of the *research question* one is trying to answer.

Observation 9.16 – Selecting the Appropriate Test: The decision of what test to use is not an arbitrary one but depends on a number of factors, such as the types and number of variables at hand, the number of groups to be tested, the sample size, and the data distribution characteristics.

The third and fourth columns contain the types of the *independent* and *dependent* variables, and the fifth the recommended *statistical test*. A decision tree chart is also provided on Figure 9.3, with the recommended statistical test at the end of each tree branch. By using these resources as a guide, the reader should be able to find a suitable statistical test for the data type and research question at hand. It must be noted that this is a just a brief introduction to the topic of statistical test suitability and selection. In addition to any decisions based on such guides, it is always helpful and advisable to consult statisticians and analysis experts before embarking on any serious analytical task.

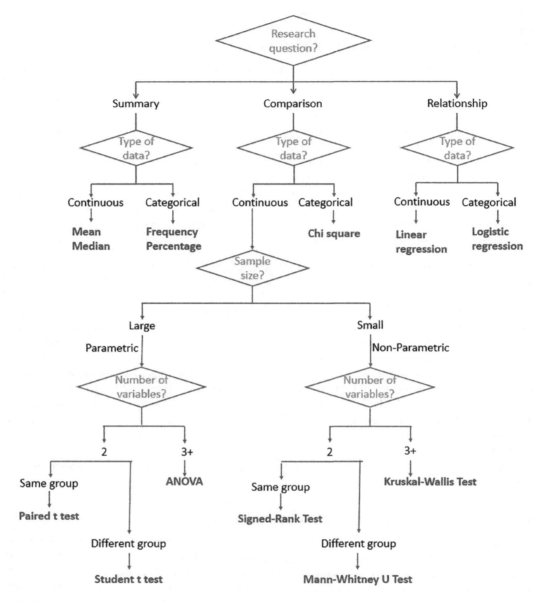

FIGURE 9.3 Choosing the right statistical analysis.

9.4 SETTING UP THE PYTHON ENVIRONMENT

General information related to the process of setting up, and operating in, the Python environment are provided in Chapter 1 of this book. Most of the essential requirements and basic programming concepts presented in these chapters are transferable and, thus, apply to the work and ideas presented here. Nevertheless, if the reader opts to focus solely on this chapter, the sections below provide a quick guide on how to set up the essential platforms, namely *Anaconda* and *Jupyter*, as well as the required libraries and modules required for the purposes of statistical analysis.

9.4.1 INSTALLING ANACONDA AND LAUNCHING THE JUPYTER NOTEBOOK

The official *Anaconda* download page allows the user to download and install the latest version of the Python platform (see Chapter 1) (Anaconda Inc., 2020). The code and examples provided in this chapter were written and tested using Python 3.9. Once Anaconda is installed, the *Anaconda Navigator* can be used to launch applications, and simple Python programs can be created and run using the *Spyder* or *Jupyter Notebook* environments.

For the purposes of this chapter, Jupyter Notebook is the platform of choice. This is due to a number of reasons. Firstly, it offers an appropriate environment for the *Pandas* library, which is required for tasks related to data exploration and modelling. Secondly, it allows for the execution of code in cells rather than running the entire file, something that can save time when it comes to debugging. Thirdly, it provides an easy way to visualize datasets and plots.

9.4.2 INSTALLING AND RUNNING THE PANDAS LIBRARY

To install Pandas, the reader can type `!pip install pandas` in the command input cell. Since Pandas is used frequently, it is common to import Pandas with a shorter name, namely pd. This is done by using the `import pandas as pd` expression:

```
!pip install pandas
import pandas as pd
```

9.4.3 REVIEW OF BASIC DATA ANALYTICS

With Pandas imported, the user can read data from local *.csv* files using the `pd.read_csv()` function and the full path directory of the file. For example, the following command can be used to read data from a local file named *purchase.csv:*

```
df = pd.read_csv('C:\Python\Example\purchase.csv', index_col=0)
```

The same applies to reading data files of other types, like Excel spreadsheets, SQL, and JSON, using the appropriate functions (i.e., `pd.read_excel()`, `pd.read_sql_query()`, and `pd.read_json()`) (The Pandas Development Team, 2020). For the purpose of importing tables from HTML webpages, Pandas uses the `pd.read_html()` function (Sharma, 2019). The following example uses the HTML dataset from a cryptocurrency website to showcase this (WorldCoinIndex, 2021). Firstly, the *requests* library is imported. After passing the website link to variable url, function request.get() attempts to connect to the web server and allocate the relevant connection information to variable `crypto_url`. If a connection is established, property `crypto_url.text` is used as an argument to the `pd.read_html` command that, in turn, passes a *dataframe* to variable `crypto_df`. This particular dataframe contains columns with unnecessary data that are discarded from the main dataset. Finally, the first five rows of the dataset are displayed:

```
1    import pandas as pd
2    import requests
3
4    # Define the url
5    url = 'https://www.worldcoinindex.com/'
6    # Request the url
7    crypto_url = requests.get(url)
8    # Read from the url to Pandas object
9    crypto_df = pd.read_html(crypto_url.text)
```

```
10  # Acquire only the relevant data form the dataset
11  dataset = crypto_df[0]
12  # Limit the displayed columns
13  df = dataset.iloc[0:102, 2:5]
14  # Print the first five rows of the dataset
15  print(df.head(5))
```

Output 9.4.3:

```
                 Name Ticker   Last price
0             Bitcoin    BTC   $ 33,839
1            Ethereum    ETH   $ 2,140.42
2       Axie Infinity    AXS   $ 40.82
3            Dogecoin   DOGE   $ 0.193697
4      Ethereumclassic    ETC   $ 47.64
```

A *dataframe* is a two-dimensional tabular data structure with labeled rows and columns. To view the dataframe, the user can simply call the name of the variable it is stored in. For instance, calling variable `crypto_df` from the *pd.read_csv* example presented above will read the entire dataframe that is stored in it. By default, the first and last five rows of a dataframe can be also retrieved using commands `df.head()` and `df.tail()` respectively. Passing a specific number to the arguments list of the `head()` function retrieves the corresponding number of rows, in this case 10.

When it comes to saving the dataframe, various different file formats can be chosen. These include, but are not limited to, the following:

1. *Plain Text CSV*: A commonly used, straightforward format.
2. *Pickle*: Python's native data storage format.
3. *HDF5*: A format designed to store large amounts of data.
4. *Feather*: A fast and lightweight binary file format that is also compatible with statistical analysis software *R*.

Depending on the requirements and nature of the task at hand, each format has its own advantages and disadvantages. The example below uses Pickle, as the process is rather straightforward: function `to_pickle()` is used to save the dataframe to file *example.pkl* and `pd.read_pickle()` to retrieve it:

```
df.to_pickle('example.pkl')
df1 = pd.read_pickle('example.pkl')
```

9.5 STATISTICAL ANALYSIS TASKS

Once the Python environment is configured and the appropriate methods and tools are determined, the reader can focus on the practical implementation of the various analytical tasks using Python. This section provides coding examples for various statistical analysis concepts and tests as well as information on the interpretation of the test results.

9.5.1 DESCRIPTIVE STATISTICS

Descriptive statistics are typically used for *summarizing* data from a sample. Depending on the type of measures used, a number of tools can be utilized for analysis and visualization (Table 9.8). If the type of measure is a *continuous* variable, functions and methods like `.describe()`, `plot(kind='hist')`, or `plt.hist()` can be used to generate summarized estimates or plot histograms (Koehrsen, 2018).

TABLE 9.8
Common Descriptive Statistical Tools for Different Types of Measures

Type of Measure	Summarized Values	Plot
Continuous Variable	Mean, Median, Standard Deviation, Range	Histogram, Box Chart and similar
Categorical Variable	Frequency, Proportion, Percentage	Pie Chart, Bar Chart, Box Chart and similar

As an example, assume a survey is conducted in order to gather personal information (i.e., *age, gender,* or *BMI*) from adults (18+) in a particular geographic area, and this information should be used to describe the *age distribution* within the sample population. The examples below show how one can generate the associated summary statistics and plot graphs:

```
1    import pandas as pd
2
3    # Define the floating numbers format
4    pd.options.display.float_format = '${:,.2f}'.format
5
6    # Define the analysis dataset
7    dataset = pd.read_csv("Survey.csv", index_col = 0)
8    print("Descriptive Statistics for Age")
9    print(dataset[["age"]].describe())
10
11   # Draw the histogram of the 'age' column
12   dataset["age"].plot(kind = 'hist', title = 'Age');
```

Output 9.5.1.a:

```
Descriptive Statistics for Age
          age
count $2,849.00
mean     $55.83
std      $16.06
min      $18.00
25%      $44.00
50%      $58.00
75%      $67.00
max     $101.00
```

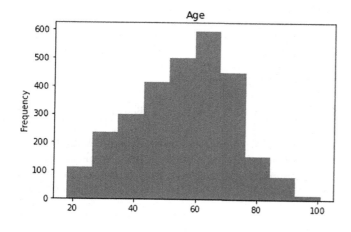

The results indicate that the mean age of this group is 55.83 years. The age ranges from 18 to 101, and the distribution is *symmetrically centred* around the mean.

For *categorical* variables one can use the .value_counts() method to generate the frequency of all values in a column, and the plot(kind='bar') function to plot the frequency using bars (Tavares, 2017). Using the same survey example, the *gender distribution* for the patient group can be calculated and plotted using the following commands:

```
1    import pandas as pd
2
3    # Define the analysis dataset
4    dataset = pd.read_csv("Survey.csv", index_col = 0)
5    print("Descriptive Statistics for Gender")
6    print(dataset[["gender"]].describe())
7
8    # Draw the bar graph for the gender column
9    dataset["gender"].value_counts().plot(kind = "bar",
10           title = "Gender", rot = 0)
```

Output 9.5.1.b:

```
Descriptive Statistics for Gender
         gender
count      2849
unique        2
top      Female
freq       1660

<AxesSubplot:title={'center':'Gender'}>
```

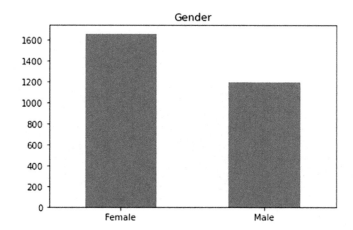

The results show that that there are 1,660 females and 1,182 males within the patient group and the related plot is generated.

As the topic of descriptive statistics is covered in detail in **Chapter 8: Data Analytics and Data Visualization**, the information provided here is only meant to function as a quick reference. Nevertheless, it is important to mention that descriptive statistics are frequently used as a way to gauge the data and provide context to many of the *inferential statistics* tasks presented in the following sections.

9.5.2 Comparison: The Mann-Whitney U Test

The *Mann-Whitney U test* is a type of *non-parametric* test for *continuous* variables. It is used to test whether the distributions of two *independent* samples are equal. This test is appropriate when the sample size is small, or the data are skewed.

As a practical example, one can consider a clinical trial comparing the treatment effects of *standard* and a *new* therapy for patients with depression. A total of ten participants are randomly allocated to the two groups (i.e., *standard therapy/new therapy*). The primary outcome of the measurements is the *depression scores*, ranging from 1 (extremely depressed) to 100 (extremely euphoric):

> **Observation 9.17 – The Mann-Whitney U Test:** A *non-parametric* test for *continuous* variables. It tests whether the distributions of two *independent* samples are equal. It is appropriate when the sample size is small or the data are skewed. Use the `mannwhitneyu()` function from the SciPy library.

Standard therapy	85	65	70	55	40	75	30	80	20	80
New therapy	75	40	60	40	50	65	35	20	25	40

The null hypothesis (H_0) is that *the depression scores of the two therapies are equal*. Since the sample size is small (<20), the Mann-Whitney U Test is the appropriate choice for analysis. To run the test, the user can use the `mannwhitneyu()` function from the SciPy library. Data arrays `data1` and `data2` contain the depression scores of the standard and new therapies. The two sets of results can be compared using the mannwhitneyu(data1, data2) function:

```
1   # Example of the Mann-Whitney U Test
2   from scipy.stats import mannwhitneyu
3   # Standard therapy
4   data1 = [85, 65, 70, 55, 40, 75, 30, 80, 20, 80]
5   # New therapy
6   data2 = [75, 40, 60, 40, 50, 65, 35, 70, 25, 40]
7   mannwhitneyu(data1, data2)
```

Output 9.5.2:

```
MannwhitneyuResult(statistic=34.0, pvalue=0.11941708700675263)
```

The results provide two values: the U statistics value (34.0) and the p-value (0.119). Since the latter is larger than the significance level of 0.05, there is no sufficient evidence to conclude that the number of bacteria in the blood between the two therapies is different. Hence, the null hypothesis can be rejected with the conclusion that *the new therapy does not improve the reduction of bacteria numbers in the blood compared to the standard therapy*.

9.5.3 Comparison: The Wilcoxon Signed-Rank Test

The *Wilcoxon Signed-Rank Test* is used to test whether the distributions of two *paired* samples are equal or not. It is a non-parametric test that can be used for both *continuous* and *ordinal* variables.

> **Observation 9.18 – The Wilcoxon Signed-Rank Test:** A *non-parametric* test for *continuous* or *ordinal* variables. It tests whether the distributions of two *paired* samples are equal. It is appropriate when the sample size is small or the data are skewed. Use the `wilcoxon()` function from the SciPy library.

As an example, one can assume a test during which depression score measurements are taken *before* and *after* a newly developed therapy for ten patients, and the goal is to find whether the therapy makes a difference:

Patient	1	2	3	4	5	6	7	8	9	10
Before therapy	85	65	70	55	40	75	30	80	20	80
After therapy	75	40	50	40	50	65	35	20	25	40

The null hypothesis (H_0) is that *there is no difference in depression scores before and after the therapy.* Since the data are taken from *pairs* and the sample size is small, the Wilcoxon Signed-Rank Test is an appropriate choice. To run the test, the user can use the `wilcoxon()` function from the SciPy library. Data arrays `data1` and `data2` contain the depression scores before and after therapy. The two sets of results can be compared using the `wilcoxon(data1, data2)` function:

```
1   # Example of the Wilcoxon Signed-Rank Test
2   from scipy.stats import wilcoxon
3   # Before therapy
4   data1 = [85, 65, 70, 55, 40, 75, 30, 80, 20, 80]
5   # After therapy
6   data2 = [75, 40, 50, 40, 50, 65, 35, 20, 25, 40]
7   wilcoxon(data1, data2)
```

Output 9.5.3:

```
WilcoxonResult(statistic=7.0, pvalue=0.037109375)
```

The test provides a p-value of 0.036 which is below the significance level of 0.05. Hence, the null hypothesis can be rejected with the conclusion that *the new therapy has a significant effect on the depression scores.*

9.5.4 Comparison: The Kruskal-Wallis Test

The *Kruskal-Wallis Test* is used to test whether the distributions (medians) of two or more *independent* samples are equal or not. It is used for *continuous* or *ordinal* variables when the sample size is small and/or data are *not normally distributed.* The test indicates whether the differences between the test groups are likely to have occurred by chance or not. It is worth noting that the Kruskal-Wallis Test is used under the assumption that the observations in each group come from populations with the same shape of distribution. Hence, if different groups have different distribution shapes (e.g., one

> **Observation 9.19 – The Kruskal-Wallis Test:** A *non-parametric* test for *continuous* or *ordinal* variables with small sample size and/or data *not normally distributed* but with a *similar skewness.* It tests whether the differences between two or more groups are by chance or not. Use the `kruskal()` function from the SciPy library.

is *right-skewed* and another *left-skewed*), the Kruskal–Wallis Test may produce inaccurate results (Fagerland & Sandvik, 2009).

As an example of how to use the test in Python, one can assume a case of three available options to alleviate depression: *standard therapy*, *new therapy*, and *new therapy plus exercise*. The purpose of the test is to determine whether there is any difference in depression scores between the three therapy options with the following depression scores:

New therapy + exercise	90	80	90	30	55	90	55	85	40	90
New therapy	85	65	70	55	40	75	30	80	20	80
Standard therapy	75	40	50	40	50	65	35	20	25	40

Since the sample size is small and the depression scores are *ordinal*, the Kruskal-Wallis Test is an appropriate choice. To run the test in Python, one can use the `kruskal()` function from the SciPy library. Data arrays `data1`, `data2` and `data3` contain the depression scores for new therapy and exercise, new therapy and standard therapy respectively. The three sets of results can be compared using the `kruskal(data1, data2, data3)` expression:

```
1    # Example of the Kruskal-Wallis Test
2    from scipy.stats import kruskal
3    # New therapy and exercise
4    data1 = [90, 80, 90, 30, 55, 90, 55, 85, 40, 90]
5    # New therapy
6    data2 = [85, 65, 70, 55, 40, 75, 30, 80, 20, 80]
7    # Standard therapy
8    data3 = [75, 40, 50, 40, 50, 65, 35, 20, 25, 40]
9    kruskal(data1, data2, data3)
```

Output 9.5.4:

```
KruskalResult(statistic=7.275735789710176, pvalue=0.026308376435655575)
```

The results show that the p-value is 0.026, which is less than the significance level of 0.05. Hence, the null hypothesis (H_0) (i.e., *the depression scores of the three therapies are equal*) can be rejected, with the conclusion that *a significant difference exists between the three treatment options.*

9.5.5   COMPARISON: PAIRED T-TEST

The *Paired t-Test*, also referred to as the *Dependent t-Test*, is used to test whether *repeated measurements* (means) taken *from the same sample* are significantly different. Since the measurements come from the same sample, the terms *paired samples, matched samples* or *repeated measures* are also commonly used for this type of test. The test is used under the assumption that the

Observation 9.20 – The Paired t-Test: A *parametric* test for *normally distributed* data with *no significant outliers.* Use the `ttest_rel()` function from the SciPy library.

measurements are *normally distributed* and *do not contain significant outliers.* If the measurements are skewed or contain significant outliers, the Wilcoxon Signed-Rank Test should be used instead.

As an example, one can assume the case of a new drug developed to assist patients by reducing blood pressure. To investigate the effectiveness of the new drug, the blood pressure of 100 patients is firstly measured prior to taking the drug and also 3 months later. Since the goal is to determine whether the new drug is effective, the null hypothesis (H_0) is that *the average blood pressure will be the same before and after taking the drug.* Assuming a dataset stored in a file named *Blood.csv,* the user can conduct the Paired t-Test in Python using the `ttest_rel()` function from the SciPy library:

```
1    import pandas as pd
2    from scipy.stats import ttest_rel
3
4    # Define the format of floating numbers
```

```
5    pd.options.display.float_format = '{:,.2f}'.format
6
7    # Define the dataset
8    dataset = pd.read_csv("Blood.csv", index_col = 0)
9    print("Descriptive Statistics for Blood before and after")
10   print(dataset[["Before", "After"]].describe())
11
12   # Prepare and display the scatter plot for the dataset
13   dataFrame = pd.DataFrame(data = dataset, columns = ["Before", "After"])
14   dataFrame.plot.scatter(x = "Before", y = "After",
15          title = "Scatter chart for Blood.csv", figsize = (7, 7))
16
17   # Calculate the Paired t-Test
18   ttest_rel(dataset[["Before"]], dataset[["After"]])
```

Output 9.5.5:

```
Descriptive Statistics for Blood before and after
        Before   After
count    80.00   80.00
mean    153.39  147.55
std      10.49   13.57
man     138.00  125.00
25%     144.75  136.00
50%     151.50  146.00
75%     159.25  157.00
max     185.00  184.00

Ttest_relResult(statistic=array([2.91731434]), pvalue=array([0.00459528]))
```

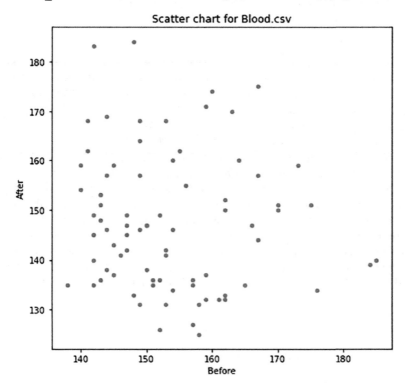

Arrays `data1` and `data2` correspond to the blood pressure scores before and after the drug therapy. The results show that the average blood pressure before taking the new drug was higher (153.38 mmHg) compared to the measurement taken after drug administration (147.55 mmHg). The test provides a p-value of 0.004, which is lower than the significance level of 0.05. Hence, the null hypothesis can be rejected with the conclusion that *a statistically significant difference in blood pressure occurs after using the new drug.*

9.5.6   COMPARISON: INDEPENDENT OR STUDENT T-TEST

The *Independent t-Test*, also known as the *Student t-Test*, is used to test whether the *means* of two *independent* samples are significantly different. To conduct Independent t-Tests in Python, the `ttest_ind()` function from the SciPy library can be used. The function accepts two arrays as parameters, corresponding to the sets of data under investigation. The reader can find more information on the official SciPy.org website (The SciPy Community, 2020).

Observation 9.21 – The Student t-Test: A *parametric* test for *normally distributed* data with *no significant outliers.* Use the `ttest _ ind()` function from the SciPy library.

Using the same survey example, one can assume a case where the user needs to know whether ages between men and women within the sample are different. In this context, the null hypothesis (H_0) *the mean ages of the two groups are equal* is used:

```
1    import pandas as pd
2    from scipy.stats import ttest_ind
3
4    # Define the format of floating numbers
5    pd.options.display.float_format = '${:,.2f}'.format
6
7    # Define the dataset
8    dataset = pd.read_csv("survey.csv", index_col = 0)
9    print("Descriptive Statistics for age grouped by gender")
10   print(dataset["age"].groupby(dataset["gender"]).describe())
11
12   # Calculate the Student t-Test
13   ttest_ind(dataset.age[dataset.gender == 'Male'],
14           dataset.age[dataset.gender == 'Female'])
```

Output 9.5.6:

```
Descriptive Statistics for age grouped by gender
          count     mean     std     min      25%     50%      75%      max
gender
Female $1,660.00 $55.27 $16.42 $18.00 $43.00 $57.00 $67.00 $101.00
Male   $1,189.00 $56.61 $15.50 $19.00 $45.00 $58.00 $68.00  $98.00

Ttest_indResult(statistic=2.1993669348926157, pvalue=0.02793196707542121)
```

The first output shows that the average age for men (56.56) is higher than that of women (55.30). The Independent t-Test is conducted in order to determine whether this difference is significant. The first statistic value is the *t score* (2.199), which is a *ratio* of the difference *between* and *within* the two groups. As a general rule, the higher the *t* score, the bigger the difference would be between groups, and vice versa. To determine whether the *t* score is high enough, one has to rely on the p-value output. In this example, the p-value is 0.0279, which is lower than the significance level of 0.05.

Thus, the null hypothesis can be rejected with the conclusion that *there is a statistically significant difference between the age of male and female individuals.*

9.5.7   COMPARISON: ANOVA

The *ANOVA (i.e., Analysis of Variance) Test* is used to compare the *means* of three or more samples. It assumes *independence of observations, homogeneity of variances,* and *normally distributed observations* within groups. In Python, the user can utilize the f_oneway() function from the SciPy library to calculate the *F-Statistic*, which, in turn, can be used to calculate the p-value. The function accepts parameters corresponding to the sample measures for each group under consideration.

Observation 9.22 – The ANOVA Test: A *parametric* test for *normally distributed, independent observations,* with *homogeneity of variances.* Use the f_oneway() function from the SciPy library.

Using the same survey data as an example, one can assume that the user needs to know whether the *Body Mass Index (BMI)* values are different across *non-smokers, former smokers* and *current smokers* (smoking status). The null hypothesis (H_0) is that *there is no difference between the means of the BMIs among people from the three different groups*:

```
1    import pandas as pd
2    from scipy.stats import f_oneway
3
4    # Define the format of floating numbers
5    pd.options.display.float_format = '{:,.2f}'.format
6
7    # Define the dataset
8    dataset = pd.read_csv("survey.csv", index_col = 0)
9    print("Descriptive statistics for survey by smokestat")
10   print(dataset.bmi.groupby(dataset.smokestat).describe(), "\n")
11
12   # Calculate the one-way ANOVA Test
13   print("Results of ANOVA by smokestat values of Never, Former, Current")
14   print(f_oneway(dataset.bmi[dataset.smokestat == "Never"], \
15        dataset.bmi[dataset.smokestat == "Former"], \
16        dataset.bmi[dataset.smokestat == "Current"]))
```

Output 9.5.7:

```
Descriptive statistics for survey by smokestat
              count   mean   std   min    25%    50%    75%    max
smokestat
Current       363.00 28.20  6.84 17.50  23.20  27.20  31.25  62.60
Former        755.00 29.22  6.24 16.80  25.05  28.20  32.40  66.20
Never       1,731.00 28.14  6.48 16.10  23.50  27.10  31.30  75.20

Results of ANOVA by smokestat values of Never, Former, Current
F_onewayResult(statistic=7.548128785289014, pvalue=0.0005377158828502398)
```

The first output shows that the former smokers have the highest mean BMI (29.22), followed by current smokers (28.30), and non-smokers (28.20). The output of the ANOVA Test shows that the F-Statistic is 6.56 and the p-value is 0.0014, indicating an overall significant effect of smoking status on BMI. However, at this point it is uncertain exactly where the difference between groups lies. To clarify this, one needs to conduct *post-hoc* tests. For more detailed information regarding post-hoc tests in Python, the reader can refer to the official documentation in Scikit-posthocs (2020).

9.5.8 COMPARISON: CHI-SQUARE

As shown, the t-Test is used to check whether means differ between two groups. The *Chi-square Test,* also known as the *Chi-squared Goodness-of-fit Test,* is the equivalent of the t-test for *categorical* variables. It tests whether categorical data from a single sample follow a specified distribution (i.e., *external* or *historical* distribution).

Observation 9.23 – The Chi-Square Test: A *parametric* test for *categorical* variables. It tests whether data from a single sample follow a specified distribution. Use the `chisquare()` function from the SciPy library.

For example, based on the example of a smoker status survey, one can assume that the proportions of *non-smokers, former smokers,* and *current smokers* are 30%, 10%, 60% respectively. The government launched a health promotion campaign in an attempt to increase smoking cession rate. To evaluate the impact of the program, the same survey was conducted for a second time a year later. The survey was completed by 500 people, and the data obtained were the following:

	Non-Smokers	Former Smokers	Current Smokers
Before programme	150	50	300
After programme	140	80	280

Since the goal is to determine the impact of the health promotion programme, the null hypothesis (H_0) assumes that *the distribution of smoking status is the same prior to, and after the implementation of the program* and, thus, the health promotion campaign has no impact. In such cases, the Chi-square Test is an appropriate choice. In Python, the test can be conducted using the `chisquare()` function from the SciPy library. The function accepts parameters corresponding to the observed frequencies in each categorical variable:

```
1   import scipy as scipy
2   from scipy.stats import chisquare
3   # Define the datasets
4   before = scipy.array([150, 50, 300])
5   print("The dataset before the program:")
6   print(before)
7   after = scipy.array([140, 80, 280])
8   print("The dataset after the program:")
9   print(after)
10
11  print("The Chi-square test results are the following:")
12  print(scipy.stats.chisquare(before, after))
```

Output 9.5.8:

```
The dataset before the program:
[150  50 300]
The dataset after the program:
[140  80 280]
The Chi-square test results are the following:
Power_divergenceResult(statistic=13.392857142857142,
pvalue=0.0012353158761688927)
```

The first value of the output (13.39) is the *Chi-square value*, followed by the p-value (0.0012). Since the p-value is less than the significance level of 0.05, the null hypothesis is rejected, indicating *that there is a significant difference in terms of the smoking status before and after the programme.*

9.5.9   RELATIONSHIP: PEARSON'S CORRELATION

Correlation is used to test whether two *continuous* variables have a *linear relationship*. The *correlation coefficient* summarizes the strength of this relationship.

As an example, the reader can assume that one needs to know whether *age* and *BMI* are correlated. The null hypothesis (H_0) for this example is that *age and BMI are not correlated*. Assuming that both age and BMI are *normally distributed* and have the same *variance*, one

> **Observation 9.24 – Pearson's Correlation:** A test used to examine whether two *normally distributed, continuous* variables have a *linear relationship*. Use the `pearsonr()` function from the SciPy library.

can use function `pearsonr()` from the SciPy library to calculate the correlation coefficient and estimate the strength of the relationship. The function accepts two arrays as parameters corresponding to the sets of data:

```
1    import pandas as pd
2    import scipy as scipy
3    import matplotlib.pyplot as plt
4    from scipy.stats import pearsonr
5
6    # Read the dataset
7    dataset = pd.read_csv("example.csv", index_col = 0)
8    print(pearsonr(dataset.age, dataset.bmi))
9
10   # Visualize the correlation with a scatter plot
11   print(plt.scatter(dataset.age, dataset.bmi, alpha = 0.5,
12        edgecolors = "none", s = 20))
```

Output 9.5.9:

```
(0.0453741864067145, 0.014235768675028503)
<matplotlib.collections.PathCollection object
at 0x000002802BD93310>
```

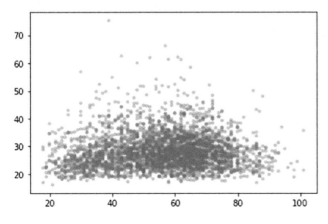

The first value of the output is the correlation coefficient (0.045), followed by the p-value (0.014). Since p-value is less than the significance level of 0.05, one can confirm that *a relationship exists between age and BMI*. Another important observation is that the correlation is *positive* (i.e., if age increases, BMI increases too), as the correlation coefficient is a *positive number*. However, the *strength* of the correlation is rather weak, as the correlation coefficient (0.045) is quite close to 0 (i.e., no correlation).

The correlation can be also visualized as a scatter plot, using the `scatter()` function as shown in the Output plot above.

9.5.10   RELATIONSHIP: THE CHI-SQUARE TEST

To test whether two *categorical* variables are *independent*, one may use the *Chi-squared Test*, also known as *Chi-squared Test of Independence* or *Pearson's Chi-square Test*.

To demonstrate the logic of the test, one can use the same survey data example and evaluate whether *gender* and *smoking status* are associated. The null hypothesis (H_0) would be that *there is no relationship between gen-*

Observation 9.25 – Pearson's Chi-Square Test: A test used to examine whether two *categorical* variables are *independent*. Use the `chi2_contingency()` function from the SciPy library.

der and smoking status. When neither of the two measurements is less than 5, one can use the `crosstab()` function from the Pandas library to create a cross table and `scipy.stats.chi2_contingency()` to conduct the Chi-square Test on the contingency/cross table. Detailed documentation for this function can be found in the official *SciPy.org* website (The SciPy Community, 2020). The following Python script makes use of both the `crosstab()` and the `chi2_contingency()` functions to provide the frequencies of the smoking status across the two gender groups and test whether there is an indication of a relationship between them:

```
1   import pandas as pd
2   import scipy as scipy
3   import matplotlib.pyplot as plt
4   from scipy.stats import chi2_contingency
5
6   # Read the dataset
7   dataset = pd.read_csv("example.csv", index_col = 0)
8   print(pd.crosstab(dataset.smokestat, dataset.gender), "\n")
9
10  # Calculate the Chi-squared Test of Independence
11  print(chi2_contingency(pd.crosstab(dataset.smokestat,
12          dataset.smokestat)))
```

Output 9.5.10.a:

```
gender       Female   Male
smokestat
Current        210    162
Former         403    367
Never         1093    683

(5835.999999999999, 0.0, 4, array([[ 47.42426319,  98.16312543,  226.41261138],
       [ 98.16312543, 203.18711446,  468.64976011],
       [ 226.41261138, 468.64976011, 1080.93762851]]))
```

The first value of the output (19.453) is the *Chi-square value*, followed by the p-value (5.96e–05), the *degrees of freedom* (2), and the *expected frequencies* as an array. Since the p-value is less than 0.05, the null hypothesis can be rejected, indicating that *a relationship between smoking status and gender exists*.

It is worth noting that if an expected frequency lower than 5 is present, the user should use the *Fisher's Exact Test* instead of the Chi-square Test. Both tests assess for independence between variables. The Chi-square Test applies an approximation assuming the sample is large, while the Fisher's Exact Test runs an exact procedure suitable for small-sized samples (Kim, 2017).

To visualize the results of the test, one can also create a *mosaic plot* using the mosaic() function from the *Statsmodels* library. The function accepts the source as a parameter and defines the names of the columns for the plot:

```
1   import pandas as pd
2   import matplotlib.pyplot as plt
3   from statsmodels.graphics.mosaicplot import mosaic
4
5   # Read the dataset
6   dataset = pd.read_csv("example.csv", index_col = 0)
7   mosaic(dataset, ["smokestat", "gender"])
8   plt.show()
```

Output 9.5.10.b:

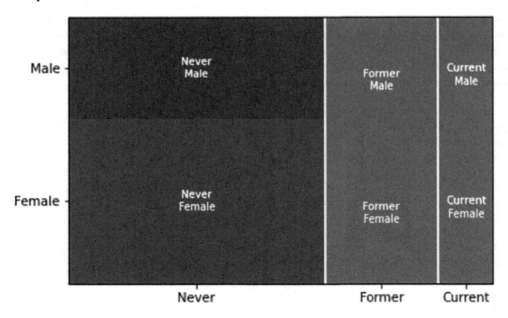

9.5.11   Relationship: Linear Regression

Linear regression is used to examine the *linear relationship* between two (i.e., univariate linear regression) or more (i.e., multivariate linear regression) variables.

To contextualize this using the previous survey example, the reader can assume a case where one wants to test the relationship between *body weight* and *BMI*, where the BMI is *normally distributed*. Additionally, predictions regarding the BMI should be made based on weight information. Since BMI is a *continuous* variable, linear regression is appropriate for

> **Observation 9.26 – Linear Regression:** A test used to examine the *linear relationship* between two (i.e., univariate) or more (i.e., multivariate) variables. Use the OLS(y, X).fit() function from the *Statsmodels* library.

the analysis. In Python, linear regression can be performed using either the *Statsmodels* or the *Scikit-learn* libraries. For this example, the test choice was function OLS(y, X).fit() from the Statsmodels library, as the Scikit-learn library is generally associated more with tasks related to *machine learning*. The related Python script and its output are provided below:

```
1    import pandas as pd
2    import matplotlib.pyplot as plt
3    import statsmodels.api as sm
4
5    # Read the dataset
6    dataset = pd.read_csv("example2.csv", index_col = 0)
7    # Independent variable
8    X = dataset.weight
9    # Dependent variable
10   y = dataset.bmi
11   # Add an intercept (beta_0) to the model
12   X = sm.add_constant(X)
13   # Function sm.OLS(dependent variable, independent variable)
14   model = sm.OLS(y, X).fit()
15   # Predictions
16   predictions = model.predict(X)
17   # Print out the statistics
18   print(model.summary())
19
20   # Plot the statistics
21   print(sm.graphics.plot_ccpr(model, "weight"))
```

Output 9.5.11.a and 9.5.11.b:

```
                           OLS Regression Results
========================================================================
Dep. Variable:               bmi    R-squared:                     0.740
Model:                       OLS    Adj. R-squared:                0.739
Method:            Least Squares    F-statistic:                   8085.
Date:           Sun, 25 Jul 2021    Prob (F-statistic):             0.00
Time:                   16:35:19    Log-Likelihood:              -7449.6
No. Observations:           2849    AIC:                       1.490e+04
Df Residuals:               2847    BIC:                       1.492e+04
Df Model:                      1
Covariance Type:       nonrobust
========================================================================
              coef    std err        t     P>|t|     [0.025     0.975]
------------------------------------------------------------------------
const       6.5712      0.251   26.188     0.000      6.079      7.063
weight      0.1218      0.001   89.918     0.000      0.119      0.124
========================================================================
Omnibus:                 268.275    Durbin-Watson:                 1.290
Prob(Omnibus):             0.000    Jarque-Bera (JB):            609.183
Skew:                      0.574    Prob(JB):                 5.22e-133
Kurtosis:                  4.953    Cond. No.                       750.
========================================================================

Notes:
[1] Standard Errors assume that the covariance matrix of the errors is
correctly specified.
Figure(432x288)
```

In this example of *linear regression,* y equals to a dependent variable, which is the variable that must be predicted or estimated. Variable x equals to a set of independent variables, which are the *predictors* of y. It must be noted that we need to add an *intercept* to the list of independent variables using `sm.add_constant(x)` before running the regression.

The output provides several pieces of information. The first part contains information about the *dependent variable*, the *number of observations*, the *model*, and the *method*. *OLS* stands for *Ordinary Least Squares*, and method *Least Squares* relates to the attempt to fit a regression line that would minimize the square of vertical distance from the data points to the regression line. Another important value presented in the first part is the *R squared* ($R^2 = 0.740$), which is the percentage of variance that the model can justify (73.9%). The larger the *R* squared value the better the model fit.

The second part of the output includes the intercept and the coefficients. The p-value is lower than .0001, indicating that there is *statistical significance in terms of the weight predicting the BMI*, with a weight increase of 1 pound leading to a respective increase in BMI by 0.1219. The linear regression equation can be also used in the following form:

$$BMI = (Intercept) + (Weight\_ coefficient) * weight$$

Once the output numbers are added, the equation would take the following form:

$$BMI = 6.5531 + 0.1219 * weight$$

Therefore, if the user knows a person's weight (e.g., 125 pounds), their BMI can be calculated as *6.5531 + 0.1219 * 125 = 21.7906*.

The user can also use the *Matplotlib* library to plot the results, as illustrated in the associated graph.

9.5.12 RELATIONSHIP: LOGISTIC REGRESSION

Logistic regression is used to describe the relationship between a *dependent, categorical* variable and one or more *independent* variables. It models the logit-transformed probability in a linear relationship with the predictor variables. For instance, using the same survey example, one can assume that the user wants to know the relationship between *smoking status* (i.e., 1=current smoker, and 0=non-smoker) and the potential predictors, such as *age, gender,* and *marital status*. In addition, the user may also want to predict the smoking status based on the predictor information. Since smoking status is a categorical variable, logistic regression is an appropriate analysis method. In Python, logistic regression can be conducted using the `Logit(y, X)` function from the *Statsmodels* library. Parameter y equals to a dependent variable, which is the variable that must be predicted or estimated. Variable X equals to a set of independent variables, which are the predictors of y:

> **Observation 9.27 – Logistic Regression:** A test used to examine the relationship between a *dependent, categorical* variable and one or more *independent* variables. Use the logit(y, X) function from the *Statsmodels* library.

```
1   # Example of Logistic Regression
2   import pandas as pd
3   import statsmodels.api as sm
4
5   # Read data
6   df = pd.read_csv("Example2.csv", index_col = 0)
7
```

```
8   x = df[["age", "gender2", "marital_divorced",
9            "marital_single", "marital_widowed"]]
10  y = df.smokestat2
11
12  # Add an intercept (beta_0) to the model
13  X = sm.add_constant(x)
14
15  logit_model = sm.Logit(y, X)
16  result = logit_model.fit()
17
18  # Print result.summary()
19  print(result.summary2())
```

Output 9.5.12:

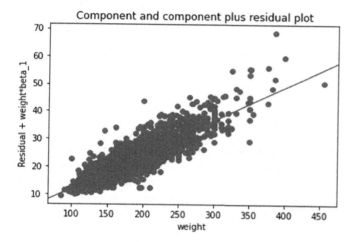

```
Optimization terminated successfully.
        Current function value: 0.373830
        Iterations 6
                    Results: Logit
=================================================================
Model:               Logit           Pseudo R-squared: 0.020
Dependent Variable:  smokestat2      AIC:              2142.0822
Date:                2021-07-27 13:21 BIC:             2177.8105
No. Observations:    2849            Log-Likelihood:   -1065.0
Df Model:            5               LL-Null:          -1086.7
Df Residuals:        2843            LLR p-value:      3.1240e-08
Converged:           1.0000          Scale:            1.0000
No. Iterations:      6.0000
-----------------------------------------------------------------
                  Coef.   Std.Err.    z     P>|z|   [0.025  0.975]
-----------------------------------------------------------------
const            -1.7107   0.2307  -7.4156  0.0000 -2.1628 -1.2585
age              -0.0109   0.0040  -2.7133  0.0067 -0.0187 -0.0030
gender2           0.1805   0.1170   1.5418  0.1231 -0.0489  0.4098
marital_divorced  0.8406   0.1422   5.9097  0.0000  0.5618  1.1194
marital_single    0.4609   0.1584   2.9096  0.0036  0.1504  0.7715
marital_widowed   0.4764   0.2229   2.1372  0.0326  0.0395  0.9133
=================================================================
```

As in linear regression, the output contains two parts. The first part provides information about the dependent variable and the number of observations, while the second part provides the intercept and the coefficients. As shown, *age* and *marital status* are significant predictors on smoking status ($p < 0.05$), while *gender* is not ($p = 0.1231$). Individuals who are divorced are 2.31 (i.e., exp(0.8406)) times more likely to be smokers than those who are married. Similar trends are also observed for those who are single (1.5855 times) and widowed (1.6102 times). In terms of age, it is observed that for every 1-year increase in age there is a decrease of approximately 1% (i.e., 1−exp(−0.0109)) in the odds of an individual being a smoker.

The output information can be also used in order to build the logistic regression as follows:
P(probability of being a smoker) =

$$\frac{\exp(-1.7107 - 0.0109 * \text{Age} + 0.1805 * \text{gender2} + 0.8406 * \text{Divorced} + 0.4609 * \text{Single} + 0.4764 * \text{Widowed})}{1 + \exp(-1.7107 - 0.0109 * \text{Age} + 0.1805 * \text{gender2} + 0.8406 * \text{Divorced} + 0.4609 * \text{Single} + 0.4764 * \text{Widowed})}$$

As such, it can be predicted that a 40-year-old divorced male will have a 24.5% probability of being a smoker:

$$\frac{\exp(-1.7107 - 0.0109 * 40 + 0.1805 * 1 + 0.8406 * 1)}{1 + \exp(-1.7107 - 0.0109 * 40 + 0.1805 * 1 + 0.8406 * 1)} = \frac{0.3244}{1 + 0.3244} = 0.2450$$

9.6   WRAP UP

This chapter focused on the introduction of basic concepts and terms related to statistics analysis and on the practical demonstration of carrying out inferential statistics analysis tasks using Python. It provided an overview of statistics and the available tools for conducting the analytical tasks. Basic statistical concepts, such as *population* and *sample*, *hypothesis*, *significance levels* and *confidence intervals*, were introduced. It also provided a practical guide for choosing the right type of statistical test for different types of tasks. The purposes and definitions of common types of statistical analysis methods were briefly discussed. Furthermore, it covered the necessary background for choosing a statistical analysis approach, such as levels and types of variables and the corresponding statistical and hypothesis tests and demonstrated how to set up the Python environment and work with various libraries specifically designed for statistical analysis. Finally, it provided a practical guide for the implementation and execution of common statistical analysis tasks in Python. Each statistical analysis method was supported by working examples, the associated Python programming code, and result interpretations.

A list of the common statistical analysis methods covered in this section, as well as the corresponding Python libraries and methods, are presented below:

Statistical Test	Library	Code
Mann-Whitney U Test	*SciPy*	`mannwhitneyu(data1, data2)`
Willcoxon Signed-rank Test	*SciPy*	`wilcoxon(data1, data2)`
Kruskal-Wallis Test	*SciPy*	`kruskal(data1, data2, data3, …)`
Paired t-Test	*SciPy*	`ttest_rel(data1, data2)`
Independent t-Test	*SciPy*	`ttest_ind(data1, data2)`
Chi-Square of goodness of Fit	*SciPy*	`chisquare (data1, data2)`
ANOVA	*SciPy*	`f_oneway(data 1, data 2, data3, …)`
Pearson's Correlation	*SciPy*	`pearsonr(var1, var2)`
Pearson's Correlation (Scatter Plot)	*Matplotlib*	`scatter(var1, var2)`

(Continued)

Statistical Test	Library	Code
Pearson's Chi-Square Test	*SciPy*	`chisquare (data1, data2)`
Pearson's Chi-Square Test (Mosaic Plot)	*Statsmodels*	`mosaic(Dataframe, ['var1', 'var2'])`
Linear Regression	*Statsmodels*	`OLS (y, X).fit()`
Logistic Regression	*Statsmodels*	`Logit (y, X)`

The basic inferential statistical tests covered in this chapter lay the foundation for other, more advanced statistical analysis tasks, such as *time to event* and *time series* analysis. Ultimately, such methods and results could be used as building blocks for even more complex system simulations, such as *Markov models, discrete-event,* and *agent-based* simulations. Although advanced statistical analysis and simulation tasks like these were not covered in this chapter, the reader should be able to explore them by building on the information and knowledge acquired. Relevant key textbooks and bibliography for the purposes of further study and self-learning can be found in the Reference List of this chapter.

9.7   EXERCISES

We conducted an experiment about *different plant species response* to *length of light* over 3 months. The data we collected are listed below:

Sample	Plant Species	Length of Daylight (Hours per Day)	Growth (cm)	Flowered or Not (1 = Yes, 0 = No)
1	A	6	4.2	0
2	B	7	3.1	1
3	A	6	4.6	1
4	A	5	3.3	0
5	B	6	2.5	0
6	A	8	5.2	1
7	B	9	3.9	1
8	B	5	2.1	0
9	A	7	3.5	1
10	B	8	3.4	1

1. The variable of *Plant Species* is:
   A. Ordinal variable
   B. Nominal variable
   C. Interval variable
   D. Ratio variable

Answer: B

2. The variable of *Length of Daylight* is:
   A. Ordinal variable
   B. Nominal variable
   C. Interval variable
   D. Ratio variable

Answer: D

3. The variable of *Growth* is:
   A. Ordinal variable
   B. Nominal variable
   C. Continuous variable
   D. Categorical variable

Answer: C

4. The variable of *Flowered or not* is:
   A. Ordinal variable
   B. Nominal variable
   C. Interval variable
   D. Ratio variable

Answer: A

5. If we want to know the correlation between *Length of Daylight* and *Growth*, which of the following statistical methods should we use?
   A. Chi-square
   B. Pearson's Correlation
   C. Logistic Regression
   D. ANOVA

Answer: B

6. The estimated correlation coefficient is 0.45. What is the strength of the correlation?
   A. Weak negative correlation
   B. Strong positive correlation
   C. Moderate positive correlation
   D. Weak positive correlation

Answer: D

7. If we want to compare the growth difference of different plant species, which statistical analysis should we use?
   A. Linear Regression
   B. Chi-square Test
   C. Student t-Test
   D. Mann-Whitney U Test

Answer: D

8. We received more data from other research teams, making the total sample size 150. Next, we would like to update our growth comparison results for different plant species. Which Python codes should we use?
   A. `mannwhitneyu(data1, data2)`
   B. `chisquare(data1, data2)`
   C. `ttest_ind(data1, data2)`
   D. `wilcoxon(data1, data2)`

Answer: C

9. Based on the total of 150 samples, we decided to investigate the relationship between *Growth* and *Length of Daylight*. What would be our *dependant variable*?
   A. Length of Daylight
   B. Growth
   C. Plant Species
   D. Flowered or not

Answer: B

10. To explore the relationship mentioned in Question 9, which statistical analysis should be used?
    A. Linear Regression
    B. Logistic Regression
    C. ANOVA
    D. Chi-square Test

Answer: A

11. Which Python code should be used to conduct the analysis used in Question 10?
    A. `ttest_rel(data1, data2)`
    B. `f_oneway(data1, data2, data3)`
    C. `OLS(y, X).fit()`
    D. `Logit(y, X)`

Answer: C

12. To explore the relationship between *Flowered or not* and *Length of Daylight*, which Python code should be used?
    A. `ttest_rel(data1, data2)`
    B. `f_oneway(data1, data2, data3)`
    C. `OLS(y, X).fit()`
    D. `Logit(y, X)`

Answer: D

REFERENCES

Anaconda Inc. (2020). *Anaconda Distribution Starter Guide.* https://docs.anaconda.com/_downloads/9ee215 ff15fde24bf01791d719084950/Anaconda-Starter-Guide.pdf.

De Winter, J. F. C., & Dodou, D. (2010). Five-point likert items: t test versus Mann-Whitney-Wilcoxon (Addendum added October 2012). *Practical Assessment, Research, and Evaluation, 15*(1), 11.

Diabetes UK. (2019). *Number of People with Diabetes Reaches 4.7 Million.* https://www.diabetes.org.uk/ about_us/news/new-stats-people-living-with-diabetes.

Fagerland, M. W., & Sandvik, L. (2009). The Wilcoxon–Mann–Whitney test under scrutiny. *Statistics in Medicine, 28*(10), 1487–1497.

Kim, H.-Y. (2017). Statistical notes for clinical researchers: Chi-squared test and Fisher's exact test. *Restorative Dentistry & Endodontics, 42*(2), 152–155.

Koehrsen, W. (2018). *Histograms and Density Plots in Python.* Towardsdatascience. com, https://towards-datascience.com/histograms-and …. https://towardsdatascience.com/histograms-and-density-plots-in-python-f6bda88f5ac0.

McDonald, J. H. (2014). Correlation and linear regression. In *Handbook of Biological Statistics* (3rd ed.). Baltimore, MD: Sparky House Publishing. https://www.biostathandbook.com/HandbookBioStatThird. pdf.

McKinney, W., & Team, P. D. (2020). Pandas-Powerful python data analysis toolkit. *Pandas—Powerful Python Data Analysis Toolkit, 1625*. https://pandas.pydata.org/docs/pandas.pdf.

McIntire, G., Martin, B., & Washington, L. (2019). Python Pandas Tutorial: A Complete Introduction for Beginners. *Learn Data Science-Tutorials, Books, Courses, and More*. https://www.learndatasci.com/tutorials/python-pandas-tutorial-complete-introduction-for-beginners/.

Minitab. (2015). Choosing between a nonparametric test and a parametric test. *State College: The Minitab Blog*. https://blog.minitab.com/blog/adventures-in-statistics-2/choosing-between-a-nonparametric-test-and-a-parametric-test.

Pandas Development Team. (2020). *pandas.read_excel*. https://pandas.pydata.org/pandas-docs/stable/reference/api/pandas.read_excel.html.

Scikit-posthocs. (2020). *The Scikit Posthocs Test*. https://scikit-posthocs.readthedocs.io/en/latest/.

SciPy Community. (2020). *scipy.stats.ttest_ind*. https://docs.scipy.org/doc/scipy/reference/generated/scipy.stats.ttest_ind.html.

Sharma, A. (2019). *Importing Data into Pandas*. https://www.datacamp.com/community/tutorials/importing-data-into-pandas#:~:targetText=To read an HTML file, to read the HTML document.

Tavares, E. (2017). *Counting and Basic Frequency Plots*. https://etav.github.io/python/count_basic_freq_plot.html.

WorldCoinIndex. (2021). *WorldCoinIndex*. https://www.worldcoinindex.com/.

10 Machine Learning with Python

Muath Alrammal
Higher Colleges of Technology
University Paris-Est (UPEC)

Dimitrios Xanthidis and Munir Naveed
University College London
Higher Colleges of Technology

CONTENTS

10.1 Introduction 409
10.2 Types of Machine Learning Algorithms 410
10.3 Supervised Learning Algorithms: Linear Regression 411
10.4 Supervised Learning Algorithms: Logistic Regression 414
10.5 Supervised Learning Algorithms: Classification and Regression Tree (CART) 418
10.6 Supervised Learning Algorithms: Naïve Bayes Classifier 430
10.7 Unsupervised Learning Algorithms: K-means Clustering 435
10.8 Unsupervised Learning Algorithms: *Apriori* 438
10.9 Other Learning Algorithms 443
10.10 Wrap Up - Machine Learning Applications 444
10.11 Case Studies 447
10.12 Exercises 447
References 447

10.1 INTRODUCTION

At the present time, *machine learning (ML)* plays an essential role in many human activities. It is applied in different areas including online shopping, medicine, video surveillance, email spam and malware detection, online customer support, and search engine result refinement. It is a subfield of *computer science* and a subset of *Artificial Intelligence (AI)*. The main focus of ML is on developing algorithms that can learn from data and make predictions based on this learning.

> **Observation 10.1 – Machine Learning:** A subfield of computer science and Artificial Intelligence that focuses on developing algorithms that can learn from data and make predictions based on their learning.

An ML program is one that learns from experience E given some tasks (T) and performance measure (P), *if it improves from that experience* (E) (Mitchell, 1997). ML behaves similarly to the growth of a child. As a child grows, its experience E in performing task T increases, which results in a higher performance measure (P).

> **Observation 10.2 – Machine Learning Process:** A Machine Learning program learns from experience (E) given some tasks (T) and performance measure (P), if it improves from that experience (E).

In ML, a computer is *trained* using a given dataset in order to predict the properties of new data. For instance, one can train a system by feeding it with 10,000 images of dogs and 10,000 more images not containing dogs, indicating in each case

DOI: 10.1201/9781003139010-10

whether a picture is a dog or not. following this training, when the system is fed with a new image it should be able to predict whether it is the image of a dog or not.

Python has an arsenal of libraries that support the implementation of ML algorithms. Some of these libraries are already discussed and used in previous chapters (e.g., *Pandas, Matplotlib*). Other libraries especially useful for ML applications are the following:

- **NumPy:** It is an array-processing library. It provides complex mathematical functions for processing multi-dimensional arrays and matrices. It is a powerful tool for handling *random numbers, Fourier transforms*, and *linear algebra*.
- **SciPy:** It is an open-source Python library used for scientific computing. It contains modules for *image optimization, signal processing, Fast Fourier transform, linear algebra,* and *ordinary differential equation (ODE)*. It is built on top of NumPy, as its underlying data structure is a multi-dimensional array.
- **Scikit-Learn:** It is built in 2010 on top of NumPy and SciPy libraries. It contains several *supervised* and *unsupervised* ML algorithms. The library is also useful in *data mining* and *data analysis*. It handles *clustering, regression, classification, model selection,* and *preprocessing.*
- **TensorFlow:** This library was developed by Google in 2015. It uses a NumPy backend for manipulating tensors.

There is an abundance of implemented ML algorithms, applying to various domains. This chapter provides an introduction to some of the most important as well as some of the most popular domain applications. This chapter concludes with a relevant case study that explores some of the main aspects of ML.

10.2   TYPES OF MACHINE LEARNING ALGORITHMS

There are three main types of ML algorithms: *supervised, unsupervised,* and *reinforcement*. A simple way to understand the difference between supervised and unsupervised ML is by introducing the concept of using some type of help to teach a computer how to map particular inputs into the relevant outputs.

In the case of supervised learning the *supervisor* uses what is referred to as *labeled* data to direct the computer into understanding how to map the input into output. As an example, assume the case of training a computer to distinguish between the images of a laptop and a desktop PC. The computer is provided with a *set of images* and a

> **Observation 10.3 – Supervised Learning:** Use *labeled* data to train a computer how to map particular input into output. If the output is in a categorical form the type is *classification*. If the output is in continuous numerical form the type is *regression*. Combining multiple supervised learning models is referred to as type of *ensembling.*

label or *flag* for each one specifying it is a laptop. The same process is repeated for the case of the desktop PC images. Although this is a simplified example, it provides a straightforward description of supervised learning.

In terms of the outputs associated with supervised learning, there are two broad types: *classification* and *regression. Classification* is related with categories, such as "sick" or "healthy" individuals, "dog" or "cat" pets, "laptop" or "desktop" PCs. *Regression* is related to outputs in the form of continuous numerical values, such as predicting an individual's height or weight, or the amount of rainfall. An additional type of supervised learning is *ensembling,* which involves combining the predictions of multiple ML models that may be too weak to stand on their own, in order to produce a more accurate prediction for a new sample.

In general, a broad statement about supervised learning is that it uses labeled data to train a computer to map inputs (X) into outputs (Y) by solving equation $Y = f(X)$ for f.

In the case of unsupervised learning there is no supervisor to train the computer in terms of mapping inputs into outputs, and no labeled training input data to model possible corresponding output variables. Essentially, the computer is left to predict the possible outputs on its own, given a set of previous inputs. There are three main types of unsupervised learning: *association, clustering,* and *dimensionality reduction.*

Association is used to discover the probability of the co-occurrence of items in a collection. It is used extensively in market-based analysis. For example, an association model might be used to predict whether a purchase of bread has an 80% probability to be connected with a purchase of eggs. *Clustering* is used to group samples in a way that ensures that objects within the same cluster

> **Observation 10.4 – Unsupervised Learning:** There is no *supervisor* to train the computer to map input into output and there is no *labeled* data for such training. The computer is trained by itself through a trial-and-error process. *Association* is used to determine the probability of the co-occurrence of items in the collection. *Clustering* is used to group samples within the same cluster. *Dimensionality reduction* is used to reduce the number of variables of the dataset.

share more similarities with each other than with objects from other clusters. *Dimensionality reduction* is used to reduce the number of variables of a dataset, while ensuring that important information is still conveyed. Dimensionality reduction can be achieved by using *feature extraction* and *feature selection* functions. The latter essentially refers to the selection of a subset of the original variables. *Feature extraction* performs data transformations from a high-dimensional space to a low-dimensional space (e.g., PCA algorithm).

Finally, *reinforcement learning* is a type of ML that allows an agent to decide the best action based on its current state, by learning behaviors that will maximize the associated rewards. It usually learns optimal actions through *trial and error.* For example, one can think of a video game in which the player needs to move to certain places at certain times in order to earn points. If a reinforcement algorithm attempts to play this game instead of a human player, it would start by moving randomly, but eventually would learn where and when it needs to move in order to maximize points accumulation through the use of an appropriate trial and error process.

10.3   SUPERVISED LEARNING ALGORITHMS: LINEAR REGRESSION

The basic idea behind *linear regression* is the quantification of the relationship between a set of inputs and their corresponding outputs. This takes the form of a line ($y = a + b.x$) where b is the slope of the regression line (the coefficient of the line) and a is the *y-axis intercept.* The goal is to have the least number of *outliers* (i.e., data with a large deviation from the line). This is measured as *the sum of the squares of all the distances of the data points from the line.* Another important parameter in linear regression is that of R^2, which suggests the possibility that the output y is affected by a related change in the input x. Obviously, like in all other statistical analysis tests, this particular test results in a *p value (statisti-*

> **Observation 10.5 – Linear Regression:** Trains a system to predict the output of a particular input by quantifying the relationship $y = a + b.x$ between a set of inputs and their corresponding outputs, where b is the *slope* of the line and a is the *y*-axis *intercept.* Use R^2 to measure the effect of the input on the possible output and p to measure the statistical significance of the test.

cal significance) that determines whether there is a statistically significant correlation between the input and output datasets.

In Python, linear regression can be implemented using the `linregress(X, y)` function of the *Stats* library. The function uses an input and an output dataset (i.e., X and y, respectively). The function output consists of five values: the *slope* of the linear regression, the *intercept*, the r value, the p value, and the *statistical error* of the test. Based on this, the overall process can be summarized in five distinct steps:

- **Step 1:** Import/read the data for the linear regression.
- **Step 2:** Define the two datasets (*X* and *y*) used to create the model.
- **Step 3:** Use `linregress()` to calculate the *slope,* the *intercept,* the *r,* and the *p* values of the linear regression.
- **Step 4 (Optional):** Use the *slope* and the *intercept* to visualize the model.
- **Step 5 (Optional):** Test the model with new data.

There are numerous real-life applications of linear regression ML algorithms. A notable example is their use in medicine and pharmaceutical research, when trying to determine the optimal dosage of a particular drug for a particular illness. Other examples include the use of such algorithms in sales and marketing, when trying to find the correct volume of promotional material (and the associated costs) for a particular product in order to maximize revenue, and the association of a student's coursework grades with their final grade in an educational context. The following Python script quantifies the relationship between the values of two columns of the *grades2.csv* dataset (*Midterm Exam* and *Final Grade*). Next, once the slope and the intercept values are calculated and the regression model is prepared for further use, both the training and the test datasets are visualized (plotted) alongside the regression line:

```
1   import pandas as pd
2   import matplotlib.pyplot as plt
3   # Request to plot inline with the rest of the results
4   # This is particularly relevant in Jupyter Anaconda
5   %matplotlib inline
6   from scipy import stats
7
8   # The function uses the calculated slope and intercept
9   # to predict the Final Grade, given the Midterm Exam grade input
10  def predictFinalGrade(X):
11          return slope * X + intercept
12
13  # Read the dataset
14  dataset = pd.read_csv("grades2.csv")
15  dataset2 = dataset[["Final Grade", "Midterm Exam"]]
16  print("The input dataset is as follows:")
17  print(dataset2)
18
19  # Define the input and output datasets
20  X = dataset2["Midterm Exam"]; y = dataset2["Final Grade"]
21
22  # Use the linregress function from the stats library
23  # to calculate slope, intercept, r, p, and std_err
24  slope, intercept, r, p, std_err = stats.linregress(X, y)
25
26  print("The slope and intercept values are: {:.2f}, \
27  {:.2f}".format(slope, intercept))
28  print("The value of R-square is: {:.2f}".format(r**2))
29  print("The value of statistical significance, p is: {:.2f}".format(p))
30
31  mymodel = list(map(predictFinalGrade, X))
32  # Plot the model of the resulting linear regression
```

```
33  plt.scatter(X, y); plt.plot(X, mymodel); plt.show()
34
35  grades = int(input("Enter the new Midterm Exam grade:"))
36  grades = predictFinalGrade(grades)
37  print("The predicted Final Grade is: {:.2f}".format(grades))
```

Output 10.3:

```
The input data set is as follows:
     Final Grade   Midterm Exam
0        67.47          70
1        75.13          82
2        66.85          40
3        54.45          44
4        76.95          82
5        45.13          50
6        73.23          62
7        81.87          84
8        62.63          64
9        58.75          52
10       49.75          62
11       44.25          42
12       62.52          68
13       47.33          52
14       68.97          70
The slope and intercept values are: 0.62,      23.96
The value of R-square is: 0.57
The value of statistical significance, p is: 0.00
```

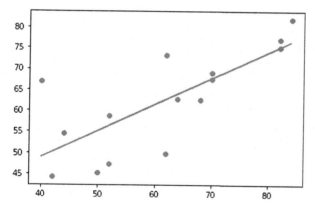

```
Enter the new Midterm Exam grade:88
The predicted Final Grade is: 78.80
```

In terms of the information provided here, the dataset is printed first with the input values used to *train* the system to quantify the regression model. The `stats.linregress()` function of the *Stats* library is used to calculate the slope and the intercept values, as well as the R^2 value, the statistical significance value (*p*) and the standard error (*std_err*). Next, the user is prompted to enter a new *Midterm Exam* grade, and the system predicts the *Final Grade* using the related function `predictFinalGrade()`.

The reader should also note that the output includes the R^2 value, which can be interpreted as *a 57% possibility that a change in the Midterm Exam will affect the Final Grade*. Another noteworthy output is that of the *p* value (i.e., statistical significance), which in this particular case is less than 0.05, suggesting that *there is a correlation between the Midterm Exam and the Final Grade*. Another value calculated during linear regression, although not displayed in the output results, is *std_err*. This value describes the maximum distance of the output values from the regression line in the form of an *error*, which is often referred to as *residual*. The script makes use of the `format()` specifier to limit the number of decimal places of the results to 2. Finally, the reader should note the inclusion of directive `%matplotlib` inline, dictating that the regression model must be plotted *inline* with the rest of the data.

10.4   SUPERVISED LEARNING ALGORITHMS: LOGISTIC REGRESSION

As shown, linear regression predictions take the form of continuous values. In the case of *logistic regression*, predictions take the form of discrete values (i.e., binary), such as whether a student will pass or fail a course, or whether it will rain or not. Its name comes from the associated logistic function: $y = 1/(1 + e^{-x})$. The plot of this function is an *S-shaped* curve. In contrast to linear regression where the output is a value directly based on

> **Observation 10.6 – Logistic Regression:** Train a system to predict the probability of an output as one of two possible values based on a given input. The function used for this purpose is the following: $y = 1/(1 + e^x)$.

the input, in logistic regression it is a *probability ranging from 0 to 1*. For example, if a value 1 represents a passing grade, an output of 0.85 means that a student is very likely to pass the course at a probability of 85%.

There are eight possible steps to follow when performing logistic regression, of which two are optional:

- **Step 1:** Import/read the data for the logistic regression.
- **Step 2:** Split the input datasets into *train* and *test* sets.
- **Step 3:** Perform *feature scaling* for the data (between 0 and 1).
- **Step 4:** Build the *logistic classifier* (with a preferred *random_state = 0* for consistent results) and fit the trained set into the classifier.
- **Step 5:** Predict the results based on the classifier.
- **Step 6:** Find the accuracy of the regression model as a percentage.
- **Step 7 (Optional):** Visualize the results of the trained set.
- **Step 8 (Optional):** Visualize the results of the test set.

The following Python script uses *Midterm Exam* and *Project* grades to create a logistic regression model and visualize its results:

```
1    # Import train_test_split to train and test the input
2    from sklearn.model_selection import train_test_split
3    # Import StandardScaler to scale the data
4    from sklearn.preprocessing import StandardScaler
5    # Import the LogisticRegression to create the classifier object
6    from sklearn.linear_model import LogisticRegression
7    # Import the accuracy_score to calculare the accuracy of the model
8    from sklearn.metrics import accuracy_score
9    # Import numpy to prepare the plot parameters
```

```
10   import numpy as np
11   # Import pyplot to create the plot
12   import matplotlib.pyplot as plt
13   # Import ListedColormap to color the data points in the plot
14   from matplotlib.colors import ListedColormap
15   # Define that results are to plotted inline
16   # This is particularly relevant in Jupyter Anaconda
17   %matplotlib inline
18
19   # Step 1: Define the input dataset. X must be a 2D list with
20   # as many rows as observations
21   X = [[60, 55], [54, 90], [70, 80], [76, 70], [64, 87], [66, 70],
22        [54, 87], [92, 70], [58, 78], [70, 71], [70, 70], [90, 76],
23        [86, 92], [72, 70],  [70, 72], [82, 87], [40, 80], [44, 90],
24        [82, 92], [50, 68]]
25   y = [0, 0, 1, 1, 1, 0, 1, 1, 0, 0, 1, 0, 1, 0, 0, 1, 1, 0, 1, 0]
26
27   # Step 2: Split set X and y into train test and test set
28   # Test size is 25% of the dataset, train size is 75%
29   # The new trained and test lists will be in random order
30   X_train, X_test, y_train, y_test = train_test_split(X, y,
31        test_size = 0.25, random_state = 0)
32   print("Trained X set:", X_train); print("Test X set:", X_test)
33   print("Trained y set:", y_train); print("Test y set:", y_test)
34
35   # Step 3: Perform feature scaling for the data (between 0 and 1)
36   sc_X = StandardScaler()
37   X_train = sc_X.fit_transform(X_train)
38   print("\nThe 2D set of trained X input:\n", X_train)
39   X_test = sc_X.transform(X_test)
40   print("\nThe 2D set of test X input:\n", X_test)
41
42   # Step 4: Build the logistic classifier
43   # Set random_state to 0 for consistent results
44   # Fit the trained set into the classifier
45   model = LogisticRegression(solver = 'liblinear',
46        random_state = 0).fit(X_train, y_train)
47   print("\n", model)
48
49   # Step 5: Predict the test results
50   y_pred = model.predict(X_test)
51   print("\nResults predicted by the model:", y_pred)
52   print("Results from the test:", y_test)
53   model.predict_proba(X)[:,1]
54
55   # Step 6: Form the confusion matrix to get the accuracy of the model
56   # Use y_test (actual output) and y_pred (predicted output)
57   accuracy = accuracy_score(y_test, y_pred)
58   print("The accuracy of the model given the test data is: ",
59        accuracy * 100, "%")
60
```

```
61  # Step 7: Visualize the training set results
62  X_set, y_set = X_train, y_train
63  X1, X2 = np.meshgrid(np.arange(start = X_set[:, 0].min() - 1,
64          stop = X_set[:, 0].max() + 1, step = 0.01),
65          np.arange(start = X_set[:, 1].min() - 1,
66          stop = X_set[:, 1].max() + 1, step = 0.01))
67  plt.contourf(X1,X2, model.predict(np.array([X1.ravel(), \
68          X2.ravel()]).T).reshape(X1.shape), alpha = 0.75,
69          cmap = ListedColormap(('red','blue')))
70
71  plt.xlim(X1.min(), X1.max())
72  plt.ylim(X2.min(), X2.max())
73  for i, j in enumerate(np.unique(y_set)):
74          plt.scatter(X_set[y_set == j, 0], X_set[y_set == j, 1])
75  plt.title('Logistic Regression: Training set')
76  plt.xlabel("Midterm Exam")
77  plt.ylabel("Project")
78  plt.show()
79
80  # Step 8: Visualize the test results
81  X_set, y_set = X_test, y_test
82  X1, X2 = np.meshgrid(np.arange(start = X_set[:, 0].min() - 1,
83          stop = X_set[:, 0].max() + 1, step = 0.01),
84          np.arange(start = X_set[:, 1].min() - 1,
85          stop = X_set[:, 1].max() + 1, step = 0.01))
86
87  plt.contourf(X1,X2, model.predict(np.array([X1.ravel(), \
88          X2.ravel()]).T).reshape(X1.shape),alpha = 0.75,
89          cmap = ListedColormap(('red','blue')))
90
91  plt.xlim(X1.min(), X1.max());plt.ylim(X2.min(), X2.max())
92  for i, j in enumerate(np.unique(y_set)):
93          plt.scatter(X_set[y_set == j, 0], X_set[y_set == j, 1])
94  plt.title('Logistic Regression: Test set')
95  plt.xlabel("Midterm Exam"); plt.ylabel("Project")
96  plt.show()
```

Output 10.4:

```
Trained X set: [[44, 90], [54, 87], [72, 70], [64, 87],
[70, 80], [66, 70], [70, 72], [70, 71], [92, 70], [40,
80], [90, 76], [76, 70], [60, 55], [82, 87], [86, 92]]
Test X set: [[82, 92], [54, 90], [50, 68], [58, 78], [7
0, 70]]
Trained y set: [0, 1, 0, 1, 1, 0, 0, 0, 1, 1, 0, 1, 0,
1, 1]
Test y set: [1, 0, 0, 0, 1]

The 2D set of trained X input:
 [[-1.69129319  1.30698109]
 [-1.01657516  1.00224457]
```

```
[ 0.19791729 -0.72459574]
[-0.34185713  1.00224457]
[ 0.06297368  0.29119268]
[-0.20691353 -0.72459574]
[ 0.06297368 -0.52143805]
[ 0.06297368 -0.62301689]
[ 1.54735334 -0.72459574]
[-1.9611804   0.29119268]
[ 1.41240974 -0.11512269]
[ 0.4678045  -0.72459574]
[-0.61174434 -2.24827836]
[ 0.87263531  1.00224457]
[ 1.14252253  1.51013878]]

The 2D set of test X input:
[[ 0.87263531  1.51013878]
 [-1.01657516  1.30698109]
 [-1.28646237 -0.92775342]
 [-0.74668795  0.088035  ]
 [ 0.06297368 -0.72459574]]


LogisticRegression(random_state=0, solver='1iblinear')

Results predicted by the model: [1 1 0 0 0]
Results from the test: [1, 0, 0, 0, 1]
The accuracy of the model given the test data is:  60.0
%
```

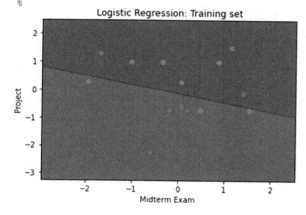

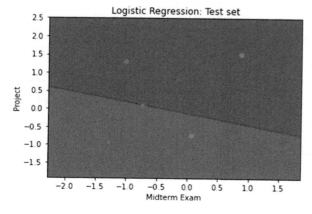

The above script and its output demonstrate the eight steps followed when using logistic regression. In Step 1 (data read), it is important to remember that input dataset X must be a two-dimensional array/list of pairs of data equal to the number of observations. In this particular case, the set includes the grades of each student for *Midterm Exam* and *Project*. The y dataset includes values 0 or 1 for each student, with 0 referring to a *fail* and 1 to a *pass*.

In the step, the script makes use of the `train_test_split()` function (*train_test_split* module) from the *Sklearn.model_selection* library. The function takes the X and y datasets, splits them to train and test subsets at a rate of 75/25 (*test_size = 0.25*), and randomizes the splitting process. The results of the function are datasets `X_train`, `X_test`, `y_train`, and `y_test`. In Step 3, the script imports the *StandardScaler* module from the *Sklearn.preprocessing* library and uses the `StandardScaler()` constructor and the `fit_transform()` function to scale output data y between 0 and 1, as required by the logistic regression model.

In Step 4, the actual logistic regression classifier is used to fit the data and execute the model using the `X_train`, `X_test`, `y_train`, and `y_test` datasets. Next, the script uses the model to predict (`.predict()`) the results of the regression (fifth step). In Step 6, the script uses function `accuracy_score()` (*Accuracy_score* module) from the *Sklearn.metrics* library to calculate the accuracy rate of the resulting regression model, as a number between 0 and 1. Finally, Steps 7 and 8 are used to visualize the training and test set results, respectively. In both cases, function `meshgrid()` is used to prepare the data for plotting and `ListedColormap()` to color the *pass* and *fail* outputs.

There are numerous different options and variations available for each of these steps, as well as for displaying and plotting the resulting data. The reader can refer to the multitude of *statistics* and/or *machine learning* textbooks and resources in order to delve deeper into the various concepts related to the interpretation and use of the results of logistic regression in various contexts.

10.5 SUPERVISED LEARNING ALGORITHMS: CLASSIFICATION AND REGRESSION TREE (CART)

A *decision tree* consists of a *root*, *nodes*, and *leaves* (Figure 10.1). The starting point of the decision tree is the *root*; each *internal node* is branching out to connect to other inputs, also in the form of nodes. Each *leaf node* is a possible output of the tree. The branching is determined by using a split function, which divides the input data into one or more branches. The leaf nodes of the tree are the outcomes.

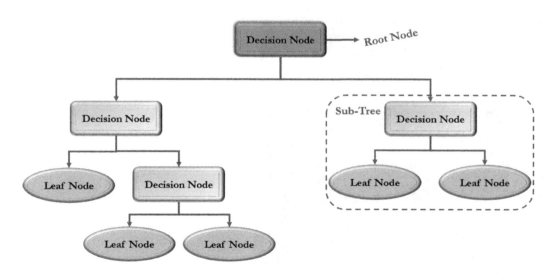

FIGURE 10.1   Decision tree.

In order to create the order (or height) of the decision tree and its features, the *decision tree algorithm* uses a function to determine the *information gain*. There are two functions serving this purpose, referred to as indices: *entropy* or *Gini* index. Their function is to measure the *impurity* of a node in the tree and, based on their value, the node is being kept or discarded. These values also determine the position of a node in the tree. There are different types of the decision tree, depending on how the indices are calculated and what choices are being made in terms of *splitting continuous values*. The most commonly used types of a decision tree are *ID3* (Quinlan, 1986), *C4.5* (Salzberg, 1994) and *CART* (Mola, 1998).

CART (Classification and Regression Tree) is one of the most important and popular types of *supervised learning algorithms*. The output can be in a form of a categorical value (e.g., it will rain or not) or a continuous value (e.g., the final price of a car). A visual representation of a decision tree is shown in Figure 10.2. The tree starts with the *Age* feature, which is a numeric attribute in a bank dataset. The values of *Age* are split into three branches: *18–23*, *24–34* and *>35*. The algorithm can split the continuous number values of the *Age* feature using a technique *that also determines the order of features within the tree*. Next, the *Age* feature (the root of the tree) is associated with three additional features (nodes): *Job*, *Marital Status*, and *Housing*.

The decision tree can be built using a training dataset. In the following example, the script makes use of a dataset of 40 bank account customer records, containing features *age, job, marital status,* and *education*. The system aims at predicting the possibility of customers

Observation 10.7 – CART: The *Classification and Regression Tree (CART)* is a decision tree with a *root*, *nodes* and *leaves* and with outputs either in a form of a *categorical* or a *continuous* value. The branching is determined by using a *split function* that divides the input data into one or more branches.

Observation 10.8 – Input and Output Datasets: The *Classification and Regression Tree (CART)* requires a 2D list/array of values as its input and output datasets. If the input and output datasets do not match, appropriate amendments are required.

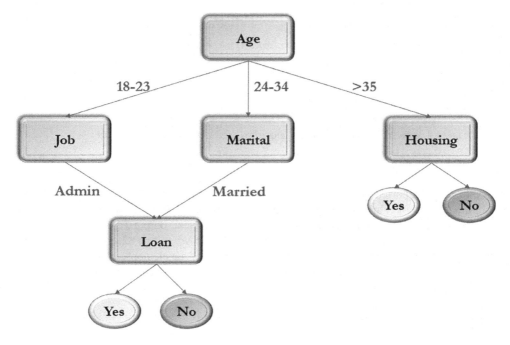

FIGURE 10.2 Example of decision tree.

making a deposit in the bank or not. In order to train the CART decision tree, these four features are used as input and the *deposit* feature as output. The possible outputs are *Yes* and *No* (depositing money or not). The script requires a number of associated libraries. Some of these

libraries are already included in the system (e.g., *Pandas* and *Numpy*), while others like *Pydoplus* and *Graphviz* must be installed explicitly. Given that the installation of any libraries depends on the particular system in use, the reader is advised to check the available `pip` install statements for specific system settings:

```
1    # Import the basic libraries
2    import pandas as pd
3    import numpy as np
4
5    # Import the DecisionTreeClassifier
6    from sklearn.tree import DecisionTreeClassifier
7    # Import the confusion_matrix, the accuracy_score, and the
8    # classification report
9    from sklearn.metrics import confusion_matrix
10   from sklearn.metrics import accuracy_score
11   from sklearn.metrics import classification_report
12
13   # Import train_test_split to split the data into train and test samples
14   from sklearn.model_selection import train_test_split
15
16   # Import the libraries for the necessary hot encoding
17   from sklearn.preprocessing import LabelEncoder
18
19   # Import the libraries to plot the graph
20   from sklearn.tree import export_graphviz
21
22   # import StringIO from sklearn.externals.six
23   from six import StringIO
24   from IPython.display import Image
25   import pydotplus
26
27   # Plot results inline
28   # This is often particularly needed in Jupyter Anaconda
29   %matplotlib inline
30
```

With the libraries imported, the next part of the script is the first step of this particular implementation. Initially, the list of values for input list *X* (2D array) is defined. Each sub-list includes the *age*, *job*, *marital status*, and *education* features of the bank customer. Next, output *Y* (single dimension list) is defined as a unidimensional list, taking values of either *Yes* or *No*. In line 82, input list *X* is converted to a *Numpy array* to facilitate a more efficient manipulation of the elements in the list. In the following line (83), the 2D array is divided into four unidimensional sub-arrays, each storing the respective elements. Finally, the data of each newly created input sub-array (*X1–X4*) and of output *Y* are printed:

```
31   #==================================================================
32   # Step 1: Define and print the input and output datasets
```

```
33   print("Step 1: Define and print the input and output datasets\n")
34
35   X = [[59, 'admin.', 'married', 'secondary'],
36        [56, 'admin.', 'married', 'secondary'],
37        [41, 'technician', 'married', 'secondary'],
38        [55, 'services', 'married', 'secondary'],
39        [54, 'admin.', 'married', 'tertiary'],
40        [42, 'management', 'single', 'tertiary'],
41        [56, 'management', 'married', 'tertiary'],
42        [60, 'retired', 'divorced', 'secondary'],
43        [37, 'technician', 'married', 'secondary'],
44        [28, 'services', 'single', 'secondary'],
45        [38, 'admin.', 'single', 'secondary'],
46        [30, 'blue-collar', 'married', 'secondary'],
47        [29, 'management', 'married', 'secondary'],
48        [46, 'blue-collar', 'single', 'tertiary'],
49        [31, 'technician', 'single', 'tertiary'],
50        [35, 'management', 'divorced', 'tertiary'],
51        [32, 'blue-collar', 'single', 'primary'],
52        [49, 'services', 'married', 'secondary'],
53        [41, 'admin.', 'married', 'secondary'],
54        [49, 'admin.', 'divorced', 'secondary'],
55        [49, 'retired', 'married', 'secondary'],
56        [32, 'technician', 'married', 'secondary'],
57        [30, 'self-employed', 'single', 'secondary'],
58        [55, 'services', 'divorced', 'tertiary'],
59        [32, 'blue-collar', 'married', 'secondary'],
60        [52, 'admin.', 'divorced', 'secondary'],
61        [38, 'unemployed', 'divorced', 'secondary'],
62        [60, 'retired', 'married', 'secondary'],
63        [60, 'retired', 'divorced', 'secondary'],
64        [30, 'admin.', 'married', 'tertiary'],
65        [44, 'unemployed', 'married', 'secondary'],
66        [32, 'blue-collar', 'married', 'secondary'],
67        [46, 'entrepreneur', 'married', 'tertiary'],
68        [34, 'management', 'married', 'secondary'],
69        [40, 'management', 'married', 'secondary'],
70        [34, 'housemaid', 'married', 'primary'],
71        [43, 'admin.', 'single', 'secondary'],
72        [52, 'technician', 'married', 'secondary'],
73        [35, 'blue-collar', 'married', 'secondary'],
74        [34, 'blue-collar', 'single', 'secondary']]
75
76   Y=['yes','yes','yes','yes','yes','yes','yes','yes','yes','yes',
77      'yes','yes','yes','yes','yes','yes','yes','yes','yes','yes',
78      'no','no','no','no','no','no','no','no','no','no',
79      'no','no','no','no','no','no','no','no','no','no' ]
80
81   # Convert the list into a numpy array for better index control
82   newX = np.array(X)
83   newX1,newX2,newX3,newX4=newX[:,0],newX[:, 1],newX[:, 2],newX[:, 3]
```

```
84   print("\nThe input of ages (X1) is :\n", newX1)
85   print("\nThe input of jobs (X2) is :\n", newX2)
86   print("\nThe input of marital status (X3) is :\n", newX3)
87   print("\nThe input of education (X4) is :\n", newX4)
88
89   print("\nThe output of deposits (Y) is :\n", Y)
90
```

Output 10.5: Step 1

```
Step 1: Define and print the input and output datasets


The input of ages (X1) is :
 ['59' '56' '41' '55' '54' '42' '56' '60' '37' '28' '38' '30' '29' '46'
 '31' '35' '32' '49' '41' '49' '49' '32' '30' '55' '32' '52' '38' '60'
 '60' '30' '44' '32' '46' '34' '40' '34' '43' '52' '35' '34']

The input of jobs (X2) is :
 ['admin.' 'admin.' 'technician' 'services' 'admin.' 'management'
 'management' 'retired' 'technician' 'services' 'admin.' 'blue-collar'
 'management' 'blue-collar' 'technician' 'management' 'blue-collar'
 'services' 'admin.' 'admin.' 'retired' 'technician' 'self-employed'
 'services' 'blue-collar' 'admin.' 'unemployed' 'retired' 'retired'
 'admin.' 'unemployed' 'blue-collar' 'entrepreneur' 'management'
 'management' 'housemaid' 'admin.' 'technician' 'blue-collar'
 'blue-collar']

The input of marital status (X3) is :
 ['married' 'married' 'married' 'married' 'married' 'single' 'married'
 'divorced' 'married' 'single' 'single' 'married' 'married' 'single'
 'single' 'divorced' 'single' 'married' 'married' 'divorced' 'married'
 'married' 'single' 'divorced' 'married' 'divorced' 'divorced' 'married'
 'divorced' 'married' 'married' 'married' 'married' 'married' 'married'
 'married' 'single' 'married' 'married' 'single']

The input of education (X4) is :
 ['secondary' 'secondary' 'secondary' 'secondary' 'tertiary' 'tertiary'
 'tertiary' 'secondary' 'secondary' 'secondary' 'secondary' 'secondary'
 'secondary' 'tertiary' 'tertiary' 'tertiary' 'primary' 'secondary•
 'secondary' 'secondary' 'secondary' 'secondary' 'secondary' 'tertiary'
 'secondary' 'secondary' 'secondary' 'secondary' 'secondary' 'tertiary'
 'secondary' 'secondary' 'tertiary' 'secondary' 'secondary' 'primary'
 'secondary' 'secondary' 'secondary' 'secondary']

The output of deposits (Y) is :
 ['yes', 'yes', 'yes', 'yes', 'yes', 'yes', 'yes', 'yes', 'yes', 'yes',
 'yes', 'yes', 'yes', 'yes', 'yes', 'yes', 'yes', 'yes', 'yes', 'yes',
 'no', 'no', 'no', 'no', 'no', 'no', 'no', 'no', 'no', 'no', 'no', 'no',
 'no', 'no', 'no', 'no', 'no', 'no', 'no', 'no']
```

In Step 2, the code addresses an important classification issue. Since models are mathematical in nature, the underlying calculations are based on textual rather than numerical data. Hence, it is necessary to *encode* the various elements of the data into numerical (*integer*) values, a process referred to as *integer encoding*. Lines 97–102 include code for finding the unique elements in each of the input sub-arrays *X1–X4*. Next, in lines *105–122*, the LabelEncoder() function (*Sklearn.preprocessing*

Observation 10.10 – Integer Encoding: The process of converting a categorical value into the numerical form necessary for the *CART* algorithm. Use the LabelEncoder() function from the *Sklearn.preprocessing* library.

library) is utilized to create the relevant objects, subsequently used by fit_transform() to produce the integer encoded sub-arrays for *X1–X4*. The same process is also applied in the case of output dataset *Y*:

```
91   #================================================================
92   # Step 2: Encode the categorical values of the input & output datasets
93   # Find and print the unique values of the categories/columns for job
94   # and marital status
95   print("\n\nStep 2: The inputs of jobs, marital status,",
96         "and education and the outputs are integer encoded")
97   jobs = np.unique(newX2)
98   print("\nThe various categories of jobs are:\n", jobs)
99   maritalStatus = np.unique(newX3)
100  print("\nThe various categories of marital status are:\n",
101        maritalStatus)
102  education = np.unique(newX4)
103  print("\nThe various categories of education are:\n", education)
104  # Integer Encode the categorical input and output values as fit()
105  # does not accept strings
106  label_encoderX2 = LabelEncoder()
107  integer_encodedX2 = label_encoderX2.fit_transform(newX2)
108  print("\nThe various categories of jobs are integer Encoded as",
109        "follows:\n", integer_encodedX2)
110  label_encoderX3 = LabelEncoder()
111  integer_encodedX3 = label_encoderX3.fit_transform(newX3)
112  print("\nThe various categories of marital status are ",
113        "integer Encoded as follows:\n",
114        integer_encodedX3)
115  label_encoderX4 = LabelEncoder()
116  integer_encodedX4 = label_encoderX4.fit_transform(newX4)
117  print("\nThe various categories of education are integer Encoded as",
118        "follows:\n", integer_encodedX4)
119  label_encoderY = LabelEncoder()
120  integer_encodedY = label_encoderY.fit_transform(Y)
121  print("\nThe various categories of output are integer Encoded as",
122        "follows:\n", integer_encodedY)
```

Output 10.5: Step 2

```
Step 2: The inputs of jobs, marital status, and education and the outputs are
integer encoded
The various categories of jobs are:
 ['admin.' 'blue-collar' 'entrepreneur' 'housemaid' 'management' 'retired'
 'self-employed' 'services' 'technician' 'unemployed']

The various categories of marital status are:
 ['divorced' 'married' 'single']

The various categories of education are:
 ['primary' 'secondary' 'tertiary']

The various categories of jobs are integer Encoded as follows:
 [0 0 8 7 0 4 4 5 8 7 0 1 4 1 8 4 1 7 0 0 5 8 6 7 1 0 9 5 5 0 9 1 2 4 4 3 0
 8 1 1]

The various categories of marital status are  integer Encoded as follows:
 [1 1 1 1 1 2 1 0 1 2 2 1 1 2 2 0 2 1 1 0 1 1 2 0 1 0 0 1 0 1 1 1 1 1 1 1 2
 1 1 2]

The various categories of education are integer Encoded as follows:
 [1 1 1 1 2 2 2 1 1 1 1 1 1 2 2 2 0 1 1 1 1 1 2 1 1 1 1 2 1 1 2 1 1 0 1
 1 1 1]

The various categories of output are integer Encoded as follows:
 [1 1 1 1 1 1 1 1 1 1 1 1 1 1 1 1 1 1 1 0 0 0 0 0 0 0 0 0 0 0 0 0 0 0 0 0 0
 0 0 0]
```

In Step 3, the code splits the datasets into *train and test input* and *train and test output*. Provided that the fit() function used in the next step needs a 2D numerical array to perform its calculations, it is necessary to combine the previously divided input sub-arrays into a single 2D array. The zip() function takes the four input sub-arrays and combines them in a single 2D array. However, since the result is still unusable for the relevant fitting calculations, the list() function is used to convert the 2D array to a suitable form (lines 127–128).

Next, function train_test_split() (*Sklearn.model_selection* library) is used with the newly created 2D array, as well as the unidimensional output array, in order to split (75/25) and randomize the datasets. This is defined explicitly by the test_size = 0.25 and the random_state = 0 arguments (lines 129–130). The test_size parameter is referring to the *hold-out validation* that splits the dataset into the *train* and *test* parts, in this case 75% and 25%.

The alternative to *hold-out validation* is the *cross-validation* technique, which selects data for training via sampling. In this approach, a block of data of fixed size is selected for training in each iteration. The technique could be also applied to smaller datasets, but the sample selection in each iteration of training can lead to heavy computation requirements and, therefore, more CPU cycles. The main types of cross-validation are *leave-p out* and *k-fold*. In the case of *k-fold*, the most commonly used selection is the ten-fold (i.e., $k = 10$). An example of a cross-validation statement is the following:

```
crossValidation = cross_validate (decisionTree, X_Train, Y_Train,
crossValidation = 10)
```

In the current context, this statement would be placed in the code just after the definition of the DecisionTreeClassifier().

The last part of this step prints the *train and test inputs* and the *train and test outputs*:

```
123  #====================================================================
124  # Step 3: Define the point to split the dataset to 3/4
125  print("\nStep 3: Define the point to split the datasets to 3/4\n")
126
127  newEncodedInput = list(zip(newX1, integer_encodedX2, integer_encodedX3,
128         integer_encodedX4))
129  X_Train, X_Test, y_Train, y_Test = train_test_split(newEncodedInput,
130         integer_encodedY, test_size = 0.25, random_state = 0)
131  print("\nTrained X set:", X_Train)
132  print("\nTest X set:", X_Test)
133  print("\nTrained y set:", y_Train)
134  print("\nTest y set:", y_Test)
135
```

Output 10.5: Step 3

```
Step 3: Define the point to split the datasets to 3/4


Trained X set: [('60', 5, 1, 1), ('34', 3, 1, 0), ('52', 8, 1, 1), ('41',
8, 1, 1), ('34', 1, 2, 1), ('44', 9, 1, 1), ('40', 4, 1, 1), ('32', 1, 2,
0), ('43', 0, 2, 1), ('37', 8, 1, 1), ('46', 1, 2, 2), ('42', 4, 2, 2),
('49', 7, 1, 1), ('31', 8, 2, 2), ('34', 4, 1, 1), ('60', 5, 0, 1), ('46',
2, 1, 2), ('56', 0, 1, 1), ('38', 9, 0, 1), ('29', 4, 1, 1), ('32', 1, 1,
1), ('32', 1, 1, 1), ('56', 4, 1, 2), ('55', 7, 0, 2), ('32', 8, 1, 1),
('49', 0, 0, 1), ('28' ,7, 2, 1), ('35', 1, 1, 1), ('55', 7, 1, 1), ('59',
0, 1, 1)]

Test X set: (('30', 6, 2, 1), ('49', 5, 1, 1), ('52', 0, 0, 1), ('54', 0,
1, 2), ('38', 0, 2, 1), ('35', 4, 0, 2), ('60', 5, 0, 1), ('30', 1, 1, 1),
('41', 0, 1, 1), ('30', 0, 1, 2)]

Trained y set: [0 0 0 1 0 0 0 1 0 1 1 1 1 1 0 1 0 1 0 1 0 0 1 0 0 1 1 0 1 1]

Test y set: [0 0 0 1 1 1 0 1 1 0]
```

In Step 4, the defined trained and test inputs and outputs are used to train and test the model (i.e., predict the possible output). This is achieved through the `DecisionTreeClassifier()` function, (*Sklearn. tree* library), which creates the *decisionTree* object model used for the output prediction (lines 144–146). The reader should note that the mathematical algorithm used in the classifier is `entropy`, `random_state = 100`, `maximum_depth = 100`, and `min_samples_leaf = 2`.

Observation 10.11 – DecisionTree Classifier(): The class used to create the decision tree model.

In terms of the *entropy* mechanism, the mathematical equation used is: $E = -\Sigma_{(i:n)}p_i log_2 p_i$. The idea is to calculate the entropy of mixed values encountered in the columns of the train dataset. If the values are heavily mixed and unequal in population, the entropy will be close to 1, otherwise it would be close to 0. Ideally, the preferred value is 0, which means that the dataset has largely homogeneous values. When visualizing the decision tree, the value of entropy suggests the *impurity* of the values in the related tree or sub-tree. The alternative to entropy is the *Gini index* mechanism, which is also used by the classifier to organize the decision tree. Its mathematical

Observation 10.12 – Entropy, Gini Index: The mathematical models used to define and organize the decision tree. They measure the level of *impurity* of the values in the dataset used for the tree.

equation is: *Gini Index = 1−Σ(P(x=k))²*. This also suggests the probabilities of uncertainty of impurity among various partitions of the dataset. In the case of this example, both mechanisms are included with that of entropy applied and the *Gini index* deactivated as a comment. Switching the activation of one over the other would showcase that the results are quite similar. For further information on either entropy or the Gini index, the reader is advised to study textbooks specifically focused on ML.

There are two more parameters specified in `DecisionTreeClassifier()` that affect the visualization of the tree: `max_depth` and `min_samples_leaf`. The former determines the maximum depth of the tree. If omitted, the tree will have no maximum depth but will grow as deep as necessary according to the calculation and the dataset. The latter will determine the minimum number of samples required to be present as leaves in the tree. If its value is 1, it will display every simple sample in the tree making the visual tree grow in size to its fullest. Increasing the value of `min_samples_leaf` will result in a reduction of the size of the visual depiction of the tree by combining the number of samples in each leaf. As mentioned, the present sample code includes two alternative versions of `DecisionTreeClassifier()` (lines 141–146): one using entropy and one the Gini index. The former uses a `min_samples_leaf` value of 1, while the latter a value of 6. Notice the difference in the size of the visual depiction of the decision tree in each case, and also how the algorithm makes decisions based on the columns of the dataset that have the greatest influence on the resulting visual depiction of the decision tree:

> **Observation 10.13 – Parameter maximum _ depth:** Used to define the depth of the decision tree (unlimited if omitted).

> **Observation 10.14 – Parameter min _ samples _ leaf:** Used to define the minimum number of samples that a leaf may have in order to be displayed in the visualization of the decision tree.

```
136  #=====================================================================
137  # Step 4: Create the classifier & train & test the input & output
138  # Create the classifier object using 4 attributes: criterion can be
139  # entropy or gini, splitter can be best or random,
140  print("\nStep 4: Define the point to split the datasets to 3/4")
141  #decisionTree = DecisionTreeClassifier(criterion = "entropy",
142  # splitter = "best", random_state = 100, max_depth = 100,
143  # min_samples_leaf = 1)
144  decisionTree = DecisionTreeClassifier (criterion = "gini",
145        splitter = "best", random_state = 100, max_depth = 100,
146        min_samples_leaf = 6)
147  # The classifier trains the input (X_Train) & the output (y_Train)
148  arrayX_Train = np.array(X_Train)
149  arrayY_Train = np.array(y_Train)
150  print("\nThe input dataset to train is:\n", arrayX_Train)
151  print("\nThe output dataset to train is:\n", arrayY_Train)
152  decisionTree.fit(arrayX_Train, arrayY_Train)
153
154  arrayY_Test1 = np.array(y_Test)
155  arrayY_Test = list(zip(arrayY_Test1, arrayY_Test1, arrayY_Test1,
156        arrayY_Test1))
157  print("\nThe output dataset to test is:\n", arrayY_Test)
158  y_Predict = decisionTree.predict(arrayY_Test)
159  print("\nThe predicted output is:\n", y_Predict)
160
```

Output 10.5: Step 4

```
Step 4: Define the point to split the datasets to 3/4

The input dataset to train is:
 [['60' '51' '1' '1']
 ['34' '3' '1' '0']
 ['52' '8' '1' '1']
 ['41' '8' '1' '1']
 ['34' '1' '2' '1']
 ['44' '9' '1' '1']
 ['40' '4' '1' '1']
 ['32' '1' '2' '0']
 ['43' '0' '2' '1']
 ['37' '8' '1' '1']
 ['46' '1' '2' '2']

 ...

The output dataset to train is:
 [0 0 0 1 0 0 0 1 0 1 1 1 1 1 0 1 0 1 0 1 0 0 1 0 0 1 1 0 1 1]

The output dataset to test is:
 [(0, 0, 0, 0), (0, 0, 0, 0), (0, 0, 0, 0), (1, 1, 1, 1), (1, 1, 1, 1),
(1, 1, 1, 1), (0, 0, 0, 0), (1, 1, 1, 1), (1, 1, 1, 1), (0, 0, 0, 0)]

The predicted output is:
 [0 0 0 0 0 0 0 0 0 0]
```

In Step 5, the code *inverts* the output to the *original column values*, it calculates the *confusion matrix* and the *accuracy score*, and provides the *classification report*. For the inversion of the output, the label encoders are used in the same way as in the case of the integer encoded arrays used in the model. Next, the confusion matrix is printed followed by the accuracy score (50%). The reader should note that, in an ideal scenario, the value of the latter approaches the 100% mark. Finally, the classification report is displayed with all the relevant details. These tasks are coded in lines 164–174. The output shows the results of Step 5.

From one training dataset, the CART algorithm can build several decision trees. The performance criteria determine which tree is preferable for the task at hand. Different metrics or performance measurement parameters are being used, the most common being accuracy, confusion matrix, precision, recall and f-score. Accuracy represents the overall accuracy of a tree. It is calculated using the correctly classified observations divided by the total number of observations, and is represented as a percentage. For example, if there are 100 observations tested and 70 of them are correctly classified, the accuracy of that tree will be 70.00. A higher accuracy suggests a better performance for the decision tree.

The confusion matrix represents the overall behavior of the tree, based on the test or train datasets. It provides more insight in terms of the performance of the tree on each class label. Therefore, the size of confusion matrix depends on the class labels, as it is always $n \times n$, where n denotes the number of the class labels. For instance, if there are three class labels in a dataset, the confusion matrix will be 3×3. In the case of the bank dataset, the confusion matrix will be 2×2, as it has only two class labels (Yes/No). The matrix will also provide a breakdown of the numbers of labels being wrongly categorized by the tree. Such information is not provided by the accuracy scores.

Precision is the measurement of the *relevance-based accuracy* (i.e., a ratio of the number of correctly predicted observations over the total number of observations) for each label. For example, assume a tree that has classified 60 customers out of 100 as *Yes*. However, only 40 out of the 60 classifications are correct. Thus, the precision will be 40/60 or 0.667.

Recall is the measure of relevance with respect to the overall classification performance in for the class labels. For example, assume a tree that predicts 60 responses of *Yes* in a dataset of 100. If 40 of these predictions are correct, while the dataset has 75 observed responses of *Yes*, the recall will be 40/75 or 0.533.

Fscore combines both the recall and the precision values into a single value. This value represents the *performance in terms of relevance* for each label. High fscore values dictate that the classifier is performing better and is more fine-tuned than one with lower values.

```
161  #=====================================================================
162  # Step 5: Invert the encoded values and calculate the confusion matrix,
163  # the accuracy score, and the classification report
164  print("\nStep 5: Invert the integer encoded results into "
165        "their original text-based")
166  invertedY_Test = label_encoderY.inverse_transform(y_Test)
167  print ("The inverted output test values are:", invertedY_Test)
168  invertedPredicted = label_encoderY.inverse_transform(y_Predict)
169  print ("The inverted predicted values of the output are:",
170        invertedPredicted)
171  confusionMatrix = confusion_matrix(invertedY_Test, invertedPredicted)
172  print("The confusion matrix for the particular case is:\n",
173        confusionMatrix)
174  accuracyScore = accuracy_score(invertedY_Test, invertedPredicted)
175  print("\nThe accuracy of the model given the test data is: ",
176        accuracyScore * 100, "%")
177  classificationReport = classification_report(y_Test, y_Predict)
178  print("\nThe classification report is as follows:\n",
179        classificationReport)
180
```

Output 10.5: Step 5

```
Step 5: Invert the integer encoded results into their original tex:-based
The inverted output test values are: ['no' 'no' 'no' 'yes' 'yes' 'yes'
'no' 'yes' 'yes' 'no']
The inverted predicted values of the output are: ['no' 'no' 'no' 'no'
'no' 'no' 'no' 'no' 'no' 'no']
The confusion matrix for the particular case is:
 [[5 0]
 [5 0]]

The accuracy of the model given the test data is:  50.0 %

The classification report is as follows:
              precision    recall  f1-score   support

           0       0.50      1.00      0.67         5
           1       0.00      0.00      0.00         5

    accuracy                           0.50        10
   Macro avg       0.25      0.50      0.33        10
weighted avg       0.25      0.50      0.33        10
```

Finally, Step 6 implements the statements used to visualize the decision tree based, on the parameters specified in the previous steps. The reader should note that the names of the features of the depicted decision tree, referred as `graphCols`, must be defined before the tree is visualized, so that proper labels are attached to the respective tree classifications:

```
181  #==================================================================
182  # Step 6: Visualizing the CART Decision Tree
183  # Define the names of the labels/features to be depicted in the
184  # decision tree
185  graphCols = ['age', 'Jobs', 'marital','education']
186
187  # Define the type of I/O to be used for the visualization of the
188  # decision tree
189  dot_data = StringIO()
190
191  # Use the export_graphviz() to prepare the visualization of the
192  # decision tree
193  export_graphviz(decisionTree, out_file = dot_data, filled = True,
194  feature_names = graphCols, rounded = True)
195
196  # Use the pydotplus library to plot the decision tree
197  graph = pydotplus.graphviz.graph_from_dot_data(dot_data.getvalue())
198
199  # Save the graph of the decision tree as a .png file in the local
200  # folder
201  graph.write_png("test.png")
202  Image(graph.create_png())
```

Output 10.5.a: Depicting the Decision Tree using gini index and min_samples_leaf = 6

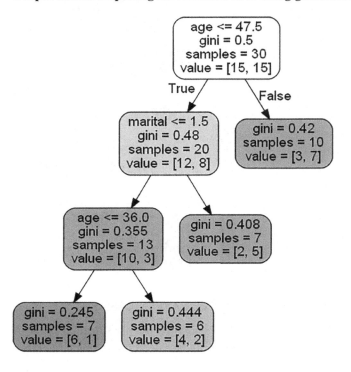

Output 10.5.b: Depicting the Decision Tree using entropy and min_samples_leaf = 1

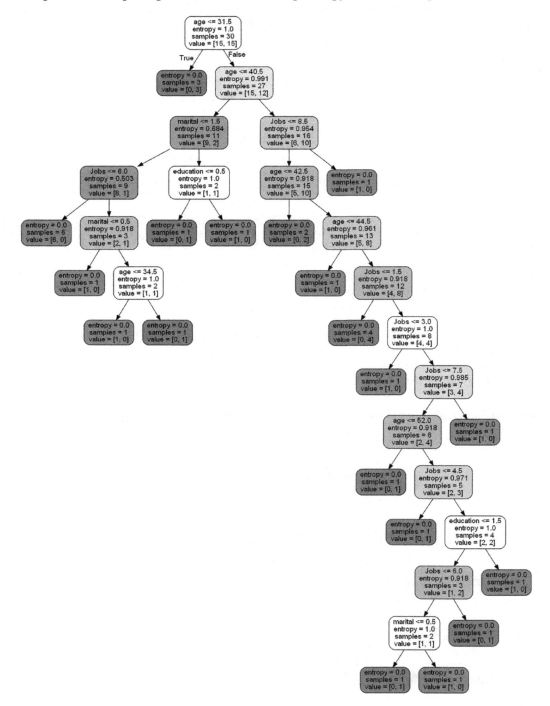

10.6   SUPERVISED LEARNING ALGORITHMS: NAÏVE BAYES CLASSIFIER

Naïve Bayes is a *probabilistic* model, which can therefore generalize the classification problem using a set of probabilities. The main concept of this model is based on the popular *Bayesian theorem.* The theorem can solve the problem of finding the probability of an event by using existing data for the conditions related to the event. For example, to find the probability of an event *A* to occur while event *B* is true is given by the equation below. This is also referred to as *posterior probability.*

$$P\left(A\,|\,B\right) = \frac{P\left(B\,|\,A\right) \cdot P\left(A\right)}{P\left(B\right)}$$

Observation 10.15 – Naïve Bayes Classifier: A supervised ML algorithm that is used to find the probability of an event given certain conditions. This probability is referred to as *posterior probability.* The known information is referred to as *prior probability.*

P(B|A) represents the known information regarding the *A* occurrence, such that *B* occurring when *A* is *True*. This probability is also called *prior probability,* as it is part of the existing knowledge. *P(A)* is the probability or *likelihood* of *A* occurring without any condition. *P(B)* represents the probability of event *B* occurring. *P(B)* is called *evidence.* Using prior probability, evidence and likelihood, a Naïve Bayes model can determine the *posterior probabilities* of each class label for a set of features, and assign a label based on these probabilities. The label with the highest or maximum *posterior probabilities* is assigned to the *current* observation.

As an example, consider the following weather data for the covering the previous 7 days, as given in Table 10.1. Based on the weather condition, the pilot instructors decide whether to run a training flight or not.

The theorem can be used to make a decision for the following weather conditions:

1. **Appearance:** Sunny
2. **Temperature:** Hot
3. **Windy:** False

To find the posterior probability for each label, calculate the probability for label *Yes*:

P(Yes) = 3/7
P(Sunny|Yes) = 1/3
P(Hot| Yes) = 1/3
P(False|Yes) = 3/3

The posterior probability for label *Yes* would be the following:

*P(Yes | (Sunny, Hot, False)) = P(Sunny | Yes) * P(Hot | Yes) * P(False | Yes) * P(Yes) =*
*= (1/3) * (1/3) * (3/3) * (3/7) = 0.047*

TABLE 10.1
Weather Data for Previous 7 Days

Appearance	Temperature	Windy	Training Flight?
Sunny	Cold	False	Yes
Cloudy	Mild	False	Yes
Sunny	Cold	True	No
Rainy	Hot	False	Yes
Rainy	Cold	True	No
Cloudy	Hot	True	No
Cloudy	Cold	False	No

Similarly, the posterior probability for label *No* for the same observation would be the following:

$P(No \mid (Sunny, Hot, False)) = P(Sunny \mid No) * P(Hot \mid No) * P(False \mid No) * P(No) =$
$(1/4) * (1/4) * (1/4) * (4/7) = 0.009$

In this case, the posterior probability of *Yes* is higher than that of *No*. Therefore, training flight will run with weather condition of *Appearance: Sunny, Temperature: Hot and Windy: False*.

Naïve Bayes may have three different implementations, depending on the data. In the case of continuous data, the *Gaussian distribution* is more suitable, whereas in the case of nominal data the *multinomial distribution* could produce better results. In the latter case (i.e., *multinomial distribution*), the implementation can be expressed in the following seven steps, with the last two being optional:

- **Step 1:** Import/read the data.
- **Step 2:** Split the input data into train and test sets.
- **Step 3:** Build the multinomial Naïve Bayes classifier.
- **Step 4:** Predict the results based on the classifier.
- **Step 5:** Find the accuracy of the regression model as a percentage.
- **Step 6 (Optional):** Visualize the results of the trained set.
- **Step 7 (Optional):** Visualize the results of the test set.

The following script uses students' *Midterm Exam* and *Project* grades to create the Naïve Bayes model and visualize the results:

```
1   # Import train_test_split to train and test the input
2   from sklearn.model_selection import train_test_split
3   # Import StandardScaler to scale the data
4   from sklearn.preprocessing import StandardScaler
5   # Import the Multinomial Naïve Bayes to create the classifier object
6   from sklearn.naive_bayes import MultinomialNB
7   # Import the accuracy_score to calculare the accuracy of the model
8   from sklearn.metrics import accuracy_score
9   # Import Numpy to prepare the plot parameters
10  import numpy as np
11  # Import Pyplot to create the plot
12  import matplotlib.pyplot as plt
13  # Import ListedColormap to color the data points in the plot
14  from matplotlib.colors import ListedColormap
15  # Plot inline
16  # This is particularly relevant in Jupyter Anaconda
17  %matplotlib inline
18
19  # Step 1: Define the input dataset. X must be a 2D list
20  # with as many rows as the observations
21  X = [[30, 75], [84, 89], [79, 84], [71, 74], [68, 71], [81, 70],
22       [61, 78], [89, 81], [58, 78], [70, 71], [70, 70], [90, 76],
23       [86, 92], [72, 70], [70, 72], [82, 87], [51, 78], [44, 71],
24       [82, 92], [50, 68]]
25  y = [0, 1, 1, 1, 0, 1, 1, 1, 0, 0, 1, 0, 1, 0, 0, 1, 1, 0, 1, 0]
26
27  # Step 2: Split the set X and y into train and test sets
28  # Test size is 25% of the dataset, Train size is 75%
29  # The new train and test lists will be in random order
30  X_train, X_test, y_train, y_test = train_test_split(X, y, \
31                          test_size = 0.25, random_state = 0)
```

```
32  print("Trained X set:", X_train); print("Test X set:", X_test)
33  print("Trained y set:", y_train); print("Test y set:", y_test)
34
35  # Step 3: Build the Naïve Bayes classifier
36  # Fit the trained set into the classifier
37  model = MultinomialNB().fit(X_train, y_train)
38  print("\n", model)
39
40  # Step 4: Predict the test results
41  y_pred = model.predict(X_test)
42  print("\nResults predicted by the model:", y_pred)
43  print("Results from the test:", y_test)
44  model.predict_proba(X)[:,1]
45
46  # Step 5: Form the confusion matrix to get the accuracy of the model
47  # Use the y_test (actual output) and the y_pred (predicted output)
48  accuracy = accuracy_score(y_test, y_pred)
49  print("The accuracy of the model given the test data is: ",
50        accuracy * 100, "%")
51  # Step 6: Visualize the training set results
52  X_set, y_set = X_train, y_train
53  X1, X2 = np.meshgrid(np.arange(start=np.array(X_set)[:, 0].min() - 1, \
54        stop = np.array(X_set)[:, 0].max() + 1, step = 0.01), \
55        np.arange(start = np.array(X_set)[:, 1].min() - 1, \
56        stop = np.array(X_set)[:, 1].max() + 1, step = 0.01))
57  plt.contourf(X1,X2, model.predict(np.array([[X1.ravel(), \
58        X2.ravel()]]).T).reshape(X1.shape), alpha = 0.75, \
59        cmap = ListedColormap(('red','blue')))
60
61  plt.xlim(X1.min(), X1.max())
62  plt.ylim(X2.min(), X2.max())
63  for i, j in enumerate(np.unique(y_set)):
64        plt.scatter(np.array(X_set)[y_set == j, 0],
65              np.array(X_set)[y_set == j, 1])
66  plt.title('Naive Bayes: Training set')
67  plt.xlabel("Midterm Exam")
68  plt.ylabel("Project")
69  plt.show()
70
71  # Step 7: Visualize the test results
72  X_set, y_set = X_test, y_test
73  X1, X2 = np.meshgrid(np.arange(start=np.array(X_set)[:, 0].min() - 1, \
74        stop = np.array(X_set)[:, 0].max() + 1, step = 0.01), \
75        np.arange(start = np.array(X_set)[:, 1].min() - 1, \
76        stop = np.array(X_set)[:, 1].max() + 1, step = 0.01))
77
78  plt.contourf(X1,X2, model.predict(np.array([[X1.ravel(), \
79        X2.ravel()]]).T).reshape(X1.shape), alpha = 0.75, \
80        cmap = ListedColormap(('red','blue')))
81
82  plt.xlim(X1.min(), X1.max());plt.ylim(X2.min(), X2.max())
83  for i, j in enumerate(np.unique(y_set)):
84        plt.scatter(np.array(X_set)[y_set == j, 0],
```

```
85                           np.array(X_set)[y_set == j, 1])
86  plt.title('Naive Bayes: Test set')
87  plt.xlabel("Midterm Exam"); plt.ylabel("Project")
88  plt.show()
```

Output 10.6:

```
Trained X set: [[44, 71], [61, 78], [72, 70], [68, 71], [79, 84], [81, 7
01, [70, 72], [70, 71], [89, 81], [51, 78], [90, 761, [71, 74], [30, 75],
[82, 87], [86, 92]]
Test X set: [[82, 92], [84, 89], [50, 68], [58, 78], [70, 70]]
Trained y set: [0, 1, 0, 0, 1, 1, 0, 0, 1, 1, 0, 1, 0, 1, 1]
Test y set: [1, 1, 0, 0, 1]

 MultinomialNB()

Results predicted by the model: [1 1 0 0 1]
Results from the test: [1, 1, 0, 0, 1]
The accuracy of the model given the test data is:   100.0 %
```

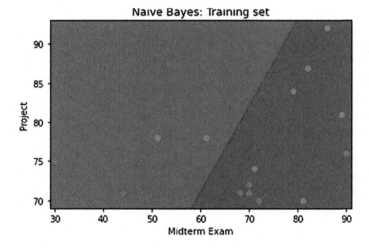

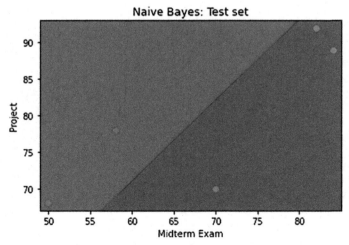

In this case, the output suggests that Naïve Bayes can predict the final grade (*Pass/Fail*) for the students with 100% accuracy. For the same data, a different implementation of Naïve Bayes may produce results with large variations (e.g., in the case of *Gaussian Naïve Bayes* function, the accuracy will be significantly lower). The reason for this is that the various Naïve Bayes functions depend on the nature of the data and are, thus, more scalable than other models.

10.7 UNSUPERVISED LEARNING ALGORITHMS: K-MEANS CLUSTERING

The *k-means clustering* algorithm is an unsupervised ML approach used to solve clustering problems in ML or data science. Its aim is to group unlabeled datasets into different clusters, where *k* is equal to the chosen number of newly created clusters. Each cluster is associated with a *centroid,* a data point representing the center of a cluster. The algorithm seeks to minimize the *sum of distances between the data point and their corresponding clusters.* Its applications may be relevant in different domains, such

> **Observation 10.16 – K-means Clustering:** An unsupervised ML algorithm that aims to group unlabeled datasets into a number (*k*) of different *clusters,* each associated with a *centroid* data point representing the center of cluster.

as *customer segmentation*, *insurance fraud detection*, and *document classification* just to name a few. Figure 10.3 presents a case of two clusters (*k* = 2) being identified in the source dataset:

K-means is, essentially, an *iterative* algorithm. First, it selects a value for *k*, that represents the number of clusters (e.g., *k* = 3 for 3 clusters). Next, it randomly assigns each data point to any of the clusters. Finally, it calculates the cluster *centroid* for each of the clusters. Once the iteration is complete a new one commences. At this stage, the algorithm reassigns each point to the *closest cluster centroid*. It then follows the same procedure to assign the points to the clusters containing the other centroids. The algorithm repeats the last two steps until there is no *switching of data points from one cluster to another,* in which case it is completed.

Implementing the k-means algorithm usually involves the following steps:

- **Step 1:** Select the number of clusters (*k*). One could also use the *elbow* function to determine the optimal number.
- **Step 2:** Select a random *centroid* for each cluster. Note that this may be other than the input dataset.

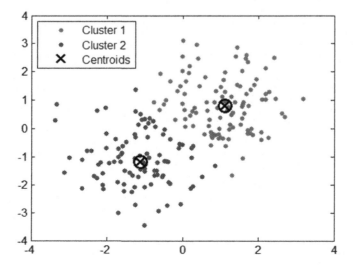

FIGURE 10.3 k-means clusters and their centroids. (See Raghupathi, 2018.)

- **Step 3:** Measure the distance (*Euclidean* function) between each point and the centroids. Assign each data point to their closest centroid.
- **Step 4:** Calculate the variance and add a new centroid for each cluster (i.e., calculate the mean of all the points for each cluster and set the new centroid).
- **Step 5:** Repeat Steps 3 and 4 until the centroid positions do not change.

The implementation of this approach in Python is rather straightforward, making it accessible to novice programmers and/or data scientists with no programming background. The following script is an example of a k-means algorithm implementation, with the objective to classify 100 customers based on their annual incomes and spending scores:

```
1   # Import Pandas
2   import pandas as pd
3   # Import Numpy as data manipulation
4   import numpy as np
5   # Import the KMeans library from the sklearn
6   from sklearn.cluster import KMeans
7   # Import the Pyplot to create the plot
8   import matplotlib.pyplot as plt
9   # Plot inline
10  # This is particularly relevant in Jupyter Anaconda
11  %matplotlib inline
12  # Import the operating system module
13  import os
14  # Import the Python data visualization library based on matplotlib
15  import seaborn as sns
16  sns.set(context = "notebook", palette = "Spectral", style = 'darkgrid',
17          font_scale = 1.5, color_codes = True)
18
19  # X is a list of 100 samples for customers, each representing the
20  # annual income and the spending score
21  X = [[15, 39], [15, 81], [16,  6], [16, 77], [17, 40], [17, 76],
22       [18,  6], [18, 94], [19,  3], [19, 72], [19, 14], [19, 99],
23       [20, 15], [20, 77], [20, 13], [20, 79], [21, 35], [21, 66],
24       [23, 29], [23, 98], [24, 35], [24, 73], [25,  5], [25, 73],
25       [28, 14], [28, 82], [28, 32], [28, 61], [29, 31], [29, 87],
26       [30,  4], [30, 73], [33,  4], [33, 92], [33, 14], [33, 81],
27       [34, 17], [34, 73], [37, 26], [37, 75], [38, 35], [38, 92],
28       [39, 36], [39, 61], [39, 28], [39, 65], [40, 55], [40, 47],
29       [40, 42], [40, 42], [42, 52], [42, 60], [43, 54], [43, 60],
30       [43, 45], [43, 41], [44, 50], [44, 46], [46, 51], [46, 46],
31       [46, 56], [46, 55], [47, 52], [47, 59], [48, 51], [48, 59],
32       [48, 50], [48, 48], [48, 59], [48, 47], [49, 55], [49, 42],
33       [50, 49], [50, 56], [54, 47], [54, 54], [54, 53], [54, 48],
34       [54, 52], [54, 42], [54, 51], [54, 55], [54, 41], [54, 44],
35       [54, 57], [54, 46], [57, 58], [57, 55], [58, 60], [58, 46],
36       [59, 55], [59, 41], [60, 49], [60, 40], [60, 42], [60, 52],
37       [60, 47], [60, 50], [61, 42], [61, 49]]
38
39  # Convert the list to an np.array for plotting the clusters
40  # of customers
```

```
41  X = np.array(X)
42
43  # Find the optimal number of clusters (elbow method)
44  from sklearn.cluster import KMeans
45  wcss = []
46  for i in range(1, 15):
47          kmeans = KMeans(n_clusters = i, init = 'k-means++', \
48                  random_state = 42)
49          kmeans.fit(X)
50          # Inertia function returns wcss for that model:
51          # WCSS is the sum of squared distance between each point
52          # and the centroid in a cluster
53          wcss.append(kmeans.inertia_)
54  # Plot the clusters and WCSS
55  plt.figure(figsize = (10,5))
56  sns.lineplot(range(1, 15), wcss, marker = 'o', color = 'red')
57  plt.title('The Elbow Method')
58  plt.xlabel('Number of clusters')
59  plt.ylabel('WCSS')
60  plt.show()
```

Output 10.7.a:

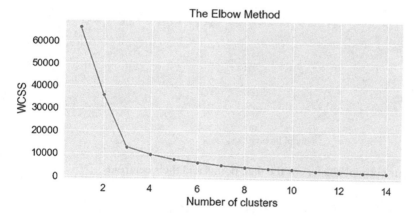

The output illustrates the identification of the optimal number of clusters that can represent the *k-means,* in this Case 4. Next, this is used to find, organize, and illustrate the respective clusters with their centroid data, as in the following script:

```
61  # Fitting K-means to the dataset
62  kmeans = KMeans(n_clusters = 4, init = 'k-means++', random_state = 42)
63  y_kmeans = kmeans.fit_predict(X)
64
65  # plot ('Annual Income (k$), Spending Score)
66  plt.figure(figsize = (15,7))
67  sns.scatterplot(X[y_kmeans == 0, 0], X[y_kmeans == 0, 1], \
68          color = 'yellow', label = 'Cluster 1', s = 50)
69  sns.scatterplot(X[y_kmeans == 1, 0], X[y_kmeans == 1, 1], \
70          color = 'blue', label = 'Cluster 2', s = 50)
```

```
71  sns.scatterplot(X[y_kmeans == 2, 0], X[y_kmeans == 2, 1], \
72         color = 'green', label = 'Cluster 3', s = 50)
73  sns.scatterplot(X[y_kmeans == 3, 0], X[y_kmeans == 3, 1], \
74         color = 'grey', label = 'Cluster 4', s = 50)
75  sns.scatterplot(kmeans.cluster_centers_[:, 0], \
76         kmeans.cluster_centers_[:, 1], color = 'red',
77         label = 'Centroids', s = 300, marker = ', ')
78  plt.grid(False)
79  plt.title('Clusters of customers')
80  plt.xlabel('Annual Income (k$)')
81  plt.ylabel('Spending Score (1-100)')
82  plt.legend()
83  plt.show()
```

Output 10.7.b: Finding and illustrating the clusters, their data points, and their centroids

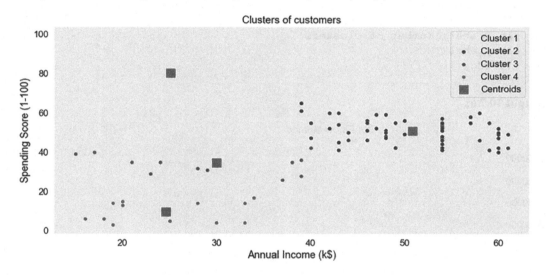

The output identifies the four optimal clusters of the data points and their centroids.

10.8   UNSUPERVISED LEARNING ALGORITHMS: *APRIORI*

The *apriori* algorithm is based on *rule mining* and is mainly used for finding the association between different items in a dataset. However, the algorithm can be also used as a classifier. It explores the data space and keeps all items in a dynamic structure. The apriori algorithm *prunes* the list of *itemsets* to keep only those that meet certain criteria. One simple criterion is the use of a *threshold* value: the most frequent item and itemset lists can be pruned using the threshold values on *support* and *confidence*. For example, if the support of an item is less than the threshold value the item is not added to the frequent items.

The association between items is determined based on two main measurements: *support* and *confidence*.

> **Observation 10.17 – Apriori:** An unsupervised ML algorithm used to find the association between different items in a dataset. It is based on the measurements of *confidence* and *support*.

> **Observation 10.18 – Support:** Calculates the likelihood of an item being in the data space and filters the reported items. Use parameter min _ support = value (0.0–1.0).

Support calculates the likelihood of an item being in the data space and *confidence* measures the relationship or association of an item with another.

For a given item (*A*) the support is calculated using the following equation (Equation 10.1):

$$\text{Support}(A) = \frac{\text{Number of observations containing } A}{\text{Total number of observations}} \tag{10.1}$$

The *confidence* is measured using the following equation (Equation 10.2) and represents the association between two items, say *A* and *B*:

$$\text{Confidence}(A \text{ to } B) = \frac{\text{Number of observations containing } A \text{ \& } B}{\text{Number of observations containing } A} \tag{10.2}$$

The `min_lift` parameter indicates the likelihood of an item being associated with another. A value of 1 indicates that the items are not associated. A *lift* value greater than 1 indicates that an item is likely to be associated with another item, while a value less than 1 means the opposite.

The `min_length` parameter defines the minimum number of items considered for the rules, and depends on the number of the available items. The association among the items can be determined up to a certain length: if the length of the association is 10, a maximum of ten items can be related to each other. Each one of these combinations is called an *itemset*. In a large dataset, the number of frequent items and itemsets could be rather substantial.

The apriori algorithm can be further explained using the dataset provided in Table 10.2. The table lists the four most recent transactions made by customers in a supermarket.

Apriori will start by calculating the support for all items as shown on Table 10.3. Next, it will apply the threshold to trim the item list and build a frequent itemset. Assume that the threshold for the support is 50%. The trimmed list of frequent items is shown on Table 10.4. Similarly, the algorithm will calculate the confidence for finding an association between two items, and trim the list using the threshold on confidence. Eventually, two rules will be selected:

> **Observation 10.19 – Confidence:** Calculates the level of confidence of the association with another item and filters the reported items. Use parameter min _ confidence = value (0.0–1.0).

> **Observation 10.20 – `min_lift`:** Defines the minimum number of items to be considered (as a combination) in the displayed rules. A value of 1 suggests an association, while a value less than 1 suggests lack of an association.

> **Observation 10.21 – `min_length`:** Defines the minimum number of items to be considered for the rules, and depends on the number of available items.

1. *If a customer buys an Apple, there are high chances the customer buys a Banana.*
2. *If a customer buys a Bread, there is a likelihood the customer will also buy Eggs.*

TABLE 10.2

Transactions at a Supermarket

Transaction ID	Items Purchased
1	Apple, Banana, Biscuits
2	Apple, Banana, Bread
3	Bread, Eggs, Cereal
4	Apple, Bread, Eggs

TABLE 10.3
Support for All Items

Item	Support
Apple	0.75
Banana	0.5
Bread	0.75
Biscuits	0.25
Cereals	0.25
Eggs	0.5

TABLE 10.4
Frequent Itemset with 50% Support

Item	Support
Apple	0.75
Banana	0.5
Bread	0.75
Eggs	0.5

The apriori implementation in Python can be described using the following four steps (the last one being optional):

- **Step 1:** Import/read the data.
- **Step 2:** Build the apriori model.
- **Step 3:** Transform the rules into a dataframe.
- **Step 4:** Create a table to display all the rules.

The following script uses the above data to create the apriori model:

```
1    # import Pandas and Numpy
2    import pandas as pd
3    import numpy as np
4
5    # import the apriori model
6    from apyori import apriori
7    # Import the accuracy_score to calculare the accuracy of the model
8    from sklearn.metrics import accuracy_score
9
10   # Step 1: Define the input dataset. X must be a 2D list with
11   # as many rows as the observations
12   X = [["Apple", "Banana", "Biscuits"], ["Apple", "Banana", "Bread"],
13       ["Bread", "Eggs", "Cereal"], ["Apple", "Bread", "Eggs"]]
14
15   # Step 2 Build the apriori model
16   rules = apriori(X, min_length = 2, min_support = 0.1, \
17       min_confidence = 0.02, min_lift = 1)
18   # rules = apriori(X, min_length = 2, min_support = 0.5,
19   # min_confidence = 0.5, min_lift = 1)
20
```

```
21  # Step3: Transform outputs in an appropriate pd.Dataframe format
22  results = list(rules)
23  results = pd.DataFrame(results)
24  print("The association rules for the particular dataset are:\n",
25          results)
26
27  # Step 4 Create an output table from the ordered statistics
28  # Note: not all tables are of the same type
29  F1 = []; F2 = []; F3 = []; F4 = []
30  C3 = results.support
31  for i in range(results.shape[0]):
32      single_list = results['ordered_statistics'][i][0]
33      F1.append(list(single_list[0]))
34      F2.append(list(single_list[1]))
35      F3.append(single_list[2])
36      F4.append(single_list[3])
37
38  # First column of the table
39  C1 = pd.DataFrame(F1)
40  # Second column of the table
41  C2 = pd.DataFrame(F2)
42  # Fourth column of the table
43  C4 = pd.DataFrame(F3,columns = ['Confidence'])
44  # Fifth column of the table
45  C5 = pd.DataFrame(F4,columns = ['Lift'])
46
47  # Concatenate all tables into one
48  table = pd.concat([C1,C2,C3,C4,C5], axis = 1)
49  print("\nImproved format of the association rules for the dataset:\n",
50          table)
```

Output 10.8.a–10.8.c:

```
The association rules for the particular dataset are:
                         items    support  \
0                       (Apple)     0.75
1                      (Banana)     0.50
2                    (Biscuits)     0.25
3                       (Bread)     0.75
4                      (Cereal)     0.25
5                        (Eggs)     0.50
6               (Apple, Banana)     0.50
7             (Apple, Biscuits)     0.25
8                (Apple, Bread)     0.50
9                (Apple, Eggs)     0.25
10           (Banana, Biscuits)     0.25
11              (Banana, Bread)     0.25
12               (Bread, Cereal)    0.25
13                (Eggs, Bread)     0.50
14               (Eggs, Cereal)     0.25
15      (Apple, Banana, Biscuits)   0.25
16         (Apple, Banana, Bread)   0.25
17          (Apple, Eggs, Bread)    0.25
18         (Eggs, Bread, Cereal)    0.25
```

```
                                    ordered_statistics
0                            [((), (Apple), 0.75, 1.0)]
1                            [((), (Banana), 0.5, 1.0)]
2                         [((), (Biscuits), 0.25, 1.0)]
3                            [((), (Bread), 0.75, 1.0)]
4                           [((), (Cereal), 0.25, 1.0)]
5                             [((), (Eggs), 0.5, 1.0)]
6     [((), (Apple, Banana), 0.5, 1.0), ((Apple), (B...
7     [((), (Apple, Biscuits), 0.25, 1.0), ((Apple),...
8                     [((), (Apple, Bread), 0.5, 1.0)]
9                    [((), (Apple, Eggs), 0.25, 1.0)]
10    [((), (Banana, Biscuits), 0.25, 1.0), ((Banana...
11                 [((), (Banana, Bread), 0.25, 1.0)]
12    [((), (Bread, Cereal), 0.25, 1.0), ((Bread), (...
13    [((), (Eggs, Bread), 0.5, 1.0), ((Bread), (Egg...
14    [((), (Eggs, Cereal), 0.25, 1.0), ((Cereal), (...
15    [((), (Apple, Banana, Biscuits), 0.25, 1.0), (...
16    [((), (Apple, Banana, Bread), 0.25, 1.0), ((Ap...
17    [((), (Apple, Eggs, Bread), 0.25, 1.0), ((Brea...
18    [((), (Eggs, Bread, Cereal), 0.25, 1.0), ((Bre...
```

```
Improved format of the association rules for the dataset:
```

	0	1	2	support	Confidence	Lift
0	Apple	None	None	0.75	0.75	1.0
1	Banana	None	None	0.50	0.50	1.0
2	Biscuits	None	None	0.25	0.25	1.0
3	Bread	None	None	0.75	0.75	1.0
4	Cereal	None	None	0.25	0.25	1.0
5	Eggs	None	None	0.50	0.50	1.0
6	Apple	Banana	None	0.50	0.50	1.0
7	Apple	Biscuits	None	0.25	0.25	1.0
8	Apple	Bread	None	0.50	0.50	1.0
9	Apple	Eggs	None	0.25	0.25	1.0
10	Banana	Biscuits	None	0.25	0.25	1.0
11	Banana	Bread	None	0.25	0.25	1.0
12	Bread	Cereal	None	0.25	0.25	1.0
13	Eggs	Bread	None	0.50	0.50	1.0
14	Eggs	Cereal	None	0.25	0.25	1.0
15	Apple	Banana	Biscuits	0.25	0.25	1.0
16	Apple	Banana	Bread	0.25	0.25	1.0
17	Apple	Eggs	Bread	0.25	0.25	1.0
18	Eggs	Bread	Cereal	0.25	0.25	1.0

The results demonstrate the apriori model at work, and also highlight the dominant associations between the items. Strong associations between *Bread* and *Eggs,* and *Apple* and *Banana* is evident.

Changing the parameter values to min_support=0.5 and min_confidence=0.5 will change the reported Output 10.8.d as follows:

The association rules for the particular dataset are:

```
        items     support                                    ordered_statistics
0       (Apple)     0.75                          [((), (Apple), 0.75, 1.0)]
1      (Banana)     0.50                          [((), (Banana), 0.5, 1.0)]
2       (Bread)     0.75                          [((), (Bread), 0.75, 1.0)]
3        (Eggs)     0.50                           [((), (Eggs), 0.5, 1.0)]
4  (Apple, Banana)  0.50    [((), (Apple, Banana), 0.5, 1.0), ((Apple), (B...
5  (Apple, Bread)   0.50                    [((), (Apple, Bread), 0.5, 1.0)]
6  (Bread, Eggs)    0.50    [((), (Bread, Eggs), 0.5, 1.0), ((Bread), (Egg...
```

Improved format of the association rules for the dataset:

```
        0        1     support   Confidence   Lift
0    Apple    None      0.75        0.75      1.0
1   Banana    None      0.50        0.50      1.0
2    Bread    None      0.75        0.75      1.0
3     Eggs    None      0.50        0.50      1.0
4    Apple  Banana      0.50        0.50      1.0
5    Apple   Bread      0.50        0.50      1.0
6    Bread    Eggs      0.50        0.50      1.0
```

Notice how filtering dramatically reduces the reported rules and output, by increasing the level of confidence and the acceptable support.

The rules extracted by apriori identify the patterns of item sales for a supermakert. The model can determine similar associations for a larger dataset and the report can be tweaked to display the top ranking associations (e.g. *Eggs* and *Bread* or *Apple* and *Banana*).

10.9   OTHER LEARNING ALGORITHMS

A number of other ML algorithms are also frequently used in real-life applications. One the most popular is *random forest* (Andrade et al., 2019; Kwon et al., 2015; Naveed & Alrammal, 2017; Naveed et al., 2020), a *supervised* ML algorithm. It can be used for both classification and regression. The main idea behind random forest is to create multiple ML decision tree models, with data-sets created using what is referred to as a *bootstrap sampling* method. According to this method, each sub-dataset is composed of random sub-samples of the original dataset. Each of the defined training datasets is used to create a different model, using the same ML algorithm and making different predictions. The best prediction is used as the result of the process.

The random forest algorithm can be described using the following four steps:

- **Step 1:** Select random samples from a given dataset.
- **Step 2:** Create a decision tree for each sample and get a prediction result for each decision tree.
- **Step 3:** Perform a vote for each of the predicted results.
- **Step 4:** Select the prediction result with the highest number votes as the final prediction.

Observation 10.22 – Random Forest: Create multiple ML decision trees from random sub-sets of the original dataset. Make predictions for each of the decision trees and vote for the best prediction.

Random forest is considered a highly accurate ML algorithm, with the larger numbers of decision trees created leading to increasingly more robust results. Since it calculates the average of all its predictions, it does not suffer from *overfitting* or *outliers* being present in the original dataset. Its main shortcomings come from the fact that it consists of multiple decision trees. Hence, it is slow in generating a final prediction as it has to get all the sub-tree predictions and vote the best one, and it is not as straightforward to interpret as a single decision tree.

The *K-Nearest Neighbors (k-NN)* algorithm uses the entire dataset as a training set, rather than splitting the dataset into a training and a test set. It assumes that similar data points are in close proximity to each other. This proximity (or distance) can be calculated using a variety of methods, such as the *Euclidean theorem*, or the *Hamming distance* (Sharma, 2020). When a new outcome is requested for a new data point, the k-NN algorithm calculates the instances between the new data point and the entire dataset, or the user-defined *k* data points that look more similar to the new data point. Next, it calculates the mean of the outcomes following a regression model, or the mode (i.e., the most frequent class).

> **Observation 10.23 – k-NN:** Use the whole data set as a training set to calculate the distances between the various *k* data points in the dataset.

The algorithm of the k-NN model follows the following six main steps:

- **Step 1:** Load the data.
- **Step 2:** Select the number (*k*) of *neighbors*.
- **Step 3:** For each new data point, calculate the distance between new and the current dataset points.
- **Step 4:** Add the distance and the index of the new data point to the current collection.
- **Step 5:** Sort the current collection of distances and indices by distance.
- **Step 6:** Pick the first *k* entries from the sorted collection, get their labels, and return the mean or mode.

The main disadvantage of k-NN is that it is becoming significantly slower as the dataset increases in size.

10.10   WRAP UP - MACHINE LEARNING APPLICATIONS

Through the use of Machine Learning (ML) algorithms, *Artificial Intelligence (AI)* has penetrated all forms of human activity. It is highly likely that the vast majority of humans has a first-hand experience of this through one of its many real-life applications. *Traffic Alerts (maps)* is such an example with several applications being used to suggestions and routes to help drivers deal with navigation and traffic. Data are collected either from other drivers currently using the same system or network and, or historical data of the various routes collected over time. Data collected when users are using the application or network include their location, average speed, and the route in which they are travelling. Figure 10.4 illustrates such an example on heavy congestion conditions (i.e., Sheikh Mohammed bin Rashid Blvd – Downtown Dubai).

Another class of examples of ML algorithms are the various *virtual personal assistants*. Such systems assist the users on various daily tasks and include advanced detection capabilities like understanding the users' voice (e.g., asking "what is my schedule for today?" will trigger the associated response). Common tasks implemented into contemporary virtual personal assistant systems include *speech recognition*, *speed-to-text* conversion, *natural language processing*, and *text-to-speech* conversion. The systems collect and refine the information based on previous interactions. They are integrated into a variety of platforms, including *smart speakers, smartphones,* and *mobile apps*.

Social media is another space where ML applications are heavily integrated and used. From personalizing news feeds to better ads targeting, social media platforms are utilizing machine learning

FIGURE 10.4 Traffic alert application.

for both corporate and end-user benefits. The list below includes some examples one may be familiar with, perhaps without even realizing that these features are nothing but the practical application of ML algorithms:

- **People You May Know:** ML works on a simple concept: understanding through experience. For example, Social Media platforms continuously monitor the friends one connects with, the most often visited profiles, one's interests, or work and personal status, or groups one belongs too. Based on continuous learning, a list of the Social Media users that one can become friends with is suggested.
- **Face Recognition:** A user uploads a personal picture with a friend and the system instantly recognizes the identity of that friend. Such systems may check the poses and projections in the picture, identify unique features, and match them with people in the user's friends or contact lists. The entire process is based on ML and is commonly referred to as *friend tagging*. It is a rather complex process taking place at the backend, but it is rather transparent on the user side, as it seems like a simple and unobtrusive feature at the front end.
- **Similar Pins:** ML is a core element in *computer vision*, a technique to extract useful information from images and videos. An example of this can be seen in platforms which use computer vision to identify the objects (or pins) in the images and recommend other related pins accordingly.

House price prediction is yet another example of ML algorithms in action. By leveraging the data collected from large numbers of houses in relation to their characteristics (e.g., square footage, number of rooms, property type), the algorithm trains the ML model to predict the price of other houses. The multiple popular online portals for searching houses or apartments (both for rental and purchase) are examples of the use of such applications.

Product recommendation is an experience most people have without even noticing. As an example, one can think of using a web browser to check a product on a specific website. It is likely that while engaging in other online activities, such as watching online videos, the same or similar products appear as an ad. In such cases, the various platforms use *smart agents* to track the user's search history and recommends ads based on it.

Recommender systems are another application of ML algorithms. Such systems use *collaborative filtering,* a method based on gathering and analyzing user behavior information and predicting what they like based on similarities with other users. Figure 10.5 provides an example of the use of collaborative filtering in an E-commerce web app. In this context one can assume a customer (Customer 1) viewing product A and other customers viewing products A, B, C, and D. Due to the similarity of interests of all the users in product A, the web app will propose products B, C and D to Customer 1.

Among the most important applications of ML is the *monitoring of video cameras.* In areas or countries utilizing excessive numbers of traffic monitoring video cameras, monitoring by human officers can be impractical and challenging. The idea of training computers to accomplish this task comes handy in such cases. Similarly, video surveillance systems powered by AI/ML make it possible to detect suspicious activity, sometimes even before it takes place. This is done by tracking unusual behavior (e.g., when one stands motionless for a long time, stumbles, or laying on public locations). The system can generate alerts sent to human attendants, who can then take appropriate actions. As activities are reported and verified, they help to improve the surveillance services even further.

In the context of *information security,* one should note the use of *spam filtering.* The term refers to processes monitoring the user's email traffic and executing appropriate preventive actions. It is crucial for such systems to ascertain that spam filters are continuously updated; this is accomplished through ML algorithms. While there are hundreds of thousands of *malware* and *security threats* detected every single day, it is generally accepted that the associated code is 90% or more similar to its predecessor. ML-based security programs can identify such coding patterns and detect new malware with slight coding variations rather easily. Similarly, ML provides great potential to secure online monetary transactions from online frauds. For instance, online payment platforms use a set of tools that helps compare millions of transactions taking place almost simultaneously and identifying suspicious of fraudulent action between buyers and sellers.

Finally, another common application of ML models can be found in the online customer support services of many e-Business or e-Commerce platforms. Such platforms frequently offer the option

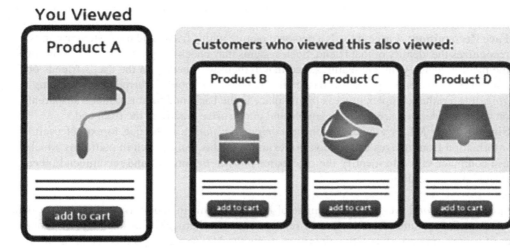

FIGURE 10.5   Product recommendations. (See Keshari, 2021.)

to chat with a customer support representative while navigating the website. While the transaction may seem like a regular conversation, it is not with a real representative but with a *chatbot*. The latter extracts information from the website and presents it to the customers in a chat-like form. Every time a new chat begins, the answer is improved based on the previously recorded answers.

The discussion on ML applications can continue further, with practical use examples like *weather prediction, distinction between animals/plants/objects,* or *customer segmentation,* just to name a few.

10.11 CASE STUDIES

Use dataset *dataset.csv* to write a Python script that predicts whether a patient will be readmitted or not within 30 days. The application should do the following:

1. Read the dataset and create a data frame with the following categories: *gender, race, age, admission type id, discharge disposition id, admission source id, max glu serum, A1Cresult, change, diabetesMed, readmitted* (categorical), *time in hospital, number of lab procedures, number of procedures, number of medications, number of outpatients, number of emergencies, number of inpatients, number of diagnoses* (numerical).
2. Apply the following ML algorithms and calculate their accuracy: *logistic regression, k-NN, SVM, Kernel SVM, Naïve Bayes, CART Decision Tree, Random Forest.*

10.12 EXERCISES

1. Use the CART example in this chapter to change the criterion from *entropy* to *Gini index* and the *max depth* to 10. How does this affect the accuracy of the model? What is the effect of changing the *max depth* to 20?
2. Test both the *BEST* and *RANDOM* splitter features on the CART example from this chapter. Explain whether the performance of a decision tree depends on the splitter feature of the classifier object.
3. Apply a smaller training dataset to the CART decision tree example to investigate whether the performance will improve or decrease (*Hint: Increase and decrease the ratio of the size of the training dataset*).
4. Find the *precision, recall* and *fscore* for a CART decision tree with *entropy* as criterion, *max dept* of *4* and *min samples leaf nodes* of *20*.
5. Use the bank dataset to train a decision tree classifier with *ten-fold cross validation* and generate the respective classification report.

REFERENCES

Andrade, E. de O., Viterbo, J., Vasconcelos, C. N., Guérin, J., & Bernardini, F. C. (2019). A model based on lstm neural networks to identify five different types of malware. *Procedia Computer Science, 159,* 182–191.

Keshari, K. (2021). *Top 10 Applications of Machine Learning: Machine Learning Applications in Daily Life.* https://www.edureka.co/blog/machine-learning-applications/.

Kwon, B. J., Mondal, J., Jang, J., Bilge, L., & Dumitraş, T. (2015). The dropper effect: Insights into malware distribution with downloader graph analytics. *Proceedings of the 22nd ACM SIGSAC Conference on Computer and Communications Security* (1118–1129), Denver, Colorado.

Mitchell, T. M. (1997). *Machine Learning* (1st ed.). New York: McGraw-Hill.

Mola, F. (1998). *Classification and Regression Trees Software and New Developments BT – Advances in Data Science and Classification* (A. Rizzi, M. Vichi, & H.-H. Bock eds.; pp. 311–318). Berlin Heidelberg: Springer.

Naveed, M., & Alrammal, M. (2017). Reinforcement learning model for classification of Youtube movie. *Journal of Engineering and Applied Science*, *12*(9), 1–7.

Naveed, M., Alrammal, M., & Bensefia, A. (2020). HGM: A Novel Monte-Carlo simulations based model for malware detection. *IOP Conference Series: Materials Science and Engineering*, *946*(1), 12003. https://doi.org/10.1088/1757-899x/946/1/012003.

Quinlan, J. R. (1986). Induction of decision trees. *Machine Learning*, *1*(1), 81–106. https://doi.org/10.1007/BF00116251.

Raghupathi, K. (2018). *10 Interesting Use Cases for the K-Means Algorithm*. DZone AI Zone. https://dzone.com/articles/10-interesting-use-cases-for-the-k-means-algorithm.

Salzberg, S. L. (1994). C4.5: Programs for machine learning by J. Ross Quinlan. Morgan Kaufmann Publishers, Inc., 1993. *Machine Learning*, *16*(3), 235–240. https://doi.org/10.1007/BF00993309.

Sharma, P. (2020). *4 Types of Distance Metrics in Machine Learning*. Analytics Vidhya. https://www.analyticsvidhya.com/blog/2020/02/4-types-of-distance-metrics-in-machine-learning/.

11 Introduction to Neural Networks and Deep Learning

Dimitrios Xanthidis
University College London
Higher Colleges of Technology

Muhammad Fahim
Higher Colleges of Technology

Han-I Wang
The University of York

CONTENTS

11.1 Introduction 449
11.2 Relevant Algebraic Math and Associated Python Methods for DL 452
    11.2.1 The Dot Method..452
    11.2.2 Matrix Operations with Python..455
    11.2.3 Eigenvalues, Eigenvectors and Diagonals 459
    11.2.4 Solving Sets of Equations with Python 460
    11.2.5 Generating Random Numbers for Matrices with Python......................461
    11.2.6 Plotting with Matplotlib..463
    11.2.7 Linear and Logistic Regression 465
11.3 Introduction to Neural Networks..466
    11.3.1 Modelling a Simple ANN with a Perceptron 467
    11.3.2 Sigmoid and Rectifier Linear Unit (ReLU) Methods 470
    11.3.3 A Real-Life Example: Preparing the Dataset.......................................473
    11.3.4 Creating and Compiling the Model 474
    11.3.5 Stochastic Gradient Descent and the Loss Method and Parameters 475
    11.3.6 Fitting and Evaluating the Models, Plotting the Observed Losses..........477
    11.3.7 Model Overfit and Underfit...482
11.4 Wrap Up...483
11.5 Case Study 484
References..484

11.1 INTRODUCTION

Deep learning is in fact a new name for an approach to artificial intelligence called neural networks, which has been going in and out of fashion for more than 70 years. Neural networks were first proposed in 1944 by Warren McCullough and Walter Pitts, two University of Chicago researchers who moved to MIT in 1952 as founding members of what's sometimes called the first cognitive science department.

(Hardesty, 2017)

DOI: 10.1201/9781003139010-11

Human intelligence is an evolutionary, biologically controlled process. Humans learn based on their experiences. Similarly, *machine* or *artificial* intelligence is subject to comparable experiences in the form of data. On a broader context, the two forms of intelligence are similar in the sense that they are subject to a common approach: "based on what I have seen and observed I think this will happen next". Once this core idea is

Observation 11.1 – Deep Learning: A specialized form of Machine Learning. It uses many layers of algorithms to process the underlying data which could be human speeches, images, text, complex objects, etc.

transferred to mathematical constructs and the associated algorithms (self-evolving), machines are observed to be capable of *learning on their own*, a process commonly referred to as *machine learning (ML)*. ML is a branch of *artificial intelligence (AI)*, an umbrella term used to describe approaches and techniques that can make machines think and act in a more rational and human-like way.

Deep learning (DL) is a specific form of ML, and therefore another branch of AI (Figure 11.1). At a basic level, DL is based on mimicking the human thinking process and developing relevant abstractions and connections. It consists of the following elements:

1. **Learning:** Facilitating the functionality to artificially obtain and process new information.
2. **Reasoning:** Offering the functionality to process information in different, and potentially overlooked, ways.
3. **Understanding:** Providing ways to showcase the results of the adopted model.
4. **Validating:** Offering the opportunity to validate the results of the model based on theory.
5. **Discovering:** Providing the mechanisms to identify new relationships within the data.
6. **Extracting:** Allowing the extraction of new meanings based on the predictors.

DL uses numerous layers of algorithms to process the underlying data, which could be spoken words, images, text, or more complex objects. The data are normally passed through interconnected layers of processing networks, as shown in Figure 11.2.

In ML, there are two types of variables: *dependent* and *independent*. One way to contextualize these variables is to think of independent variables as the *inputs* of the ML process and dependent as the *outputs*. For example, one can predict a person's *weight* by knowing that person's *height*.

Another notion the reader should be familiar with is that of *data plotting*. Essentially, plotting is a way to visualize the data in an effort to identify underlying patterns and groupings. As data can be *scattered,* when plotting them the goal is to find a *line* that represents the best *fit* for a given dataset. A simple equation can define such a process: $Y = F(X) + B$ where Y is the dependent variable (predicted weight) and X the independent variable (an individual's height).

In ML, there are mainly two types of predictions:

1. **Linear Regression:** Linear regression is focused on predicting *continuous* values. This topic is thoroughly discussed in Chapter 10: Machine Learning with Python. It is highly recommended that the reader goes through the basic discussions on that chapter before proceeding to the next sections of the present one, as they offer a useful foundation for understanding many aspects of DL.

FIGURE 11.1 Scope of data-based learning technologies.

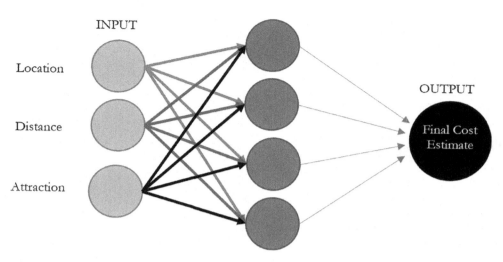

FIGURE 11.2 DL processing and layering structure.

2. **Logistic Regression:** Logistic regression is focused on predicting values *classified* as 0 or 1, and is one of the cornerstones of DL.

DL is applied in cases of learning based on *unlabelled* data with *unknown features*. Thus, *feature extraction (FE)* is a vital aspect of DL. FE uses algorithms to construct the meaning of the features, so the training and testing processes can be applied.

This chapter covers the following:

1. An introduction to the theory and mathematical constructs of DL fundamentals, supported by the associated mathematical equations, and working examples and related Python scripts.
2. An introductory discussion on *Neural Networks (NN)* and DL algorithms implementing NN with working examples and scripts.
3. Examples of building a DL model using NN.

It should be noted that, since there are several mathematical concepts involved in the DL processes, it is possible to face compatibility issues when working with more than one libraries. In such cases, it is, often, quite useful to know if a particular library is installed in the system and, if so, which version. In that case, the following statements may come handy:

```
1   # scipy
2   import scipy
3   print('scipy: %s' % scipy.__version__)
4   # numpy
5   import numpy
6   print('numpy: %s' % numpy.__version__)
7   # matplotlib
8   import matplotlib
9   print('matplotlib: %s' % matplotlib.__version__)
10  # pandas
11  import pandas
12  print('pandas: %s' % pandas.__version__)
```

```
13  # statsmodels
14  import statsmodels
15  print('statsmodels: %s' % statsmodels.__version__)
16  # scikit-learn
17  import sklearn
18  print('sklearn: %s' % sklearn.__version__)
```

Output 11.1:

```
scipy: 1.4.1
numpy: 1.19.5
matplotlib: 3.2.2
pandas: 1.1.5
statsmodels: 0.10.2
sklearn: 1.0.1
```

In addition to *Pandas, MatplotLib, Nympy*, and *SciPy* libraries already covered in previous chapters, there are a few more that are essential in DL scripts. Some of these must be installed prior to their import and use in the script. However, given the variety of installations depending on the operating systems and configurations, it is deemed impractical to cover all those in the present chapter. The reader is advised to seek instructions in the many online available sites. A list of these libraries, with a brief description, follows:

1. **TensorFlow:** It is used for backpropagation and passes the data for training and prediction.
2. **Theano:** It helps with defining, optimizing and evaluating mathematical equations on multi-dimensional arrays. It is very efficient when performing symbolic differentiation.
3. **Pytorch:** It helps with tensor computations with GPU and Neural Networks based data modeling.
4. **Caffe:** It helps with implementing DL frameworks using improved expressions and speed.
5. **Apache mxnet:** As a core component, it comes with a dynamic dependency scheduler that provides parallelism for both symbolic and imperative operations.

11.2   RELEVANT ALGEBRAIC MATH AND ASSOCIATED PYTHON METHODS FOR DL

There are some essential mathematical concepts that must be explained and their Python implementations described before delving into the introduction of DL with Python. The most fundamental are the *dot()* method, the matrix operations, *eigenvalues/eigenvectors* and diagonals, solving equations through sets, generating random numbers, and linear and logistic regression.

11.2.1   THE DOT METHOD

A method often used in DL that is not covered in previous chapters is the *dot* method. It implements the math equation that sums the products of two arrays:

$$x.y = x^T b = \sum_{n=1}^{N} x_n y_n$$

The *dot* method is important in the context of DL, as the main method of the latter is to accept multiple inputs

Observation 11.2 – The Dot Method:
Calculates the sum of vectors, provided in the form of matrices.

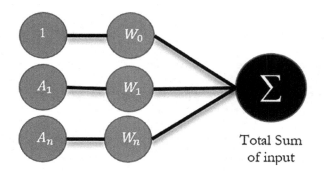

FIGURE 11.3 The dot method in DL.

from various *neurons* and calculate their sum. Since the inputs are always in the form of *vectors* (i.e., pairs of values like *course grade* and its *weight*), the *dot* method is an effective means for this calculation. Figure 11.3 illustrates the functionality of the dot method:

Consider the following Python script:

```
1    import numpy as np
2
3    # 1x2 and 1x3 arrays
4    x1, y1 = np.array([1, 2]), np.array([3, 4])

5    x2, y2 = np.array([1, 2, 3]), np.array([4, 5, 6])
6    print("The two arrays x1 and y1 are:\n", x1, y1)
7    print("The two arrays x2 and y2 are:\n", x2, y2)
8
9    # Product of 2 arrays calculated as xi*yi (for each of the 2 elements)
10   print("\nCreate a new list as products of the elements of the two \
11          arrays (x1 * y1):", x1 * y1)
12   print("\nCreate a new list as products of the elements of the two \
13          arrays (x2 * y2):", x2 * y2)
14
15   # Loop calculates the dot method of the 2 arrays (x1, y1 & x2, y2)
16   Dot = 0
17   for i in range(len(x1)):
18       Dot += x1[i] * y1[i]
19   print("\nUsing a regular loop to calculate the dot value for \
20          the 1x2 arrays:", Dot)
21   Dot = 0
22   for i in range(len(x2)):
23       Dot += x2[i] * y2[i]
24   print("Using a regular loop to calculate the dot value for \
25          the 1x3 arrays:", Dot)
26   # The zip method with parallel iterations calculates
27   # the dot for x1, y1 and x2, y2
28   Dot = 0
29   for g, h in zip(x1, y1):
30           Dot += g * h
31   print("\nUsing the zip method for parallel iterations:", Dot)
32
```

```
33  Dot = 0
34  for g, h in zip(x2, y2):
35        Dot += g * h
36  print("Using the zip method for parallel iterations:", Dot)
37
38  # The sum method calculates the dot for two arrays
39  print("\nThe sum of the products of the elements of the two arrays \
40  (np.sum(x1 * y1)):", np.sum(x1 * y1))
41  print("\nThe sum of the products of the elements of the two arrays \
42  (np.sum(x2 * y2)):", np.sum(x2 * y2))
43
44  # A different version of the sum method calculates the dot of 2 arrays
45  print("\nThe sum of the products of the elements of the two arrays \
46        ((x1 * y1).sum()):", (x1 * y1).sum())
47  print("The sum of the products of the elements of the two arrays \
48        ((x1 * y1).sum()):", (x2 * y2).sum())
49
50  # The dot method on two arrays
51  print("\nUse the dot method on the elements of the two arrays \
52  (np.dot(x1, y1)):", np.dot(x1, y1))
53  print("Use the dot method on the elements of the two arrays \
54        (np.dot(x2, y2)):", np.dot(x2, y2))
55
56  # A different version of the dot method on two arrays
57  print("\nAnother way to use the dot method on the elements \
58        of the two arrays (x1.dot(y1)):", x1.dot(y1))
59
60  print("Another way to use the dot method on the elements \
61  of the two arrays (x2.dot(y2)):", x2.dot(y2))
62
63  # Direct use of the dot notation on two arrays
64  print("\nAnother way to use the dot method (x1 @ y1):", x1 @ y1)
65  print("\nAnother way to use the dot method (x2 @ y2):", x2 @ y2)
```

Output 11.2.1:

```
The two arrays x1 and y1 are:
 [1 2] [3 4]
The two arrays x2 and y2 are:
 [1 2 3] [4 5 6]

Create a new list as products of the elements of the two arrays (x1 * y1)
: [3 8]
Create a new list as products of the elements of the two arrays (x2 * y2)
: [ 4 10 18]
Using a regular loop to calculate the dot value for the 1x2 arrays: 11
Using a regular loop to calculate the dot value for the 1x3 arrays: 32

Using the zip method for parallel iterations: 11
Using the zip method for parallel iterations: 32
```

```
The sum of the products of the elements of the two arrays (np.sum(x1 * y1))
: 11
The sum of the products of the elements of the two arrays (np.sum(x2 * y2))
: 32
The sum of the products of the elements of the two arrays ((x1 * y1).sum())
: 11
The sum of the products of the elements of the two arrays ((x1 * y1).sum())
: 32
Use the dot method on the elements of the two arrays (np.dot(x1, y1)): 11
Use the dot method on the elements of the two arrays (np.dot(x2, y2)): 32

Another way to use the dot method on the elements of the two arrays (x1.dot
(y1)): 11
Another way to use the dot method on the elements of the two arrays (x2.dot
(y2)): 32

Another way to use the dot method (x1 @ y1): 11

Another way to use the dot method (x2 @ y2): 32
```

This script calculates and presents the sum of the products of the elements of two arrays (based on their indices) in varying ways and presents their results. For illustration purposes, it uses two types of arrays (i.e., 1×2 elements and 1×3 elements). The reader should notice the various forms that the *dot* method can take. The method is quite useful and becomes handy in the examples provided in the following sections.

11.2.2 MATRIX OPERATIONS WITH PYTHON

Another algebraic concept that is quite useful in DL is that of *matrix multiplication*. Broadly speaking, this process requires that the size of the *second dimension* of the *first matrix* must be the same as the size of the *first dimension* of the *second matrix*. In other words, the number of columns in the *first matrix* must be equal to the number of rows in the *second matrix*. The resulting matrix has the size of the *first dimension* of the first matrix (or its number of rows) and the size of the *second dimension* of the second matrix (or its number of columns). For the calculation of the various elements of the new matrix the *dot* method is used.

As an example, one can assume the following two matrices:

$$npArray = \begin{bmatrix} 1 & 2 \\ 5 & 6 \end{bmatrix}$$

$$newMatrix = \begin{bmatrix} 3 & 4 & 5 \\ 1 & 2 & 3 \end{bmatrix}$$

The first array (*npArray*) has two columns, whereas the second (*newMatrix*) has two rows. Hence, it is possible to have a new matrix as the product of these two matrices. The resulting matrix will be calculated as follows:

$$\begin{bmatrix} (1*3+2*1) & (1*4+2*2) & (1*5+2*3) \\ (5*3+6*1) & (5*4+6*2) & (5*5+6*3) \end{bmatrix} = \begin{bmatrix} 5 & 8 & 11 \\ 21 & 32 & 43 \end{bmatrix}$$

Another mathematical Python method that often comes handy when using matrices is `exp()` from the *Numpy* library. The method accepts an array of elements (an algebraic matrix) as an argument and creates a new matrix as a result of $e^\wedge x_i y_i$. Using the previous example of matrix *npArray*, the resulting matrix will be as follows:

$$\begin{bmatrix} e^\wedge 1 & e^\wedge 2 \\ e^\wedge 5 & e^\wedge 6 \end{bmatrix} =: \begin{bmatrix} 2.71828283 & 7.3890561 \\ 148.4131591 & 403.42879349 \end{bmatrix}$$

Observation 11.3 – The `exp()` Method: Creates a new matrix as a result of $e^\wedge x_i y_i$ of the elements of the original matrix.

Observation 11.4 – Inverse Matrix: A matrix which, if multiplied by the original, gives the *identity* matrix.

Another concept often used in DL is that of the *inverse matrix*. If such a matrix is multiplied by the original, it will result into the *identity matrix*. If, in turn, the latter is multiplied by the original matrix, *it will not change it*. This is similar to integer 1, which when multiplied by any other integer it does not incur any value changes. The identity matrices for 2×2, 3×3, and 4×4 matrices can be expressed as follows:

$$\begin{bmatrix} 1 & 0 \\ 0 & 1 \end{bmatrix}$$

$$\begin{bmatrix} 1 & 0 & 0 \\ 0 & 1 & 0 \\ 0 & 0 & 1 \end{bmatrix}$$

$$\begin{bmatrix} 1 & 0 & 0 & 0 \\ 0 & 1 & 0 & 0 \\ 0 & 0 & 1 & 0 \\ 0 & 0 & 0 & 1 \end{bmatrix}$$

This pattern can continue in a similar fashion for larger square matrices. It is important to note that there are two requirements for a matrix to have a corresponding inverse: it must be a *square* matrix and its *determinant* value must be *non-zero*.

The *determinant* is a special number, either integer or real, calculated from a matrix. Its most important role is precisely to determine whether a matrix can have an inverse one, in which case the determinant is non-zero. If not, it will have a value of 0 or *extremely close* to 0. It must be noted that even a number like $2.3e{-}23$ is considered as 0 and, therefore, such a determinant would suggest that it is not feasible to have an inverse matrix.

The determinant is calculated by *subtracting the product* of the diagonal elements of the matrix. For example, in the case of matrix $\begin{bmatrix} 1 & 2 \\ 5 & 6 \end{bmatrix}$ the deter-

Observation 11.5 – Identity Matrix: A matrix that has all its first diagonal elements with a value of 1, which causes no change to the corresponding values when multiplied by the original matrix.

Observation 11.6 – Determinant: A special number, integer or real, calculated from the *diagonals* of a matrix. It determines whether a matrix has an inverse (value is non-zero) or not (value is 0).

minant is calculated as $1*6 - 5*2 = 6 - 10 = -4$. However, in the case of $\begin{bmatrix} 1 & 3 & 2 \\ 5 & 4 & 8 \\ 7 & 6 & 9 \end{bmatrix}$ things are more complicated. In this case the determinant is calculated as *1*((4*9) − (6*8)) − 3*((5*9) − (7*8)) + 2*((5*6) − (7*4)) = 1*(36−48) − 3*(45−56) + 2*(30−28) = − 12 − 3*(−11) + 2*2 = −12 + 33 + 4 = 25.* The pattern for *3 × 3* or larger matrices is as follows:

- Multiply the first element of the first row with the determinant of the matrix that is not in the same row or column.
- Similarly, calculate the same values for all the elements of the first row of the matrix.
- Calculate the final determinant as *first result − second result + third result − fourth result* and so forth.

The reader should note that the determinant can be calculated only for *square* matrices.
The following script briefly demonstrates the above concepts:

```
1   import numpy as np
2
3   # Create a 2-dimensonal array (2x2) using the array function (Numpy)
4   npArray = np.array([[1, 2], [5, 6]])
5   # Show the entire array and the 2nd element of the 1st dimension
6   # in 2 different ways
7   print("\nThe nparray's array's contents:\n", npArray)
8   print("The 2nd element of the 1st dimension of the array:",
9           npArray[0][1])
10  print("The same result from a different syntax:", npArray[0, 1])
11  print("\nThe elements of the 2nd dimension:", npArray[:, 0])
12  print("\nShow the result of the e^x for each element of the input \
13          array:\n", np.exp(npArray))
14
15  # Create a 2-dimensonal array (2x3) using the array function (Numpy)
16  newMatrix = np.array([[3, 4, 5], [1, 2, 3]])
17  print("\nThe 2x3 matrix newMatrix is:\n", newMatrix)
18  # Multiply the arrays npArray and newMatrix applying the .dot method
19  print("\nThe product of npArray and newMatrix using the .dot method \
20          is:\n", npArray.dot(newMatrix))
21
22  # Create a 2-dimensional array (3x3) using the array function (Numpy)
23  newMatrix2 = np.array([[1, 3, 2], [5, 4, 8], [7, 6, 9]])
24  print("\nThe 3x3 matrix newMatrix2 is:\n", newMatrix2)
25  # Determinant values for npArray & newMatrix2. The matrices are squares
26  print("\nThe determinant for the npArray is: ", np.linalg.
    det(npArray))
27  print("The determinant for the newMatrix is: ",
28          np.linalg.det(newMatrix2))
29  # Calculate and display the inverse matrix for npArray and newMatrix2
30  inverseNpArray = np.linalg.inv(npArray)
31  print("\nThe inverse matrix for the npArray is:\n", inverseNpArray)
32  inverseNewMatrix2 = np.linalg.inv(newMatrix2)
33  print("\nThe inverse matrix for the newMatrix2 is:\n",
34          inverseNewMatrix2)
35  # Multiplying  original npArray & newMatrix2 matrices with their
36  # inverse produces the identity matrix
37  print("\nThe product of the npArray and its inverse matrix is:\n",
```

```
38            inverseNpArray.dot(npArray))
39  print("\nThe product of the newMatrix2 and its inverse matrix is:\n",
40            inverseNewMatrix2.dot(newMatrix2))
```

Output 11.2.2:

```
The nparray's array's contents:
 [[1 2]
 [5 6]]
The 2nd element of the 1st dimension of the array: 2
The same result from a different syntax: 2

The elements of the 2nd dimension: [1 5]

Show the result of the e^x for each element of the input array:
 [[  2.71828183    7.3890561 ]
 [148.4131591   403.42879349]]

The 2x3 matrix newMatrix is:
 [[3 4 5]
 [1 2 3]]

The product of npArray and newMatrix using the .dot method is:
 [[ 5  8 11]
 [21 32 43]]

The 3x3 matrix newMatrix2 is:
 [[1 3 2]
 [5 4 8]
 [7 6 9]]


The determinant for the npArray is:  -3.999999999999999
The determinant for the newMatrix is:  25.000000000000007

The inverse matrix for the npArray is:
 [[-1.5   0.5 ]
 [ 1.25 -0.25]]

The inverse matrix for the newMatrix2 is:
 [[-0.48 -0.6   0.64]
 [ 0.44 -0.2   0.08]
 [ 0.08  0.6  -0.44]]

The product of the npArray and its inverse matrix is:
 [[ 1.00000000e+00 -2.22044605e-16]
 [-5.55111512e-17  1.00000000e+00]]

The product of the newMatrix2 and its inverse matrix is:
 [[ 1.00000000e+00  6.66133815e-16  9.99200722e-16]
 [-2.08166817e-16  1.00000000e+00 -1.24900090e-16]
 [ 7.21644966e-16  1.11022302e-16  1.00000000e+00]]
```

The results showcase the output of the calculations. Note that the rather complicated calculations for the determinant lead to the respective values not being whole numbers. In addition, the product of *newMatrix2* and its inverse matrix is the identity matrix of *3 × 3*, although some of its elements appear to be non-zero values, but are quite close to that.

11.2.3 EIGENVALUES, EIGENVECTORS AND DIAGONALS

Another concept related to matrix operations is that of *eigenvalues* and *eigenvectors,* which determine whether a particular matrix changes direction when multiplied by a specified vector. As an example, consider a square matrix A. Its eigenvector and eigenvalue will be the ones that make the following equation true: $AV = \lambda V$ where A is the original matrix, V is the eigenvector and λ is the eigenvalue. It is beyond the scope of this chapter to cover algebraic mathematics in any sort of detail. The reader can find such information on the multitude of related books and resources. For the purposes of this chapter, it should suffice to mention that the concept of eigenvalues and eigenvectors is useful in several transformation processes, including but not limited to computer graphics, physics applications, and predictive modelling.

Another notion that must be mentioned is that of a *diagonal*. It is often useful to find the diagonals above or below the *main diagonal* of a matrix. In the case of the former, a positive integer is suggested, whereas in the case of the latter a negative one.

The following script is a demonstration of how the concepts of eigenvalue, eigenvector, and diagonals are calculated and/or identified:

> **Observation 11.7 – Eigenvalue, Eigenvector:** Mathematical concepts that suggest whether a particular matrix changes direction when multiplied by a specified vector ($AV = \lambda V$).

```
1   import numpy as np
2
3   # Create a 2x2 array using the array function (Numpy) and
4   # display its contents
5   npArray = np.array([[1, 2], [5, 6]])
6   print("\nThe nparray's array's contents:\n", npArray)
7
8   # Create a 3x3 array using the array function (Numpy) and
9   # display its contents
10  newMatrix = np.array([[1, 3, 2], [5, 4, 8], [7, 6, 9]])
11  print("\nThe 3x3 matrix newMatrix2 is:\n", newMatrix)
12
13  # Display the diagonal for both arrays
14  print("The diagonal of the npArray is: ", np.diag(npArray))
15  print("The diagonal of the npArray above the main diagonal is: ",
16        np.diag(npArray, 1))
17  print("The diagonal of the npArray below the main diagonal is: ",
18        np.diag(npArray, -1))
19  print("The diagonal of the newMatrix is: ", np.diag(newMatrix))
20  print("The diagonal of the newMatrix above the main diagonal is: ",
21        np.diag(newMatrix, 1))
22  print("The diagonal of the newMatrix below the main diagonal is: ",
23        np.diag(newMatrix, -1))
24
25  # Calculate and display the Eigenvalue and Eigenvector for both arrays
26  eigenValueNpArray, eigenVectorNpArray = np.linalg.eig(npArray)
```

```
27  print("\nThe eigenvalues of the npArray are: \n", eigenValueNpArray)
28  print("\nThe eigenvectors of the npArray are: \n", eigenVectorNpArray)
29  eigenValueNewMatrix, eigenVectorNewMatrix = np.linalg.eig(newMatrix)
30  print("\nThe eigenvalues of the newMatrix are: \n",
31          eigenValueNewMatrix)
32  print("\nThe eigenvectors of the newMatrix are: \n",
33          eigenVectorNewMatrix)
```

Output 11.2.3:

```
The nparray's array's contents:
 [[1 2]
 [5 6]]


The 3x3 matrix newMatrix2 is:
 [[1 3 2]
 [5 4 8]
 [7 6 9]]
The diagonal of the npArray is:  [1 6]
The diagonal of the npArray above the main diagonal is:  [2]
The diagonal of the npArray below the main diagonal is:  [5]
The diagonal of the newMatrix is:  [1 4 9]
The diagonal of the newMatrix above the main diagonal is:  [3 8]
The diagonal of the newMatrix below the main diagonal is:  [5 6]

The eigenvalues of the npArray are:
 [-0.53112887  7.53112887]

The eigenvectors of the npArray are:
 [[-0.79402877 -0.2928046 ]
 [ 0.60788018 -0.9561723 ]]

The eigenvalues of the newMatrix are:
 [15.86430285+0.j         -0.93215143+0.84080839j -0.93215143-0.84080839j]

The eigenvectors of the newMatrix are:
 [[ 0.22516436+0.j          0.76184671+0.j          0.76184671-0.j         ]
 [ 0.60816639+0.j         -0.24748842+0.39196634j -0.24748842-0.39196634j]
 [ 0.76120605+0.j         -0.36476897-0.26766596j -0.36476897+0.26766596j]]
```

11.2.4 Solving Sets of Equations with Python

Python provides a convenient way to solve sets of equations by treating them as *matrices*. The idea behind this is to take a set of equations, produce the relevant matrices (i.e., one with the variable coefficients and one with the resulting values for each equation), and call the `solve()` method (*Numpy* library). Consider the following example of a set of three equations:

$$5x - 3y + 2z = 10$$

$$-4x - 3y - 9z = 3$$

$$2x + 4y + 3z = 6$$

Firstly, the following matrix of the variable coefficients is produced:

$$\begin{bmatrix} 5 & -3 & 2 \\ -4 & -3 & -9 \\ 2 & 4 & 3 \end{bmatrix}$$

Observation 11.8 – The `solve()` Method: A method that solves a set of equations using relevant, appropriately processed matrices.

This is followed by the matrix for their solutions:

$$\begin{bmatrix} 10, & 3, & 6 \end{bmatrix}$$

Finally, the `solve()` method is called, producing the respective solutions for x, y, and z:

```
1    import numpy as np
2
3    # Assume the following set of equations:
4    # 5x - 3y + 2z = 10
5    # -4x - 3y - 9z = 3
6    # 2x + 4y + 3z = 6
7    # Use solve() to solve the equations
8
9    # Create a 3x3 matrix based on the equations and and display contents
10   equations = np.array([[5, -3, 2], [-4, -3, -9], [2, 4, 3]])
11   results = np.array([10, 3, 6])
12   print("\nThe solution for x, y, and z is:\n",
13           np.linalg.solve(equations, results))
```

Output 11.2.4:

```
The solution for x, y, and z is:
 [ 3.90225564  1.46616541 -2.55639098]
```

11.2.5 GENERATING RANDOM NUMBERS FOR MATRICES WITH PYTHON

Sometimes it is useful to generate matrices with random numbers in order to evaluate models prior to using actual data. Through the *Numpy* library, Python provides several methods that offer such functionality. The following script can be divided into three distinct parts. In the first part, a 3×4 matrix is generated and filled with *0 s*. Next, another two matrices are generated and filled with *1 s and 20 s*, respectively. Finally, a 4×4 identity matrix is generated. In the second part, the script uses the `rand()` and `randn()` methods to generate numbers for the matrices, either through the regular random numbers generator or from the *Normal Gaussian Distribution* that has a mean of 0. In the third part, the script demonstrates the use of basic statistics methods from *Numpy,* including `mean()`, `var()`, and `std()` to calculate the mean, the statistical variance, and the standard deviation of the data, respectively:

Observation 11.9 – `rand()`, `randn()`, `mean()`, `var()`, `std()`: Some of the methods of the *Random* package of the *Numpy* library that provide basic descriptive statistical calculations on matrices.

```
1    import numpy as np
2
3    # Generate 3x4 matrices of zeroes, ones, 20s, and a 4x4 identity matrix
4    print("Generate a 3x4 matrix of zeroes\n", np.zeros((3, 4)))
5    print("\nGenerate a 3x4 matrix of ones\n", np.ones((3, 4)))
6    print("\nGenerate a 3x4 matrix of 20s\n", 20 * np.ones((3, 4)))
7    print("\nGenerate an Identify matrix 4x4\n", np.eye(4))
8
9    # Generate a random number, a 3x4 matrix of random numbers,
10   # a 3x4 matrix of random numbers from the Normal (Gaussian)
11   # Distribution (i.e., mean = 0), and a 4x4 matrix of random
12   # numbers between 5 and 15 from the Normal Distribution
13   print("\nGenerate a random number\n", np.random.random())
14   print("\nGenerate an array 3x4 with random numbers\n",
15         np.random.random((3, 4)))
16   print("\nGenerate an array 3x4 with random numbers from the Normal \
17         Distribution\n", np.random.randn(3, 4))
18   print("\nGenerate an array 4x4 with random numbers between 5 and 15\n",
19         np.random.randint(5, 15, size = (4, 4)))
20
21   # Generate an array of 10 items with random numbers from the
22   # Normal (Gaussian). Distribution and use it as a source for performing
23   # basic statistics
24   npArray = np.random.randn(10)
25   print("\nGenerate an array of 10 random numbers from the Normal \
26         Distribution\n", np.random.randn(10))
27   # Print the mean of the new array
28   print("\nThe mean of the new array is: ", npArray.mean(), )
29   # Print the variance of the new array
30   print("The variance of the new array is: ", npArray.var())
31   # Print the standard deviation (i.e., the square root of the variance)
32   print("The stdDev of the new array is: ", npArray.std())
```

Output 11.2.5:

```
Generate a 3x4 matrix of zeroes
 [[0. 0. 0. 0.]
 [0. 0. 0. 0.]
 [0. 0. 0. 0.]]

Generate a 3x4 matrix of ones
 [[1. 1. 1. 1.]
 [1. 1. 1. 1.]
 [1. 1. 1. 1.]]

Generate a 3x4 matrix of 20s
 [[20. 20. 20. 20.]
 [20. 20. 20. 20.]
 [20. 20. 20. 20.]]
```

```
Generate an Identify matrix 4x4
 [[1. 0. 0. 0.]
  [0. 1. 0. 0.]
  [0. 0. 1. 0.]
  [0. 0. 0. 1.]]

Generate a random number
 0.8435542056822151

Generate an array 3x4 with random numbers
 [[0.35570211 0.27618855 0.0541145  0.58001638]
  [0.20641101 0.48294052 0.92104823 0.61556587]
  [0.19491554 0.5713989  0.63918665 0.81824177]]

Generate an array 3x4 with random numbers from the Normal Distribution
 [[-0.24286997 -1.00451518  0.06104505 -1.85966171]
  [-0.47202171  0.01079039  0.03526387  0.44499205]
  [ 2.2395344   0.42076315  0.6505322  -0.6350833 ]]

Generate an array 4x4 with random numbers between 5 and 15
 [[ 7 13  7 12]
  [ 7  9  5  5]
  [12 12 10  8]
  [12  7  9 13]]

Generate an array of 10 random numbers from the Normal Distribution
 [ 0.80516765 -0.34184534 -1.01860459  1.55026532  1.52091946  0.68490906
  -0.07417641  1.35254549  0.21432432  0.29326124]

The mean of the new array is:  0.009347564051776013
The variance of the new array is:  0.6866073562925792
The stdDev of the new array is:  0.8286177383405325
```

11.2.6 PLOTTING WITH MATPLOTLIB

As it is already discussed in Chapter 8 on Data Analytics and Data Visualization and Chapter 9 on Statistics, Python offers libraries that effectively and efficiently address all types of charts that might be required by the analysis of data at hand. These include *Matplotlib* and *Scipy* and are widely used for Deep Learning as well. The following two scripts are a quick refresh of how to use these libraries to visualize/plot the results of the mathematical methods of the previous sections:

```
1    # Import the Numpy and Matplotlib libraries
2    import numpy as np
3    import matplotlib.pyplot as plt
4    # Plot inline alongside the rest of the results
5    # This is particularly relevant in Jupyter Anaconda
6    %matplotlib inline
7
8    # Plot a line as the sin of the values between 0 and 40
9    # with 4 different types of intervals
10   for i in range(1, 5):
11           A = np.linspace(0, 40, 20*i)
```

```
12          B = np.sin(A) + 0.2 * A
13          plt.plot(A, B)
14          plt.xlabel("Input"); plt.ylabel("Output")
15          titleShow = "Basics of Charts. Number of samples: " + str(20*i)
16          plt.title(titleShow); plt.show()
```

Output 11.2.6.a–11.2.6.d:

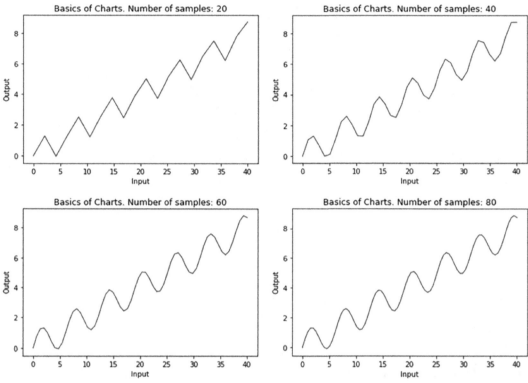

```
1    # Import the Scipy and Matplotlib libraries
2    from scipy.stats import norm
3    import matplotlib.pyplot as plt
4    # Plot inline alongside the rest of the results
5    # This is particularly relevant in Jupyter Anaconda
6    %matplotlib inline
7
8    # Create data points between -10 and 10, with 2000 intervals
9    x = np.linspace(-10, 10, 2000)
10   # loc is the mean and scale is the standard deviation
11   # Calculate the probability density function (Norm module/Scipy)
12   fx = norm.pdf(x, loc = 0, scale = 1)
13   # Plot the chart
14   plt.plot(x, fx); plt.show()
15   # Calculate the cumulative distribution function (Norm module/Scipy)
16   fx2 = norm.cdf(x, loc = 0, scale = 1)
17   # Plot the chart
```

```
18  plt.plot(x, fx2); plt.show()
19  # Calculate the log of the probability density function (Norm
20  # module/Scipy)
21  fx3 = norm.logpdf(x, loc = 0, scale = 1)
22  plt.plot(x, fx3); plt.show()
23  # Calculate the log of the cumulative distribution function
24  # (Norm module/Scipy)
25  fx4 = norm.logcdf(x, loc = 0, scale = 1)
26  plt.plot(x, fx4); plt.show()
```

Output 11.2.6.e–11.2.6.h:

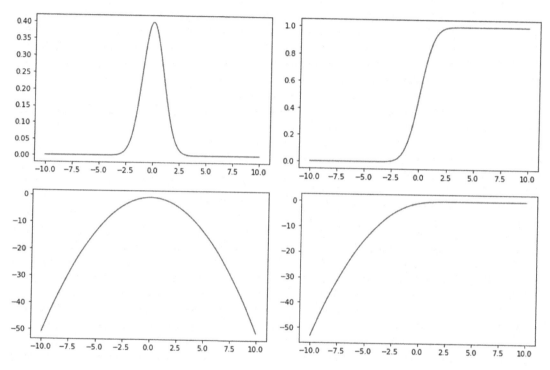

11.2.7 LINEAR AND LOGISTIC REGRESSION

Regression can involve either categorical or continuous variables. The input could be continuous, categorical, or discrete. If y shows the outcome and x shows the input, the model can be written as follows:

$y = F(x)$, where F is the *DL* model that suggests the relationship between input and output.

In the case of Linear Regression this model reveals a directly proportional relationship between input and output with some possible Regression coefficients (γ) of the various inputs (x) and the possibility of an error (φ) of the model calculations. Eventually, in the case of Linear Regression, the model can be written as follows:

$$y = F(x) = \gamma_0 + \gamma_1 x_1 + \cdots + \gamma_n x_n + \varphi$$

In the case of Logistic Regression (*LR*), the backbone of a *DL* Neural Network, the *DL* algorithm is used to classify the possible outputs as accurately as possible. The categories are encoded as either

0 or 1 and a sigmoid method is used to output a number between 0 and 1. The output is interpreted as a probability that the data is to be categorized as 1.

11.3 INTRODUCTION TO NEURAL NETWORKS

"Neural networks reflect the behavior of the human brain, allowing computer programs to recognize patterns and solve common problems in the fields of AI, machine learning, and deep learning."

(IBM Cloud Education, 2020)

The *artificial neural networks (ANN)* technique was inspired by the basics of human functioning. The main idea behind it is to interpret data through a series of multiple ML-based *perceptrons* (covered in detail in the next section), and *label* or *cluster* the input as required. Real world data such as images, sounds, time series, or other complex data are translated into numbers using *vectors*. ANN is quite helpful in classifying and clustering *raw* data even if they are unidentified and unlabelled. This is because it groups data based on similarities it observes or learns in its deeper layers, thus, transforming them into labelled training data, in a similar way the human brain does.

A *deep neural network* consists of one or more perceptrons in two or more layers (input and output). The perceptrons of each different layer are fed by the previous layer, using the same input but with different *weights*. The target of DL in ANN is to find correlations and map inputs to outputs. At a basic level, it extracts unknown features from the input data that can be fed to other algorithms, while also creating components of larger ML applications that may include classification, regression and reinforcement learning. It approximates the unknown method ($f(x) = y$) for any input x and output y. During learning, ANN finds the right method by evolving into a tuned transformation of x into y. In simple terms, this could represent methods like $f(x) = 7x + 18$ or $f(x) = 8x - 0.8$.

ANN performs particular well in clustering. It falls into the category of *unsupervised learning*, as it does not require labels to perform its tasks. It consists of the *input layer*, the *hidden layer(s)*, the *output layer*, the adjustable *weights* for model training and learning for all layers, and the *activation method*.

The *neuron* is the basic building block of a neural network. It is also known as the *linear unit* of the neural network system Figure 11.4.

In Figure 11.4 above, X is the input to the neuron and w is the weight. In its most basic form, the key for a neuron to be able to learn is the modification of value w. Y is

> **Observation 11.10 – Neuron:** The basic building block of a *neural network*, also called the *linear unit*. It learns by modifying the values of the *weights* of the inputs and adding up the sum of *inputs × weights* and the possible *bias* of the model.

the output and b the *bias* of the model. The bias is independent of the input and its value is provided with the model. The neuron sums up all the input values to come up with the equation that describes its model like a slope equation in linear algebra: $Y = wX + b$.

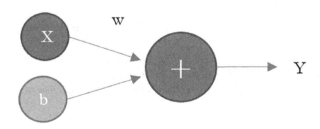

FIGURE 11.4 A typical neuron.

11.3.1 MODELLING A SIMPLE ANN WITH A PERCEPTRON

Figure 11.5 illustrates the method of a single neuron in a single layer (i.e., a *perceptron*). Its fundamental functionality is to mimic the behavior of the human brain's neuron. The idea is to take the inputs of the model ($x_1, x_2,..., x_n$) and multiply each by their respective weights ($w_1, w_2,..., w_n$), in order to produce the relevant k values ($k_1, k_2,..., k_n$). Often, a constant bias value multiplied by its associated weight is also added to this sum. Next, the sum of the k values is calculated and applied to the selected *sigmoid activation method*. Finally, the result is frequently *normalized* using some type of method as the unit step. A perceptron is also called a *single-layer* neural network because its output is decided based on the outcome of a single activation method associated with a single neuron. Figure 11.5 illustrates this model.

Class *FirstNeuralNetwork* presented below implements a basic perceptron (i.e., single-layer ANN). The implementation includes the following steps:

> **Observation 11.11 – Perceptron:** A single-layer neural network as its output is decided on a single activation method associated with a single neuron.

1. Generate and initialize a new object (named *ANN*) based on the *FirstNeuralNetwork* class, to initiate the perceptron model (lines 46 and 5–10). Instead of reading the weights from a data file, these are randomly generated as an array of *3 × 1* values, ranging from *−1* to *1*. The calculation uses the following formula: *(max−min) * randomset (lines × columns) + min*. Hence, in this case, the formula will be *(1−(−1)) * np. random.random((3, 1)) + (−1) = 2 * np.random.random(3, 1)−1*. The reader should keep in mind that by using the `seed()` method with a particular parameter, in this case *1*, the random sequence of numbers will always be the same. If it is preferred to have a different sequence of numbers every time the script runs, the seeding line should not be included.
2. Instead of reading the training inputs and outputs from a dataset, these are given as arrays of values (lines 49–52). Since the *dot* method will be used on the inputs and weights to calculate their sum, it is necessary that the number of columns of the former must match the number of lines of the latter (in this case *3*).

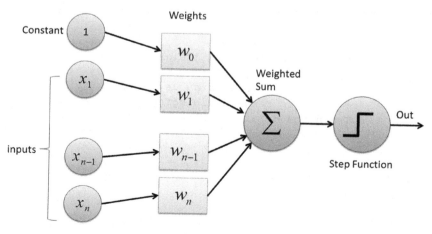

Fig : Perceptron

FIGURE 11.5 Perceptron.

3. Call the `Training()` method to train the model (line 56). For optimum training results, it is necessary to define the number of required iterations. The number is rather subjective; however, empirical experience suggests that a number of iterations between 10,000 and 15,000 is sufficient.
4. Use the `dot()` method to calculate the weighted sum of the inputs and their weights (lines 38–42).
5. Use the `Sigmoid()` method (lines 12–15) to calculate the output based on the result of the `dot()` method in step 4 (lines 41–42).
6. An optional step would be to calculate the training process error as the result of the training output (originally provided) – the calculated output. There are various ways to calculate this error, depending on the required level of accuracy. In this case, the error is calculated based on the last iteration of the training process (lines 28–36).
7. Another optional step would be to adjust the weights vector, based on the error calculated in the previous step (line 34).

```
1    import numpy as np
2
3    class FirstNeuralNetwork():
4
5        def __init__(self):
6            # Create a random number using the seed method
7            np.random.seed(1)
8            # Convert weights to a 3x1 matrix with values from -1 to 1 and
9            # a mean of 0 multiplied by 2
10           self.weights = 2 * np.random.random((3, 1)) -1
11
12       def Sigmoid(self, x):
13           # Use the sigmoid method to calculate the output
14           sigmoid =  1 / (1 + np.exp(-x))
15           return sigmoid
16
17       def SigmoidDerivative(self, x):
18           derivative =  x * (1 - x)
19           return derivative
20
21       def Training(self, trainingInputs, trainingOutputs,
22                       trainingIterations):
23           # Train the model for continuous adjustment of the weights
24           for iteration in range(trainingIterations):
25               # Train the data through the neuron
```

```
26                    output = self.NeuronThinking(trainingInputs)
27
28                    # Compute the error rate for back-propagation
29                    theError = trainingOutputs - output
30
31                    # Perform weight adjustments during the training phase
32                    theAdjustments = np.dot(trainingInputs.T,
33                        theError * self.SigmoidDerivative(output))
34                self.weights += theAdjustments
35            print("\nThe calculated error vector of the training process \
36                is: \n", theError)
37
38        def NeuronThinking(self, inputs):
39            # Pass the inputs through the neuron
40            inputs = inputs.astype(float)
41            output = self.Sigmoid(np.dot(inputs, self.weights))
42            return output
43
44  if __name__ == "__main__":
45        # Create an object based on the FirstNeuralNetwork neuron class
46        ANN = FirstNeuralNetwork()
47        print("Randomly Generated Weights:\n", ANN.weights)
48
49        # Train the data with 4 input values and 1 output
50        trainingInputs = np.array([[0,0,1], [1,1,1], [1,0,1], [0,1,1]])
51        print("\nThe training inputs:\n", trainingInputs)
52        trainingOutputs = np.array([[0],[1],[1],[0]])
53        print("\nThe training output:\n", trainingOutputs)
54
55        # Call the Training method to train the model
56        ANN.Training(trainingInputs, trainingOutputs, 15000)
57        print("\nThe adjusted weights vector is:\n", ANN.weights)
58
59        firstInput = str(input("\nProvide first input: "))
60        secondInput = str(input("Provide second input: "))
61        thirdInput = str(input("Provide third input: "))
62        print("The three inputs are: ", firstInput, secondInput,
63            thirdInput)
64        print("The new data is projected to be: ")
65        print(ANN.NeuronThinking(np.array([firstInput, secondInput,
66            thirdInput])))
```

Output 11.3.1: Test it with 1, 0, 0 and 0, 1, 0

Output test 1

```
Randomly Generated Weights:
  [[-0.16595599]
   [ 0.44064899]
   [-0.99977125]]

The training inputs:
  [[0 0 1]
   [1 1 1]
   [1 0 1]
   [0 1 1]]

The training output:
  [[0]
   [1]
   [1]
   [0]]
```

```
The calculated error vector
of the training process is:
  [[-0.00786416]
   [ 0.00641397]
   [ 0.00522118]
   [-0.00640343]]

The adjusted weights vector is:
  [[10.08740896]
   [-0.20695366]
   [-4.83757835]]

Provide first input: 1
Provide second input: 0
Provide third input: 0
The three inputs are:  1 0 0
The new data is projected to be:
[0.9999584]
```

Output test 2

```
Randomly Generated Weights:
  [[-0.16595599]
   [ 0.44064899]
   [-0.99977125]]

The training inputs:
  [[0 0 1]
   [1 1 1]
   [1 0 1]
   [0 1 1]]

The training output:
  [[0]
   [1]
   [1]
   [0]]
```

```
The calculated error vector
of the training process is:
  [[-0.00786416]
   [ 0.00641397]
   [ 0.00522118]
   [-0.00640343]]

The adjusted weights vector is:
  [[10.08740896]
   [-0.20695366]
   [-4.83757835]]

Provide first input: 0
Provide second input: 1
Provide third input: 0
The three inputs are:  0 1 0
The new data is projected to be:
[0.44844546]
```

11.3.2 SIGMOID AND RECTIFIER LINEAR UNIT (RELU) METHODS

Both *sigmoid* and *rectifier linear unit (ReLU)* are *activation methods* used in DL.

The *sigmoid* method is defined as: $\sigma(x) = \dfrac{1}{1+e^{-x}}$.
One of the drawbacks of the sigmoid method is that it slows down the DL process in case of big data inputs, as it takes time to make the necessary calculations. This is especially true when the input is a large number. For this reason, it is mostly used when its output is expected to fall in the range between 0 and 1, much like a probability output.

Observation 11.12 – The Sigmoid Method: It takes input values in a range and calculates the relevant output values given a specific formula. The output is always probabilistic ranging from 0 to 1. The method is slow with big data, and particularly with large numbers.

In most cases, the *ReLU* method is used instead. The concept of this method is simple: if the input value is higher than or equal to 0, it is returned as output unchanged; if it is lower, the method returns 0 as output. The method is particularly useful as it is rather fast, regardless of the input. The obvious problem with ReLU is that it ignores the negative input values, thus, not mapping them into the output.

The following script creates a sequence of input floats ranging from *−10* to *10*. Next, it calculates the outputs for each of the inputs using the sigmoid method and the outputs using ReLU. Finally, it plots the results of the inputs and outputs for both cases:

Observation 11.13 – The Rectifier Linear Unit (ReLU) Method: It takes input values in a range. For each input higher than or equal to 0 it results in the same value as the input. For each input value lower than 0, it results in 0. An important restriction with this method is that *it ignores negative values.*

```
1    # Import matplotlib, numpy and math
2    import matplotlib.pyplot as plt
3    import numpy as np
4    import math
5
6    # linspace(start, end) creates a sequence of integer input numbers
7    x = np.linspace(-10, 10)
8    print("The generated array of floats is: \n", x)
9
10   # Use the sigmoid function to calculate the output
11   sigmoid = 1/(1 + np.exp(-x))
12   print("\nThe calculated array of sigmoids is: \n", sigmoid)
13
14   # Create the Numpy array for the ReLU results & initialize with zeros
15   relu = np.zeros(len(x))
16   # Use the ReLU function to calculate the ReLU output based on the input
17   for i in range(len(x)):
18       if x[i] > 0:
19           relu[i] = x[i]
20       else:
21           relu[i] = 0.0
22   print("\nThe resulting array of ReLU is: \n", relu)
23
24   plt.plot(x, sigmoid)
25   plt.xlabel("x")
26   plt.ylabel("Sigmoid(X)")
27   plt.title("The sigmoid function for inputs -10 to 10")
28   plt.show()
29
30   plt.plot(x, relu)
31   plt.xlabel("x")
32   plt.ylabel("ReLU(X)")
33   plt.title("The ReLU function for inputs -10 to 10")
34   plt.show()
```

Output 11.3.2:

```
The generated array of floats is:
 [-10.           -9.59183673   -9.18367347   -8.7755102    -8.36734694
  -7.95918367   -7.55102041   -7.14285714   -6.73469388   -6.32653061
  -5.91836735   -5.51020408   -5.10204082   -4.69387755   -4.28571429
  -3.87755102   -3.46938776   -3.06122449   -2.65306122   -2.24489796
  -1.83673469   -1.42857143   -1.02040816   -0.6122449    -0.20408163
   0.20408163    0.6122449     1.02040816    1.42857143    1.83673469
   2.24489796    2.65306122    3.06122449    3.46938776    3.87755102
   4.28571429    4.69387755    5.10204082    5.51020408    5.91836735
   6.32653061    6.73469388    7.14285714    7.55102041    7.95918367
   8.36734694    8.7755102     9.18367347    9.59183673   10.          ]

The calculated array of sigmoids is:
 [4.53978687e-05 6.82792246e-05 1.02692018e-04 1.54446212e-04
 2.32277160e-04 3.49316192e-04 5.25297471e-04 7.89865942e-04
 1.18752721e-03 1.78503502e-03 2.68237328e-03 4.02898336e-03
 6.04752187e-03 9.06814944e-03 1.35769169e-02 2.02816018e-02
 3.01959054e-02 4.47353464e-02 6.58005831e-02 9.57904660e-02
 1.37437932e-01 1.93321370e-01 2.64947903e-01 3.51547277e-01
 4.49155938e-01 5.50844062e-01 6.48452723e-01 7.35052097e-01
 8.06678630e-01 8.62562068e-01 9.04209534e-01 9.34199417e-01
 9.55264654e-01 9.69804095e-01 9.79718398e-01 9.86423083e-01
 9.90931851e-01 9.93952478e-01 9.95971017e-01 9.97317627e-01
 9.98214965e-01 9.98812473e-01 9.99210134e-01 9.99474703e-01
 9.99650684e-01 9.99767723e-01 9.99845554e-01 9.99897308e-01
 9.99931721e-01 9.99954602e-01]

The resulting array of ReLU is:
 [ 0.           0.           0.           0.           0.           0.
   0.           0.           0.           0.           0.           0.
   0.           0.           0.           0.           0.           0.
   0.           0.           0.           0.           0.           0.
   0.           0.20408163   0.6122449    1.02040816   1.42857143   1.83673469
   2.24489796   2.65306122   3.06122449   3.46938776   3.87755102   4.28571429
   4.69387755   5.10204082   5.51020408   5.91836735   6.32653061   6.73469388
   7.14285714   7.55102041   7.95918367   8.36734694   8.7755102    9.18367347
   9.59183673  10.          ]
```

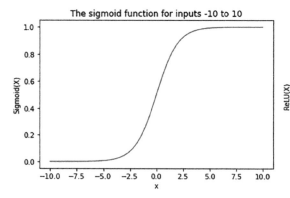

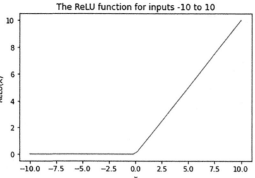

11.3.3 A Real-Life Example: Preparing the Dataset

The basic tasks when creating a multi-layer NN is to *create*, *compile* and *fit* the model, if necessary, *plot* the associated observations and data, and *evaluate* it. Among the most important concepts in DL are the *sequential model*, the *dense class*, the *activation class*, and adding *layers* to the model. A detailed analysis of these topics is beyond the scope of this chapter and the reader is encouraged to consider related sources specializing in DL. Nevertheless, a relatively common real-life example is examined in order to showcase and introduce some of the basic associated notions. This is split into a number of distinct steps, presented in the following sections.

> **Observation 11.14 – The `sample()` Method:** Use this *Pandas* method with the `frac` and `random_state` parameters, to define a sample from the original set to be used in the DL process.

The first step involves reading a dataset from a *CSV* file (*diabetes.csv*) and taking a random sample (i.e., 70%) of its rows to use as a training dataset (*frac* parameter). For the same input, the sample will also be the same, as a result of the `random_state = 0` parameter. Next, the index of the dataset is dropped, in order to keep only the remaining columns. Finally, the NN is optimized by scaling the dataset values to a range between 0 and 1:

```
1   import pandas as pd
2   import numpy as np
3
4   # Step 1: Read the csv file
5   MyDataFrame = pd.read_csv('diabetes.csv')
6   MyDataSource = MyDataFrame.to_numpy()
7
8   X = MyDataSource[:,0:8]
9   y = MyDataSource[:,8]
10
11  # Step 2: Use frac to split dataset to the train & test parts (70/30)
12  # Use random state to return the sample rows in every iteration
13  # Remove the index column from the dataset and print the first 4 rows
14  # Scale the dataset values to [0, 1] to optimize the NN
15  My_train = MyDataFrame.sample(frac = 0.7, random_state = 0)
16  My_test = MyDataFrame.drop(My_train.index)
17  print(My_train.head(4))
18  maxTo = My_train.max(axis = 0)
19  minTo = My_train.min(axis = 0)
20  My_train = (My_train - minTo) / (maxTo - minTo)
21  My_test = (My_test - minTo) / (maxTo - minTo)
22
23  # Split the features and the target
24  Xtrain = My_train.drop('Outcome', axis = 1)
25  Xtest = My_test.drop('Outcome', axis = 1)
26  Ytrain = My_train['Outcome']
27  Ytest = My_test['Outcome']
28  print("\nThe dataset contains", Xtrain.shape[0], "rows and",
29        Xtrain.shape[1], "columns")
```

Output 11.3.3:

```
     Pregnancies  Glucose  •••  Age  Outcome
661            1      199  •••   22        1
122            2      107  •••   23        0
113            4       76  •••   25        0
14             5      166  •••   51        1

[4 rows x 9 columns]

The dataset contains 538 rows and 8 columns
```

Number 8 in the output indicates the number of inputs, as the number of features in the dataset.

11.3.4 CREATING AND COMPILING THE MODEL

The next step involves the creation of four different models as a way to examine different scenarios. Firstly, the *Keras and Layers* libraries (*TensorFlow* package) are imported. These libraries are necessary in order to create the DL model and define its details. Next, the four models are created. *SimpleModel* consists of only the input and the output layers, with the former having just 12 neurons. *MakeItWider* doubles the number of neurons keeping the same basic layers. *MakeItDeeper* keeps the number of neurons the same as in the case of *SimpleModel,* but adds a third hidden layer between the input and the output. Finally, *FinalModel* defines a significant number of neurons per layer (a rather common case) and adds two layers between the input and the output.

In all four cases, the newly created DL models are created following the *sequential* approach. This simply means that each layer builds upon the input from the previous layer, thus connecting all layers to each other. The minimum number of layers in any DL model is 2: the input and the output. Any other layer is a *hidden* layer. There is no consensus as to what is the correct number of neurons per layer, although there are some suggested mathematical formulae on how to determine this number. As a rough guide, the reader should note that a number between 500 and 1,000 neuros per layer is commonly used. It must be also noted that the various layers in the NN do not have to consist of the same number of neurons.

The *activation* parameter defines the type of *stochastic gradient descent* used to optimize the weights of the model. In all four cases of this example, the *ReLU* method is selected. The optional `input_shape` parameter defines the number of features in the NN model (i.e., in this case *8*). This number defines the columns of the data set excluding the index (which is not used) and the output (i.e., the *outcome* column).

> **Observation 11.15 – Sequential Approach:** Each of the layers of the NN builds on the input from its previous layer, ensuring that all layers connected to each other.

Once the models are created, they must be *compiled*. Compilation basically deals with training and adjusting weights, and is often known as *backend processes*. It determines the best network representation for train/test and makes predictions on the specified hardware (i.e., either GPU or CPU). It also supports distributed computing such as *Hadoop/MapReduce*. At the moment of writing, *Theano* and *TensorFlow* are among the most commonly used libraries. In terms of the associated methods/parameters used in all four cases of this example, the `loss` method of choice is `mae`, the `optimizer` is `adam`, and the `metric` is `accuracy`. These methods/parameters are discussed in more detail in the following section.

The additionial part of the script is the following:

```
1    from tensorflow import keras
2    from tensorflow.keras import layers
3
4    # Step 3: Prepare the models for testing and compiling
5    # Prepare a simple model
6    SimpleModel = keras.Sequential([layers.Dense(12,
7        activation = 'relu'), layers.Dense(1)])
8    SimpleModel.compile(loss = 'mae', optimizer = 'adam',
9        metrics = ['accuracy'])
10   # Make the model wider by doubling the neuros of the layer
11   MakeItWider = keras.Sequential([layers.Dense(24, activation = 'relu'),
12       layers.Dense(1)])
13   MakeItWider.compile(loss = 'mae', optimizer = 'adam',
14       metrics = ['accuracy'])
15   # Make the model deeper by adding another layer
16   MakeItDeeper = keras.Sequential([layers.Dense(12, activation = 'relu'),
17       layers.Dense(12, activation = 'relu'),
18       layers.Dense(1)])
19   MakeItDeeper.compile(loss = 'mae', optimizer = 'adam',
20       metrics = ['accuracy'])
21   # Prepare the final model with many neuros and adding another layer
22   FinalModel = keras.Sequential([
23       layers.Dense(600, activation = 'relu', input_shape = [8]),
24       layers.Dense(600, activation = 'relu'),
25       layers.Dense(600, activation = 'relu'),
26       layers.Dense(1)])
27   FinalModel.compile(loss = 'mae', optimizer = 'adam',
28       metrics = ['accuracy'])
```

Notice that there is no output for the above script which serves as a preparation step.

11.3.5 STOCHASTIC GRADIENT DESCENT AND THE LOSS METHOD AND PARAMETERS

Stochastic gradient descent (SGD) is a family of algorithms aiming to optimize the weights for the best possible mapping of inputs to outputs. The selected algorithm is defined by the `optimizer` paramenter/method, which at present is most often `adam`.

The loss parameter/method deals with the measurement of the integrity of the NN predictions. In simple terms, it measures the *disparity* between *predicted* values and *desired* values. Several `loss` method options are available, including *mean square error (MSE), root mean square (RMS)*, and *mean absolute error (MAE)*.

MSE is amongst the most well-known methods of calculating the average (mean) of the differences between

Observation 11.16 – Stochastic Gradient Descent (SGD): A family of algorithms aiming to optimize the weights for the best mapping of inputs to outputs.

Observation 11.17 – Method `loss` Parameters: Select from a number of available mathematical methods to calculate the loss resulting from the process (e.g., *mean square error, root mean square,* and *mean absolute error*).

the *real observations* and the *predictions*. The mathematical equation for this particular method is the following:

$$\text{MSE} = \sum_{k=1}^{K} \frac{(x_i - x_i')^2}{K}$$

RMS is one of the most popular and, possibly, most accurate methods. It calculates the square root of the MSE. Its mathematical equation is the following:

$$\text{RMSE} = \sqrt{\sum_{k=1}^{K} \frac{(x_i - x_i')^2}{K}}$$

Finally, *MAE* is calculated as the mean of the absolute errors between the real and the predicted observations as in the following formula (i.e., x_k=true observations, x_k=predictions):

$$\text{MAE} = \frac{1}{K} \sum_{k=1}^{K} |x_k - x_k'|$$

The following script showcases the use of all three loss measuring methods discussed above:

```
1    import numpy as np
2
3    # Define the actual and the predicted values as np arrays
4    actual = np.array([1.8, 2, 1.9])
5    print("The actual observations are: \n", actual)
6    predicted = np.array([2, 1.7, 1.7])
7    print("\nThe predicted observations are: \n", predicted)
8    # Array calculated on the differences between the 2 sets of values
9    difference = predicted - actual
10   print("\nThe differences in the observations are: \n", difference)
11
12   # Calculate the array based on the squares of the differences
13   squareOfDifferences = difference ** 2
14   print("\nThe squares of the differences of the observations: \n",
15        squareOfDifferences)
16
17   # Calculate the mean square error for the observations
18   MSE = squareOfDifferences.mean()
19   print("\nThe Mean Square Error is calculated as: ", MSE)
20
21   # Calculate the mean of the square of the differences
22   meanSquareDifferences = squareOfDifferences.mean()
23   RMSE = np.sqrt(meanSquareDifferences)
24   print("\nThe root mean of square of differences is: ", RMSE)
```

```
25
26   # Calculate the mean of the absolute error of the differences
27   absoluteDifferences = np.absolute(difference)
28   meanAbsoluteDifference = absoluteDifferences.mean()
29   print("\nThe mean of the absolute differences of the observations \
30         is: ", meanAbsoluteDifference)
```

Output 11.3.5:

```
The actual observations are:
 [1.8 2.  1.9]

The predicted observations are:
 [2.  1.7 1.7]

The differences in the observations are:
 [ 0.2 -0.3 -0.2]

The squares of the differences of the observations:
 [0.04 0.09 0.04]

The Mean Square Error is calculated as:  0.056666666666666664

The root mean of square of differences is:  0.23804761428476165

The mean of the absolute differences of the observations is:
0.2333333333333333
```

11.3.6 FITTING AND EVALUATING THE MODELS, PLOTTING THE OBSERVED LOSSES

The next step involves the *fitting* of the various models, as well as the *plotting* of the relevant observations. The reader can follow the implementation of this step in the following script, taking note of the following:

1. For practical reasons, the number of iterations during model training is set to 5 (as defined by the epochs parameter). It must be noted that this is a quite small number to be truly efficient, but it is sufficient for demonstration purposes. In reality, this number is expected to be at least three digits long (i.e., between 100 and 1,000).

2. The fitting process investigates the training of the models with 300 rows of train data (shown in the batch_size).

3. The observations from the four different models are plotted together using the plot method (*Matplotlib.pyplot* library).

Observation 11.18 – The epochs Parameter: Used to define the number of iterations of the training set during the training/fitting step. Usually, the number is in the hundreds.

Observation 11.19 – The batch_size Parameter: Used to define the number of rows to be observed during the training/fitting step.

```
1   import matplotlib.pyplot as plt
2
3   # Step 4: Fit the models and plot the observations
4   # Fit the SimpleModel
5   print("\nThe observation epochs for the simple model: \n")
6   Observations1 = SimpleModel.fit(Xtrain, Ytrain, validation_data =
7       (Xtest, Ytest), batch_size = 300,  epochs = 5)
8   # Prepare the dataframe from the SimpleModel observation history
9   Observation1DataFrame = pd.DataFrame(Observations1.history)
10
11  # Fit the MakeItWider model
12  print("\nThe observation epochs for the wider model: \n")
13  Observations2 = MakeItWider.fit(Xtrain, Ytrain, validation_data =
14      (Xtest, Ytest), batch_size = 300,  epochs = 5)
15  # Prepare the dataframe from the MakeItWider observation history
16  Observation2DataFrame = pd.DataFrame(Observations2.history)
17
18  # Fit the MakeItDeeper model
19  print("\nThe observation epochs for the deeper model: \n")
20  Observations3 = MakeItDeeper.fit(Xtrain, Ytrain, validation_data =
21      (Xtest, Ytest), batch_size = 300,  epochs = 5)
22  # Prepare the dataframe from the MakeItDeeper observation history
23  Observation3DataFrame = pd.DataFrame(Observations3.history)
24
25  # Fit the FinalModel model
26  print("\nThe observation epochs for the final model: \n")
27  Observations4 = FinalModel.fit(Xtrain, Ytrain, validation_data =
28      (Xtest, Ytest), batch_size = 300,  epochs = 5)
29  # Prepare the dataframe from the FinalModel observation history
30  Observation4DataFrame = pd.DataFrame(Observations4.history)
31
32  # Plot the observations from the 4 models
33  plt.xlabel("Epochs")
34  plt.ylabel("Loss")
35  plt.title("History of observations of loss")
36  Observation1DataFrame['loss'].plot(label = "Simple model")
37  Observation2DataFrame['loss'].plot(label = "Make it wider")
38  Observation3DataFrame['loss'].plot(label = "Make it deeper")
39  Observation4DataFrame['loss'].plot(label = "Final model")
40  plt.legend()
41  plt.grid()
```

Output 11.3.6:

```
The observation epochs for the simple model:

Epoch 1/5
2/2 [==============================] - 3s 946ms/step - loss: 0.4762 - accuracy: 0.6450 - val_loss: 0.4424 - val_accuracy: 0.6652
Epoch 2/5
2/2 [==============================] - 0s 118ms/step - loss: 0.4627 - accuracy: 0.6450 - Val_loss: 0.4290 - val_accuracy: 0.6652
Epoch 3/5
2/2 [==============================] - 0s 122ms/step - loss: 0.4491 - accuracy: 0.6450 - val_loss: 0.4161 - val_accuracy: 0.6652
Epoch 4/5
2/2 [==============================] - 0s 106ms/step - loss: 0.4359 - accuracy: 0.6450 - val_loss: 0.4037 - val_accuracy: 0.6652
Epoch 5/5
2/2 [==============================] - 0s 96ms/step - loss: 0.4231 - accuracy: 0.6450 - val_loss: 0.3924 - val_accuracy: 0.6652

The observation epochs for the wider model:

Epoch 1/5
2/2 [==============================] - 2s 590ms/step - loss: 0.4248 - accuracy: 0.6450 - val_lose: 0.3855 - val_accuracy: 0.6652
Epoch 2/5
2/2 [==============================] - 0s 65ms/step - loss: 0.4069 - accuracy: 0.6450 - val_loss: 0.3767 - val_accuracy: 0.6652
Epoch 3/5
2/2 [==============================] - 0s 90ms/step - loss: 0.3956 - accuracy: 0.6450 - val_loss: 0.3717 - val_accuracy: 0.6652
Epoch 4/5
2/2 [==============================] - 0s 92ms/step - loss: 0.3890 - accuracy: 0.6450 - val_loss: 0.3705 - val_accuracy: 0.6652
Epoch 5/5
2/2 [==============================] - 0s 45ms/step - loss: 0.3856 - accuracy: 0.6450 - val_loss: 0.3715 - val_accuracy: 0.6652
```

The observation epochs for the deeper model:

Epoch 1/5
2/2 [==============================]- 1s 404ms/step - loss: 0.4124 - accuracy: 0.6506 - val_loss: 0.3940 - val_accuracy: 0.6652
Epoch 2/5
2/2 [==============================]- 0s 124ms/step - loss: 0.4054 - accuracy: 0.6487 - val_loss: 0.3866 - val_accuracy: 0.6652
Epoch 3/5
2/2 [==============================]- 0s 81ms/step - loss: 0.3985 - accuracy: 0.6487 - val_loss: 0.3796 - val_accuracy: 0.6652
Epoch 4/5
2/2 [==============================]- 0s 71ms/step - loss: 0.3927 - accuracy: 0.6468 - val_loss: 0.3731 - val_accuracy: 0.6696
Epoch 5/5
2/2 [==============================]- 0s 56ms/step - loss: 0.3855 - accuracy: 0.6468 - val_loss: 0.3672 - val_accuracy: 0.6696

The observation epochs for the final model:

Epoch 1/5
2/2 [==============================]- 2s 751ms/step - loss: 0.4379 - accuracy: 0.6450 - val_loss: 0.3437 - val_accuracy: 0.6652
Epoch 2/5
2/2 [==============================]- 0s 53ms/step - lose: 0.3767 - accuracy: 0.6450 - val_loss: 0.3955 - val_accuracy: 0.6609
Epoch 3/5
2/2 [==============================]- 0s 54ms/step - loss: 0.3973 - accuracy: 0.6506 - val_loss: 0.3819 - val_accuracy: 0.6609
Epoch 4/5
2/2 [==============================]- 0s 56ms/step - loss: 0.3785 - accuracy: 0.6468 - val_loss: 0.3543 - val_accurcy: 0.6652
Epoch 5/5
2/2 [==============================]- 0s 55ms/step - loss: 0.3521 -accuracy: 0.6450 - val_lose: 0.3510 - val_accuracy: 0.6652

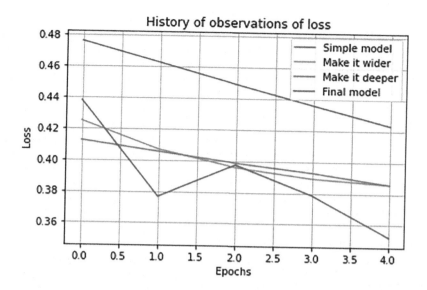

In the last step, the script evaluates the four models using the input and output testing datasets defined in the beginning of the process using the evaluate() method, and prints the relevant accuracy results for the models:

Observation 11.20 – The evaluate() Method: Used to calculate the accuracy of the suggested model through the test input and output datasets.

```
1   # Step 5: Evaluate the accuracy of the models
2   # Evaluate the accuracy of the Simple model
3   Accuracy1 = SimpleModel.evaluate(Xtest, Ytest)
4   print('Measured Accuracy for SimpleModel: %2.2f' % (Accuracy1[1]*100))
5
6   # Evaluate the accuracy of the MakeItWider model
7   Accuracy2 = MakeItWider.evaluate(Xtest, Ytest)
8   print('Measured Accuracy for MakeItWider: %2.2f' % (Accuracy2[1]*100))
9
10  # Evaluate the accuracy of the MakeItDeeper model
11  Accuracy3 = MakeItDeeper.evaluate(Xtest, Ytest)
12  print('Measured Accuracy for MakeItDeeper: %2.2f' % (Accuracy3[1]*100))
13
14  # Evaluate the accuracy of the FinalModel model
15  Accuracy4 = FinalModel.evaluate(Xtest, Ytest)
16  print('Measured Accuracy for FinalModel: %2.2f' % (Accuracy4[1]*100))
```

Output 11.3.6.d:

```
8/8 [==================] - 0s 4ms/step - loss: 0.3924 - accuracy: 0.6652
Measured Accuracy for SimpleModel: 66.52
8/8 [==================] - 0s 2ms/step - loss: 0.3715 - accuracy: 0.6652
Measured Accuracy for MakeItWider: 66.52
8/8 [==================] - 0s 2ms/step - loss: 0.3672 - accuracy: 0.6696
Measured Accuracy for MakeItDeeper: 66.96
8/8 [==================] - 0s 4ms/step - loss: 0.3510 - accuracy: 0.6652
Measured Accuracy for FinalModel: 66.52
```

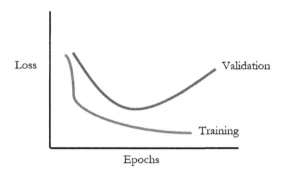

FIGURE 11.6 Validation vs. training.

11.3.7 MODEL OVERFIT AND UNDERFIT

Too much or too little learning is not good for any model. The optimum fit between the former (i.e., *overfit*), and the latter (i.e., *underfit*) must be found, although all models suffer from either one or the other to some extent. In simple terms, a model is *underfit* when *loss* is not low enough, which means that it does not have enough data to learn. The opposite is *overfit*, meaning that the model learned using too much data, which is wrongfully taken as good data. Figure 11.6 illustrates this concept:

> **Observation 11.21 – Overfit, Underfit:** A model is *overfit* when there is too much data used by the model to learn and much of it is not good. A model is *underfit* when there is not enough data to learn, and the *loss* is not low enough.

Finally, the reader must be aware of a common problem that must be addressed. Every DL model has a characteristic called *capacity*. This is the indication of size and complexity of the associated patterns. In the case of neural networks this relates to the number of neurons and layers. In order to overcome underfitting, the capacity of the model has to be increased.

TensorFlow provides a *callback* method for *early stopping*. As the name indicates, the method runs after every *epoch* during the training process.

The next script.

```
1   from tensorflow.keras.callbacks import EarlyStopping
2
3   # Define the minimum amount of change and the waiting time (epochs)
4   # before stopping
5   EarlyStopping1 = EarlyStopping(min_delta = 0.001, patience = 10,
6   restore_best_weights = True)
7
8   print("\nThe observation epochs for the new model: \n")
9   NewObservation = FinalModel.fit(Xtrain, Ytrain, validation_data =
10      (Xtest, Ytest), batch_size = 300, epochs = 400,
11      callbacks = [EarlyStopping1], verbose = 0)
12  Observation5DataFrame = pd.DataFrame(NewObservation.history)
13  Observation5DataFrame.loc[:, ['loss', 'val_loss']].plot();
```

Output 11.3.7:

The observation epochs for the new model:

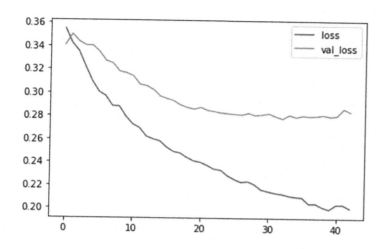

11.4 WRAP UP

While the concepts presented in the previous sections should suffice to provide a general idea of the basic structural components of neural networks, in reality, DL utilizes a number of different network types and topologies that expand and extend their complexity and functionality. Examples include, but are not limited to, the following:

- **Convolutional Neural Network (CNN):** A *regularized* variation of the multi-layer ANN, using a hierarchical approach to produce increasingly complex patterns. CNNs are commonly used for image processing.
- **Recurrent Neural Network (RNN):** As the name suggests, RNNs make use of *recurrent* loops and connections across the network components. This provides the ability to store data within the network and use them as a feedback mechanism to inform events. RNNs are particularly popular in machine translation applications.
- **Self-Organizing Map (SOM):** SOMs differ to other ANNs in terms of the approach taken towards data training (i.e., *competitive learning* instead of *error-correction learning*). The aim of such types of networks is to reduce the *dimensionality* of the input data.
- **Auto-Encoder:** Autoencoders aim at encoding data efficiently by ignoring *noise* that may be present in the signal or data. They are used for various applications like information retrieval, image processing and reconstruction, and popularity prediction.

As this chapter was meant to method as an introduction to some of the basic concepts behind neural networks, it does not explicitly cover the various types and topologies these can take within the DL context. The reader can find a wealth of information on these topics on the broad and extensive associated bibliography and online resources covering the theory and application of ML and DL algorithms.

11.5   CASE STUDY

1. Phase 1: Build the necessary logical gates (i.e., *AND, OR, XOR*) that will act as building blocks for the perceptron modeling in the ANN. Follow the guidelines below:
   a. User real numbers instead of integers.
   b. Implement $_iD$ vectors for weights (w) and inputs (x) as follows: $z = \vec{w}.\ \vec{x}$.
   c. Feed the sum of the vectors to the sigmoid activation method.
2. Phase 2: Implement the following tasks:
   a. Task 1: (Perceptron Engineering):
      – Write a sigmoid method.
      – Write a method to send values to the weights.
   b. Task 2: (Validating):
      – Develop a method/method to insert values for the weights.
      – Develop a method/method to provide a sample to the network.
      – Test the network with the XOR gate-based weights.
   c. Task 3: Implement a multi-layer perceptron (i.e., Class).
   d. Task 4: Develop a backpropagation algorithm by following the next steps:

      – Feed a sample to the network: $y = \begin{bmatrix} 0 \\ 1 \end{bmatrix}$.

      – Calculate the MSE: $MSE = \dfrac{1}{n}\sum_{i=0}^{n-1}\left(y_i - \text{Output}_i\right)^2$

      – Calculate the error terms of each Neuron's output:

      $$\sigma_k = \text{Output}_k *\left(1 - \text{Output}_k\right)*\left(y_k - \text{Output}_k\right)$$

      $\text{Output}_k *\left(1 - \text{Output}_k\right)$, is derivative of sigmoid function

      – Repeatedly compute the error terms in the hidden layers:

      $$\sigma_{\text{Hid}} = \text{Output}_{\text{Hid}} *\left(1 - \text{Output}_{\text{Hid}}\right)* \sum_{k\in\text{outputs}} w_{k\text{Hid}}\sigma_k$$

      – Apply the *delta* rule:

      $$\Delta w_{ij} = \gamma *\sigma_i *x_{ij}$$

      – Adjust the weights for the best model outcome:

      $$w_{ij} = w_{ij} + \Delta w_{ij}$$

   e. Task 5: Validate the class.

REFERENCES

Hardesty, L. (2017). *Explained: Neural Networks – Ballyhooed Artificial-Intelligence Technique Known as "Deep Learning" Revives 70-Year-Old Idea.* MIT News Office. https://news.mit.edu/2017/explained-neural-networks-deep-learning–0414.

IBM Cloud Education. (2020). *Neural Networks.* IBM. ibm.https://wwwcom/cloud/learn/neural-networks.

12 Virtual Reality Application Development with Python

Christos Manolas
The University of York
Ravensbourne University London

Ourania K. Xanthidou
Brunel University London

Dimitrios Xanthidis
University College London
Higher Colleges of Technology

CONTENTS

12.1   Introduction...485
12.2   3D Video Game Engines and VR Development Platforms 487
12.3   Motion Trackers and Head Mounted Displays VS Keyboards, Mice and Display Screens.....489
12.4   The Vizard Environment and Creating the Graphics Window 490
12.5   Creating the 3D World 491
12.6   Collisions and Gravity 492
12.7   Creating Additional 3D Objects..495
12.8   3D (Cartesian) Coordinates and Basic Object Positioning 496
12.9   Euler Angles and Object Orientation...499
12.10   Absolute vs Relative Positioning...500
12.11   Creating and Positioning Multiple Objects through Lists 504
12.12   Using Prefabricated Animations... 510
12.13   Basic Movement.. 514
12.14   Basic Interaction 519
12.15   Integrating VR Hardware and Exporting a Standalone.exe File 523
12.16   Conclusion...525
12.17   Case Study 526
References...526

12.1   INTRODUCTION

The idea of artificially created immersive environments like the ones used in *Virtual Reality (VR)* has been around for a surprisingly long time, but the various incarnations of the technological systems used over the years have been largely limited to highly specialized and inaccessible to the general public contexts, such as simulation for training purposes and scientific research (Carlson, 2017; Rosen, 2008; Smith, 2010). Nevertheless, the rapid technological advances of the past decades made the technologies required for the exploration of this idea accessible and affordable to millions of people. Powerful computers, high quality 3D graphics, VR headsets, accurate motion trackers and sensors, and development software suitable for VR are no longer restricted to specialized computer

DOI: 10.1201/9781003139010-12

laboratories or simulation training facilities. Such technologies and tools are becoming increasingly affordable to anyone interested in exploring the possibilities offered by their use and integration, both as an end user and a VR developer.

A significant contributing factor to the rapid expansion of VR development over the past few years is also the fact that much of the development work can be done in existing 3D graphics and video game platforms (*game engines*), with little or no need for serious modifications of their core components and workflows. It is, thus, unsurprising that VR development shares a lot with 3D graphics and video games development, and it is true that someone with experience in the latter will find it a lot easier to make the transition compared to someone starting from scratch. At the same time, VR development is a unique medium with its own technical and creative characteristics and challenges. Therefore, although VR developers are bound to have significant exposure to video game platforms and 3D graphics technologies on their way towards mastering their craft, it is perfectly feasible to start their VR development journey without previous experience in these areas. The path of least resistance for such an entry to VR development would likely have the form of a platform that combines a simple development interface and a solid and computationally powerful base that can support the substantial demands of VR applications and hardware. *Vizard*, the software package used in this chapter is exactly this: a VR development platform offering a Python scripting interface that is built on top of the required low-level core classes and libraries used for the rendering and manipulation of the VR environment (WorldViz, 2019). This allows the aspiring VR developer to get exposure to fundamental concepts, principles, and techniques without having to engage with the more challenging and difficult to master programming languages like C++ or C# that are used in some of the leading commercial VR development platforms (Epic Games, 2019; Unity Technologies, 2019).

It must be noted that while this chapter focuses on the basic introduction of concepts related to the technical implementation of VR applications with Python, *the most important element of the VR experience is arguably an engaging and meaningful storyline or a key feature acting as the main focal point*. The understanding of the technological aspects of VR development without a matching understanding of the conceptual aspects of the process would be more suitable as an exercise rather than a viable VR development approach. It is, thus, recommended that once a basic understanding of the main aspects of the technical implementation is established, the reader should refer to the rich literature covering the conceptual and aesthetic implications of the emergence of VR technologies, and the challenges it has brought about for content developers.

This chapter firstly introduces some basic concepts and technologies related to VR and 3D games development, and then the focus shifts to the introduction of basic VR development tasks through Python scripting. In addition to providing an introduction to some of the basic Python commands used in the Vizard platform, the main aim of this chapter is to expose the reader to some essential conceptual and technical aspects of VR development and to the programming logic one may wish to adopt for this type of work.

A basic understanding of object-oriented programming principles and logic is assumed for the reader. As such, concepts like *class* and *function/method* structures, *instantiation, inheritance,* and *polymorphism* are not covered here, although they certainly are as relevant in this context as in any other object-oriented programming situation. Nevertheless, as Python scripting commands in the Vizard environment are relatively easy to use and self-explanatory, even readers without a solid understanding of these concepts should be able to follow the programming ideas presented in this chapter and make a start in VR development.

For clarity and compatibility purposes, the Python scripts developed in this chapter are loosely following some of the examples provided in the official Vizard documentation and online tutorials. This is in order a) to support an easy transition to the very detailed Vizard guides and API reference if the reader decides to delve deeper into the platform, and b) to utilize the basic collection of *prefabricated 3D objects* that come with the standard Vizard installation, thus avoiding possible inconsistencies, confusion, and folder structure or file format issues a random choice of assets could cause. The code presented in this chapter was written and tested in the 64-bit version of Vizard 7 (WorldViz, 2019).

12.2   3D VIDEO GAME ENGINES AND VR DEVELOPMENT PLATFORMS

Before starting to write the first lines of Python code, one needs to establish a basic understanding of the platforms and technologies that make VR possible. Although a thorough introduction of a complex and convoluted topic such as the technological foundation of VR systems is beyond the scope of this chapter, the reader may benefit from conceptualizing some basic ideas related to it. This brief introduction is especially aimed at those who have an interest in learning more about VR development but do not necessarily have a background on programming in the context of 3D graphics or video games development. Readers familiar with these concepts can start at later sections of this chapter, as required.

At the moment of writing, many VR developers use existing game development platforms that provide the essential tools and functionality for the creation and operation of the 3D environments, instead of building new ones from the ground up. This is especially true on a commercial level, where VR applications are commonly developed using advanced, industry-standard 3D game development

Observation 12.1 – 3D Game Engine: A software platform providing specialized tools, libraries, and interfaces for the development of video games and interactive audiovisual content.

platforms, or *game engines*, like Epic Games' *Unreal Engine* (Epic Games, 2019) and *Unity 3D* (Unity Technologies, 2019). Such platforms provide pre-built classes, functions/methods and tools specifically designed for working with 3D graphics, and specialized methods like *physics modelling* and *3D graphics rendering* (Dunn & Parberry, 2011). As 3D game engines support the manipulation of fully animated 3D graphics, they are also suitable for the development of applications for other types of media that rely on such features like VR (Glover & Linowes, 2019). Indeed, the line between 3D video games and VR is becoming increasingly blurred, as a large volume of commercial VR application development occurs in the context of video games.

A typical 3D game development environment provides access to both the visual model of the 3D world under development and the necessary tools and interfaces to work with it. For readers who have not worked with a game engine before, a screenshot of a project in the Unity 3D environment is provided in Figure 12.1. In this example, the main screen displays the *3D world* or *map* from a given angle (i.e., the *camera* or *viewpoint*), a list of all the objects that reside on the 3D world (top-left), hierarchical views of the file and folder structures (bottom and bottom-left), and a panel providing access to detailed settings and parameters for given objects within the 3D world (right). In video games and VR development, an *object* refers to anything that can be added to the 3D world, irrespectively of whether it is visible or not. For instance, in Figure 12.1, the hills and the ground are one single object (i.e., the 3D world), the vegetation is a series of separate objects based on the same template, and the selected little white symbol at the lower center is an audio object that can be triggered when the user enters the area outlined by the faint spherical outlines. The reader should note that everything that exists within the 3D environment is treated like a separate object, irrespectively of its size, complexity, and attributes.

Although game or VR development platforms usually provide extensive and elaborate tools for addressing many of the common development tasks through a visual interface, a significant amount of coding is also required for developing viable applications. As such, a programming interface exposing the required programming classes and functions/methods is also provided in these platforms. In the vast majority of cases, this is done by means of a scripting interface that supports internal or external coding *editors* and *compilers*. As an example, Figure 12.2 shows a script added as a *component* to an object in Unity 3D and opened for editing

Observation 12.2 – Camera or Viewpoint: In the 3D environment of a 3D engine, a camera or *viewpoint* represents the angle and position from which the user observes the environment. A given 3D environment may have many camera/viewpoint objects, but the user utilizes one of them at any given time. A good example of this is the typical 3D multiplayer video game, in which different users observe the same 3D environment through their own, individual cameras/viewpoints.

FIGURE 12.1 The Unity 3D development environment. (See Unity Technologies, 2019.)

FIGURE 12.2 Script created in Unity 3D and opened for editing in Visual Studio. (See Microsoft, 2019.)

using an *external editor* (Microsoft, 2019; Unity Technologies, 2019). The programming languages used for scripting in each development platform varies, depending on the underlying architecture and structure, and the intended audience. Nevertheless, irrespectively of the programming language and tools used, what is important to note is that one of the functions of the game engine is to integrate the compiled scripts into the 3D environment, and to assign them to the intended objects and functionality modules.

Despite the fact that 3D game development platforms are currently some of the biggest players in VR development, the increasing interest in VR as a separate and unique medium has led to the appearance of a number of dedicated tools specifically focusing on VR development. As with the 3D game development platforms discussed earlier, the structure and target audience of these platforms vary, and so do the supported programming languages. Although it is generally true that lower-level languages like C++ are better suited for computationally heavy tasks, such as working with animated 3D graphics, some of the dedicated VR development platforms aim at those who need to use them without delving into the details of low-level programming and the mathematical principles behind 3D animation or physics modelling. Vizard, the platform used in this chapter, is such a platform, providing a more accessible interface for VR development. This is where Python comes into the picture, as Vizard is one of the few platforms that use it explicitly for all the associated scripting tasks. Through the Vizard interface, Python scripts can be used to manipulate and control the appearance, structure, and behavior of the 3D environment, and the interaction of the user with the VR world. This is achieved through specialized libraries and methods that do all the necessary work for the translation of the Python commands to the appropriate lower-level languages and platforms.

12.3   MOTION TRACKERS AND HEAD MOUNTED DISPLAYS VS KEYBOARDS, MICE AND DISPLAY SCREENS

Two of the most obvious differences between VR applications and those utilizing regular screen displays, keyboards, and mice are a) the way input from the user is being received, and b) the method of delivery of the visual content. A *motion tracker* generally refers to a sensor that tracks the user's movements and translates them to numerical coordinates that are passed to the application. This essentially replaces traditional input devices like handheld game controllers, keyboards, and mice. Nevertheless, at a primary level, there is not much difference between receiving the necessary coordinate numbers by a motion tracker or a keyboard and mouse. It must be noted here that although working natively with motion trackers and through a keyboard or mouse are two distinct and frequently different processes, for exploratory and prototyping tasks like the ones presented in this chapter such differences are less crucial.

As with input devices, video game and VR applications also differ in the way the output is handled. In VR, a *Head Mounted Display (HMD)* projects the 3D environment images to each eye of the user using specialized short distance projectors instead of a screen (Glover & Linowes, 2019). Although the study of the mechanisms of vision and stereoscopic image projection are beyond the scope of this chapter, interested readers can find more information on seminal textbooks on the subject of *stereoscopic media*, such as Ray Zone's *Stereoscopic Cinema and the Origins of 3-D Film* (Zone, 2007) and Bernard Mendiburu's *3D Movie Making, Stereoscopic Digital Cinema from Script to Screen* (Mendiburu, 2012). As with motion trackers, the differences between monitoring the development work on a screen or through an HMD should be of secondary importance when it comes to demonstration and prototyping tasks, although admittedly there are significant differences in terms of how animators or graphic designers work when professional work is being produced. Although there is certainly a point where the requirement for displaying content through an HMD and receiving input from motion trackers becomes essential, for many of the basic steps of a VR project one can work using a regular computer screen, a keyboard, and a mouse. In the current context, all the concepts presented can be explored and tested without a strict requirement for working with specialized HMDs or motion trackers. Nevertheless, the process of connecting such devices through the dedicated interface provided by Vizard (*Vizconnect*) is briefly covered in the latter parts of this chapter. This is in order to allow the reader to potentially integrate VR hardware devices of their choice to the scripts developed over this chapter with minimal changes and amendments to the code.

Observation 12.3 – Head Mounted Display (HMD): A wearable, head-worn device projecting visual content directly to the eyes of the user from a close distance. It is usually equipped with *motion trackers* that provide positional information, so the display can adapt to the user's head and eye movements and render the visual content accordingly.

12.4 THE VIZARD ENVIRONMENT AND CREATING THE GRAPHICS WINDOW

Vizard is available on three different types of licenses that can be downloaded from the official company website at *WorldViz.com* (WorldViz, 2019). It offers a Python scripting interface and a selection of libraries and tools allowing for the navigation and management of the VR project(s). The Python scripts reside at the top right of the main window, while a project/file explorer window is provided on the left side, and a debugging and information panel on the bottom (Figure 12.3):

Unlike the game development platform examples presented earlier, the reader will notice that in Vizard there is no permanently visible 3D environment. This is because the 3D environment is *rendered in real time* on a separate window when the appropriate Python command is executed. As such, instead of the VR developer working directly on an existing 3D graphics design window

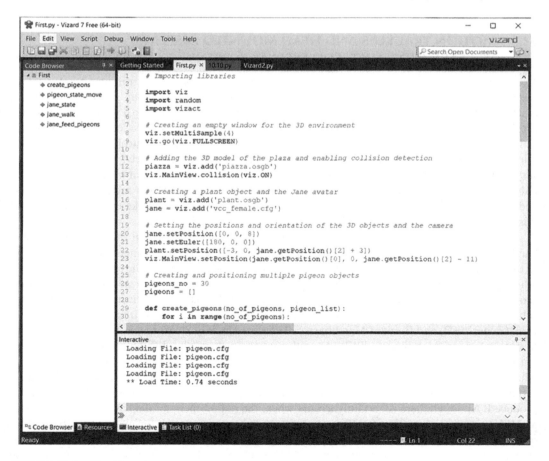

FIGURE 12.3 The Vizard scripting environment. (See WorldViz, 2019.)

as in Figure 12.1, a Python script must be run in order for the actual VR environment to show up. A new script can be created using the *File > New Vizard File* command from the application menu. The Python libraries required for the creation and manipulation of the VR environment are provided by Vizard. As in most other programming environments, such libraries need to be explicitly declared and imported to the script in order for the user to get access to their classes and objects. In Vizard, the viz keyword is used to declare the main library that enables the instantiation of the 3D world:

> **Observation 12.4 – The viz Library:** Import the viz library to a project to gain access to the methods needed to initiate and run the 3D world.

```
import viz
```

Once the library is imported, an empty 3D world can be created by using the viz.go() method:

```
viz.go()
```

Several aspects of the appearance and characteristics of the window within which the VR environment will be displayed can be also controlled by using the appropriate *flags*. For instance, the following command will create a full screen window as opposed to the default one (800×600 pixels):

> **Observation 12.5 – The viz.go() Method:** Use the viz.go() method to create the 3D world. Use the FULLSCREEN flag to create a full screen window.

```
viz.go(viz.FULLSCREEN)
```

Similarly, using the PROMPT flag will allow the user to select the presentation mode prior to launching the VR environment window:

```
viz.go(viz.PROMPT)
```

An exhaustive list of all the available options can be found in the Vizard reference and API documentation (WorldViz, 2019). At this stage, the developed script should look like the following:

```
1    # Import the viz library
2    import viz
3
4    # Create an empty window for the 3D environment
5    viz.go(viz.FULLSCREEN)
```

Pressing the *Run* button on the Vizard toolbar or *F5* on the keyboard will execute the script. As this is the first time the script is being executed, it must be saved before proceeding. Scripts can be saved on any location, but as with any other project, keeping the file and folder structure reasonably organized certainly helps in the long term. After running the script, the window hosting the VR application (the *graphics window*) will be launched. At this stage, an empty black screen will be displayed. The reason for this is that the script in its current form creates an empty 3D space *but does not fill this space with any visual content*.

12.5   CREATING THE 3D WORLD

The 3D environment within which the action will take place can be referred to as the *map* or the *3D world*. A map may consist of various integrated 3D objects, or designed as a single object irrespectively of how large and complex this object may be. It must be noted that, in most cases, the 3D objects and graphics used for 3D games and VR development are created outside the development platforms, in specialized 3D software like *3D Studio Max* and *Blender* (Autodesk, 2020; Blender,

2020). As such, the structure, appearance, and characteristics of the 3D objects are decided during the design phase by 3D graphics designers and animators. Once the design is completed, the objects are exported to suitable formats, ready for use in the 3D game and VR development platforms. Such predesigned objects are often referred to as *prefabricated* objects or *prefabs*. In this section, a prefabricated 3D model of a plaza will be used as the visual environment for the map of the VR application. The plaza 3D object is stored as a *.osgb* file, a 3D object filetype supported by Vizard (WorldViz, 2019). In Vizard, new 3D objects can be created or imported into the 3D environment using the `viz.add()` and `viz.addChild()` methods. For instance, the following command will load the prefabricated plaza object to the, initially empty, map:

```
piazza = viz.add('piazza.osgb')
```

The name of the *.osgb* file of the prefabricated object (`piazza.osgb`) must be declared in the argument list and passed to a newly created variable (`piazza`). For someone without a basic understanding of variables and instantiation, the structure of this command may be slightly confusing. This concept is discussed further in Section 12.7 Creating Additional 3D Objects. For the time being, it should suffice to mention that when new 3D objects are imported or created within the 3D world, *they must be allocated to newly created variables*, in this instance the variable named `piazza`.

> **Observation 12.6 – Map or 3D World:** The 3D environment where the action takes place.

> **Observation 12.7 – Prefabricated Objects (Prefabs):** Pre-designed 3D objects, such as environments or character models that the user of a game engine can import to a project instead of creating them from scratch.

By default, newly added 3D objects are positioned with their center point aligned to the center of the map. The camera (viewpoint) through which the user observes the map is also initially positioned at the center (Figure 12.4).

With the graphics window active, the user can move within the map by pointing the mouse towards the desired direction, while holding down the left mouse button. Alternative modes of movement and orientation adjustments are also available through the mouse by holding down a) the right mouse button or, b) both the left and right mouse buttons. These are the default navigation controls in Vizard. As expected, these controls can be modified and reallocated, and this topic is covered in more detail in Section 12.15 Integrating VR Hardware and Exporting a

> **Observation 12.8 – The add() and addChild() Methods:** Use the add() and addChild() methods to add new 3D objects to the map.

> **Observation 12.9 – Movement within the Map:** In Vizard, the default way of moving within the map is through the mouse.

Standalone.*exe* File. At this point, the default controls should be sufficient for testing the script and familiarizing with the 3D environment. Below is a version of the complete script developed in this section:

```
# Import the viz library
import viz

# Create an empty window for the 3D environment
viz.go(viz.FULLSCREEN)

# Add a prefabricated 3D model of the chosen environment
piazza = viz.add('piazza.osgb')
```

12.6 COLLISIONS AND GRAVITY

While navigating, the reader will notice that the camera can move freely through the boundaries of the 3D world (i.e., the walls of the buildings and the floor of the plaza). This is because the 3D model graphics that were imported to the empty 3D environment are just that: graphics. At this

FIGURE 12.4 3D plaza model through the main camera/viewpoint at the center of the map. (See WorldViz, 2019.)

stage, it has not been specifically declared that when the camera overlaps with the graphics it should be prevented from moving further in order to emulate a more realistic behavior. In 3D games and VR development, this is known as a *collision*, an event occurring at the space and time where one object (e.g., the camera) starts overlapping with another (e.g., the walls and/or the floor). The process of detecting when and where a collision occurs is commonly referred to as *collision detection*, and it is one of the key tools in the VR developer's toolkit. From a programming perspective, when two objects collide, it is frequently required that specific actions are performed, such as the objects bouncing off each other or a specific reaction or animation being triggered (e.g., a bomb exploding when touching the ground). Although there are many ways to deal with collisions through code, the Vizard environment simplifies things by providing methods and commands that take care of the collision detection tasks. One of the collision-related tasks that must be controlled in almost every VR application is *the interaction between the camera and the graphics of the 3D world*. In Vizard, one can do this by calling the `collision()` method of the camera object (`viz.MainView`):

> **Observation 12.10 – Collision and Collision Detection:** Terms referring to the detection of objects overlapping within the 3D world. Collision detection is used to trigger actions and events, and thus determine the behaviour of 3D objects as necessary.

```
viz.MainView.collision()
```

The reason for accessing collision by specifically including the path of `MainView` in the `viz` library (i.e., `viz.MainView`) is that there can be multiple different views in a single VR application and the VR developer may not want to enable collision detection for all of them at once. If the reader is not familiar with classes and object-oriented programming structures, it is sufficient to think of the `MainView` as the visual environment that is currently active (i.e., the 3D plaza) and `viz` as the window that hosts this environment.

The `collision()` method accepts as an argument either the `viz.ON` or `viz.OFF` flag and sets the state of collision detection between the camera and other objects in the 3D world accordingly:

```
viz.MainView.collision(viz.ON)
```

Once collision detection is switched on, the camera will not be able to pass through the graphics of the 3D world anymore. In this case, most of the hard work has been handled by the classes behind the `viz.MainView.collision()` method. Such tasks would typically include the precise measurement of the location of the surfaces of the 3D object, as well as the boundaries and

> **Observation 12.11 – Camera Collision On/Off:** Once collision detection is switched on, the camera will not be able to pass through the graphics of the 3D world anymore.

the geometry of each surface. However, at the level of abstraction the typical VR developer would normally operate, this is of academic interest rather than something that would be required on a practical level. What really matters is that one is familiar with the concept of collision detection and the use of relevant commands and methods like `viz.MainView.collision()`, as well as of the fact that the default state of collision detection for the camera in Vizard is `OFF` when a new 3D world is created.

The second issue the user will notice while navigating the 3D world at this stage of development is that the camera could be placed at any given point statically, both vertically and horizontally. An additional consideration one has to deal with when designing the behavior of the various objects is the presence (or absence) of *gravity*. This is especially important in VR development, as it is often the case that *the camera represents the viewpoint of the user*. The `viz.MainView.collision(viz.ON)` command, in addition to providing collision detection between the camera and the 3D objects, automatically sets the gravity feature for the camera on. By default, gravity is set at the rate of the gravitational power of the earth (i.e., 9.8 m/s²). The camera becomes subject to the gravitational pull, so it is prevented from floating statically, although the user can still navigate vertically while the mouse button is held down. If the mouse button is released while navigating above ground, the camera will start falling at a speed of 9.8 m/s² and will stop only when it collides with the ground surface. This is, of course, unless the user intentionally navigates outside the limits of the plaza and past the buildings, in which case the camera will keep on falling indefinitely into the void space outside the map.

> **Observation 12.12 – The `gravity()` Method:** Use the `gravity()` method to set up the gravity pull applied to objects within the 3D environment.

Although the automatic adjustment of the gravitational pull is a convenient feature, there are many instances in which the apparent gravity of the 3D world should be defined exclusively. For instance, if the 3D world is supposed to be located at a floating space station with minimal gravitational powers at play, the VR developer may wish to switch off the gravity altogether. Similarly, if the VR world is set on the moon, the gravity needs to be set at a lower level (i.e., 1.62 m/s²) than the default settings of 9.8 m/s² for the gravity on earth. In its simplest form, setting up the gravity for the 3D world is achieved by passing the desired gravity pull as a numerical argument (in m/s²) to the `viz.MainView.gravity()` method. For example:

```
viz.MainView.gravity(0)
```

will set off gravity, as it will be equal to zero, while:

```
viz.MainView.gravity(9.8)
```

will set gravity back to the rate expected on earth. Accordingly:

```
viz.MainView.gravity(1.62)
```

will emulate the effect of gravity the user would experience on the moon. Below is a version of the complete script for this section:

```
1   # Import the viz library
2   import viz
3
4   # Create an empty window for the 3D environment
5   viz.go(viz.FULLSCREEN)
6
7   # Add a prefabricated 3D model of the chosen environment
8   piazza = viz.add('piazza.osgb')
9
10  # Enable collision detection and gravity for the created 3D world
11  viz.MainView.collision(viz.ON)
```

Running this script should allow the user to move around the 3D world using the mouse controls without being able to pass through solid objects, while being also subjected to the standard gravitation pull of the earth.

12.7 CREATING ADDITIONAL 3D OBJECTS

For someone without prior knowledge of object-oriented programming, the concept of an object can be confusing. However, in the Vizard environment, objects should be much easier to recognize, as they usually represent tangible elements of the 3D world. As shown, the script used in the previous sections already includes the creation of an object:

```
piazza = viz.add('piazza.osgb')
```

In this instance, a new variable with the name piazza is declared and the *piazza.osgb* 3D model (prefabricated outside Vizard) is allocated to it through the viz.add() method. Thus, the 3D object that was created has the name piazza and is currently the only object that was explicitly declared and created in the script. Additional objects can be created and added to the 3D world using the same structure but different variable names. For example, the following command will create a new object named plant using the prefabricated *plant.osgb* model:

```
plant = viz.add('plant.osgb')
```

Objects of different file types can be imported in a similar way, irrespectively of how simple or complex they are in terms of their design. For instance, the following command:

```
jane = viz.add('vcc_female.cfg')
```

will allocate a new object to variable jane, using the *vcc_female.cfg* model. Files of type.*cfg* are used for *avatars*, 3D objects that represent characters and can have prefabricated behaviors and animations attached to them by design. Below is a complete version of the script developed in the section with comments:

```
1    # Import the viz library
2    import viz
3
4    # Create an empty window for the 3D environment
5    viz.go(viz.FULLSCREEN)
6
7    # Add the 3D model of the plaza and enable collision detection
8    piazza = viz.add('piazza.osgb')
9    viz.MainView.collision(viz.ON)
10
11   # Create a plant object based on the 'plant.osgb' model
12   plant = viz.add('plant.osgb')
13
14   # Create a female character object (Jane) based on the
15   # 'vcc_female.cfg' model
16   jane = viz.add('vcc_female.cfg')
```

Running the script will instantiate a new 3D environment, add the plaza model (3D world), turn on collisions and gravity for the camera, and add two new objects: a plant and a female character named Jane. Note that the new objects may not be instantly visible, as they are all created at the same place as the camera (i.e., the center point of the 3D world). Navigating slightly backwards will allow the user to see the objects.

12.8   3D (CARTESIAN) COORDINATES AND BASIC OBJECT POSITIONING

In the previous examples, the position of the plaza model, as well as the positions of the newly created objects and the camera itself, were determined by Vizard and automatically placed at the center point of the 3D environment. However, in most occasions, the VR developer would require objects to be instantiated at specific points. As VR development is done on a 3D environment, the position of objects in space is best described by a *three-dimensional (Cartesian) coordinate system* along three axes: *x*, *y*, and *z* (DQ, 2019; Dunn & Parberry, 2011). Vizard uses what is known as a *left-handed coordinate space*, as x defines the left-to-right axis, y the bottom-to-top axis, and z the back-to-front axis in relation to the center of the 3D world (Figure 12.5).

Any given position within the 3D world can be described using the three numbers corresponding to the x, y, and z axes. In programming terms, methods or commands handling the positioning of objects would normally accept these numbers as arguments. In Vizard, the structure of the Python command for positioning an object to a specific point is the following:

> **Observation 12.13 – 3D Cartesian Coordinate System:** A mathematical coordinate system describing the positioning of a given object in 3D space, using three mutually perpendicular axes (x, y, z).

```
<object>.setPosition([0,0,0])
```

The <object> keyword is a placeholder for the object of choice. For example, the following command:

```
jane.setPosition([0, 0, 8])
```

> **Observation 12.14 – The setPosition() Method:** Use the setPosition() method on any object to place it in a particular position relative to the center of the map.

will position Jane eight units to the front (axis z) of the camera when the 3D world is instantiated. Similarly, adding the following command:

```
plant.setPosition([-3, 0, 10])
```

FIGURE 12.5   3D Cartesian coordinates in Vizard. (See WorldViz, 2019.)

will position the plant three units to the left (axis x) and 10 to the front (axis z). Note that the starting point of the axes is the center of the 3D world. As such, a negative value on axis x (e.g., −*3*) will move the object to the left. Negative and positive values on other axes work in the same manner. Any number can be used for positioning if and when necessary, but one needs to be careful with the initial positioning of objects, as *collision detection and gravity features will not prevent an object from being created outside the boundaries of the 3D world.* In this case, the object would fall into the void. As an example of this, setting axis y to a negative value in the current script will position the object in the empty space under the floor of the plaza.

It must be noted that, although not explicitly defined, another object is automatically created when the script is run: the camera itself. In object-oriented programming, it is often the case that when something is essential for the functionality of the program it is created automatically behind the scenes, although manual access and control over this may be also given to the programmer. In the current context, this means that the camera can be also controlled and positioned, as any other object. Accessing different types of objects within the `viz` library hierarchy can take some time to master. Without getting into the details of how one works with multiple cameras and different views, Vizard allows the VR developer to access the default camera using the `viz.MainView.setPosition()` method. Adding the following line to the script will initiate the camera 10 units behind the center point of the 3D world on the z axis:

```
viz.MainView.setPosition([0, 0, -10])
```

By doing this, upon loading the 3D world, the user will appear to be 10 units to the rear of the center point, facing Jane and the plant near the fountain (Figure 12.6).

FIGURE 12.6   3D objects and camera starting positions. (See WorldViz, 2019.)

Below is a version of the complete script developed in this section:

```
1    # Import the viz library
2    import viz
3
4    # Create an empty window for the 3D environment
5    viz.go(viz.FULLSCREEN)
6
7    # Add the 3D model of the plaza and enable collision detection
8    piazza = viz.add('piazza.osgb')
9    viz.MainView.collision(viz.ON)
10
11   # Create a plant object and Jane
12   plant = viz.add('plant.osgb')
13   jane = viz.add('vcc_female.cfg')
14
15   # Set the positions and orientation of the 3D objects and the camera
16   jane.setPosition([0, 0, 8])
17   plant.setPosition([-3, 0, 10])
18   viz.MainView.setPosition([0, 0, -10])
```

12.9   EULER ANGLES AND OBJECT ORIENTATION

The orientation of objects within the 3D world is defined and controlled using a similar logic to positioning. Objects can be rotated around each of the three axes (x, y, z) of the 3D coordinate system. In mathematics, this is known as the *Euler Angles*, which describe the rotations around the three axes as the *yaw, pitch*, and *roll* (Ardakani & Bridges, 2010). The command used to adjust the Euler angles in Vizard is the following:

Observation 12.15 – Euler Angles and the setEuler() Method: *Euler Angles* are mathematical definitions that describe the rotations around the x, y, and z axes of a 3D coordinate system. These rotations are commonly referred to as the *yaw, pitch*, and *roll*. Use the setEuler(x, y, z) method to adjust the rotation of an object based on Euler Angles.

```
<object>.setEuler([0, 0, 0])
```

Note that the orientation is defined in terms of degrees, with each degree 'measuring 1/360th of a complete revolution' (Dunn & Parberry, 2011). Thus, the arguments passed to the command should be between *0* and *360*. Note that the degrees can also have negative values (i.e., *−360 to 0*). This dictates whether the rotation is calculated on a *clockwise* or *counter-clockwise* basis. Based on this, adding the following command:

```
jane.setEuler([180, 0, 0])
```

will cause Jane to turn by 180° clockwise on the y axis (yaw) and face the camera rather than the fountain at the center of the plaza. Similarly, positioning the camera 10 units to the left of the original position ([-10, 0, 0]) and adjusting its y axis rotation by 45° ([45, 0, 0]) will cause it to move closer to the corner of the plaza, while facing the area where the two 3D objects are from a certain angle:

```
viz.MainView.setPosition([-10, 0, 0])
viz.MainView.setEuler([45, -10, 0])
```

As mentioned, the use of positive or negative values dictates whether the angle is calculated on a clockwise or counter-clockwise basis. Changing the argument values of the setPosition() commands for the plant and the camera to the ones below:

```
plant.setPosition([-8, 0, 2])
viz.MainView.setPosition([-10, 0, 0])
```

and the values of the setEuler() command for the camera to the following:

```
viz.MainView.setEuler([45, 15, 0])
```

will cause the plant to be initiated two units in front of the camera, with the camera facing directly to it at a 15° downward angle (Figure 12.7).

Below is a version of the complete script developed in this section:

```
1    # Import the viz library
2    import viz
3
4    # Create an empty window for the 3D environment
5    viz.go(viz.FULLSCREEN)
6
7    # Add the 3D model of the plaza and enable collision detection
```

FIGURE 12.7 Viewpoint positioning and orientation. (See WorldViz, 2019.)

```
8    piazza = viz.add('piazza.osgb')
9    viz.MainView.collision(viz.ON)
10
11   # Create a plant object and Jane
12   plant = viz.add('plant.osgb')
13   jane = viz.add('vcc_female.cfg')
14
15   # Set the positions and orientation of the 3D objects and the camera
16   jane.setPosition([0, 0, 8])
17   jane.setEuler([180, 0, 0])
18   plant.setPosition([-8, 0, 2])
19   viz.MainView.setPosition([-10, 0, 0])
20   viz.MainView.setEuler([45, 15, 0])
```

12.10 ABSOLUTE VS RELATIVE POSITIONING

Positioning objects explicitly through static numbers may work in simple situations, but as the projects and tasks become more complex the VR developer will need to calculate positioning in relation to other objects rather than by means of hard-coded values. As an example based on the current version of the script, if the camera was to point at Jane from a short distance, the commands used for its positioning and orientation could have been changed to something like the following:

```
viz.MainView.setPosition([0, 0, 5])
viz.MainView.setEuler([0, 15, 0])
```

This would position the camera three units in front of Jane who is currently at [0, 0, 8] (i.e., *8−5 = 3*) and facing slightly downwards (15°). However, this way of working assumes that the VR developer knows the exact 3D coordinates for all objects at all times (or is willing to spend a significant amount of time trying different numbers like the author of this chapter frequently did while writing). This is obviously not an efficient way of working and, more often than not, the positioning of objects needs to be decided on the fly. Thus, the reader must become familiar with another key concept in VR and 3D content development: the distinction between *absolute* and *relative* positioning. So far, the values used as arguments for the setPosition() method were absolute values, in the sense that they referred to absolute positions within the 3D world. Relative positioning ignores the global 3D world coordinates and uses another object as a *reference point*, in relation to which the coordinate values are calculated. In order to do this, one needs to have a way for reading

Observation 12.16 – Absolute versus Relative Positioning: Terms describing what the reference for measuring an object's position within the 3D world is. *Absolute* positioning describes the position of the object within the 3D environment in absolute terms (i.e., in relation to the center of the 3D environment), while *relative* positioning describes its position in relation to another object.

Observation 12.17 – The get-Position() Method: Use the getPosition() method to read the current position of an object in terms of its Cartesian coordinates.

the *current position* of any given object. In Vizard, this can be done through the getPosition() method. As with setPosition(), getPosition() returns three numerical values corresponding to the three coordinate axes (x, y, z). However, as in this instance the values are being *read* instead of *assigned*, no argument passing is needed. The syntax of the command is the following:

```
<object>.getPosition()
```

and it returns all three coordinate numbers of the chosen object automatically. The values returned by getPosition() could be stored in a variable or passed on to another method or object. For those unfamiliar with basic programming concepts like storing values to variables or passing them to methods, it should suffice to mention that in Python a new variable that can hold the type of information returned by getPosition() can be created simply by declaring it with a unique name. In the current context, examples of this are the cases where Jane and the plant are created:

```
jane_pos = jane.getPosition()
```

This line of code will store the coordinates of Jane's current position in the newly created jane_pos variable. Note that this command should be added in the script after Jane has been placed on the desired position, otherwise it will return inaccurate numbers. For testing and debugging purposes, the reader can use the print() method to monitor the values stored in the variable:

```
jane_pos = jane.getPosition()
print("Jane's current position is", jane_pos)
```

The line will be printed on the *Interactive* window at the bottom panel of the Vizard environment (Figure 12.3).

One could also use the contents of jane_pos as a direct argument to another object. The following line will position the plant at the same place as Jane:

```
plant.setPosition(jane_pos)
```

TABLE 12.1

Coordinate Table and Indices (WorldViz, 2019)

Coordinates List	−10	0	5
Index = 0	−10		
Index = 1		0	
Index = 2			5

as the x, y, z coordinates passed on will correspond to Jane's. The assignment of all three coordinates at once may be handy at times, but the VR developer will frequently need to access and use the values of the three axes individually. Although familiarity with concepts like *arrays* and *indexing* is assumed for the reader, a brief explanation is provided below for those who may not feel entirely comfortable with them. The `jane_pos` variable stores the three numbers corresponding to the three axes as an *array* or, in Python terms, a *list,* an *organized collection of data values*. The values stored in a list can be accessed individually by using an *index*, an integer number defining the position of the stored value within the array (Table 12.1).

Note that, in programming terms, numbering frequently starts at *0* instead of *1*. As such, if one wanted to access the first value within the list, an index value of *0* should be used. In Python, indices can be used to pick values from lists using *square brackets* after the name of the variable that contains the list. For instance:

```
jane_pos[0]
```

will return the first of the three coordinate numbers stored in `jane_pos`. As with values being returned by `getPosition()` in earlier examples, the values acquired in this manner need to be stored in corresponding variables, or passed on as arguments to another command, object, or method. For example:

```
jane_x = jane_pos[0]
jane_z = jane_pos[2]
```

will store the x and z coordinate values to two newly created and appropriately named variables. Once the variables are created, replacing the plant positioning command with the following:

```
plant.setPosition([jane_x, 0, jane_z])
```

will position the plant on the same x and z axes as Jane. Calculations can be also performed directly on the variables, either before they are passed to `setPosition()` or in its arguments list. For instance, the following lines:

```
jane_pos = jane.getPosition()
jane_z = jane_pos[2]
cam_z = jane_z -3
viz.MainView.setPosition([0, 0, cam_z])
```

will position the camera three units in front of Jane, irrespectively of where Jane is positioned within the 3D world. After replacing the camera positioning commands with the lines above, the reader can try placing Jane back to the center point by changing the z axis value in the original positioning command:

```
jane.setPosition([0, 0, 0])
```

Running the script with these settings will cause Jane and the camera to move together in relation to each other. The same can be also achieved by skipping the janeZ variable allocation and doing the calculation directly in the arguments section:

```
jane_pos = jane.getPosition()
cam_z = jane_pos[2] - 3
viz.MainView.setPosition([0, 0, cam_z])
```

or by skipping the intermediate variable allocations altogether and working directly with the jane_pos command:

```
jane_pos = jane.getPosition()
viz.MainView.setPosition([jane_pos[0], 0, jane_pos[2] - 3])
```

Ultimately, one could do all the work in one single line:

```
viz.MainView.setPosition(jane.getPosition()[0], 0, jane.getPosition()[2] -3)
```

as the index value for the z axis for Jane can be read straight from jane.getPosition() without being passed to a variable. Any of the above variations could be used to achieve the same goal, and it is just a matter of programming experience, style, and preference which option should be chosen. Although the general consensus is that code should be as concise as possible (in which case the last version should be preferred), such decisions are always based on convenient compromises between efficiency, ease-of-use, and readability.

The above example demonstrates where the real power of relative positioning lies: the VR developer can lock the camera to an object without the need to know where it may be at any given moment. This is especially important in situations where objects are moving at randomly generated positions and directions within the 3D world during run-time. In such cases, it is impossible for one to know the position of the objects at the time the code is written.

Once the relationships between objects are established, elements of the 3D world can be moved around by modifying a minimal number of values, both during the design phase and at run-time. For instance, the relative distance of the camera from Jane can be modified simply by changing the number deducted from Jane's z axis position in the camera positioning command:

```
viz.MainView.setPosition(jane.getPosition()[0], 0, jane.getPosition()[2] -11)
```

Similarly, changing the coordinate values of the x and z axes when Jane's position is firstly defined in the script will not affect her relative distance from the camera. In this particular occasion this is rather handy, as changing the z axis coordinate during Jane's positioning to *8* again:

```
jane.setPosition([0, 0, 8])
```

will move her back to the fountain while the camera is automatically placed three units behind the center point (i.e., *8–11 = −3*). This would also create enough space for the new objects that will be created in the next section.

Below is a version of the complete script developed in this section:

```
1    # Import the viz library
2    import viz
3
4    # Create an empty window for the 3D environment
5    viz.go(viz.FULLSCREEN)
6
```

```
7   # Add the 3D model of the plaza and enable collision detection
8   piazza = viz.add('piazza.osgb')
9   viz.MainView.collision(viz.ON)
10
11  # Create a plant object and Jane
12  plant = viz.add('plant.osgb')
13  jane = viz.add('vcc_female.cfg')
14
15  # Set the positions and orientation of the 3D objects and the camera
16  jane.setPosition([0, 0, 8])
17  jane.setEuler([180, 0, 0])
18  plant.setPosition([-3, 0, jane.getPosition()[2] + 3])
19  viz.MainView.setPosition(jane.getPosition()[0], 0,
20  jane.getPosition()[2] - 11)
```

12.11   CREATING AND POSITIONING MULTIPLE OBJECTS THROUGH LISTS

Familiarity with the basic programming `for` loop structure is assumed for the reader. If the reader has no previous exposure to the use of loops in Python, Chapter 2 covers this topic in detail.

In its most basic form, the `for` loop repeats a task for a set number of times. For instance, the following statement:

```
for i in range(5):
    print('Iteration no.', i)
```

will print the number of the iteration the `for` loop goes through at run-time. The `range` keyword is just a way to tell the `for` loop to do five iterations (starting at *0*). When it comes to 3D objects, a `for` loop can be also used for their instantiation and positioning. For example, the following statement:

```
for i in range(5):
    pigeon = viz.add('pigeon.cfg')
    pigeon.setPosition([i, 0, 0])
```

will create five pigeons and position them one unit apart on the x axis along the center point of the map. Note that the pigeons will appear on the right side in relation to the camera once the 3D world is instantiated. The *pigeon.cfg* avatar used for the above example comes with the standard installation of Vizard, similarly to the *vcc_female.cfg* female character used for Jane.

Although using the `range` keyword can be certainly convenient on many occasions, it may not be the optimal choice for other common VR development tasks, such as the positioning of objects in *asymmetric* or *random* points within the 3D world. For dealing with such tasks, a slightly different way of using Python `for` loops can be utilized. Instead of using a predefined range, the conditional arguments can be structured as a *list*, exactly like the one used previously for positioning and orientation purposes. As with the previous list examples, the `for` loop uses an index to access the various values within the list. In this instance, this is not based on selection rather than on a *sequence*. For example, the following block of code:

```
for i in [0, 1, 2, 3, 4]:
    pigeon = viz.add('pigeon.cfg')
    pigeon.setPosition([i, 0, 0])
```

will have exactly the same result as the previous one that utilized the `range` keyword. Using the `for` loop has the disadvantage of requiring the conditional arguments to be passed on manually, so it is not optimal for generating or manipulating large numbers of objects. However, it provides the

VR developer with direct control over the values passed as arguments. For example, modifying the above statement in the following manner:

```
for i in [-5, -2, 1, 2, 6]:
    pigeon = viz.add('pigeon.cfg')
    pigeon.setPosition([i, 0, 0])
```

will position the pigeons at asymmetric, but *explicitly defined* positions on the x axis along the center point.

Another aspect one needs to deal with frequently when managing numerous instances of an object is the use and management of unique names or identifiers. Although in the example above the five pigeons were successfully instantiated and positioned, they all appear to have the same name: `pigeon`. This can be confusing for the VR developer. Although behind the scenes each pigeon has a unique identifier indeed, in the current form of the script there seems to be no direct way of manipulating individual pigeons after they are created. This type of control is essential, especially when it comes to making the 3D world look more realistic at subsequent stages of development. One of the tricks that enable one to deal with such issues is to create lists of 3D objects rather than separate ones. In Python, a new variable can be designated as a list by allocating square brackets to it during the declaration:

```
my_empty_list = []
```

This command will create an empty list that can be later populated with different types of items, including 3D objects. Pre-defined arguments can be also supplied to the list during the variable declaration:

```
my_list = [-8, 0, 6]
```

New values can be added to an existing list using the `append()` method. The appended value is passed as an argument. For instance, the following command:

> **Observation 12.18 – 3D Object Lists:** In the context of 3D engines or VR development platforms, 3D objects operate as regular OOP objects. As such, the user can create collections of objects using regular OOP structures, such as lists and arrays.

```
my_empty_list.append(7)
```

will add number 7 to `my_empty_list`. Similarly:

```
my_list.remove(6)
```

will remove number 6 from `my_list`. The reader can find more information on Python lists in Chapter 2. In the current context, the list structure will enable the dynamic manipulation of multiple objects, during or after their instantiation. For example, an empty list holding multiple instances of the pigeon can be created as a new list variable named `pigeons`. The `pigeons` variable will host as many pigeon objects as desired (within reason). Replacing the `for` loop of the previous example with the following lines:

```
pigeons = []

pigeon = viz.add('pigeon.cfg')
pigeon.setPosition([-1, 0, 0])
pigeons.append(pigeon)

pigeon = viz.add('pigeon.cfg')
pigeon.setPosition([1, 0, 0])
pigeons.append(pigeon)
```

will create two new pigeons, position them two units apart along the x axis (*−1* and *1*) and at the center point on the z axis (*0*), and add them to the `pigeons` list. This time, however, each of the pigeons can be explicitly accessed at any point after instantiation, using the *corresponding index value* on the `pigeons` list. For example, adding the following command to the previous example:

```
pigeons[0].setPosition([0, 0, 5])
```

will place the first of the two newly created pigeons closer to Jane along the z axis (as mentioned, an index value of *0* corresponds to the first object in the list).

A more efficient and condensed way of dealing with such tasks would be to use a `for` loop, both for the creation and positioning of multiple pigeons within the 3D world and for adding them to the list. For instance, replacing the last example with the following block of code:

```
pigeons = []

for i in [-5, -3, 2, 4, 5]:
    pigeon = viz.add('pigeon.cfg')
    pigeon.setPosition([i, 0, 0])
    pigeons.append(pigeon)
```

will create five pigeons and position them at asymmetrical but predefined points along the x axis from the center point of the 3D world, before finally adding them to the `pigeons` list. As discussed, access to individual pigeons within the `pigeons` list is now available for further manipulation. Adding the following lines to the script will reposition pigeons *1*, *4*, and *5* to the specified coordinates:

```
pigeons[0].setPosition([-1, 0, 6])
pigeons[3].setPosition([2, 0, 3])
pigeons[4].setPosition([4, 0, 4])
```

Finally, the positioning of the pigeons can be made to appear slightly more realistic by generating random values for the coordinates of the pigeons, rather than trying to do this explicitly like in the previous example. Python provides several methods for the generation of random values. For this example, the `uniform()` method from the `random` library is used. This method generates values within a given range. For instance:

Observation 12.19 – Randomization: In the context of interactive 3D content development, randomization commonly refers to the process of allocating artificially generated random sequences of actions to specific objects and/or items within object lists and arrays. This provides a sense of realism to the artificial 3D worlds.

```
import random
print('Random value', random.uniform(1, 10))
```

will print a random value between *1* and *10*. In terms of the pigeons example, `random.uniform()` can be used to randomly generate coordinate numbers for any of the three axes. For example, the following commands:

```
pigeons[1].setPosition([random.uniform(-5, 5), 0, 0])
pigeons[3].setPosition([0, 0, random.uniform(2, 7)])
pigeons[4].setPosition([random.uniform(-5, 5), 0, random.uniform(2, 7)])
```

will position the second pigeon from the pigeons list (*index = 1*) between *−5* and *5* units on the x axis from the center point, the fourth pigeon (*index = 3*) between *2* and *7* units on the z axis, and the fifth one (*index = 4*) somewhere within the area defined by *x=−5* to *5* and *y=2* to *7*. In reality, such tasks need to be automated further, especially in situations involving large numbers

Observation 12.20 – The `random.uniform()` Method: Use the `uniform()` method from the *random* library to generate random values within a given range.

of objects. The following example will create five pigeons, randomly position them along the x and z axes within the limits passed as arguments to `random.uniform()`, and append each pigeon to the pigeons list for further use:

```
import random

[...]

pigeons = []

for i in range(5):
    pigeon = viz.add('pigeon.cfg')
    pigeon.setPosition([random.uniform(-5, 5), 0, random.uniform(2, 7)])
    pigeons.append(pigeon)
```

The real value of this structure lies on the fact that the VR developer could use the exact same lines of code to create and position any number of pigeons. The only difference would be the value passed as an argument to the `range()` method.

As the script starts becoming slightly lengthier, this may be a good time to quickly bring up the notion of organizing the code through *functions*. This will be also necessary for the tasks covered in the next section. A basic level of understanding of functions in Python is assumed for the reader. However, a brief example is provided below for those not familiar with this aspect. A Python function can be declared by using the `def` keyword followed by a (unique) name and a colon. For instance:

```
def create_pigeons():
```

will declare the `create_ pigeons()` function. Commands can be added to the function in a similar manner as in the `for` loop case shown earlier. For instance, if one would like to group all the pigeon instantiation and positioning commands developed earlier under the `create_pigeons()` function, the structure could look like the one below:

```
def create_pigeons():
    pigeons = []
    for i in range(5):
        pigeon = viz.add('pigeon.cfg')
        pigeon.setPosition([random.uniform(-5, 5), 0, random.uniform(2, 7)])
        pigeons.append(pigeon)
```

Using this structure, instead of having to type all the commands needed to create new pigeons, the VR developer can merely call `create_ pigeons()` whenever needed. As an example, the following commands will create 15 pigeons (i.e., 3×5):

```
create_pigeons()
create_pigeons()
create_pigeons()
```

Alternatively, one could also use a `for` loop for the same task to further automate the workflow:

```
for i in range(3):
    create_pigeons()
```

However, this structure intentionally contains a logical error. As the variables declared inside a Python function have a *local* scope, the pigeons list will not be visible

Observation 12.21 – Passing 3D Objects as Arguments: In line with universal OOP programming principles, 3D objects and object lists can be passed to methods and functions as arguments. This allows VR developers to build programming structures that use these objects dynamically, based on the requirements of the VR application.

to the rest of the script. In other words, while the pigeons will be created and positioned as expected, the VR developer will not have control over individual pigeons outside the `create_pigeons()` function, which was one of the original reasons for doing all this work. To address this, one can create the `pigeons` list *outside* the function, so it is visible globally, and then *pass it to the function as an argument*:

```python
pigeons = []

def create_pigeons (pigeon_list):
    for i in range(4):
        pigeon = viz.add('pigeon.cfg')
        pigeon.setPosition([random.uniform(-5, 5), 0, random.uniform(2, 7)])
        pigeon_list.append(pigeon)

for i in range(3):
    create_pigeons(pigeons)
```

This is an important notion, as it dictates whether a variable should be declared inside or outside the function. More details on this topic can be found in Chapters 2 and 3, and in other sources covering the Python language in detail (Ascher & Lutz, 1999). Back to the current example, any other parameter that may need to be modified within the function can be passed as an argument too. For example, if the number of pigeons created inside the `create_pigeons()` function needs to be adjusted frequently, it can be passed as an argument like in the following example:

```python
pigeons = []

def create_pigeons(no_of_pigeons, pigeon_list):
    for i in range(no_of_pigeons):
        pigeon = viz.add('pigeon.cfg')
        pigeon.setPosition([random.uniform(-5, 5), 0, random.uniform(2, 7)])
        pigeon_list.append(pigeon)

create_pigeons(30, pigeons)
```

This structure will generate 30 randomly positioned pigeons (Figure 12.8). In a similar manner, the VR developer could have created any number of pigeons just by passing the desired number as a numeric argument to the `create_pigeons()` function.

In order to be able to manipulate the number of pigeons easily when the script becomes more complex, the pigeon number can be passed to a variable at the beginning of the block (or the script). This will allow for the adjustment of the number of pigeons globally, removing thus the need to check the details of each function or block of code if the number of pigeons needs to be changed later on:

```python
pigeons_no = 30

[...]

pigeons = []

def create_pigeons(no_of_pigeons, pigeon_list):
    for i in range(no_of_pigeons):
        pigeon = viz.add('pigeon.cfg')
        pigeon.setPosition([random.uniform(-5, 5), 0, random.uniform(2, 7)])
        pigeon_list.append(pigeon)

create_pigeons(pigeons_no, pigeons)
```

FIGURE 12.8 Multiple 3D object creation and positioning. (See WorldViz, 2019.)

As with most programming tasks, different ways of achieving the same result are available. In general terms, one should adopt the structuring and programming style that best suits their style or the project requirements, but this is something that is being developed over time rather than decided based on a few examples in a book chapter. Nevertheless, it must be mentioned that various alternative methods to achieve similar results with the examples above can be found on the tutorial pages of Vizard. Comparing different approaches will help the reader to start generating different programming ideas about the same task and developing a more creative mindset towards programming in general.

Below is a version of the complete script developed in this section, resulting in the 3D environment shown in Figure 12.8:

```
1    # Import libraries
2    import viz
3    import random
4
5    # Create an empty window for the 3D environment
6    viz.go(viz.FULLSCREEN)
7
8    # Add the 3D model of the plaza and enable collision detection
9    piazza = viz.add('piazza.osgb')
10   viz.MainView.collision(viz.ON)
11
12   # Create a plant object and Jane
13   plant = viz.add('plant.osgb')
14   jane = viz.add('vcc_female.cfg')
15
```

```
16   # Set the positions and orientation of the 3D objects and the camera
17   jane.setPosition([0, 0, 8])
18   jane.setEuler([180, 0, 0])
19   plant.setPosition([-3, 0, jane.getPosition()[2] + 3])
20   viz.MainView.setPosition(jane.getPosition()[0], 0,
21   jane.getPosition()[2] - 11)
22
23   # Create and position multiple pigeon objects
24   pigeons_no = 30
25   pigeons = []
26
27   def create_pigeons(no_of_pigeons, pigeon_list):
28       for i in range(no_of_pigeons):
29           pigeon = viz.add('pigeon.cfg')
30           pigeon.setPosition([random.uniform(-5, 5), 0,
31           random.uniform(2, 7)])
32           pigeon_list.append(pigeon)
33
34   create_pigeons(pigeons_no, pigeons)
```

12.12   USING PREFABRICATED ANIMATIONS

The 3D objects created so far in this chapter are completely motionless. Although the complexities of detailed object animation are best left to 3D animators and game physics programmers, there are a number of ways the VR developer can inject some life to the 3D world. Arguably, the easiest way to do this is by using the prefabricated animations that are commonly attached to the 3D objects during the design phase. In the simplest scenarios, the VR developer can access these animations simply by calling specific methods and commands. In the case of avatar objects like Jane, such animations can take the form of object *states*. The Vizard platform provides access to the various states of an object through the <object>.

Observation 12.22 – Prefabricated Animations: Prefabricated 3D objects can be static or animated. Although game engines provide the necessary tools for animating initially static objects, it is common for prefabricated objects to include certain animation behaviors by default. These are determined by the original creators of the objects (e.g., 3D artists, animators).

state() method. Different states for a given avatar are defined by a predetermined, unique numerical index value that is passed as an argument. For example, adding the following command to the script:

```
jane.state(4)
```

will make Jane applaud the user for bringing her to life. Similarly, changing Jane's state to the following:

```
jane.state(1)
```

will make her look around idly, waiting for further instructions.

As mentioned, in the case of avatar objects like Jane and the pigeons, animations can be stored as properties of the prefabricated object when it is designed. Object animations can be also programmed by the VR developer, but this topic exceeds the scope of this chapter. The reader can find more information on this topic on the detailed Vizard API reference (WorldViz, 2019).

Observation 12.23 – The state() Keyword: In Vizard, the state() keyword is used to describe, handle, and trigger the various different animations allocated to a given 3D object.

Nevertheless, a basic level of realism can be achieved by automating the states of an object, in order to make it behave in a seemingly natural manner. One way of performing such tasks is by using a timer to trigger specific states at predetermined intervals. Vizard offers two timer methods through the `vizact` library. In its simplest form, the timer will trigger an *event* at a predefined time. In this particular instance, this event will be Jane's *state*. Replacing the commands of the previous example with the following lines:

```
import vizact

[...]

def jane_state():
    jane.state(1)

jane.state(4)
jane_state_timer = vizact.ontimer2(3, 1, jane_state)
```

will automate Jane's actions by using a timed trigger. Jane will firstly appear to be applauding and three seconds later will switch to the waiting state. This structure firstly imports the `vizact` library that is necessary for using the `ontimer()` method. Next, it creates a function named `jane_state()` that changes Jane's state to the idle animation (*state 1*), and sets Jane's initial state to the applauding animation (*state 4*). Finally, it uses the `ontimer2()` method from the `vizact` library to trigger the jane_state function after three seconds. Note that the `ontimer()` method is allocated to the `jane_state_timer` variable. This is in order to have control over this particular timer if more timers are added to the script. The second argument in the arguments list of the `ontimer()` method dictates how many times the underlying task must be executed. Note that if an action is to be repeated indefinitely rather than for a set number of times, the `ontimer()` method can be used instead of `ontimer2()`. The `ontimer()` method behaves exactly like `ontimer2()`, but the second argument is missing, as the timer will repeat indefinitely. As mentioned earlier, changing an object's state indefinitely may be required in order to try and create a sense of realism. For instance, the VR developer may want to do so in order to create the illusion that computer-controlled characters have some sort of intelligence and make informed decisions. In the existing script, changing the last structure to the following:

```
def jane_state():
    x = int(random.choice([1, 9]))
    jane.state(x)

jane.state(1)
jane_state_timer = vizact.ontimer((random.randint(10, 20)), jane_state)
```

will make Jane's states randomly change between waiting (*state 1*) and looking for something or someone (*state 9*). This is decided by a variation of `random.choice()`, the command used in earlier examples that picks an element of the supplied list at random on every iteration. Similarly, if the `random.randint()` is used as an argument for the timer it will pick a random integer from the range provided in its argument list (e.g., 10–20 seconds). Note that Jane's state is initialized to *state 1* before the timer is called. This way, Jane starts at a preset animation rather than motionless.

Observation 12.24 – Timer: In OOP, and in programming in general, a *timer* represents an object that allows the programmer to execute specific commands at predefined times and/or intervals. In the context of game engines, the timer is a particularly important programming component, as it allows one to dictate and refine the apparent movements and actions of 3D objects within the 3D world.

The same logic can be also used in order to change the state of specific objects from an object list. For example, a similar structure could be used to change the state of a given pigeon from the `pigeon` list:

> **Observation 12.25 – The `ontimer()` and `ontimer2()` Methods:** Use the `ontimer()` or `ontimer2()` methods from the `vizact` library to repeatedly trigger a particular action.

```
def pigeon_state():
    x = int(random.choice([1, 3]))
    pigeons[4].state(x)
```

```
pigeon_state_timer = vizact.ontimer((random.randint(10, 20)), pigeon_state)
```

In this case, it is likely that the animation states should be changed for random pigeons from the list on every iteration of the timer rather than a given, preselected one. In order to do this, choosing a pigeon from the list needs to be randomized too:

```
def pigeon_state():
    x = int(random.choice([1, 3]))
    pigeons[random.randint(0, pigeons_no)].state(x)
```

```
pigeon_state_timer=vizact.ontimer((random.randint(10, 20)), pigeon_state)
```

Adding the above commands to the script will gradually make the pigeons change their state in a random order and at random times (Figure 12.9).

The `ontimer()` method offers another way of doing this. Instead of randomizing the selection from the list inside the function, the selected item can be passed as an argument. Replacing the last structure with the following:

FIGURE 12.9 Adding animation and movement to 3D objects. (See WorldViz, 2019.)

```
def pigeon_state(rand_pigeon):
    x = int(random.choice([1, 3]))
    rand_pigeon.state(x)
```

```
pigeon_state_timer=vizact.ontimer(1,pigeon_state,vizact.choice(pigeons))
```

will pick a random pigeon from the `pigeon` list when the `ontimer()` is executed, and will pass it to the `pigeon_state()` function as an argument with the name `rand _pigeon`. The `vizact.choice()` method automatically selects a random object from a list. Note that the `ontimer()` method allows the user to send arguments to the chosen function (`pigeon_state`) by adding them to the argument list. The advantage of this version of the structure over the previous ones is that the VR developer does not have to change anything in the code if the pigeons list is altered elsewhere in the script. For example, if one decides to create 300 pigeons rather than 30 when the `pigeons` list is firstly created, the `vizact.choice()` will adapt to the size of the list. Writing code that dynamically adjusts to possible changes of variables in other parts of the program may take some more work and thinking, but it is certainly worth the effort.

Having introduced the `random.choice()` method, it may be worth making a small amendment to the `create_pigeons` function discussed earlier. One can use this method to generate random states for the pigeons *when they are instantiated for the first time*. This will prevent the pigeons from being instantiated in a static state and having to wait until they are randomly animated by the `pigeon_state` function later on. Adding the following line after setting the position of a pigeon inside the `create_pigeons` function:

[...]

```
pigeon.state(random.choice([1, 3]))
```

[...]

will instantiate the pigeons on an animated state. At this phase, many pigeons may still move simultaneously in a rather artificial manner, but the amendments that will be made on the next section should help to significantly moderate this effect.

Below is a version of the complete script developed in this section:

> **Observation 12.26 – The `choice()` Method**: Use the `choice()` method from the `vizact` library to randomly select an object from a given object list.

```
1   # Import libraries
2   import viz
3   import random
4   import vizact
5
6   # Create an empty window for the 3D environment
7   viz.setMultiSample(4)
8   viz.go(viz.FULLSCREEN)
9
10  # Add the 3D model of the plaza and enable collision detection
11  piazza = viz.add('piazza.osgb')
12  viz.MainView.collision(viz.ON)
13
14  # Create a plant object and Jane
15  plant = viz.add('plant.osgb')
16  jane = viz.add('vcc_female.cfg')
```

```
17
18 # Set the positions and orientation of the 3D objects and the camera
19 jane.setPosition([0, 0, 8])
20 jane.setEuler([180, 0, 0])
21 plant.setPosition([-3, 0, jane.getPosition()[2] + 3])
22 viz.MainView.setPosition(jane.getPosition()[0], 0,
23 jane.getPosition()[2] - 11)
24
25 # Create and position multiple pigeon objects
26 pigeons_no = 30
27 pigeons = []
28
29 def create_pigeons(no_of_pigeons, pigeon_list):
30     for i in range(no_of_pigeons):
31         pigeon = viz.add('pigeon.cfg')
32         pigeon.setPosition([random.uniform(-5, 5), 0,
33         random.uniform(2, 7)])
34         pigeon.state(random.choice([1, 3]))
35         pigeon_list.append(pigeon)
36
37 create_pigeons(pigeons_no, pigeons)
38
39 # Initialize and randomize the animation states of Jane
40 def jane_state():
41     x = int(random.choice([1, 9]))
42     jane.state(x)
43
44 jane.state(1)
45 jane_state_timer = vizact.ontimer((random.randint(10, 20)),
46 jane_state)
47
48 # Initialize & randomize the animation states & movement of the pigeons
49 def pigeon_state(rand_pigeon):
50     x = int(random.choice([1, 3]))
51     rand_pigeon.state(x)
52
53 pigeon_state_timer = vizact.ontimer(1, pigeon_state,
54 vizact.choice(pigeons))
```

12.13   BASIC MOVEMENT

In addition to changing avatar object states, the vizact library offers an easy way to create complex avatar object movements, such as walking, turning, or moving different parts of the body of a character in a controlled manner. In spite of their technical and developmental complexity, such actions are similar to the positioning and orientation tasks discussed in previous sections, as they also use the Cartesian coordinates system to determine their direction and angle changes over time. For instance, the following command:

```
walk = vizact.walkTo([0, 0, 2])
```

Observation 12.27 – Animation vs Movement: Although prefabricated animations *animate* the object (e.g., provide a walking movement to a character's body), they may not *move* the object within the 3D world (e.g., the character gets a walking animation movement but remains at the same point in space). In most cases, animations and movement commands must be combined to create a realistic object behavior.

can be used to instruct a character to perform a (prefabricated) walking movement from their current position towards a specific point in the 3D world (i.e., two units from the center point on the z axis). Similarly, the following command:

```
turn = vizact.turn(150)
```

can be used to make a character change their orientation by 150° on the z axis (clockwise), with *0* being the *north* of the 3D world. Note that the designated `vizact` movement commands are passed to the `walk` and `turn` variables. In order for the movement to take place, these variables must be added to the desired object using the `<object>.addAction()` method. The following command:

```
jane.addAction(walk)
```

will cause Jane to perform the walking movement that has been passed to the `walk` variable. The following statements will make Jane walk to the left of the camera and turn in order to look at the position the `MainView` camera is located during instantiation:

```
jane_walk = vizact.walkTo([-3, 0, 2])
jane.addAction(jane_walk)
jane_turn = vizact.turn(150)
jane.addAction(jane_turn)
```

Movement actions can be also randomly determined and triggered. As an example, a new function named `pigeon_walk` can be created, based on the `pigeon_state` function:

> **Observation 12.28 – The `walkTo()`, `turn()` and `addAction()` Methods**: Use the `walkTo()` and `turn()` methods to prepare an object to move and/or turn in a specified manner. Use the `addAction()` method to enable the move or turn action.

```
def pigeon_walk(rand_pigeon):
    walk = vizact.walkTo([random.randint(-5, 5), 0, random.randint(0, 8)])
    rand_pigeon.addAction(walk)
```

The `pigeon_walk` function allocates a walking action to the `walk` variable. The coordinates of the movement are determined by generating random values for the x (*−5* to *5*) and z (*0* to *8*) axes. This is in order to create the illusion that the pigeons are moving randomly but, at the same time, to contain them within a predetermined area around the center of the 3D world. A new instance of `ontimer()` can be used to call the `pigeon_walk` function at random times and for random pigeons from the `pigeons` list:

```
pigeon_walk_timer = vizact.ontimer(random.randint(1, 3),
pigeon_walk, vizact.choice(pigeons))
```

The above structure will cause the pigeons to start moving around one by one and at random times.

Although the latest amendments will make Jane and the pigeons move randomly, a more careful examination will unveil that the walking animation occasionally stops while the characters are still in motion. This is more noticeable when the random times between changing states are shorter or when the number of objects is lower. The reader can set the number of pigeons to a very low number (e.g., *2–3*) and the `ontimer()` interval for the `jane_state_timer` to one or two seconds in order to observe this issue more comfortably:

[...]

```
pigeons_no = 30
```

[...]

```
jane_state_timer = vizact.ontimer(1, jane_state)
```

[...]

(*Note: Remember to change the values back to their original settings after this test otherwise Jane will behave erratically.*)

If a state is triggered while Jane or the pigeons are in walking mode they will appear to be floating. The reason for this inconsistency is that the `ontimer()` commands that were originally created for controlling the states of Jane and the pigeons through the `jane_state` and `pigeon_state` functions, are occasionally overriding the prefabricated walking animation *while the walking movement is still in place*. In order to address such issues, the logical order and the structure of the script must be controlled. For example, in the case of Jane's movement, the random state allocation can be disabled by stopping the respective timer:

[...]

```
jane_state_timer.setEnabled(0)
jane_walk = vizact.walkTo([-3, 0, 2])
jane.addAction(jane_walk)
jane_turn = vizact.turn(150)
jane.addAction(jane_turn)
```

[...]

The `setEnabled(0)` command stops `jane_state_timer` from triggering different states for Jane. It can be put back to action by passing value *1* as an argument.

Another way of addressing such overlaps would be to control the animation states and the movements of an avatar object under the same timer and structure rather than separate ones. In the current script, in the case of the pigeons, one may choose to unify the randomization of the states and the movement in one function. This will allow a more precise control over the state of each object at any given

> **Observation 12.29 – The `setEnabled()` Method**: Use the `setEnabled()` method to deactivate (0) or activate (1) the timer associated with the actions of an object.

time and will prevent the changing state from overriding a simultaneous movement of any given pigeon. For this sort of control, the VR developer will need to utilize a simple logical control structure using the *if...else* statement (Chapter 2). Modifying the existing `pigeon_state` function to the following:

```
def pigeon_state_move(rand_pigeon):
    random_switch = random.choice([1, 2])
    if (random_switch == 1):
        rand_pigeon.clearActions()
        x = int(random.choice([1, 3]))
        rand_pigeon.state(x)
    else:
        walk = vizact.walkTo([random.randint(-5, 5), 0, random.randint(0, 8)])
        rand_pigeon.addAction(walk)

pigeon_state_timer = vizact.ontimer(1, pigeon_state_move,
vizact.choice(pigeons))
```

will allow for the control of the state and movement of any given pigeon in a mutually exclusive manner, avoiding state/movement overlapping. A random value generator has been added to the start of the function. This is in order to determine whether the current iteration of the function will alter the state a) or the movement b) of the pigeon. Note that the `clearActions()` command has been added to the beginning of the if...else statement. This clears all previous actions, such as walking animations, that may have been added to this pigeon on previous iterations. This way, the pigeon will either change state or move to a new location, but the possibility of changing states while walking is now controlled. To improve clarity, the above structure can be placed just under the pigeon creation section.

The above solutions to this logical issue may not be the optimal or the most appropriate ones. However, they are provided in order to encourage the reader to start thinking creatively when it comes to structuring and controlling the order and the general logic of the actions of the various objects. This is an important part not only of VR development, but of coding in general, and it is too broad a subject to be analyzed in a single book chapter.

> **Observation 12.30 – The `clearActions()` Method**: Use the `clearActions()` method to cancel all previous actions associated with a particular object.

As a last comment for this section, once adding animation and movement to the 3D world, the user may notice some rendering *artefacts* and *aliasing* along the edges of the objects. Vizard provides the `viz.setMultiSample()` method in order to smoothen out such effects. Adding the following line:

```
viz.setMultiSample(4)
```

before the `viz.go()` command at the beginning of the script, will provide *anti-aliasing* processing while ren-

> **Observation 12.31 – Anti-Aliasing:** As the 3D environment consists of audiovisual reconstructions based on *samples* rather than *continuous* signals, distortions and artefacts can frequently occur in the form of *aliasing*. *Anti-aliasing* refers to the various techniques and tools addressing distortions and artefacts caused by sampling.

dering the 3D world. The number passed as an argument dictates the resolution of the anti-aliasing correction, and it can be adjusted according to the seriousness of the effect and the computational power of the system.

Below is a version of the complete script for this section:

```
1   # Import libraries
2   import viz
3   import random
4   import vizact
5
6   # Create an empty window for the 3D environment
7   viz.setMultiSample(4)
8   viz.go(viz.FULLSCREEN)
9
10  # Add the 3D model of the plaza and enable collision detection
11  piazza = viz.add('piazza.osgb')
12  viz.MainView.collision(viz.ON)
13
14  # Create a plant object and the Jane avatar
15  plant = viz.add('plant.osgb')
16  jane = viz.add('vcc_female.cfg')
17
18  # Set the positions and orientation of the 3D objects and the camera
```

```
19  jane.setPosition([0, 0, 8])
20  jane.setEuler([180, 0, 0])
21  plant.setPosition([-3, 0, jane.getPosition()[2] + 3])
22  viz.MainView.setPosition(jane.getPosition()[0], 0,
23  jane.getPosition()[2] - 11)
24
25  # Create and position multiple pigeon objects
26  pigeons_no = 30
27  pigeons = []
28
29  def create_pigeons(no_of_pigeons, pigeon_list):
30      for i in range(no_of_pigeons):
31          pigeon = viz.add('pigeon.cfg')
32          pigeon.setPosition([random.uniform(-5, 5), 0,
33          random.uniform(2, 7)])
34          pigeon.state(random.choice([1, 3]))
35          pigeon_list.append(pigeon)
36
37  create_pigeons(pigeons_no, pigeons)
38
39  # Initialize & randomize the animation states & movement of the pigeons
40  def pigeon_state_move(rand_pigeon):
41      random_switch = random.choice([1, 2])
42      if (random_switch == 1):
43          rand_pigeon.clearActions()
44          x = int(random.choice([1, 3]))
45          rand_pigeon.state(x)
46      else:
47          walk = vizact.walkTo([random.randint(-5, 5), 0,
48          random.randint(0, 8)])
49          rand_pigeon.addAction(walk)
50
51  pigeon_state_timer = vizact.ontimer(1, pigeon_state_move,
52  vizact.choice(pigeons))
53
54  # Randomize the animation states of Jane
55  def jane_state():
56      x = int(random.choice([1, 9]))
57      jane.state(x)
58
59  jane.state(1)
60  jane_state_timer = vizact.ontimer((random.randint(10, 20)), jane_state)
61
62  # Jane walking and turning actions
63  jane_state_timer.setEnabled(0)
64  jane_walk = vizact.walkTo([-3, 0, 2])
65  jane.addAction(jane_walk)
66  jane_turn = vizact.turn(160)
67  jane.addAction(jane_turn)
```

12.14 BASIC INTERACTION

At this point, the 3D world created over the previous sections should be showing some signs of life. However, one of the main elements that make VR special is still missing. As opposed to traditional linear immersive media, one of the fundamental characteristics of VR is that it is *interactive*. Interaction applies to many facets of the VR experience, such as the interaction between the user and the 3D world, between the various different 3D objects, or the way specific elements of the environment react to the user's input. Providing a concrete and well-designed interaction system to the user is a rather complicated process involving thorough planning, technical aptitude and understanding, and extensive prototyping and testing. In this section, a few basic, related ideas are presented, in order to help the reader understand the logic and the challenges behind the design of interaction systems for VR. More detailed information on such topics can be found in works that cover VR development more extensively, such as Sherman and Craig's *Understanding Virtual Reality: Interface, Application and Design* (Sherman & Craig, 2018) and Glover and Linowes' *Complete Virtual Reality and Augmented Reality Development with Unity* (Glover & Linowes, 2019).

In its most basic form, an interaction system may consist of simple keyboard or mouse commands, or user movements that trigger specific *events* and *behaviors*. In Vizard, one of the commands allowing the VR developer to handle such tasks comes in the form of the onkeydown() method from the `vizact` library. The method determines if a specific key has been pressed and executes any associated actions accordingly. As an example, one could use the keyboard to instruct Jane to dance. Before asking Jane to do so, it is a good idea to remove the following lines that automate Jane's walking and turning movements from the end of the previously created script:

Observation 12.32 – Interactive versus Linear Systems: One of the main characteristics of VR systems is that they are *interactive*. Essentially, this means that the interaction between the user and the 3D environment can affect the user experience, as opposed to *linear* system productions like film or television where the experience is largely pre-determined in terms of the audiovisual content.

```
jane_state_timer.setEnabled(0)
jane_walk = vizact.walkTo([-3, 0, 2])
jane.addAction(jane_walk)
jane_turn = vizact.turn(160)
jane.addAction(jane_turn)
```

Changing the deleted lines with the following:

Observation 12.33 – The onkeydown() and onkeyup() Methods: Use the onkeydown() and onkeyup() methods to trigger particular actions when a specified key is pressed and/or released.

```
vizact.onkeydown('d', jane.state, 5)
```

will make Jane dance when button '*d*' is pressed on the keyboard. Jane will revert to one of the randomized states dictated by the next `jane_state_timer` trigger (`jane_state` function), as the `jane_state_timer.setEnabled(0)` command used to stop this from happening is deleted.

The above is a rather trivial task and does not add much to the VR experience. One may want to replace the dance command once more with the following lines:

```
jane_state_timer.setEnabled(0)
walk_to_camera = vizact.walkTo([viz.MainView.getPosition()[0], 0,
viz.MainView.getPosition()[2] + 3])
jane.addAction(walk_to_camera)
```

This block of code will firstly stop `jane_state_timer` and then make Jane to walk towards the camera. The positional x and y coordinates for the `walkTo()` method are acquired through, and calculated based on, the relevant `getPosition()` list elements. If the VR developer prefers Jane to follow the camera only when a specific event takes place (e.g., if a specific button is pressed), the above lines can be put into a function and called on demand:

```
def jane_walk():
    jane_state_timer.setEnabled(0)
    walk_to_camera = vizact.walkTo([viz.MainView.getPosition()[0], 0,
    viz.MainView.getPosition()[2] + 3])
    jane.addAction(walk_to_camera)

vizact.onkeydown('w', jane_walk)
```

An element of independence in the way Jane behaves and reacts to the user's input can be also added by randomizing Jane's willingness to move to the point dictated by the user, adding a sense of realism to her behavior. This idea could be further expanded into a more elaborate interaction system based on Jane's seemingly randomized mood and the ability to control the level of politeness and patience of the user through input commands. This is an extremely brief introduction to another element the reader needs to be aware of: one can use simplistic or elaborate *artificial intelligence* techniques to create the illusion of realism in VR projects. Nevertheless, the creation of elaborate, highly realistic VR experiences of this level exceeds the scope of this chapter. For the time being, Jane's response to the user's walking request can be based simply on random chance:

```
def jane_walk():
    jane_state_timer.setEnabled(0)
    random_switch = random.choice([1, 2])
    if random_switch == 1:
        walk_to_camera = vizact.walkTo([viz.MainView.getPosition()[0], 0,
        viz.MainView.getPosition()[2] + 3])
        jane.addAction(walk_to_camera)
    else:
        jane.clearActions()
        jane_state_timer.setEnabled(1)

vizact.onkeydown('w', jane_walk)
```

In the `jane_walk` function, the timer that causes Jane to switch states is firstly disabled in order to avoid interfering with the user's input. Next, the `random.choice()` method is used to randomly generate a value of either *1* or *2*. If *1* is generated, Jane walks towards the camera, otherwise she refuses to follow the user's instruction and switches back to the normal cycle of changing states, as the `jane_state_timer` is enabled again. Note that `clearActions()` is used in order to stop Jane from walking before starting the timer again. If this was not done, the state changes while she is still in a walking movement would have caused her to appear floating.

In line with the concept and the theme of the 3D world under development, one more feature that could be added to the script is Jane's ability to feed the pigeons *upon request*. This type of task is interesting, as it involves both the interaction of the user with Jane and of Jane with the pigeons. Adding the following function to the end of the script:

```
def jane_feed_pigeons(no_of_pigeons):
    jane_state_timer.setEnabled(0)
    pigeon_state_timer.setEnabled(0)
    jane.clearActions()
    jane.state(15)
```

```
        for i in range(no_of_pigeons):
            walk = vizact.walkTo([jane.getPosition()[0], 0, jane.
            getPosition()[2]])
            pigeons[i].addAction(walk)

vizact.onkeydown('f', jane_feed_pigeons, pigeons_no)
vizact.onkeyup('f', jane_state)
```

will disable the timers controlling the random allocation of states and movements for Jane and the pigeons, clear any existing actions in place (e.g., walking), change Jane's state to *feeding* (*state 15*), and make the pigeons walk to Jane as long as she is feeding them. Note that an additional onkeyup() command has been added after onkeydown(), with the same key identifier (i.e., '*f*'). This allows the user to feed the pigeons as long as key '*f*' is being pressed, and reverts back to a regular state when it is released. Finally, adding the following line:

```
pigeon_state_timer.setEnabled(1)
```

to the beginning of the jane_state function will ensure that when Jane stops feeding the pigeons they will also return to their regular behavior. This is due to the timer turning on again.

Below is a version of the complete script developed in this section:

```
1    # Import libraries
2    import viz
3    import random
4    import vizact
5
6    # Create an empty window for the 3D environment
7    viz.setMultiSample(4)
8    viz.go(viz.FULLSCREEN)
9
10   # Add the 3D model of the plaza and enable collision detection
11   piazza = viz.add('piazza.osgb')
12   viz.MainView.collision(viz.ON)
13
14   # Create a plant object and the Jane avatar
15   plant = viz.add('plant.osgb')
16   jane = viz.add('vcc_female.cfg')
17
18   # Set the positions and orientation of the 3D objects and the camera
19   jane.setPosition([0, 0, 8])
20   jane.setEuler([180, 0, 0])
21   plant.setPosition([-3, 0, jane.getPosition()[2] + 3])
22   viz.MainView.setPosition(jane.getPosition()[0], 0,
23   jane.getPosition()[2] - 11)
24
25   # Create and position multiple pigeon objects
26   pigeons_no = 30
27   pigeons = []
28
29   def create_pigeons(no_of_pigeons, pigeon_list):
30       for i in range(no_of_pigeons):
31           pigeon = viz.add('pigeon.cfg')
32           pigeon.setPosition([random.uniform(-5, 5), 0,
```

```
33              random.uniform(2, 7)])
34              pigeon.state(random.choice([1, 3]))
35              pigeon_list.append(pigeon)
36
37  create_pigeons(pigeons_no, pigeons)
38
39  # Initialize & randomize the animation states & movement of the pigeons
40  def pigeon_state_move(rand_pigeon):
41      random_switch = random.choice([1, 2])
42      if random_switch == 1:
43          rand_pigeon.clearActions()
44          x = int(random.choice([1, 3]))
45          rand_pigeon.state(x)
46      else:
47          walk = vizact.walkTo([random.randint(-5, 5), 0,
48          random.randint(0, 8)])
49          rand_pigeon.addAction(walk)
50
51  pigeon_state_timer = vizact.ontimer(1,
52  pigeon_state_move, vizact.choice(pigeons))
53
54  # Randomize the animation states of Jane
55  def jane_state():
56      x = int(random.choice([1, 9]))
57      jane.state(x)
58
59  jane.state(1)
60  jane_state_timer = vizact.ontimer((random.randint(10, 20)), jane_state)
61
62  # Jane walking and behaviour
63  def jane_walk():
64      jane_state_timer.setEnabled(0)
65      random_switch = random.choice([1, 2])
66      if random_switch == 1:
67          walk_to_camera = vizact.walkTo([viz.MainView.getPosition()[0],
68          0, viz.MainView.getPosition()[2] + 3])
69          jane.addAction(walk_to_camera)
70      else:
71          jane.clearActions()
72          jane_state_timer.setEnabled(1)
73
74  vizact.onkeydown('w', jane_walk)
75
76  # Jane feeding the pigeons
77  def jane_feed_pigeons(no_of_pigeons):
78      jane_state_timer.setEnabled(0)
79      pigeon_state_timer.setEnabled(0)
80      jane.clearActions()
81      jane.state(15)
82      for i in range(no_of_pigeons):
83          walk = vizact.walkTo([jane.getPosition()[0], 0,
84          jane.getPosition()[2]])
```

```
85                pigeons[i].addAction(walk)
86
87  vizact.onkeydown('f', jane_feed_pigeons, pigeons_no)
88  vizact.onkeyup('f', jane_state)
```

12.15 INTEGRATING VR HARDWARE AND EXPORTING A STANDALONE.EXE FILE

It may seem slightly odd that the integration of some of the most essential elements of any VR experience, the *HMDs* and the *motion tracking controllers*, were left for last. Partially, this is due to the fact that a large part of the development, testing, and prototyping work presented here can be completed without them. This also allows anyone without access to such equipment to get exposed to VR development and follow the work done in this chapter. From a developmental perspective, assigning actions to the VR hardware is not hugely different to assigning actions to other input devices like a keyboard or a mouse. Similarly, experiencing the 3D world through the HMD, although certainly more immersive and enveloping for the user, does not show anything that is not already there while using the screen.

From a strategic perspective, connecting the VR hardware to an already developed VR prototype after the basic building blocks, functionality, and logic are put in place is also a valid option. Working long hours with an HMD on trivial tasks can cause increased discomfort and fatigue, and utilizing the motion sensors to test basic interaction and collision tasks makes no big difference compared to using keyboard and mouse commands. Once a basic structure is in place and the reader desires to start looking at the various development aspects in more detail, different pieces of VR hardware can be integrated to the existing projects, as required. When the VR application is intended to be used with various different VR hardware products, each of these products and their integration needs to be tested individually. This is another reason for choosing to start the exploration of VR development with a hardware-agnostic approach and to leave the hardware related decisions for the latter stages of the development cycle.

From a practical perspective, integrating the VR hardware to VR applications developed in platforms like Vizard should not pose a big challenge to the VR developer. Most modern VR development platforms provide ample support and tools for connecting devices from various VR hardware manufacturers. The connection of the devices is usually through a corresponding *Software Development Kit (SDK)* and the related collections of libraries, tools and middleware, and/or through a dedicated *Graphical User Interface (GUI)*. In all cases, the connection of the VR hardware to the development plat-

> **Observation 12.34 – VR Hardware:** Although an application designed for VR could be also run on a conventional computer system, the main point of creating it is for the user to be able to experience it as an *immersive* 3D experience. In order to do so, the application needs to be deployed to specialized VR hardware systems. These systems consist of *HMDs* or *3D display* setups, *motion trackers*, and other specialized input/output controllers.

> **Observation 12.35 – Software Development Kit (SDK):** As with most other OOP environments, VR development platforms rely on *SDKs* to add specialized functionality to the applications and make them compatible with third party software and hardware. This is particularly relevant during integration and deployment, in order to make the VR application compatible with the various different VR systems that are available at any given time.

form is not as troublesome and experimental as it used to be in the past, and VR hardware manufacturers provide detailed tutorials, manuals, and support for device connectivity. As such, the reader should have no major issues integrating their preferred VR hardware to their projects with minimal code adjustments. In Vizard, the connection of the VR hardware to the development platform is managed through a dedicated visual interface called *Vizconnect* (Figure 12.10).

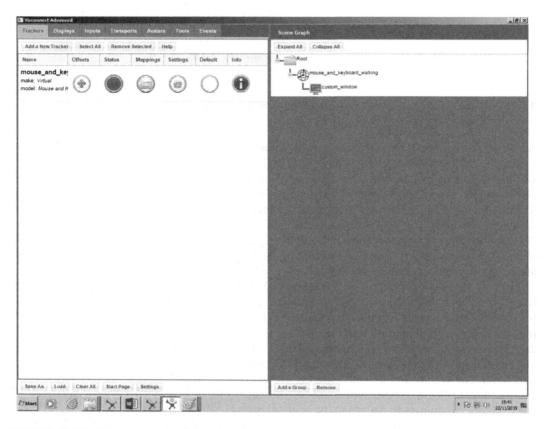

FIGURE 12.10 Vizconnect visual interface and hierarchical view of input and output devices. (See WorldViz, 2019.)

Vizconnect allows the user to assign the required HMDs, trackers and controllers, test functionality, and adjust or calibrate various system parameters. As a simple example, below are the first steps of basic display and tracker integration, in line with the official Vizard documentation (WorldViz, 2019):

- Open Vizconnect (*Tools > Vizconnect*).
- Vizconnect will ask the user to save the new configuration file. Save it under the name *viz_config_file.py*, on the same folder as the original script developed throughout this chapter.
- On the start-up screen select *Option 3 (Advanced Configuration)*.
- On the *Displays* tab create a new display and select *Custom Window*. Apply and Exit.
- On the *Trackers* tab select *Mouse and Keyboard Walking*. Apply and Exit.
- On the *Scene Graph* panel (right side), drag the *Custom Window* icon onto the *Mouse and Keyboard Walking* icon. This should add it as a child object (Figure 12.10).
- Exit Vizconnect and return to the original script.

Once the hardware is set, it can be accessed from the script by importing the `vizconnect` library:

```
import vizconnect
```

and replacing the `viz.go()` command with a link to the configuration file created through Vizconnect:

```
vizconnect.go('viz_config_file.py')
```

Observation 12.36 – The `vizconnect` Library: Use the `vizconnect` library to access the newly configured HMDs through the VR environment.

Once these changes take place, the first few lines of the script should look like the following:

```
# Import libraries
import viz
import random
import vizact
import vizconnect

# Create an empty window for the 3D environment
viz.setMultiSample(4)
vizconnect.go('viz_config_file.py')
```

[…]

The above adjustments will allow navigation in the 3D world using the directional arrows on the keyboard, and x-y-z rotation using the mouse. Although keyboards and mice are not VR hardware controllers in the strict sense, the basic process for integrating them to a script is the same as for dedicated VR hardware. The reader is encouraged to refer to the detailed Vizard tutorials to explore the multitude of options offered through the Vizconnect system. The ultimate decisions are down to VR hardware availability and compatibility, and the requirements of the VR project at hand.

Once the selected VR hardware is connected and integrated, the project can be exported as a standalone application (*.exe*) using the built-in wizard (*File > Publish as EXE*). The wizard collects and tests all the dependencies and necessary components and files for the project, so no extra adjustments, coding, or conversions are required. Note that adding VR hardware is not a requirement for exporting the application, so projects can be exported without such an integration, if necessary.

12.16   CONCLUSION

In a topic as complex and multifaceted as VR development, a book chapter like this can only scratch the surface and point the reader to some general, broad directions. Creating a polished and meaningful VR experience takes a lot more than just animating and randomizing a few objects and their behaviors. Thorough planning is required in order to understand the requirements and lay down all the details of the project long before a single line of code is written. Most importantly, *the VR experience needs a storyline, or a key feature that keeps the user interested and engaged.* Without this, even the most spectacular VR environment will lose its shine once the user is accustomed to the technological novelty.

The script developed in this chapter is meant to function as an introduction to some of the basic concepts and challenges VR developers deal with on a regular basis. In its current form, it creates a 3D world (map), adds 3D objects and avatars, provides movement functionality via prefabricated animations, and some basic interaction between the user (camera) and the 3D objects. At this stage, the script is nowhere near being completed or even being fully functional in a technical sense. Major issues are still present that need to be addressed, such as avatars missing collision detection and being able to walk through the 3D world, the 3D world being completely silent, the pigeons concentrating on the exact same spot during feeding, interaction choices being very limited, and a multitude of other movement, structural, and logical issues. Some of these are intentionally left unresolved so the reader can try to address them in the exercises provided below. For the more ambitious readers, Vizard provides access to numerous specialized Python methods and commands, allowing for a much more efficient and organized way to deal with tasks like the ones covered in this chapter. Automating tasks through *sequences, directors,* and *threads*, using time and execution management methods like `waittime()` and `waitkey()` instead of simple timers, using the elaborate *OPAL physics* system, and consolidating and structuring code in *classes* and *custom actions* are just some of the areas the reader may wish to explore (WorldViz, 2019). The abundance of tools and libraries and the extensive documentation provided by Vizard should be more than enough for starting the exploration of the finer details and challenges of VR application development.

12.17 CASE STUDY

Improve the existing script by adding the following features:

1. Allow Jane to feed the pigeons at random times without being instructed by the user. The feeding must stop automatically after a random amount of time has elapsed, and Jane should return to her normal *states* cycle.
2. Allow Jane to walk to random positions without the user's instruction. Make sure to restrict Jane's walking area so she does not get outside the 3D world boundaries or pass-through 3D objects.
3. Make the pigeons spread around while eating instead of concentrating at the same single point as Jane.
4. Make the pigeons walk faster towards Jane when feeding is triggered.
5. Make the pigeons change back to their normal states at random times after they have walked towards Jane for feeding.
6. Add a general ambience background sound to the 3D world.
7. Make Jane whistle once when each round of the pigeon feed commences. The whistle sound must emanate from the position Jane is at any given moment rather than being omnipresent. Note that for this sort of task the audio file needs to be in MONO rather than STEREO.
8. Add pigeon chirping sounds to the pigeons. The sounds must be allocated to random pigeons. As above, for this sort of task the audio file needs to be in MONO rather than STEREO.
9. Add some variety to the pigeons' chirping by adding two more chirping sounds and randomly switch between the three different sounds at run-time. Obviously, the three chirping sounds need to be distinguishably different to each other in order for this exercise to have practical value.
10. Add some randomness to the intervals at which the different pigeon chirps are triggered to improve realism.

REFERENCES

Ardakani, H. A., & Bridges, T. J. (2010). *Review of the 3-2-1 euler angles: a yaw-pitch-roll sequence.* Tech. Rep. Guildford: University of Surrey.
Ascher, D., & Lutz, M. (1999). *Learning Python.* Sebastopol, CA: O'Reilly.
Autodesk. (2020). *3DS Max -3D modelling and rendering software for design visualisation, games and animation.* https://www.autodesk.co.uk/products/3ds-max/overview?plc=3DSMAX&term=1-YEAR&support=ADVANCED&quantity=1.
Blender. (2020). *Our mission.* https://www.blender.org/about/.
Carlson, W. E. (2017). *Computer graphics and computer animation: a retrospective overview.* Columbus: Ohio State University.
DQ, N. (2019). *Cartesian coordinates.* http://mathinsight.org/cartesian_coordinates.
Dunn, F., & Parberry, I. (2011). *3D Math Primer for Graphics and Game Development.* Boca Raton, FL: Taylor and Francis.
Epic Games. (2019). *What is Unreal Engine 4?* https://www.unrealengine.com/en-US/what-is-unreal-engine-4.
Glover, J., & Linowes, J. (2019). *Complete Virtual Reality and Augmented Reality Development with Unity: Leverage the Power of Unity and Become a Pro at Creating Mixed Reality Applications.* Birmingham: Packt Publishing Ltd.
Mendiburu, B. (2012). *3D Movie Making: Stereoscopic Digital Cinema from Script to Screen.* New York: Routledge.
Microsoft. (2019). *Visual studio: best in-class tools for any developer.* https://visualstudio.microsoft.com/.
Rosen, K. R. (2008). The history of medical simulation. *Journal of Critical Care, 23*(2), 157–166.
Sherman, W. R., & Craig, A. B. (2018). *Understanding Virtual Reality: Interface, Application, and Design.* Cambridge, MA: Morgan Kaufmann.
Smith, R. (2010). The long history of gaming in military training. *Simulation & Gaming, 41*(1), 6–19.
Unity Technologies. (2019). *Unity 3D.* https://unity3d.com/unity.
WorldViz. (2019). *Vizard 6 (64-bit).* https://www.worldviz.com/vizard-virtual-reality-software.
Zone, R. (2007). *Stereoscopic cinema and the origins of 3-D film, 1838–1952.* Lexington: University Press of Kentucky.

Appendix
Case Studies Solutions

CHAPTER 2 – INTRODUCTION TO PROGRAMMING WITH PYTHON

DESCRIPTION

Write a *Python application* that displays the following menu and runs the associated functions based on the user's input:

1. Body mass index calculator.
2. Check customer credit.
3. Check a five-digit for palindrome.
4. Convert an integer to the binary system.
5. Initialize a list of integers and sort it.
6. Exit.

Specifics on the components of the application:

- **Body Mass Index Calculator:** Read the user's weight in kilos and height in meters, and calculate and display the user's body mass index. The formula is: BMI=(weightKilos)/ (heightMeters×heightMeters). If the BMI value is less than 18.5, display the message *"Underweight: less than 18.5"*. If it is between 18.5 and 24.9, display the message *"Normal: between 18.5 and 24.9"*. If it is between 25 and 29.9, display the message *"Overweight: between 25 and 29.9"*. Finally, if it is more than 30, display the message *"Obese: 30 or greater"*.

- **Check Department-Store Customer Balance:** Determine if a department-store customer has exceeded the credit limit on a charge account. For each customer, the following facts are to be entered by the user:
  - a. Account number.
  - b. Balance at the beginning of the month.
  - c. Total of all items charged by the customer this month.
  - d. Total of all credits applied to the customer's account this month.
  - e. Allowed credit limit.

    The program should accept input for each of the above from as integers, calculate the new balance (= beginning balance+charges−deposits), display the new balance, and determine if the new balance exceeds the customer's credit limit. For customers whose credit limit is exceeded, the program should display the message "Credit limit exceeded".

- A palindrome is a number or a text phrase that reads the same backward as forward (e.g., 12321, 55555). Write an application that reads a five-digit integer and determines whether or not it is a palindrome. If the number is not five digits long, display an error message indicating the issue to the user. When the user dismisses the error dialog, allow them to enter a new value.

- **Convert Decimal to Binary:** Accept an integer between 0 and 99 and print its binary equivalent. Use the modulus and division operations, as necessary.

- **List Manipulation and Bubble Sort:** Write a script that does the following:
  a. Initialize a list of integers of a Maximum size, where Maximum is entered by the user
  b. Prompt the user to select between automatic or manual entry of integers to the list
  c. Fill the list with values, either automatically or manually, depending on the user's selection
  d. Sort the list using Bubble Sort
  e. Display the list if it has less than 100 elements.

The above should be implemented using a single Python script. Avoid adding statements in the main body of the script unless necessary. Try to use functions to run the various tasks of the application. Have the application/menu run continuously until the user enters the value associated with exiting.

SOLUTION

```
1    # The package random allows to generate different types of random
2    # numbers
3    import random
4
5    def BMI():
6        # Collect inputs needed for the calculation
7        weight = float(input("Enter your weight in Kilogram: "))
8        height = float(input("Enter your height in Meters: "))
9        # Call the function calculateBMI sending the inputs as parameters
10       calcualteBMI(weight, height)
11
12   def CustomerCredit():
13       # Collect inputs needed for the calculation
14       balanceStart = int(input("Enter your balance at the begining \
15   of the month: "))
16       monthCharges = int(input("Enter the total amount of your \
17   charges this month: "))
18       monthCredits = int(input("Enter the total amount of your \
19   credits this month: "))
20       limit = int(input("Enter your Credit Limit: "))
21       # Call the function calculateCustomerCredit by sending the inputs
22       # as parameters
23       calculateCustomerCredit(balanceStart, monthCharges,
24           monthCredits, limit )
25
26   def Palindrome():
27       # Collect string input needed for the calculation
28       number = input("Enter a five-digit number:")
29       if (len(number) != 5):
30           print("The number entered is not with 5 digits ... \
31               Please try again")
32       calculatePalindrome(number)
33
34   def ConvertToBinary():
35       # Collect the integer number to be converted into binary
36       number = int(input("Enter an integer number between 0 and 99: "))
37       if (number < 0) or (number > 99):
```

```
38            print("The number entered is not within the range ... \
39                    Please try again")
40        calculateBinary(number)
41
42  def ArraySorting():
43      # Collect the needed input to sort the array's elements
44      maxSize = int(input("Enter the maximum size of your array: "))
45      filling = input("Fill the array automatically (Y/N)? ")
46      if (filling != 'Y' and filling != 'y' and filling != 'n' and
47          filling != 'N'):
48          print("The entered answer is not valid ... Please try again")
49      else:
50          bubbleSort(maxSize, filling)
51
52  # --------- Functions Section ------------
53  # ---------------------------------------
54  # ********* BMI Function ****************
55  def calcualteBMI(W, H):
56      bmi = W/(H*H)
57      if (bmi<18.5):
58          print("Underweight: less than 18.5")
59      elif (bmi>=18.5) and (bmi<=24.9):
60          print("Normal: between 18.5 and 24.9")
61      elif (bmi>=25) and (bmi<=29.9):
62          print("Overweight: between 25 and 29.9")
63      elif (bmi>=30):
64          print("Obese: 30 or greater")
65      else:
66          print ("BMI's calculation failed .. ")
67
68  # ********** Customer Credit Function ****
69  def calculateCustomerCredit(blnc, charges, credits, limit):
70      newBalance = blnc + charges - credits
71      print ("Your new balance is: ", newBalance)
72      if (newBalance>limit):
73          print("You exceeded your credit limit this month")
74
75  # ********** Palindrome Function ****
76  def calculatePalindrome(Nbr):
77      # To be a palindrome, the 5th digit should be equal to the first
78      # digit and the 4th digit should be equal to the second
79      if (Nbr[4] == Nbr[0]) and (Nbr[3] == Nbr[1]):
80          print ("Your number is a palindrome")
81      else:
82          print ("Your number is not a palindrome")
83
84  # ********** Binary Conversion Function ****
85  def calculateBinary(Nbr):
86      # Keep dividing by 2 and keeping the remainders
87      result = Nbr
88      rem = ""
89      while (result != 0):
```

```
 90            result = Nbr//2
 91            rem = rem + str(Nbr%2)
 92            Nbr = result
 93       # Reverse the string rem to get the exact binary number
 94       binary = ""
 95       l = len(rem)
 96       for i in range(0,l):
 97            binary = binary + rem[l-i-1]
 98       print("The binary number is: ", binary)
 99
100  # ********** Array Sorting Function ****
101  def bubbleSort(Max, Fill):
102       Array = []
103       if (Fill == 'y' or Fill=='Y'):
104            for i in range(0, Max):
105                 number = random.randrange(0,1000)
106                 Array.append(number)
107       else:
108            for i in range(0, Max):
109                 number = int(input("Enter an element of the array: "))
110                 Array.append(number)
111       if (len(Array) <= 100):
112            print("The original array is:", Array)
113       # bubble sort
114       for k in range(Max-1,0,-1):
115            for i in range(k):
116                 if Array[i] > Array[i+1]:
117                      temp = Array[i]
118                      Array[i] = Array[i+1]
119                      Array[i+1] = temp
120       if (len(Array) <= 100):
121            print("The sorted array is:", Array)
122
123  # --------- MENU ---------------------------------------------
124  # ------------------------------------------------------------
125  def DisplayMenu():
126       print("-------------------------------------------")
127       print("----------      M E N U      ------------")
128       print("-------------------------------------------")
129       print("    1- Body Mass Index Calculator")
130       print("    2- Customer Credit")
131       print("    3- Five-Digit Palindrome")
132       print("    4- Integer Binary Conversion")
133       print("    5- Array Integers Sorting")
134       print("-------------------------------------------")
135       rep = int(input(" -----  Choose an option (1-5)  --- : "))
136       return rep
137
138  while (True):
139       selection = DisplayMenu()
140       if (selection == 1):
141            BMI()
142       elif (selection == 2):
```

```
143              CustomerCredit()
144         elif (selection == 3):
145              Palindrome()
146         elif (selection == 4):
147              ConvertToBinary()
148         elif (selection == 5):
149              ArraySorting()
```

Output: Case Study Chapter 2

```
-------------------------------------------
----------          M E N U      ------------
-------------------------------------------
     1- Body Mass Index Calculator
     2- Customer Credit
     3- Five-Digit Palindrome
     4- Integer Binary Conversion
     5- Array Integers Sorting
-------------------------------------------

-----    Choose an option (1-5)   --- :

```

```
-----    Choose an option (1-5)   --- : 1
Enter your weight in Kilogram: 78
Enter your height in Meters: 1.72
Overweight: between 25 and 29.9
-------------------------------------------
```

```
-----    Choose an option (1-5)   --- : 2
Enter your balance at the begining of the month: 1137
Enter the total amount of your charges this month: 876
Enter the total amount of your credits this month: 1600
Enter your Credit Limit: 5000
Your new balance is:   413
-------------------------------------------
```

```
-----    Choose an option (1-5)   --- : 3
Enter a five-digit number:12321
Your number is a palindrome
-------------------------------------------
```

```
-----    Choose an option (1-5)   --- : 4
Enter an integer number between 0 and 99: 77
The binary number is:   1001101
-------------------------------------------
```

```
-----    Choose an option (1-5)   --- : 5
Enter the maximum size of your array: 5
Fill the array automatically (Y/N)? Y
The original array is: [332, 758, 537, 61, 862]
The sorted array is: [61, 332, 537, 758, 862]
-------------------------------------------
```

CHAPTER 3 – OBJECT ORIENTED PROGRAMMING WITH PYTHON

Description

Sherwood real estate requires an application to manage properties. There are two types of properties: apartments and houses. Each property may be available for rent or sale. The following are the requirements:

- Both types of properties (apartment or house) are described using a reference number, address, built up area, number of bedrooms, number of bathrooms, number of parking slots, pool availability and gym availability.
- A house requires extra attributes such as the number of floors, plot size and house type (villa or townhouse).
- An apartment requires additional attributes such as floor and number of balconies.
- Each type of property (house or apartment) may be available for rent or sale.
- A rental property should include attributes such as deposit amount, yearly rent, furnished (yes or no), includes maids' room (yes or no).
- A property available for sale has attributes such as sale price and estimated annual service charge.
- All properties include an agent commission of 2% that is fixed for all properties. Both types of sale properties have a tax of 4% which is fixed for all properties.
- All properties require a method to display the details of that property.
- All properties should include a method to compute the agent commission. For rental properties agent commission is calculated by using the yearly rental amount, whereas for purchase properties it is calculated using the sale price.
- Both types of purchase properties should include a method to compute the tax amount. Tax amount is computed on sale price.

Implement a Python application that creates the four types of properties – RentalApartment, RentalHouse, SaleApartment, SaleHouse - by using multiple inheritance and abstract classes. Implement class attributes and instance attributes using encapsulation. All numeric attributes such as price should be validated for inputs with a suitable minimum and maximum price.

Define the methods in the abstract class and implement it in the respective classes. Override the print function to display each property details.

Create a main application to offer the above functionalities and test them by creating new properties of each type and calling the respective methods.

Solution

First it is necessary to create the parent class from which all the rest will somehow inherit its functions/methods and attributes. This is an *abstract* class called "Property" which uses the ABC Abstract Base Class.

```
1    # Import the ABC Abstract Base Classes
2    from abc import ABC, abstractmethod
3
4    # Define abstract class Property
5    class Property(ABC):
6
7        # Define class attributes
8        agentCommission = 0.02
```

```
9
10        # Define the constructor of the class and its attributes
11        def __init__(self, refNumber, owner, address, builtUpArea,
12                    bedNumber, bathNumber, parkNumber, pool, gym):
13          self.refNumber = refNumber
14          self.owner = owner
15          self.address = address
16          self.builtUpArea = builtUpArea
17          self.bedNumber = bedNumber
18          self.bathNumber = bathNumber
19          self.parkNumber = parkNumber
20          self.pool = pool
21          self.gym = gym
22
23        # Define abstract method setPropertyAttributes
24        @abstractmethod
25        def setPropertyAttributes(self):
26          self.owner = input("Owner: ")
27          self.address = input("Address: ")
28          self.builtUpArea = input("Area ")
29          self.bedNumber = input("Bedrooms: ")
30          self.bathNumber = input("Bathrooms: ")
31          self.parkNumber = input("Parking Slots: ")
32          self.pool = input("Pool: ")
33          self.gym = input("Gym: ")
34
35        # Define abstract method displayPropertyDetails
36        @abstractmethod
37        def displayPropertyDetails(self):
38          print("Reference Number: " + self.refNumber)
39          print("Owner: " + self.owner)
40          print("Address: " + self.address)
41          print("Built Up Area: " + self.builtUpArea)
42          print("Beds: " + str(self.bedNumber))
43          print("Bathrooms: " + str(self.bathNumber))
44          print("Parking Slots: " + str(self.parkNumber))
45          print("Pool: " + self.pool)
46          print("Gym: " + self.gym)
```

Then, the "Apartment" class is implemented that inherits from the parent "Property" class as follows:

```
1     from Property import Property
2
3     # Define sub-class Apartment
4     class Apartment(Property):
5
6         # Define class attributes
7         floor = 0
8         balconyNumber = 0
9
10        # Implement abstract method calculateCommission
```

```
11       def calculateCommission():
12         pass
13
14       # Implement abstract method setPropertyAttributes
15       def setPropertyAttributes(self):
16         Property.setPropertyAttributes(self)
17
18       # Implement abstract method displayPropertyDetails
19       def displayPropertyDetails(self):
20         Property.displayPropertyAttributes(self)
21
22       # Set extra attributes for Apartments
23       def setApartmentAttributes(self):
24         self.floor = input("Enter apartment floor: ")
25         self.balconyNumber = input("Enter number of balconies: ")
26
27       # Display Property details
28       def displayPropertyDetails(self):
29         Property.displayPropertyDetails(self)
30         print("Floor: " + str(self.floor))
31         print("Number of Balconies: " + str(self.balconyNumber))
```

The third class, the "RentalProperty", provides rental details related to the apartment to be rented:

```
1    # Define sub-class RentalProperty
2    class RentalProperty():
3
4       # Define class attributes
5       depositAmount = 0
6       yearlyRent = 0
7       isFurnished = "No"
8       maidsRoom = "No"
9       agentCommission = 0.02
10
11       # Set extra attributes for rental properties
12       def setRentalAttributes(self):
13         self.depositAmount = int(input("Enter deposit amount: "))
14         self.yearlyRent = int(input("Enter yearly rent: "))
15         self.isFurnished = input("Property furnished (Yes/No): ")
16         self.maidsRoom = input("Maids room (Yes/No): ")
17
18       # Display rental property details
19       def displayRentalDetails(self):
20         print("Deposit Amount: " + str(self.depositAmount))
21         print("Yearly Rent: " + str(self.yearlyRent))
22         print("Property furnished: " + str(self.isFurnished))
23         print("Maids room: " + str(self.maidsRoom))
24         print("Agent Commission: " + str(self.agentCommission))
25
26       # Calculate Agent Commission
27       def calcCommission(self):
28         self.agentCommission = self.agentCommission * self.yearlyRent
```

Finally, the "RentalApartment" class inherits from both the "Apartment" and the "RentalProperty" classes to create the objects with the details of the apartments available as well as their rental details. The class has no particular attributes or functions/methods by itself and looks as follows:

```
1    from Apartment import Apartment
2    from RentalProperty import RentalProperty
3
4    # Define sub-class RentalApartment
5    class RentalApartment(Apartment, RentalProperty):
6        pass
```

Likewise with the case of the "RentalApartment" the "SaleApartment" class inherits from both the "Apartment" (already listed above) and the "SaleProperty" classes to create the objects with the details of the apartments available as well as their details "for sale". The "SaleProperty" class is shown below:

```
1    # Define sub-class SaleProperty
2    class SaleProperty():
3
4        # Define class attributes
5        salePrice = 0
6        annualServiceCharge = 0
7        tax = 0.04
8
9        # Set extra attributes for sale properties
10       def setSaleAttributes(self):
11           self.salePrice = input("Enter sale price: ")
12           self.annualServiceCharge=input("Enter annual service charge: ")
13
14       # Display sale property details
15       def displaySaleDetails(self):
16           print("Sale price: " + str(self.salePrice))
17           print("Annual service charge: " + str(self.annualServiceCharge))
18           print("Tax: " + str(self.tax))
```

Similarly, the "SaleApartment" class is given below:

```
1    from Apartment import Apartment
2    from SaleProperty import SaleProperty
3
4    # Define sub-class SaleApartment
5    class SaleApartment(Apartment, SaleProperty):
6        pass
```

Having completed the classes for the cases of apartments to rent or for sale, the same pattern and similar classes will be needed for the cases of houses to rent or for sale. Once again, the parent class is the one given before, i.e., "Property". The "House" class follows:

```
1    from Property import Property
2
3    # Define sub-class House
```

```
4    class House(Property):
5
6        # Define class attributes
7        numberOfFloors = 1
8        plotSize = 200
9        houseType = ""
10
11       # Implement abstract method setPropertyAttributes
12       def setPropertyAttributes(self):
13           Property.setPropertyAttributes(self)
14
15       # Implement abstract method displayPropertyDetails
16       def displayPropertyDetails(self):
17           Property.displayPropertyAttributes(self)
18
19       # Set extra attributes for Houses
20       def setHouseAttributes(self):
21           self.numberOfFloors = input("Enter number of floors: ")
22           self.plotSize = input("Enter plot size: ")
23           self.houseType=input("Enter type of house (Villa/Townhouse): ")
24
25       # Display Property details
26       def displayPropertyDetails(self):
27           Property.displayPropertyDetails(self)
28           print("Number of Floors: " + str(self.numberOfFloors))
29           print("Plot Size: " + str(self.plotSize))
30           print("House Type: " + str(self.houseType))
```

The "SaleHouse" class is given below:

```
1    from House import House
2    from SaleProperty import SaleProperty
3
4    # Define sub-class RentalHouse
5    class SaleHouse(House, SaleProperty):
6        pass
```

The next class is the "RentalHouse":

```
1    from House import House
2    from RentalProperty import RentalProperty
3
4    # Define sub-class RentalHouse
5    class RentalHouse(House, RentalProperty):
6        pass
```

Finally, the main script that will run the application and all the classes, i.e., "Chapter3CaseStudyApplication, is given below:

```
1    from RentalApartment import RentalApartment
2    from SaleApartment import SaleApartment
3    from RentalHouse import RentalHouse
4    from SaleHouse import SaleHouse
5
6    print("Select any of the following options")
7    print("===================================")
8    print("1. Apartments to rent")
9    print("2. Apartments for sale")
10   print("3. Houses to rent")
11   print("4. Houses for sale")
12   print("===================================")
13   selection = input("Enter your choice: ")
14
15   # Enter the general details of the property
16   propertyCode = input("Enter the code of the property:")
17   propertyOwner = input("Enter the name of the owner:")
18   propertyAddress = input("Enter the address of the property:")
19   propertyArea = input("Enter the location/area of the property:")
20   propertyBedNum = int(input("Enter the number of bedrooms (1-10):"))
21   propertyBathNum = int(input("Enter the number of bathrooms (1-5):"))
22   propertyParkNum = int(input("Enter the number of parking slots (1-5):"))
23   propertyPool = input("Does the property include a swimming pool \
24   (Yes/No):")
25   propertyGym = input("Does the property include a gym (Yes/No):")
26   if (selection == "1"):
28       # Create the object with the details of the property
29       newRentalApartment = RentalApartment(propertyCode,
30           propertyOwner, propertyAddress, propertyArea,
31           propertyBedNum, propertyBathNum, propertyParkNum,
32           propertyPool, propertyGym)
33
34       # Enter the specific details of the apartment for rent
35       # and report all property details
36       newRentalApartment.setApartmentAttributes()
37       newRentalApartment.setRentalAttributes()
38       print("\nReporting property details")
39       print("==========================")
40       newRentalApartment.displayPropertyDetails()
41       print("\nReporting rental details")
42       print("==========================")
43       newRentalApartment.calcCommission()
44       newRentalApartment.displayRentalDetails()
45
46   if (selection == "2"):
47       # Create the object with the details of the property
48       newSaleApartment = SaleApartment(propertyCode, propertyOwner,
49           propertyAddress, propertyArea, propertyBedNum, propertyBathNum,
```

```
50              propertyParkNum, propertyPool, propertyGym)
51
52      # Enter the specific details of the apartment for rent
53      # and report all property details
54      newSaleApartment.setApartmentAttributes()
55      newSaleApartment.setSaleAttributes()
56      print("\nReporting property details")
57      print("==========================")
58      newSaleApartment.displayPropertyDetails()
59      print("\nReporting sale details")
60      print("==========================")
61      newSaleApartment.displaySaleDetails()
62
63  if (selection == "3"):
64      # Create the object with the details of the property
65      newRentalHouse = RentalHouse(propertyCode, propertyOwner,
66          propertyAddress, propertyArea, propertyBedNum, propertyBathNum,
67          propertyParkNum, propertyPool, propertyGym)
68
69      # Enter the specific details of the house for rent
70      # and report all property details
71      newRentalHouse.setHouseAttributes()
72      newRentalHouse.setRentalAttributes()
73      print("\nReporting property details")
74      print("==========================")
75      newRentalHouse.displayPropertyDetails()
76      print("\nReporting rental details")
77      print("==========================")
78      newRentalHouse.calcCommission()
79      newRentalHouse.displayRentalDetails()
80
81  if (selection == "4"):
82      # Create the object with the details of the property
83      newSaleHouse = SaleHouse(propertyCode, propertyOwner,
84          propertyAddress, propertyArea, propertyBedNum, propertyBathNum,
85          propertyParkNum, propertyPool, propertyGym)
86
87      # Enter the specific details of the house for sale
88      # and report all property details
89      newSaleHouse.setHouseAttributes()
90      newSaleHouse.setSaleAttributes()
91      print("\nReporting property details")
92      print("==========================")
93      newSaleHouse.displayPropertyDetails()
94      print("\nReporting sale details")
95      print("==========================")
96      newSaleHouse.displaySaleDetails()
```

Output: Case Study Chapter 3

```
Select any of the following options
=====================================
1. Apartments to rent
2. Apartments for sale
3. Houses to rent
4. Houses for sale
=====================================
Enter your choice: 1
Enter the code of the property:VS7256S
Enter the name of the owner:Alex Fora
Enter the address of the property:1550 Wilder Avenue
Enter the location/area of the property:Downtown
Enter the number of bedrooms (1-10):4
Enter the number of bathrooms (1-5):4
Enter the number of parking slots (1-5):3
Does the property include a swimming pool (Yes/No):Yes
Does the property include a gym (Yes/No):Yes
Enter apartment floor: 16
Enter number of balconies: 2
Enter deposit amount: 2500
Enter yearly rent: 24000
Property furnished (Yes/No): Yes
Maids room (Yes/No): No
```

```
Reporting property details
============================
Reference Number: VS7256S
Owner: Alex Fora
Address: 1550 Wilder Avenue
Built Up Area: Downtown
Beds: 4
Bathrooms: 4
Parking Slots: 3
Pool: Yes
Gym: Yes
Floor: 16
Number of Balconies: 2

Reporting rental details
============================
Deposit Amount: 2500
Yearly Rent: 24000
Property furnished: Yes
Maids room: No
Agent Commission: 480.0
```

CHAPTER 4 – GRAPHICAL USER INTERFACE
PROGRAMMING WITH PYTHON

DESCRIPTION

Enhance the "Countries" application in order to include the following functionality:

- Add one more listbox to display more content for each country (e.g., size, population, etc.)
- Add a combobox to allow the user to select the font name of the contents of the listboxes.
- Add a combobox to allow the user to select the font size of the contents of the listboxes.
- Add a combobox to change the background color of the content in the listboxes.

SOLUTION

```
1    import tkinter as tk
2    from tkinter import *
3    from tkinter import ttk
4    from tkinter import messagebox
5
6    countries = ['E.U.', 'U.S.A.', 'Russia', 'China', 'India', 'Brazil']
7    capital = ['Brussels', 'Washinghton', 'Moscow', 'Beijing', 'New Delhi',
8               'Brazilia']
9    population = ['450m', '330m', '145m', '1,400m', '1,350m', '210m']
10   fontName = ['Arial', 'Garamond', 'Times New Roman', 'Courier']
11   fontSize = [8, 10, 12, 14]
12   fontColor = ['cyan', 'grey', 'green', 'red']
13
14   global newCountry, newCapital, newPopulation
15   global FontNameSelection, FontSizeSelection, FontColorSelection
16   global CountriesFrame, CapitalFrame, PopulationFrame, SettingsFrame
17   global checkButton1, checkButton2, checkButton3
18   global radioButton
19   global CountriesList, CapitalList, PopulationList
20   global FontNameList, FontSizeList, FontColorList
21   global CountriesScrollBar, CapitalScrollBar, PopulationScrollBar
22
23   # Create the interface for the listboxes
24   def drawListBoxes():
25       global FontNameSelection, FontSizeSelection, FontColorSelection
26       global CountriesList, CapitalList, PopulationList
27       global CountriesFrame, CapitalFrame, PopulationFrame
28       global CountriesScrollBar, CapitalScrollBarPopulationScrollBar
29
30       color = FontColorSelection.get()
31       size = FontSizeSelection.get()
32       name = FontNameSelection.get()
33
34       # Create the CountriesFrame labelframe & place the
35       # CountriesList widget in it
36       CountriesFrame = tk.LabelFrame(winFrame, text = 'Countries')
37       CountriesFrame.config(bg = 'light grey', fg = 'blue', bd = 2,
```

```
38          width = 13, relief = 'sunken')
39      # Create a scrollbar widget to attach to the CountriesList
40      CountriesScrollBar = Scrollbar(CountriesFrame, orient = VERTICAL)
41      CountriesScrollBar.pack(side = RIGHT, fill = Y )
42      # Create the listbox in the CountriesFrame
43      CountriesList = tk.Listbox(CountriesFrame, bg = color,
44          font = (name, size), yscrollcommand = CountriesScrollBar,
45          width = 13, height = 8)
46      CountriesList.pack(side = LEFT, fill = BOTH)
47      # Associate the scrollbar command with its parent widget,
48      # i.e., the CountriesList yview
49      CountriesScrollBar.config(command = CountriesList.yview)
50      # Place the Countries frame and its parts onto the interface
51      CountriesFrame.pack();
52      CountriesFrame.place(relx = 0.02, rely = 0.05)
53      CountriesList.bind('<Double-Button-1>',
54          lambda event: alignList('countries'))
55
56      # Create the CapitalFrame labelframe and place the CapitalList
57      # widget in it
58      CapitalFrame = tk.LabelFrame(winFrame, text = 'Capitals')
59      CapitalFrame.config(bg = 'light grey', fg = 'blue', bd = 2,
60          width = 14, relief = 'sunken')
61      # Create a scrollbar widget to attach to the CapitalFrame
62      CapitalScrollBar = Scrollbar(CapitalFrame, orient = VERTICAL)
63      CapitalScrollBar.pack(side = RIGHT, fill = Y)
64      # Create the listbox in the CapitalFrame
65      CapitalList = tk.Listbox(CapitalFrame,
66          yscrollcommand = CapitalScrollBar, bg = color,
67          font = (name, size), width = 13, height = 8)
68      CapitalList.pack(side = LEFT, fill = BOTH)
69      # Associate the scrollbar command with its parent widget,
70      # i.e., the CapitalList yview
71      CapitalFrame.pack()
72      CapitalFrame.place(relx = 0.25, rely = 0.05)
73      CapitalList.bind('<Double-Button-1>',
74          lambda event: alignList('capital'))
75
76      # Create the PopulationFrame labelframe and place the
77      # PopulationList widget in it
78      PopulationFrame = tk.LabelFrame(winFrame, text = 'Populations')
79      PopulationFrame.config(bg = 'light grey', fg = 'blue',
80          bd = 2, width = 14, relief = 'sunken')
81      # Create a scrollbar widget to attach to the PopulationFrame
82      PopulationScrollBar = Scrollbar(PopulationFrame, orient = VERTICAL)
83      PopulationScrollBar.pack(side = RIGHT, fill = Y)
84      # Create the listbox in the PopulationFrame
85      PopulationList = tk.Listbox(PopulationFrame, bg = color,
86          width = 13, height = 8, font = (name, size),
87          yscrollcommand = PopulationScrollBar )
88      PopulationList.pack(side = LEFT, fill = BOTH)
89      # Associate the scrollbar command with its parent widget,
```

```
90          # i.e., the PopulationList yview
91          PopulationFrame.pack()
92          PopulationFrame.place(relx = 0.50, rely = 0.05)
93          PopulationList.bind('<Double-Button-1>',
94              lambda event: alignList('population'))
95
96      # Create the interface for the new entries
97      def drawNewEntries():
98          global newCountry, newCapital, newPopulation
99
100         # Create the labelframe & place the newCountry Entry widget in it
101         NewCountryFrame = tk.LabelFrame(winFrame, text = 'New Country')
102         NewCountryFrame.config(bg = 'light grey', fg = 'blue',
103             bd = 2, width = 14, relief = 'sunken')
104         NewCountryFrame.pack()
105         NewCountryFrame.place(relx = 0.02, rely = 0.5)
106         newCountry = tk.StringVar()
107         newCountry.set('')
108         NewCountryEntry = tk.Entry(NewCountryFrame,
109             textvariable = newCountry, width = 13)
110         NewCountryEntry.config(bg = 'dark grey', fg = 'red',
111             relief = 'sunken')
112         NewCountryEntry.grid(row = 0, column = 0)
113
114         # Create the labelframe & place the newCapital Entry widget in it
115         NewCapitalFrame = tk.LabelFrame(winFrame, text = 'New Capital')
116         NewCapitalFrame.config(bg = 'light grey', fg = 'blue',
117             bd = 2, width = 14, relief = 'sunken')
118         NewCapitalFrame.pack()
119         NewCapitalFrame.place(relx = 0.25, rely = 0.5)
120         newCapital = tk.StringVar()
121         newCapital.set('')
122         NewCapitalEntry = tk.Entry(NewCapitalFrame,
123             textvariable = newCapital, width = 13)
124         NewCapitalEntry.config(bg = 'dark grey', fg = 'red',
125             relief = 'sunken')
126         NewCapitalEntry.grid(row = 0, column = 0)
127
128         # Create the labelframe & place the newPopulation Entry widget in it
129         NewPopulationFrame = tk.LabelFrame(winFrame,
130             text = 'New Population')
131         NewPopulationFrame.config(bg = 'light grey', fg = 'blue',
132             bd = 2, width = 14, relief = 'sunken')
133         NewPopulationFrame.pack()
134         NewPopulationFrame.place(relx = 0.50, rely = 0.5)
135         newPopulation = tk.StringVar()
136         newPopulation.set('')
137         NewPopulationEntry = tk.Entry(NewPopulationFrame, width = 13,
138             textvariable = newPopulation)
139         NewPopulationEntry.config(bg = 'dark grey', fg = 'red',
140             relief = 'sunken')
141         NewPopulationEntry.grid(row = 0, column = 0)
```

```
142
143   # Create the interface for the action buttons
144   def drawButtons():
145       # Create the labelframe to place the buttons in it
146       ButtonsFrame = tk.LabelFrame(winFrame, text = "Actions")
147       ButtonsFrame.config(bg = 'light grey', fg = 'blue',
148           bd = 2, width = 14, relief = 'sunken')
149       ButtonsFrame.pack()
150       ButtonsFrame.place(relx = 0.75, rely = 0.05)
151
152       newRecordButton = tk.Button(ButtonsFrame,
153           text = 'Insert\nnew record', width = 8, height = 2)
154       newRecordButton.grid(row = 0, column = 0)
155       newRecordButton.bind('<Button-1>', lambda event,
156           a = 'insertRecord': buttonsClicked(a))
157
158       deleteRecordButton = tk.Button(ButtonsFrame,
159           text = 'Delete\n record', width = 8, height = 2)
160       deleteRecordButton.grid(row = 0, column = 1)
161       deleteRecordButton.bind('<Button-1>', lambda event,
162           a = 'deleteRecord': buttonsClicked(a))
163
164       clearRecordsButton = tk.Button(ButtonsFrame,
165           text = 'Clear\n records', width = 8, height = 2)
166       clearRecordsButton.grid(row = 1, column = 0)
167       clearRecordsButton.bind('<Button-1>', lambda event,
168           a = 'clearAllRecords': buttonsClicked(a))
169
170       changeSettingsButton = tk.Button(ButtonsFrame,
171           text = 'Change\n settings', width = 8, height = 2)
172       changeSettingsButton.grid(row = 1, column = 1)
173       changeSettingsButton.bind('<Button-1>', lambda event,
174           a = 'changeSettings': buttonsClicked(a))
175
176       exitButton = tk.Button(ButtonsFrame, text = 'Exit', width = 8,
177           height = 2)
178       exitButton.grid (columnspan = 2, row = 2, column = 0)
179       exitButton.bind('<Button-1>', lambda event : winFrame.destroy())
180       exit()
181
182   # Create the interface for the checkbuttons
183   def drawCheckButtons():
184       global checkButton1, checkButton2, checkButton3
185
186       # Create the labelframe to place the checkbuttons in it
187       CheckButtonsFrame = tk.LabelFrame(winFrame, text = "Enable/Disable")
188       CheckButtonsFrame.config(bg = 'light grey', fg = 'blue', bd = 2,
189           relief = 'sunken')
190       CheckButtonsFrame.pack()
191       CheckButtonsFrame.place(relx = 0.75, rely = 0.45)
192
193       checkButton1 = IntVar(value = 1)
```

```
194     CountriesCheckButton = tk.Checkbutton(CheckButtonsFrame,
195         text = 'Countries', width = 16, height = 1, bg = 'light blue',
196         variable = checkButton1, onvalue = 1, offvalue = 0,
197         command = checkClicked).grid(row = 0, column = 0)
198
199     checkButton2 = IntVar(value = 1)
200     CapitalCheckButton = tk.Checkbutton(CheckButtonsFrame,
201         text = 'Capitals', width = 16, height = 1, bg = 'light blue',
202         variable = checkButton2, onvalue = 1, offvalue = 0,
203         command = checkClicked).grid(row = 1, column = 0)
204
205     checkButton3 = IntVar(value = 1)
206     PopulationCheckButton = tk.Checkbutton(CheckButtonsFrame,
207         text = 'Populations', width = 16, height = 1, bg = 'light blue',
208         variable = checkButton3, onvalue = 1, offvalue = 0,
209         command = checkClicked).grid(row = 2, column = 0)
210
211 # Create the interface for the radiobuttons
212 def drawRadioButtons():
213     global radioButton
214
215     # Create the labelframe to place the radiobuttons
216     RadioButtonsFrame = tk.LabelFrame(winFrame, text = "Containers")
217     RadioButtonsFrame.config(bg = 'light grey', fg = 'blue',
218         bd = 2, relief = 'sunken')
219     RadioButtonsFrame.pack()
220     RadioButtonsFrame.place(relx = 0.75, rely = 0.7)
221
222     radioButton = IntVar()
223     visibleRadioButton = tk.Radiobutton(RadioButtonsFrame,
224         text = 'Visible', width = 8, height = 1, bg = 'light green',
225         variable = radioButton, value = 1,
226         command = radioClicked).grid(row = 0, column = 0)
227
228     invisibleRadioButton = tk.Radiobutton(RadioButtonsFrame,
229         text = 'Invisible', width = 8, height = 1, bg = 'light green',
230         variable = radioButton, value = 2,
231         command = radioClicked).grid(row = 0, column = 1)
232
233     radioButton.set(1)
234
235 # Create the interface for the settings comboboxes
236 def drawSettingsCombos():
237     global FontNameSelection, FontSizeSelection, FontColorSelection
238     global FontNameList, FontSizeList, FontColorList
239     global SettingsFrame
240
241     # Create the labelframe and place the settings combos in it
242     SettingsFrame = tk.LabelFrame(winFrame, text = 'Settings')
243     SettingsFrame.config(bg = 'light grey', fg = 'blue', bd = 2,
244         width = 20, relief = 'sunken')
```

```
245        SettingsFrame.pack()
246        SettingsFrame.place(relx = 0.02, rely = 0.65)
247
248        # Create the label in the entry frame
249        FontNameLabel = tk.Label(SettingsFrame,
250            text = 'Select font name', width = 17)
251        FontNameLabel.config(bg = 'light grey', fg = 'red', bd = 3,
252            relief = 'flat', font = 'Arial 14 bold')
253        FontNameLabel.grid(column = 0, row = 0)
254        # Create the combobox to select the font name from the combo
255        FontNameCombo = ttk.Combobox(SettingsFrame,
256            textvariable = FontNameSelection, width = 14)
257        FontNameCombo['values'] = fontName
258        FontNameCombo.current(0); FontNameCombo.grid(column = 0, row = 1)
259
260        # Create the label in the entry frame
261        FontSizeLabel = tk.Label(SettingsFrame,
262            text = 'Select font size', width = 17)
263        FontSizeLabel.config(bg = 'light grey', fg = 'red', bd = 3,
264            relief = 'flat', font = 'Arial 14 bold')
265        FontSizeLabel.grid(column = 1, row = 0)
266        # Create the combobox to select the font name from the combo
267        FontSizeCombo = ttk.Combobox(SettingsFrame,
268            textvariable = FontSizeSelection, width = 14)
269        FontSizeCombo['values'] = fontSize
270        FontSizeCombo.current(0); FontSizeCombo.grid(column = 1, row = 1)
271
272        # Create the label in the entry frame
273        FontColorLabel = tk.Label(SettingsFrame,
274            text = 'Select font color', width = 17)
275        FontColorLabel.config(bg = 'light grey', fg = 'red', bd = 3,
276            relief = 'flat', font = 'Arial 14 bold')
277        FontColorLabel.grid(column = 2, row = 0)
278        # Create the combobox to select the font name from the combo
279        FontColorCombo = ttk.Combobox(SettingsFrame,
280            textvariable = FontColorSelection, width = 14)
281        FontColorCombo['values'] = fontColor
282        FontColorCombo.current(0); FontColorCombo.grid(column = 2, row = 1)
283
284  # Define the method 'indexSelectedListbox' that will identify
285  # the row selected in any of the listboxes
286  def alignList(a):
287      global CountriesList, CapitalList, PopulationList
288      global selectedIndex
289
290      if (a == 'countries'):
291          selectedIndex = int(CountriesList.curselection()[0])
292          CapitalList.selection_set(selectedIndex)
293          PopulationList.selection_set(selectedIndex)
294
295      if (a == 'capital'):
```

```
296          selectedIndex = int(CapitalList.curselection()[0])
297          CountriesList.selection_set(selectedIndex)
298          PopulationList.selection_set(selectedIndex)
299
300      if (a == 'population'):
301          selectedIndex = int(PopulationList.curselection()[0])
302          CountriesList.selection_set(selectedIndex)
303          CapitalList.selection_set(selectedIndex)
304
305  # Define the checkClicked function that will control the state
306  # of the containers
307  def checkClicked():
308      global checkButton1, checkButton2
309
310      # Control the state of the containers as NORMAL or DISABLED
311      # based on the checkbuttons' state
312      if (checkButton1.get() == 1):
313          CountriesList.config(state = NORMAL)
314      else:
315          CountriesList.config(state = DISABLED)
316
317      if (checkButton2.get() == 1):
318          CapitalList.config(state = NORMAL)
319      else:
320          CapitalList.config(state = DISABLED)
321
322      if (checkButton3.get() == 1):
323          PopulationList.config(state = NORMAL)
324      else:
325          PopulationList.config(state = DISABLED)
326
327  # Define the radioClicked function that will display or hide the frames
328  # of the containers
329  def radioClicked():
330      global CountriesFrame, CapitalFrame
331      global radioButton
332
333      # Use the destroy() function to destroy the frames of the
334      # containers. The lists are not destroyed
335      CountriesFrame.destroy()
336      CapitalFrame.destroy()
337      PopulationFrame.destroy()
338
339      if (radioButton.get() == 1):
340          drawListBoxes()
341          populate()
342
343  # Populate the listboxes
344  def populate():
345      global CountriesList, CapitalList, PopulationList
346      global FontNameSelection, FontSizeSelection, FontColorSelection
```

```
347
348        color = FontColorSelection.get(); size = FontSizeSelection.get()
349        name = FontNameSelection.get()
350
351        for i in range (int(len(countries))):
352            CountriesList.insert(i, countries[i])
353
354        for i in range (int(len(capital))):
355            CapitalList.insert(i, capital[i])
356
357        for i in range (int(len(population))):
358            PopulationList.insert(i, population[i])
359
360        PopulationList.config(bg = color, font = (name, size))
361        CountriesList.config(bg = color, font = (name, size))
362        CapitalList.config(bg = color, font = (name, size))
363
364    # Define the method 'buttonsClicked' that will trigger the code
365    # to be executed when any of the buttons is clicked
366    def buttonsClicked(a):
367        global CountriesList, PopulationCombo, CapitalList
368        global FontNameSelection, FontSizeSelection, FontColorSelection
369        global newCountry, newPopulation, newCapital, populationSelection
370        global selectedIndex
371
372        if (a == "insertRecord"):
373            if (newCountry != '' and newCapital != ''
374                and newPopulation != ''):
375                countries.append(newCountry.get())
376                CountriesList.delete('0', 'end')
377                capital.append(newCapital.get())
378                CapitalList.delete('0', 'end')
379                population.append(newPopulation.get())
380                PopulationList.delete('0', 'end')
381                # Call the function populate() to re-populate the containers
382                # with the renewed lists
383                populate()
384
385        if (a == 'deleteRecord'):
386            # Use the messagebox.askyesno() to pop a message to ask
387            # confirmation for deleting the elements
388            deleteElementOrNot = messagebox.askokcancel(
389                title = "Delete element",
390                message = "Are you ready to delete the elements?",
391                icon = 'info')
392            if (deleteElementOrNot == True):
393                # Use the pop() method to remove the selected elements
394                # from the lists
395                countries.pop(selectedIndex)
396                capital.pop(selectedIndex)
397                population.pop(selectedIndex)
```

```
398              CountriesList.delete('0', 'end')
399              CapitalList.delete('0', 'end')
400              PopulationList.delete('0', 'end')
401              # Call the function populate() to re-populate the containers
402              # with the renewed lists
403              populate()
404
405      if (a == 'clearAllRecords'):
406          # Use the messagebox.askyesno() to pop a message to ask
407          # confirmation for clearing the lists              ·
408          clearListsOrNot = messagebox.askokcancel(
409              title = "Clear all elements",
410              message = "Are you ready to clear the lists?",
411              icon = 'info')
412          if (clearListsOrNot == True):
413              countries.clear(); capital.clear()
414              population.clear()
415              CountriesList.delete('0', 'end')
416              CapitalList.delete('0', 'end')
417              PopulationList.delete('0', 'end')
418              # Call the function populate() to re-populate the containers
419              # with the renewed lists
420              populate()
421
422      if (a == 'changeSettings'):
423          CountriesList.delete('0', 'end'); CapitalList.delete('0', 'end')
424          PopulationList.delete('0', 'end')
425          # Call the function populate() to re-populate the containers
426          # with the renewed lists and change the interface settings
427          populate()
428
429  # Create the frame for the 'Enhanced Countries' program and configure
430  # its size and background color
431  winFrame = tk.Tk()
432  winFrame.title ('Countries'); winFrame.geometry("650x350")
433  winFrame.config (bg = 'light grey'); winFrame.resizable(False, False)
434
435  FontNameSelection = tk.StringVar(); FontSizeSelection = tk.IntVar()
436  FontColorSelection = tk.StringVar()
437
438  FontNameSelection.set(fontName[0]); FontSizeSelection.set(fontSize[0])
439  FontColorSelection.set(fontColor[0])
440
441  # Create the Graphical User Interface
442  drawListBoxes(); drawNewEntries(); drawButtons()
443  drawCheckButtons(); drawRadioButtons(); drawSettingsCombos()
444
445  # Call populate() method to populate the listboxes and comboboxes
446  populate()
447
448  winFrame.mainloop()
```

Output: Case Study Chapter 4

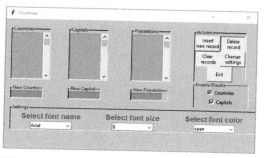

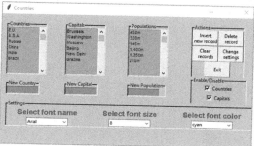

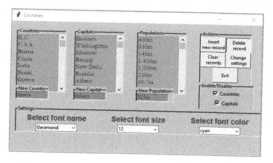

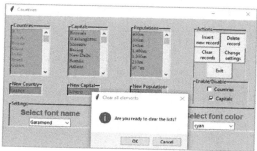

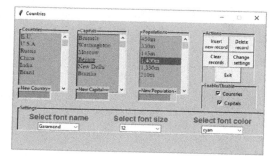

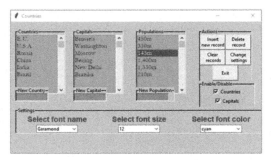

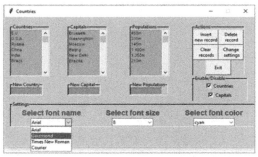

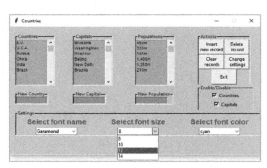

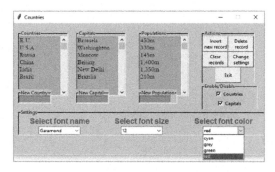

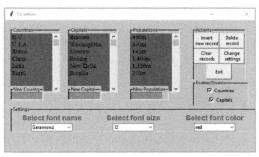

CHAPTER 5 – APPLICATION DEVELOPMENT WITH PYTHON

DESCRIPTION

Complete the integration of the *Basic Widgets* Python script from Chapters 4 with a full menu system in an object-oriented application, using all three types of menus (i.e., regular, toolbar, popup), as described in this chapter. The menu system should include the following options: *Color dialog, Open File dialog, Separator, Basic Widgets, Save As, Separator, About, Exit*.

SOLUTION

Firstly, the main application must be created that will include the main interface with all the three types of menus and the call to the *Basic Widgets* application and the APIs for the *Color Dialog, Open File Dialog, Save As,* and the *About* window. The script for the application follows:

```
1    import tkinter as tk
2    from tkinter import ttk
3
4    from tkinter import filedialog
5    from tkinter import colorchooser
6    from tkinter import Menu
7    from tkinter import *
8    from tkinter import messagebox as mbox
9    # Import from PIL the necessary image processing classes
10   from PIL import Image, ImageTk
11
12   import Chapter5CaseStudyBasicWidgets
13
14   class CaseStudy:
15       global openFileToolTip, saveAsToolTip
16       global colorsDialogToolTip, exitToolTip
17       global photo1, photo2, photo3, photo4
18       global openFileButton, saveAsButton, colorsButton, exitButton
19
20       # Show the messagebox for the info of the particular application
21       def _msgBox(self):
22           mbox.showinfo("About", "The Case Study for the Chapter 5: \
23           \nApplication Development with Python")
24
25       # Exit the application
26       def _quit(self):
27           MsgBox = mbox.askquestion("Exit Application",
28               "Exit the application?", icon = "warning")
29           if MsgBox == 'yes':
30               self.Main_winFrame.quit()
31               self.Main_winFrame.destroy()
32               exit()
33           else:
34               msg2 =mbox.showinfo("Return", "You will now return to the \
35               application screen")
36
37       # Create the main project window
```

```
38      def __init__(self,Main_winFrame):
39
40          global openFileToolTip, saveAsToolTip
41          global colorsDialogToolTip, exitToolTip
42          global photo1, photo2, photo3, photo4
43          global openFileButton, saveAsButton, colorsButton, exitButton
44          global popupmenu
45
46          self.Main_winFrame = Main_winFrame
47          self.Main_winFrame.title("Application Development with Python")
48          self.Main_winFrame.config(bg = 'linen')
49          self.Main_winFrame.resizable(False, False)
50          self.Main_winFrame.geometry("800x600")
51
52          # Add a menu to the main frame
53          menuBar = tk.Menu(self.Main_winFrame)
54          self.Main_winFrame.config(menu = menuBar)
55          fileMenu = Menu(menuBar, tearoff = 0)
56          fileMenu.add_command(label = "Basic Widget",
57              command = self.BasicWidget_app)
58          fileMenu.add_separator()
59          fileMenu.add_command(label = "Exit (Ctrl+Q)",
60              command = self._quit,underline = 1, accelerator = "Ctrl+Q")
61          menuBar.add_cascade(label = "Basic Apps", menu = fileMenu)
62
63          helpMenu = Menu(menuBar, tearoff=0)
64          helpMenu.add_command(label = "About", command = self._msgBox)
65          menuBar.add_cascade(label = "Help", menu = helpMenu)
66
67          # Create the toolbar and invoke the bindButton function to
68          # bind them
69          self.images()
70          toolbar = tk.Frame(self.Main_winFrame, bd = 1, relief = RAISED)
71          toolbar.pack(side = TOP, fill = X)
72          openFileButton=tk.Button(toolbar, image=photo1, relief=FLAT)
73          saveAsButton=tk.Button(toolbar, image = photo2, relief = FLAT)
74          colorsButton=tk.Button(toolbar, image = photo3, relief = FLAT)
75          exitButton = tk.Button(toolbar, image = photo4, relief = FLAT)
76          self.bindButtons()
77          openFileButton.pack(side = LEFT, padx = 0, pady = 0)
78          saveAsButton.pack(side = LEFT, padx = 0, pady = 0)
79          colorsButton.pack(side = LEFT, padx = 0, pady = 0)
80          exitButton.pack(side = LEFT, padx = 0, pady = 0)
81
82          # Create the Popup menu
83          popupmenu = tk.Menu(self.Main_winFrame, tearoff = 0)
84          popupmenu.add_command(label="Open File dialog", image = photo1,
85              compound = LEFT, command = self.openDialog)
86          popupmenu.add_command(label = "Save As dialog", image = photo2,
87              compound = LEFT, command = self.saveAsDialog)
88          popupmenu.add_command(label = "Color dialog", image = photo3,
89              compound = LEFT, command = self.colorDialog)
```

```
90          popupmenu.add_separator()
91          popupmenu.add_command(label = "Exit", image = photo4,
92              compound = LEFT, command = self._quit)
93          self.Main_winFrame.bind('<Button-1>',
94              lambda event: self.popupMenu(event))
95
96      # Methods to be called using the interactive interface
97      # ----------------------------------------------------
98      def BasicWidget_app(self):
99          app1Frame = tk.Tk
100         app1 = Chapter5CaseStudyBasicWidgets.BasicWidgets(app1Frame)
101
102     # ToolBar Menu images
103     def images(self):
104         global photo1, photo2, photo3, photo4
105         image1 = Image.open("images/OpenFile.gif")
106         image1 = image1.resize((24, 24), Image.ANTIALIAS)
107         photo1 = ImageTk.PhotoImage(image1)
108         image2 = Image.open("images/SaveAs.gif")
109         image2 = image2.resize((24, 24), Image.ANTIALIAS)
110         photo2 = ImageTk.PhotoImage(image2)
111         image3 = Image.open("images/ColorsDialog.gif")
112         image3 = image3.resize((24, 24), Image.ANTIALIAS)
113         photo3 = ImageTk.PhotoImage(image3)
114         image4 = Image.open("images/Exit.gif")
115         image4 = image4.resize((24, 24), Image.ANTIALIAS)
116         photo4 = ImageTk.PhotoImage(image4)
117
118     # ToolBar Menu Application 1
119     def colorDialog(self):
120         # Assign the user's selection of the color to a set of variables
121         (rgbSelected, colorSelected) = colorchooser.askcolor()
122         # Use the color part of the set of variables to change the
123         # color of the form
124         self.Main_winFrame.config(background = colorSelected)
125
126     # ToolBar Menu Application 2
127     def openDialog(self):
128         filedialog.askopenfile(title = "Open File Dialog")
129
130     # ToolBar Menu Application 3
131     def saveAsDialog(self):
132         filedialog.asksaveasfilename(title = "Save As Dialog")
133
134     # ToolBar Menu Application 4
135     def quit(self):
136         self.winFrame.destroy()
137         exit()
138
139     # Call the Popup Menu
140     def popupMenu(self, event):
141         global popupmenu
```

```
142              popupmenu.tk_popup(event.x_root, event.y_root)
143
144      # Create the Tooltips to show and hide
145      def showToolTips(self, a):
146          global openFileToolTip, saveAsToolTip
147          global colorsDialogToolTip, exitToolTip
148          if (a == 1):
149              openFileToolTip = tk.Label(self.Main_winFrame,
150              relief = FLAT, text = "Open the Open File dialog",
151              background = 'cyan')
152
153              openFileToolTip.place(x = 25, y = 30)
154          if (a == 2):
155              saveAsToolTip = tk.Label(self.Main_winFrame, bd = 2,
156                  relief = FLAT, text = "Open the Save As Dialog",
157                  background = 'cyan')
158              saveAsToolTip.place(x = 50, y = 30)
159          if (a == 3):
160              colorsDialogToolTip = tk.Label(self.Main_winFrame,
161                  relief = FLAT, bd = 2,
162                  text = "Open the Colors Dialog", background = 'cyan')
163              colorsDialogToolTip.place(x = 75, y = 30)
164          if (a == 4):
165              exitToolTip = tk.Label(self.Main_winFrame, relief = FLAT,
166                  bd = 2, text = "Click to exit the application",
167                  background = 'cyan')
168              exitToolTip.place(x = 100, y = 30)
169
170      def hideToolTips(self, a):
171          global openFileToolTip, saveAsToolTip
172          global colorsDialogToolTip, exitToolTip
173
174          if (a == 1):
175              openFileToolTip.destroy()
176          if (a == 2):
177              saveAsToolTip.destroy()
178          if (a == 3):
179              colorsDialogToolTip.destroy()
180          if (a == 4):
181              exitToolTip.destroy()
182
183      # Define the bindButtons function to bind the buttons with the
184      # various events.
185      def bindButtons(self):
186          global openFileButton, saveAsButton, colorsButton, exitButton
187
188          openFileButton.bind('<Button-1>',
189              lambda event: self.openDialog())
190          openFileButton.bind('<Enter>',
191              lambda event: self.showToolTips(1))
192          openFileButton.bind('<Leave>',
193              lambda event: self.hideToolTips(1))
```

```
194          saveAsButton.bind('<Button-1>',
195             lambda event: self.saveAsDialog())
196          saveAsButton.bind('<Enter>',
197             lambda event: self.showToolTips(2))
198          saveAsButton.bind('<Leave>',
199             lambda event: self.hideToolTips(2))
200          colorsButton.bind('<Button-1>',
201             lambda event: self.colorDialog())
202          colorsButton.bind('<Enter>',
203             lambda event: self.showToolTips(3))
204          colorsButton.bind('<Leave>',
205             lambda event: self.hideToolTips(3))
206          exitButton.bind('<Button-1>',
207             lambda event: self._quit())
208          exitButton.bind('<Enter>',
209             lambda event: self.showToolTips(4))
210          exitButton.bind('<Leave>',
211             lambda event: self.hideToolTips(4))
212
213  # Create the main window frame
214  Main_winFrame = tk.Tk()
215  app = CaseStudy(Main_winFrame)
216  # Call the main window form of the application
217  Main_winFrame.mainloop()
```

Then, the main *Basic Widgets* application must be created and called from the main application. The application is the same as in the previous chapter with the addition of the manipulation of the color, the size, and the font name of the entry box text. The script is as follows:

```
1   import tkinter as tk
2   from tkinter import ttk
3   from tkinter import colorchooser
4   from tkinter.font import Font
5
6   class BasicWidgets():
7       global fonts
8       fonts = ['Arial', 'Tahoma', 'Verdana', 'Silom', 'Herculanum',
9               'Courier']
10      global fontSize, fontName, fontColor, colorSelected, fontDetails
11
12      # Declare the global variables
13      global tempText, winText
14      global textVar
15      global textEntryBox
16      global LfontName
17
18      # The method that hides the contents of the EntryBox
19      def hideEntryContents(self):
20          self.tempText = self.textEntryBox.get()
21          self.textEntryBox.delete(0, "end")
22
23      # The method that shows the contents of the Entry Box
```

```
24        def showEntryContents(self):
25            if self.tempText != '':
26                self.textEntryBox.insert(0,self.tempText)
27
28        # The method to enable or disable the Entry box
29        def enableDisableEntryWidget(self, a):
30            if (a == 'e'):
31                self.textEntryBox.config(state = 'normal')
32            elif (a == 'd'):
33                self.textEntryBox.config(state = 'disable')
34
35        # The method to turn the font to bold and back to normal
36        def boldContentsofEntryWidget(self, a):
37            global fontDetails
38
39            if (a == 'b'):
40                self.textEntryBox.config(font = 'bold')
41            elif (a == 'n'):
42                self.setFormatting()
43
44        # The method to change the font of the content to password and back
45        def passwordEntryWidget(self, a):
46            if (a == 'p'):
47                self.textEntryBox.config(show = '*')
48            elif (a == 'n'):
49                self.textEntryBox.config(show = '')
50
51        # The method to control the color options
52        def setColor(self):
53            global colorSelected, fontColor
54            (rgbSelected,self.colorSelected ) = colorchooser.askcolor()
55            self.fontColor = self.colorSelected
56            self.setFormatting()
57
58        # The method to change the formatting option
59        def change(self, event):
60            global LfontName, fontName, fonts, fontIndex
61            fontIndex = LfontName.current()
62            self.fontName.set(fonts[fontIndex])
63            self.setFormatting()
64
65        # The method to select the font size based on the onScale selection
66        def onScale(self,val):
67            global fontSize
68            self.fontSize = str(val)
69            self.setFormatting()
70
71        # The method to finalize the format of the text
72        def setFormatting(self):
73            global fontSize, fontColor, fontName, fontDetails
74
75            fontDetails = "" + str(self.fontName.get()) + " " + \
```

```
76                    str(self.fontSize)
77            self.textEntryBox.config(foreground = self.fontColor,
78                font = fontDetails)
79
80        # Initialize the widgets of the form
81        def __init__(self,winFrame):
82
83            global textVar
84            global winText, tempText
85            global fontSize, fontName, fontColor, fontDetails
86            global LfontName
87
88            self.tempText = ''
89
90            winFrame = tk.Tk()
91            self.winFrame = winFrame
92            self.winFrame.title("Basic Widgets")
93            self.winFrame.config(bg = 'light grey')
94            self.winFrame.resizable(True, True)
95            self.winFrame.geometry('335x290')
96
97            # The label widget
98            winLabel = tk.Label(self.winFrame,
99                text = "Enter your Text", width = 15)
100           winLabel.config(bg = 'linen', font = "Arial 14 bold")
101           winLabel.grid(column = 0, row = 0)
102
103           # The textEditbox widget
104           textVar = tk.StringVar()
105           self.textEntryBox = ttk.Entry(self.winFrame,
106               width = 18, textvariable = textVar)
107           self.textEntryBox.grid(column = 1, row = 0)
108
109           # Button 1: Show the Entry widget
110           btn1_show_entry_w = tk.Button(self.winFrame,
111               text = "Show the\n Entry Widget", fg = 'red')
112           btn1_show_entry_w.config(width = 18,
113               font = 'Arial 11 bold', borderwidth = 8 ,
114               command = self.textEntryBox.grid)
115           btn1_show_entry_w.grid(column = 0, row = 1)
116
117           # Button 2: Hide the Entry widget
118           btn2_hide_entry_w = tk.Button(self.winFrame,
119               text = "Hide the\n Entry Widget", fg = 'red')
120           btn2_hide_entry_w.config(width = 28,
121               font = "Arial 11 bold", borderwidth = 8,
122               command = self.textEntryBox.grid_remove)
123           btn2_hide_entry_w.grid(column = 1, row = 1)
124
125           # Button 3: Hide the contents of the entry widget
126           btn3_Hide_EntryContent = tk.Button(self.winFrame,
127               text = "Hide the contents\n of the Entry Widget",
```

```
128             fg = 'steelblue')
129         btn3_Hide_EntryContent.config(width = 18,
130             font = "Arial 11 bold", borderwidth = 8 ,
131             command = self.hideEntryContents)
132         btn3_Hide_EntryContent.grid(column = 0, row = 2)
133
134         # Button 4: Show the contents of the entry widget
135         btn4_show_EntryContent = tk.Button(self.winFrame,
136             text = "Show the contents\n of the Entry Widget",
137             fg = 'steelblue')
138         btn4_show_EntryContent.config(width = 28,
139             font = "Arial 11 bold", borderwidth = 8,
140             command = self.showEntryContents)
141         btn4_show_EntryContent.grid(column = 1, row = 2)
142
143         # Button 5: Enable the entry widget
144         btn5_ebable_EntryW = tk.Button(self.winFrame,
145             text = "Enable the\n Entry Widget", fg = 'green')
146         btn5_ebable_EntryW.config(width = 18,
147             font = "Arial 11 bold", Borderwidth = 8)
148         btn5_ebable_EntryW.grid(column = 0, row = 3)
149         btn5_ebable_EntryW.bind('<Button-1>', lambda event,
150             a = 'e': self.enableDisableEntryWidget(a))
151
152         # Button 6: Disable the entry widget
153         btn6_Disable_EntryW = tk.Button(self.winFrame,
154             text = "Disable the\n Entry Widget", fg = 'green')
155         btn6_Disable_EntryW.config(width = 28,
156             font = "Arial 11 bold", borderwidth = 8)
157         btn6_Disable_EntryW.grid(column = 1, row = 3)
158         btn6_Disable_EntryW.bind('<Button-1>', lambda event,
159             a = 'd': self.enableDisableEntryWidget(a))
160
161         self.fontBold = False
162
163         # Button 7: Change the content of the entry widget to Bold
164         btn7_Bold_EntryW = tk.Button(self.winFrame,
165             text = "Bold  contents of\n the Entry Widget",
166             fg = 'orchid')
167         btn7_Bold_EntryW.config(width = 18,
168             font = "Arial 11 bold", borderwidth = 8)
169         btn7_Bold_EntryW.grid(column = 0, row = 4)
170         btn7_Bold_EntryW.bind('<Button-1>', lambda event,
171             a = 'b': self.boldContentsofEntryWidget(a))
172
173         # Button 8: Return the contents of the entry widget back to
174         # normal
175         btn8_normalFont_EntryW = tk.Button(self.winFrame,
176             text = "Return to normal \nthe Entry Widget",
177             fg = 'orchid')
178         btn8_normalFont_EntryW.config(width = 28,
```

```
179                 font = "Arial 11 bold", borderwidth = 8)
180         btn8_normalFont_EntryW.grid(column = 1, row = 4)
181         btn8_normalFont_EntryW.bind('<Button-1>', lambda event,
182             a = 'n': self.boldContentsofEntryWidget(a))
183
184         # Button 9: Change the contents of the entry widget to password
185         btn9_Password_EntryW = tk.Button(self.winFrame,
186             text = "Show Entry Contents\n as Password Text",
187             fg = 'purple')
188         btn9_Password_EntryW.config(width = 18,
189             font = "Arial 11 bold", borderwidth = 8)
190         btn9_Password_EntryW.grid(column = 0, row = 5)
191         btn9_Password_EntryW.bind('<Button-1>', lambda event,
192             a = 'p': self.passwordEntryWidget(a))
193
194         # Button 10: Change the contents of the entry widget back
195         # to normal
196         btn10_noPassword_EntryW = tk.Button(self.winFrame,
197             text = "Show Entry Contents\n as Normal Text",
198             fg = 'purple')
199         btn10_noPassword_EntryW.config(width = 28,
200             font = "Arial 11 bold", borderwidth = 8)
201         btn10_noPassword_EntryW.grid(column = 1, row = 5)
202         btn10_noPassword_EntryW.bind('<Button-1>', lambda event,
203             a = 'n': self.passwordEntryWidget(a))
204
205         # Button 11: Select the preferred font size
206         btn11_FontSize = tk.Button(self.winFrame,
207             text = "Use the scale to\nselect the font size",
208             fg = 'dark grey')
209         btn11_FontSize.config(width = 18, font = "Arial 11 bold",
210             borderwidth = 8)
211         btn11_FontSize.grid(column = 0, row = 6)
212
213         self.fontSize = 8
214         fontScale = tk.Scale(self.winFrame, length = 200 ,
215             from_ = 8, to = 16 )
216         fontScale.config(resolution = 1, activebackground = 'darkblue',
217             orient = 'horizontal', command = self.onScale,
218             bg = 'linen', fg = 'DarkMagenta', troughcolor = 'Thistle')
219         fontScale.grid(column = 1 , row = 6)
220
221         # Button 12: Select the preferred font name
222         btn12_FontName = tk.Button(self.winFrame, fg = 'brown',
223             text = "Use the combobox to\nselect the font name")
224         btn12_FontName.config(width = 18, font = "Arial 11 bold",
225             borderwidth = 8)
226         btn12_FontName.grid(column = 0, row = 7)
227
228         self.fontName = tk.StringVar()
229         self.fontName.set(fonts[0])
```

```
230            LfontName = ttk.Combobox(self.winFrame,
231                textvariable = self.fontName, width = 18)
232            LfontName['values'] = fonts; LfontName.current(1)
233            LfontName.grid(column = 1 , row = 7)
234            LfontName.bind("<<ComboboxSelected>>", self.change)
235
236            # Button 13: Create the Font Color Selector
237            self.fontColor = "#00fcff"
238            btn13_ColorSelector = tk.Button(self.winFrame,
239                Text = 'Select the preferred\ntext color',
240                width = 18, Fg = 'orange',font = "Arial 11 bold",
241                command = self.setColor)
242            btn13_ColorSelector.grid(column = 0, row = 8)
243
244            # Start the form
245            self.winFrame.mainloop()
```

Output: Case Study Chapter 5

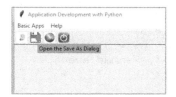

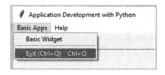

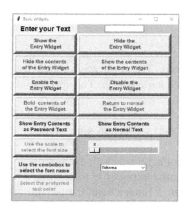

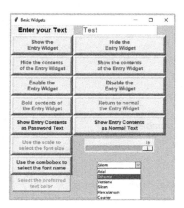

CHAPTER 6 – DATA STRUCTURES AND ALGORITHMS WITH PYTHON

DESCRIPTION

Create an application that implements the specified algorithms and tasks. The application should use a *GUI interface* in the form of a *tabbed notebook*, using *one tab for each algorithm*. The application requirements are the following:

1. Implement the Following Static Sorting Algorithms: bubble sort, insertion sort, shaker sort, merge sort.
2. Ask the user to enter a regular arithmetic expression in a form of a phrase, with each of the operators limited to *single digit integer numbers*. Convert the infix expression to postfix.
3. Ask the user to enter a sequence of integers, insert them into a binary search tree and implement the BST ADS algorithm with both inorder and postorder traversals.

SOLUTION

```
1    # Import the necessary libraries
2    import tkinter as tk
3    from tkinter import ttk
4
5    # Import the random module to generate random numbers
6    import random
7    import time
8
9    # Declare and/or initialise the global variables and widgets
10   global tab1, tab2, tab3, tab4, tab5
11   global i, j, k, mergeComp
12   global textVar
13   global winLabel
14   global tempText
15   global list
16
17   #----------- CASE STUDY - PART A: SORTING -----------
18   #------Bubble Sort------
19   def bubbleSort(size):
20       global list
21       comparisons = 0
22
23       # Start the timer
24       startTime = time.process_time()
25
26       # The Bubble sort algorithm
27       for i in range (size - 1):
28           for j in range (size - 1):
29               comparisons += 1
30               if (list[j] > list[j + 1]):
31                   temp = list[j]
32                   list[j] = list[j + 1]
33                   list[j+1] = temp
34
```

```
35          # End the timer
36          endTime = time.process_time()
37
38          listToString = ' '.join(map(str, list))
39          winLabel1 = tk.Label(tab1,
40              text = "The sorted list is: " + listToString)
41          winLabel1.grid(column = 1, row = 1)
42          winLabel2 = tk.Label(tab1,
43              text = "The number of comparisons is = " + str(comparisons))
44          winLabel2.grid(column = 1, row = 3)
45          winLabel3 = tk.Label(tab1, text="The elapsed time in seconds = " +
46              str(endTime - startTime))
47          winLabel3.grid(column = 1, row = 5)
48
49  #------ Insertion Sort ------
50  def insertionSort(size):
51      global list
52      comparisons = 0
53
54      # Start the timer
55      startTime = time.process_time()
56
57      # The Insertion sort algorithm
58      for i in range(1, size):
59          temp = list[i]
60          loc = i
61          while ((loc > 0) and (list[loc - 1] > temp)):
62              comparisons += 1
63              list[loc] = list[loc - 1];
64              loc = loc - 1
65          list[loc] = temp
66
67      # End the timer
68      endTime = time.process_time()
69
70      listToString = ' '.join(map(str, list))
71      winLabel1 = tk.Label(tab2,
72          text = "The sorted list is: " + listToString)
73      winLabel1.grid(column = 1, row = 1)
74      winLabel2 = tk.Label(tab2, text="The number of comparisons is = " +
75          str(comparisons))
76      winLabel2.grid(column = 1, row = 3)
77      winLabel3 = tk.Label(tab2, text="The elapsed time in seconds = " +
78          str(endTime - startTime))
79      winLabel3.grid(column = 1, row = 5)
80
81  #------ Shaker Sort ------
82  def shakerSort(size):
83      global list
84      comparisons = 0
85
86      # Start the timer
```

```
87      startTime = time.process_time()
88
89      # The Shaker Sort algorithm
90      swapped = True; start = 0; end = size - 1
91
92      # Keep running the Shaker Sort while there are swaps taking place
93      while (swapped == True):
94          # Set swap to false to start the new loop
95          swapped = False;
96
97          # Loop from left to right using Bubble sort
98          for i in range(start, end):
99              comparisons += 1
100             if (list[i] > list[i + 1]):
101                 temp = list[i]; list[i] = list[I + 1]
102                 list[I + 1] = temp; swapped = True;
103         # If there were no swaps, then the list is sorted
104         if (swapped == False):
105             break
106         # If there was at least one swap, then reset swap to false
107         # and continue
108         else:
109             swapped = False
110         # Decrease the end of the list to -1 since one more largest
111         # element moved to the right
112         end -= 1
113
114         # Loop from right to left using Bubble sort
115         for i in range (end, start, -1):
116             comparisons += 1
117             if (list[i] < list[i - 1]):
118                 temp = list[i]; list[i] = list[i - 1]
119                 list[i-1] = temp; swapped = True
120
121         # Increase the start of the list by 1 since one more smallest
122         # element moved to the left
123         start += 1
124
125     # End the timer
126     endTime = time.process_time()
127
128     # Print output
129     listToString = ' '.join(map(str, list))
130     winLabel1 = tk.Label(tab3, text = "The sorted list is: " +
131         listToString)
132     winLabel1.grid(column = 1, row = 1)
133     winLabel2 = tk.Label(tab3, text="The number of comparisons is = " +
134         str(comparisons))
135     winLabel2.grid(column = 1, row = 3)
136     winLabel3 = tk.Label(tab3, text="The elapsed time in seconds = " +
137         str(endTime - startTime))
138     winLabel3.grid(column = 1, row = 5)
```

```
139
140  #------ Merge Sort ------
141  def merge(first, middle, last):
142      global list
143      global i, j, k, mergeComp
144      size1 = middle - first + 1; size2 = last - middle
145
146      # Create temporary lists
147      leftList = []; rightList = []
148
149      # Copy data of the original list to temporary lists leftList
150      # and rightList
151      for i in range(0 , size1):
152          leftList.append(list[first + i])
153      for j in range(0 , size2):
154          rightList.append(list[middle + 1 + j])
155
156      # Merge the temporary lists leftList and rightList into the
157      # original list until one of the sub-lists is empty
158      i = 0; j = 0; k = first
159      while (i < size1 and j < size2):
160          if (leftList[i] <= rightList[j]):
161              list[k] = leftList[i]; i += 1; mergeComp += 1
162          else:
163              list[k] = rightList[j]; j += 1; mergeComp += 1
164          k += 1
165
166      # If the leftList becames empty, copy its remaining elements to the
167      # original list
168      while (i < size1):
169          list[k] = leftList[i]; i += 1; k += 1
170
171      # If the rightList becames empty, copy its remaining elements to
172      # the original list
173      while (j < size2):
174          list[k] = rightList[j]; j += 1; k += 1
175
176  # The mergeSort algorithm
177  def mergeSort(first, last):
178      global list
179      # Start the timer
180      startTime = time.process_time()
181
182      # The recursive step
183      if (first <= last-1):
184          middle = (first + last)//2
185          mergeSort(first, middle)
186          mergeSort(middle + 1, last)
187          merge(first, middle, last)
188
189      # End the timer
190      endTime = time.process_time()
```

```
191
192      # Print output
193      listToString = ' '.join(map(str, list))
194      winLabel1 = tk.Label(tab4,
195          text = "The sorted list is: " + listToString)
196      winLabel1.grid(column = 1, row = 1)
197      winLabel2 = tk.Label(tab4,
198          text = "The number of comparisons is = " + str(mergeComp))
199      winLabel2.grid(column = 1, row = 3)
200      winLabel3 = tk.Label(tab4,
201          text = "The elapsed time in seconds = " +
202          str(endTime - startTime))
203      winLabel3.grid(column = 1, row = 5)
204
205  #----- Sorting Function Calls -----
206  def multiSorting():
207      global list
208      global i, j, k, mergeComp
209      list = []
210      i, j, k, mergeComp = 0, 0, 0, 0
211      size = 10
212
213      # Use the randint() method of the random class to generate
214      # random integers
215      for i in range (size):
216          newNum = random.randint(-100, 100)
217          list.append(newNum)
218
219      # Call sorting functions
220      bubbleSort(size)
221      insertionSort(size)
222      shakerSort(size)
223      mergeSort(0, size - 1)
224
225  #-------------------------------------------------------------------
226  #------ CASE STUDY PART B: INFIX TO POSTFIX ------
227  #------ Infix to Postfix ------
228  # Source: https://cppsecrets.com/users/
229  # 258265898665726650506471776573764667797/
230
231  # INFIX-TO-POSTFIX-CONVERSION-USING-STACK.php
232
233  def infixToPostfix(infixExpression):
234      global textVar
235      global winLabel
236
237      # Initialize set of operators and priorities dictionary
238      OPERATORS = set(['+', '-', '*', '/', '(', ')', '^'])
239      PRIORITY = {'+':1, '-':1, '*':2, '/':2, '^':3}
240
241      # Initialize stack and output expression
242      stack = []
```

```
243        Output = ''
244
245        # Infix to Postfix conversion
246        for ch in infixExpression:
247            if ch not in OPERATORS:
248                Output += ch
249            elif ch == '(':
250                stack.append('(')
251            elif ch == ')':
252                while stack and stack[-1] != '(':
253                    Output += stack.pop()
254                stack.pop()
255            else:
256                while stack and stack[-1] != \
257                    '(' and PRIORITY[ch] <= PRIORITY[stack[-1]]:
258                    Output += stack.pop()
259                stack.append(ch)
260        while stack:
261            Output += stack.pop()
262
263        textVar.set(Output)
264
265   # Create GUI
266   def createInfixPostfixGUI():
267        global textVar
268        textVar = tk.StringVar()
269
270        # Create and initialize labels, buttons and text boxes
271        winLabel1 = tk.Label(tab5, text = 'Enter Infix Expression:')
272        winLabel1.grid(column = 1, row = 1)
273
274        winLabel2 = tk.Label(tab5, textvariable = textVar)
275        winLabel2.grid(column = 2, row = 3)
276
277        textVarLocal = tk.StringVar()
278        textVarLocal.set('(A * B) + C - D')
279        winText = ttk.Entry(tab5, textvariable = textVarLocal)
280        winText.grid(column = 2, row = 1)
281
282        winButtonConvert = tk.Button(tab5, text = 'Convert to Postfix')
283        winButtonConvert.grid(column = 1, row = 3)
284        winButtonConvert.bind('<Button-1>',
285            lambda event: infixToPostfix(textVarLocal.get()))
286
287   #-------------------------------------------------------------------
288   #------ CASE STUDY PART C: BST POSTORDER ------
289   # BST class initialization
290   class BinarySearchTree:
291        def __init__(self, key):
292            self.left = None
293            self.right = None
294            self.data = key
```

```
295
296   # BST class initialization
297   def insert(root, newData):
298
299       if (root == None):
300           return BinarySearchTree(newData)
301       else:
302           if root.data == newData:
303               return root
304           elif root.data < newData:
305               root.right = insert(root.right, newData)
306           else:
307               root.left = insert(root.left, newData)
308       return root
309
310   # Inorder Traversal
311   def traverseInorderBST(root):
312       global tempText
313
314       # If the BST current node is not a leaf traverse the left subtree.
315       # If it is a leaf, print its data & then traverse the right subtree
316       if (root):
317           traverseInorderBST(root.left)
318           tempText = tempText + ' ' + str(root.data)
319           traverseInorderBST(root.right)
320
321   # Postorder Traversal
322   def traversePostorderBST(root):
323       global tempText
324
325       # If the BST current node is not a leaf traverse the left subtree.
326       # If it is a leaf, print its data & then traverse the right subtree
327       if (root):
328           traversePostorderBST(root.left)
329           traversePostorderBST(root.right)
330           tempText = tempText + ' ' + str(root.data)
331
332   # Create GUI and call traversal functions
333   def createInorderPostorderGUI():
334       global tempText
335
336       newData = random.randint(-100, 100)
337       winLabel1 = tk.Label(tab6, text = 'Root: ')
338       winLabel1.grid(column = 1, row = 1)
339       winLabel2 = tk.Label(tab6, text = str(newData))
340       winLabel2.grid(column = 2, row = 1)
341       bst = BinarySearchTree(newData)
342
343       for i in range (10):
344           newData = random.randint(-100, 100)
```

```
345            bst = insert(bst, newData)
346
347     tempText = ''
348     traverseInorderBST(bst)
349     winLabel3 = tk.Label(tab6, text = 'Inorder Traversal: ')
350     winLabel3.grid(column = 1, row = 3)
351     winLabel4 = tk.Label(tab6, text = tempText)
352     winLabel4.grid(column = 2, row = 3)
353
354     tempText = ''
355     traversePostorderBST(bst)
356     winLabel5 = tk.Label(tab6, text = 'Postorder Traversal: ')
357     winLabel5.grid(column = 1, row = 5)
358     winLabel6 = tk.Label(tab6, text = tempText)
359     winLabel6.grid(column = 2, row = 5)
360
361 #------------------------------------------------------------------------
362 #----------------------- MAIN BODY --------------------------------------
363 winFrame = tk.Tk()
364 winFrame.title("Chapter 6 - Case Study")
365 winFrame.resizable(True, True)
366 winFrame.geometry('500x150')
367
368 # Create the notebook with the tab pages
369 tabbedInterface = ttk.Notebook(winFrame)
370 tab1 = ttk.Frame(tabbedInterface)
371 tabbedInterface.add(tab1, text = "Bubble Sort")
372 tab2 = ttk.Frame(tabbedInterface)
373 tabbedInterface.add(tab2, text = "Insertion Sort")
374 tab3 = ttk.Frame(tabbedInterface)
375 tabbedInterface.add(tab3, text = "Shaker Sort")
376 tab4 = ttk.Frame(tabbedInterface)
377 tabbedInterface.add(tab4, text = "Merge Sort")
378 tab5 = ttk.Frame(tabbedInterface)
379 tabbedInterface.add(tab5, text = "Infix to Postfix")
380 tab6 = ttk.Frame(tabbedInterface)
381 tabbedInterface.add(tab6, text = "Postorder Traversal")
382 tabbedInterface.pack()
383
384 # Invoke the Sorting functions
385 multiSorting()
386
387 # Invoke the Infix to Postfix functions
388 createInfixPostfixGUI()
389
390 # Invoke the Inorder to Postorder functions
391 createInorderPostorderGUI()
392
393 winFrame.mainloop()
394 #------------------------------------------------------------------------
```

Output: Case Study Chapter 6

Chapter 6 - Case Study

| Bubble Sort | Insertion Sort | Shaker Sort | Merge Sort | Infix to Postfix | Postorder Traversal |

The sorted list is: -84 -73 -58 -57 -42 -13 -5 10 45 72
The number of comparisons is = 81
The elapsed time in seconds = 0.0

Chapter 6 - Case Study

| Bubble Sort | Insertion Sort | Shaker Sort | Merge Sort | Infix to Postfix | Postorder Traversal |

The sorted list is: -96 -93 -42 -25 -14 33 38 67 72 82
The number of comparisons is = 0
The elapsed time in seconds = 0.0

Chapter 6 - Case Study

| Bubble Sort | Insertion Sort | Shaker Sort | Merge Sort | Infix to Postfix | Postorder Traversal |

The sorted list is: -91 -84 -52 -43 3 23 39 51 72 96
The number of comparisons is = 9
The elapsed time in seconds = 0.0

Chapter 6 - Case Study

| Bubble Sort | Insertion Sort | Shaker Sort | Merge Sort | Infix to Postfix | Postorder Traversal |

The sorted list is: -72 -67 -40 -33 -28 18 30 31 62 63
The number of comparisons is = 19
The elapsed time in seconds = 0.015625

Chapter 6 - Case Study

| Bubble Sort | Insertion Sort | Shaker Sort | Merge Sort | Infix to Postfix | Postorder Traversal |

Enter Infix Expression: (A+B*C-A*B)+(A-B)
Convert to Postfix   ABC*+AB*-AB-+

Chapter 6 - Case Study

| Bubble Sort | Insertion Sort | Shaker Sort | Merge Sort | Infix to Postfix | Postorder Traversal |

Root: -13
Inorder Traversal: -97 -62 -58 -45 -28 -13 -12 28 30 43 90
Postorder Traversal: -97 -58 -28 -45 -62 -12 30 28 90 43 -13

CHAPTER 7 – DATABASE PROGRAMMING WITH PYTHON

DESCRIPTION

Create an application that provides the following functionality:

a. Prompt the user for their credentials and the name of the MySQL database to connect to. Display a list of the tables that are available in the connected database in a *status bar* form at the bottom of the application window (Hint: A *label* can be used for this purpose).

b. Allow the user to define a new table and set the number of its attributes. Based on user selection, create the interface required for the specifications of the attributes in the new table (i.e., *attribute name*, *type*, and *size*, *primary* or *foreign key* designation). The interface should be created on-the-spot.

The application must use a GUI interface and the MySQL facilities for the database element.

SOLUTION

```
1    import mysql.connector
2    import tkinter as tk
3    from tkinter import ttk
4
5    global LoginF, DescF, columnDetailFrame
6    global userVar, passVar, hostVar, dbVar, statusVar, tableVar
7    global tablesCombo, columnsCombo
8    global connect, cursor, config; global numOfColumns
9    global tables, columns, dataTypes, primaryKey
10   #-------------------------------------------------------------------
11   def initializeVarObjects():
12       global userVar, passVar, hostVar, dbVar, statusVar, tableVar
13       global numOfColumns; global tables, columns, dataTypes, primaryKey
14
15       userVar = tk.StringVar(); userVar.set("Enter the username here")
16       passVar = tk.StringVar(); passVar.set("Enter the password here")
17       hostVar = tk.StringVar(); hostVar.set("Enter the host here")
18       dbVar = tk.StringVar(); dbVar.set("Enter the database name here")
19       tableVar = tk.StringVar(); tableVar.set("")
20       statusVar = tk.StringVar(); statusVar.set("")
21       numOfColumns = 0; dataTypes = ["Char", "Integer"]
22       primaryKey = [True, False]; tables = []; columns = []
23   #-------------------------------------------------------------------
24   # Define the function to control the Column Size Scale widget change
25   def onScale(val):
26       global numOfColumns
27       v = int(val)
28       numOfColumns = v
29   #-------------------------------------------------------------------
30   def createGUI():
31       global userVar, passVar, hostVar, dbVar, statusVar
32       global tablesCombo; global connect, cursor, config
33       global LoginF, DescF, StatusF
34
```

```
35        # Create the frames for the GUI
36        LoginF = tk.LabelFrame(winFrame, text = 'Login',
37            bg = 'light grey', fg = 'red')
38        LoginF.grid(column = 0, row = 0)
39
40        # Create the labels and entry boxes
41        userL = tk.Label(LoginF, text = "Username:",
42            bg = "light grey").grid(column = 0, row = 0)
43        userT = ttk.Entry(LoginF, textvariable = userVar,
44            width = 46).grid(column = 1, row = 0)
45        passL = tk.Label(LoginF, text = "Password:",
46            bg = "light grey").grid(column = 0, row = 1)
47        passT = ttk.Entry(LoginF, textvariable = passVar,
48            width = 46).grid(column = 1, row = 1)
49        hostL = tk.Label(LoginF, text = "Hostname:",
50            bg = "light grey").grid(column = 0, row = 2)
51        hostT = ttk.Entry(LoginF, textvariable = hostVar,
52            width = 46).grid(column = 1, row = 2)
53        dbL = tk.Label(LoginF, text = "DB Name:",
54            bg = "light grey").grid(column = 0, row = 3)
55        dbT = ttk.Entry(LoginF, textvariable = dbVar,
56            width = 46).grid(column = 1, row = 3)
57        connectB = tk.Button(LoginF, text = "Connect", fg = 'blue')
58        connectB.bind("<Button-1>", lambda event: dbConnect())
59        connectB.grid(columnspan = 2, column = 0, row = 4)
60
61        # The frame to display the status of the operations
62        StatusF = tk.LabelFrame(winFrame, text = 'Status',
63            bg = 'light grey', fg = 'red')
64        StatusF.grid(column = 0, row = 3)
65        statusT = ttk.Entry(StatusF, textvariable = statusVar,
66            width = 55).grid(column = 1, row = 0)
67    #-------------------------------------------------------------------
68    def dbCreateTableGUI():
69        global NewTableFrame; global tableNoColsScale; global tableNameVar
70
71    # The frame for the creation of the new table and its attributes
72        NewTableFrame = tk.LabelFrame(winFrame, text = 'New Table',
73            bg = 'light grey', fg = 'red')
74        NewTableFrame.grid(column = 0, row = 1)
75        # Create the label, and entry for the name of the new table
76        tableNameLabel = tk.Label(NewTableFrame, text = "Name:",
77            bg = "light grey")
78        tableNameLabel.grid(column = 0, row = 0)
79        tableNameVar = tk.StringVar()
80        tableNameVar.set("Enter name")
81        tableNameText = ttk.Entry(NewTableFrame,
82            textvariable = tableNameVar, width = 26)
83        tableNameText.grid(column = 1, row = 0)
84        tableNoColsLabel = tk.Label(NewTableFrame,
85            text = "No. of columns:", bg = "light grey")
86        tableNoColsLabel.grid(column = 2, row = 0)
87        tableNoColsVar = tk.IntVar()
```

```
88      tableNoColsScale = tk.Scale(NewTableFrame, length = 100,
89          from_ = 0, to = 40)
90      tableNoColsScale.config(resolution = 1, orient = 'horizontal')
91      tableNoColsScale.config(bg = 'light grey', fg = 'red',
92          troughcolor = 'cyan', command = onScale)
93      tableNoColsScale.grid(column = 3, row = 0)
94      # Create button to finalize the number of columns in the new table
95      tableNoColsButton = tk.Button(NewTableFrame,
96          text = "Finalize columns", fg = "blue")
97      tableNoColsButton.bind("<Button-1>",
98          lambda event: dbFinalizeColumns())
99      tableNoColsButton.grid(column = 1, row = 1)
100     # Create the button to execute the statement that will create
101     # the new table
102     tableNoColsButton = tk.Button(NewTableFrame, text = "Create table",
103         fg = "blue")
104     tableNoColsButton.bind("<Button-1>", lambda event: dbCreateTable())
105     tableNoColsButton.grid(column = 2, row = 1)
106 #-----------------------------------------------------------------
107 # Define the function to finalize the number of columns in the new table
108     def dbFinalizeColumns():
109         global columnTypeCombo, columnPrimaryKeyCombo
110         global tableNoColsScale, columnSizeScale
111         global connect
112         global columnNameText
113         global tables, columns, dataTypes, primaryKey
114         global statusVar
115         global tablesCombo, columnsCombo; global ColumnDetailFrame
116
117     numOfColumns = int(tableNoColsScale.get())
118     statusVar.set("Decision is for " + str(numOfColumns) + \
119         " attributes/columns")
120
121     columnNameLabel = []*numOfColumns
122     columnTypeLabel = []*numOfColumns
123     columnSizeLabel = []*numOfColumns
124     columnPrimaryKeyLabel = []*numOfColumns
125     columnNameText = []*numOfColumns
126     referenceTableLabel = []*numOfColumns
127     columnTypeCombo = []*numOfColumns
128     columnPrimaryKeyCombo = []*numOfColumns
129     columnSizeScale = []*numOfColumns
130
131     # The frame for the definition of each of the columns of the table
132     ColumnsFrame = tk.LabelFrame(winFrame,
133         text = 'The Attributes of the table')
134     ColumnsFrame.config(bg = 'light grey', fg = 'red', bd = 2,
135         relief = 'sunken')
136     ColumnsFrame.grid(column = 0, row = 2)
137
138     ColumnDetailFrame = [None]*numOfColumns
139     for i in range(numOfColumns):
140         # Create the label frame to accept the attribute details
```

```
141        ColumnDetailFrame[i] = tk.LabelFrame(ColumnsFrame,
142            text = "Column " + str(i+1) + " details")
143        ColumnDetailFrame[i].config(bg = 'light grey', fg = 'red', )
144        ColumnDetailFrame[i].grid(column = 0, row = i)
145
146        # Create the label to prompt for the name of the new column
147        newLabel = tk.Label(ColumnDetailFrame[i],
148            text = "Name:", bg = "light grey")
149        columnNameLabel.append(newLabel)
150        columnNameLabel[i].grid(column = 0, row = 0)
151
152        # Create the text to accept the name of the new column
153        newColumnNameText = ttk.Entry(ColumnDetailFrame[i], width = 8)
154        columnNameText.append(newColumnNameText)
155        columnNameText[i].grid(column = 1, row = 0)
156
157        # Create the combobox to select the data type of the new column
158        newColumnTypeLabel = tk.Label(ColumnDetailFrame[i],
159            text = "Type:", bg = "light grey")
160        columnTypeLabel.append(newColumnTypeLabel)
161        columnTypeLabel[i].grid(column = 2, row = 0)
162        newColumnTypeCombo = ttk.Combobox(ColumnDetailFrame[i],
163            width = 4)
164        newColumnTypeCombo['values'] = dataTypes
165        newColumnTypeCombo.current(0)
166        columnTypeCombo.append(newColumnTypeCombo)
167        columnTypeCombo[i].grid(column = 3, row = 0)
168
169        # Create the scale to select the size of the column
170        newColumnSizeLabel = tk.Label(ColumnDetailFrame[i],
171            text = "Size:", bg = "light grey")
172        columnSizeLabel.append(newColumnSizeLabel)
173        columnSizeLabel[i].grid(column = 4, row = 0)
174        newColumnSize = tk.Scale(ColumnDetailFrame[i],
175            length = 61, from_ = 0, to = 30)
176        newColumnSize.config(resolution = 1, orient = 'horizontal')
177        newColumnSize.config(bg = 'light grey', fg = 'red',
178                troughcolor = 'cyan', command = onScale)
179        newColumnSize.grid(column = 5, row = 0)
180        columnSizeScale.append(newColumnSize)
181
182        # Create the combobox to decide if the current column is part
183        # a primary key
184        newColumnPrimaryKeyLabel = tk.Label(ColumnDetailFrame[i],
185            text = "Primary\nKey?", bg = "light grey")
186        columnPrimaryKeyLabel.append(newColumnPrimaryKeyLabel)
187        columnPrimaryKeyLabel[i].grid(column = 6, row = 0)
188        newColumnPrimaryKeyCombo = ttk.Combobox(
189            ColumnDetailFrame[i], width = 4)
190        newColumnPrimaryKeyCombo['values'] = primaryKey
191        newColumnPrimaryKeyCombo.current(0)
192        columnPrimaryKeyCombo.append(newColumnPrimaryKeyCombo)
```

```
193              columnPrimaryKeyCombo[i].grid(column = 7, row = 0)
194  #-------------------------------------------------------------------
195  # Define the function to create the new table
196  def dbCreateTable():
197      global tableNameVar, statusVar; global connect
198      global columnNameText
199      global columnTypeCombo, columnPrimaryKeyCombo
200      global columnSizeScale, tableNoColsScale
201
202      cursor = connect.cursor()
203      numOfColumns = int(tableNoColsScale.get())
204
205      sqlString = "Create table " + tableNameVar.get() + "("
206      for i in range (0, numOfColumns):
207          sqlString = sqlString + str(columnNameText[i].get()) + " " + \
208              str(columnTypeCombo[i].get()) + "(" + \
209          str(columnSizeScale[i].get()) + ")"
210          if (str(columnPrimaryKeyCombo[i].get()) == "True"):
211              sqlString = sqlString + " Primary key"
212          if (i == numOfColumns-1):
213              sqlString = sqlString + ")"
214          else:
215              sqlString = sqlString + ", "
216              statusVar.set(sqlString)
217
218      try:
219          cursor.execute(sqlString)
220          statusVar.set("Table " + tableNameVar.get() + " is created")
221      except:
222          statusVar.set("There was a problem with creating the table")
223  #-------------------------------------------------------------------
224  # Define the function to connect to the database
225  def dbConnect():
226      global userVar, passVar, hostVar, dbVar, statusVar
227      global connect, cursor, config; global tables, columns
228
229      GUIDB = 'GuiDB'
230      config = {'user': userVar.get(), 'password': passVar.get(),
231              'host': hostVar.get(), 'database': dbVar.get()}
232
233      connect = mysql.connector.connect(**config)
234      cursor = connect.cursor()
235      try:
236          cursor.execute("Show tables")
237          tables = cursor.fetchall()
238          cursor.execute("Desc " + str(tables[0][0]))
239          columns = cursor.fetchall()
240          statusVar.set("DB tables are: " + str(tables))
241      except:
242          statusVar.set("There was a problem with the connection")
243      dbCreateTableGUI()
244  #-------------------------------------------------------------------
```

```
245  # Create the basic window frame with the tk.Tk() constructor and
246  # give a title
247  winFrame = tk.Tk()
248  winFrame.config(bg = "grey")
249  winFrame.title("Data Definition Language: Create")
250
251  initializeVarObjects()
252  createGUI()
253  winFrame.mainloop()
```

Output: Case Study Chapter 7

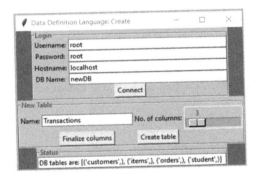

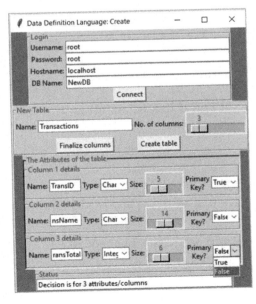

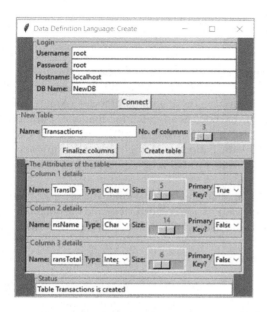

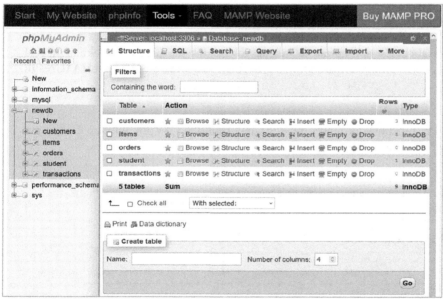

CHAPTER 8 – DATA ANALYTICS AND DATA VISUALIZATION

Description

Readmission is considered a quality measure of hospital performance and a driver of healthcare costs. Studies have shown that patients with diabetes are more likely to have higher early readmissions (readmitted within 30 days of discharge), compared to those without diabetes (American Diabetes Association, 2018; McEwen & Herman, 2018). To reduce early readmission, one solution is to provide additional assistance to patients with high risk of readmission. For this purpose, the US Department of Health would like to know how to identify the patients with high risk of readmission using the collected clinical records of diabetes patients from 130 US hospitals between 1999 and 2008.

As an attempt to assist the US Department of Health in understanding the data, you are asked to explore, analyse (descriptively), and visualise the data of readmission (**readmitted**) and the potential risk factors, such as time in hospital (**time_in_hospital**) and hemoglobin A1c results (**HA1Cresult**), using techniques covered in this chapter.

More specifically, your work should cover the following:

1. Data Acquisition: Import the related data file (i.e., Diabetes.csv).
2. Data Exploration: Report the number of records/samples and the number of columns/variables in the dataset.
3. Descriptive Statistics: Use suitable techniques to summarise or describe the three variables we are interested in: **readmitted**, **time_in_hospital**, and **HA1Cresult**.
4. Data Visualisation: Use appropriate techniques to visualise the three variables and the relationships between **readmitted** and **time_in_hospital**, and **readmission** and **HA1Cresult**.

Solution

1. Use the *read_csv* functions to import (read in) data from the CSV file into the data frame. To view the first few rows of the dataset, use the *head()* function.

```
1    import pandas as pd
2    dataset = pd.read_csv('Diabetes.csv')
3    dataset.head()
```

Output: Case Study Chapter 8.a

	encounter_id	patient_nbr	race	gender	age	weight	admission_type_id	discharge_disposition_id	admission_source_id	time_in_hospital	...	ci
0	2278392	8222157	Caucasian	Female	[0-10)	?	6	25	1	1	...	
1	149190	55629189	Caucasian	Female	[10-20)	?	1	1	7	3	...	
2	64410	86047875	AfricanAmerican	Female	[20-30)	?	1	1	7	2	...	
3	500364	82442376	Caucasian	Male	[30-40)	?	1	1	7	2	...	
4	16680	42519267	Caucasian	Male	[40-50)	?	1	1	7	1	...	

5 rows × 50 columns

2. Use the *len()* and the *shape()* functions to report the number of records and columns. Use the *columns* function to get a list of the available columns in the dataset.

```
1    import pandas as pd
2    dataset = pd.read_csv('Diabetes.csv')
```

```
3      print("Number of samples", len(dataset))
4      print ("Number of records", dataset.shape[0])
5      print ("Number of columns", dataset.shape[1])
6      print ("List of columns", dataset.columns)
```

Output: Case Study Chapter 8.b

```
Number of samples 101766
Number of records 101766
Number of columns 50
List of columns Index(['encounter_id', 'patient_nbr', 'race', 'gender', 'age', 'weight',
        'admission_type_id', 'discharge_disposition_id', 'admission_source_id',
        'time_in_hospital', 'payer_code', 'medical_specialty',
        'num_lab_procedures', 'num_procedures', 'num_medications',
        'number_outpatient', 'number_emergency', 'number_inpatient', 'diag_1',
        'diag_2', 'diag_3', 'number_diagnoses', 'max_glu_serum', 'HA1Cresult',
        'metformin', 'repaglinide', 'nateglinide', 'chlorpropamide',
        'glimepiride', 'acetohexamide', 'glipizide', 'glyburide', 'tolbutamide',
        'pioglitazone', 'rosiglitazone', 'acarbose', 'miglitol', 'troglitazone',
        'tolazamide', 'examide', 'citoglipton', 'insulin',
        'glyburide-metformin', 'glipizide-metformin',
        'glimepiride-pioglitazone', 'metformin-rosiglitazone',
        'metformin-pioglitazone', 'change', 'diabetesMed', 'readmitted'],
       dtype='object')
```

3. The *describe()* function can be used to investigate the individual variables. The function generates descriptive statistics that summarise the central tendency and dispersion of a continuous variable, and the number of levels and the most frequent level of a categorical variable (excluding NaN – Not a Number values).

```
1      import pandas as pd
2      dataset = pd.read_csv('Diabetes.csv')
3      print(dataset.readmitted.describe(), "\n")
4      print(dataset.time_in_hospital.describe(), "\n")
5      print(dataset.HA1Cresult.describe(), "\n")
```

Output: Case Study Chapter 8.c

```
count       101766
unique           3
top             NO
freq         54864
Name: readmitted, dtype: object

count    101766.000000
mean          4.395987
std           2.985108
min           1.000000
25%           2.000000
50%           4.000000
75%           6.000000
max          14.000000
Name: time_in_hospital, dtype: float64

count       101766
unique           4
top           None
freq         84748
Name: HA1Cresult, dtype: object
```

Based on the results of the *describe()* function, it is clear that both **readmitted** and **HA1Cresult** are categorical variables, as they have unique values (levels of the categorical variable) of 2 and 4, respectively. To describe categorical variables, the *value_counts()* function can be used. The *dropna=False* attribute can be applied to include the counts of NaN, while the *normalize=True* attribute can be used to obtain the percentages.

```
1    import pandas as pd
2    dataset = pd.read_csv('Diabetes.csv')
3    print(dataset.readmitted.value_counts(dropna=False), "\n")
4    print(dataset.HA1Cresult.value_counts(dropna=False), "\n")
5    print(dataset.readmitted.value_counts(dropna=False,
6          normalize=True), "\n")
7    print(dataset.readmitted.value_counts(dropna=False,
8          normalize=True), "\n")
```

Output: Case Study Chapter 8.d

```
NO      54864
>30     35545
<30     11357
Name: readmitted, dtype: int64

None    84748
>8       8216
Norm     4990
>7       3812
Name: HA1Cresult, dtype: int64

NO      0.539119
>30     0.349282
<30     0.111599
Name: readmitted, dtype: float64

NO      0.539119
>30     0.349282
<30     0.111599
Name: readmitted, dtype: float64
```

On the other hand, based on the results of the *describe()* function, it is clear that **time_in_hospital** is a continuous variable. This is because the results display the mean, standard deviation (sd), median, 25th and 75th quintiles, and minimum and maximum values. Functions like *mean ()*, *median ()* and *mode ()* can be used to derive mean, median, and mode respectively. Functions like *max ()*, *min ()*, *quantile (0.25)*, *quantile (0.75)*, *std ()*, *skew ()* and *kurtosis ()* are useful to obtain information about the maximum, minimum, 25th percentile, 75th percentile, standard deviation, skewness and kurtosis of a continuous variable, respectively.

```
1     import pandas as pd
2     dataset = pd.read_csv('Diabetes.csv')
3     print('Mean:', dataset.time_in_hospital.mean(), "\n")
4     print('Median:', dataset.time_in_hospital.median(), "\n")
5     print('Mode:', dataset.time_in_hospital.mode(), "\n")
6     print('Max:', dataset.time_in_hospital.max(), "\n")
7     print('Min:', dataset.time_in_hospital.min(), "\n")
8     print('25th percentile:', dataset.time_in_hospital.quantile(0.25), "\n")
9     print('75th percentile:', dataset.time_in_hospital.quantile(0.75), "\n")
10    print('Standard deviation:', dataset.time_in_hospital.std(), "\n")
11    print('Skewness:', dataset.time_in_hospital.skew(), "\n")
12    print('Kurtosis:', dataset.time_in_hospital.kurtosis(), "\n")
```

Output: Case Study Chapter 8.e

```
Mean: 4.395986871843248

Median: 4.0

Mode: 0    3
dtype: int64

Max: 14

Min: 1

25th percentile: 2.0

75th percentile: 6.0

Standard deviation: 2.9851077674705677

Skewness: 1.133998719333879

Kurtosis: 0.8502508404660913
```

The mean (3.87), median (3.0) and mode (2) values, alongside the skewness estimate (1.36), suggest that the time in hospital (**time_in_hospital**) is not normally distributed, but *highly* and *positively* skewed. In addition, since Kurtosis is 1.69 (>0), the distribution of **time_in_hospital** is leptokurtic, meaning that the data are heavily-tailed, with profusion of outliers.

4. To visualise individual variables, bar and pie charts can be used for categorical variables (e.g., **readmitted** and **HA1Cresult**), while histograms and box plots can be used for continuous variables (e.g., **time_in_hospital**). In this instance, we use the *plt.pie* function from the *matplotlib.pyplot* package to plot the pie charts for the categorical variables.

```
1    import pandas as pd
2    import matplotlib.pyplot as plt
3    dataset = pd.read_csv('Diabetes.csv')
4
5    # labels
6    readmit_label = dataset.readmitted.unique()
7
8    # Count the frequencies
9    readmit_count = dataset.readmitted.value_counts()
10
11   # Plot the pie chart
12   plt.pie(readmit_count,labels = readmit_label,
13       autopct = "%1.1f%%", startangle = 90)
14   plt.axis("equal")
15   plt.legend(title = "Early readmission")
16   plt.title("Early readmission")
```

Output: Case Study Chapter 8.f

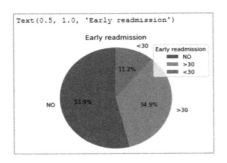

```
1     import pandas as pd
2     import matplotlib.pyplot as plt
3     dataset = pd.read_csv('Diabetes.csv')
4
5     # labels
6     HA1C_label = dataset.HA1Cresult.unique()
7
8     # Count the frequencies
9     HA1C_count = dataset.HA1Cresult.value_counts()
10
11    # Plot the pie chart
12    plt.pie(HA1C_count, labels = HA1C_label, autopct = "%1.1f%%",
13        startangle = 90)
14    plt.axis("equal")
15    plt.legend(title = "HA1C result")
16    plt.title("HA1C result")
```

Output: Case Study Chapter 8.g

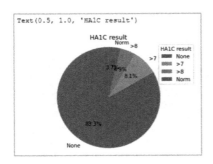

To illustrate the distribution of continuous variable **time_in_hospital**, the *plot.hist()* function was used to plot the histogram.

```
1     import pandas as pd
2     import matplotlib.pyplot as plt
3     dataset = pd.read_csv('Diabetes.csv')
4     Plt = dataset.time_in_hospital.plot.hist(legend = True)
```

Output: Case Study Chapter 8.h

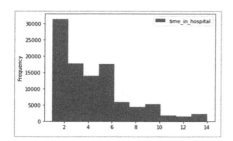

Apart from the visualisation of individual variable, one can also visualise the relationship between two variables. For the relationship between **readmitted** and **HA1Cresult**, the nested or compound bar chart can be used. In this instance, we used the *plot.bar ()* function to display the frequency of **HA1Cresult** by **readmitted**.

```
1    import pandas as pd
2    import matplotlib.pyplot as plt
3    dataset = pd.read_csv('Diabetes.csv')
4
5    # create a crosstable
6    dataset1 = pd.crosstab(dataset.HA1Cresult, dataset.readmitted)
7    fig, ax = plt.subplots(1,2)
8
9    # draw the first chart
10   plt.subplot(1,2,1)
11   plot1 = dataset1['<30'].plot.bar(figsize = (10, 7), legend = True,
12                                     sharey = True, rot = 0)
13   plot1.set_title('Early admission')
14   plot1.set_ylabel('Frequencies')
15   plot1.set_xlabel('HA1C result')
16
17   # draw the second chart
18   plt.subplot(1,2,2)
19   plot2 = dataset1['>30'].plot.bar(figsize = (10, 7), legend = True,
20                                     sharey = True, rot = 0)
21   plot2.set_title('Readmitted')
22   plot2.set_ylabel('Frequencies')
23   plot2.set_xlabel('HA1C result')
```

Output: Case Study Chapter 8.i

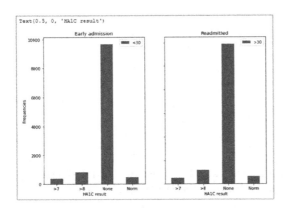

For the relationship between **readmitted** and **time_in_hospital**, box plots can be used. In this instance, we used the *boxplot ()* function to display the box plots of **time_in_hospital** by **readmitted**.

```
1      import pandas as pd
2      import matplotlib.pyplot as plt
3      dataset = pd.read_csv('Diabetes.csv')
4      dataset.boxplot(column = ['time_in_hospital'],
5          by = ['readmitted'], grid = False)
```

Output: Case Study Chapter 8.k

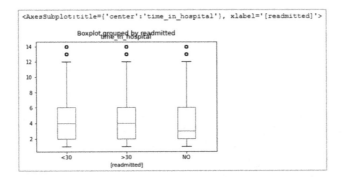

Based on the plot, it appears that the distributions of **time_in_hospital** did not differ significantly between early readmissions and other admissions.

CHAPTER 9 – STATISTICAL ANALYSIS WITH PYTHON

Description

Readmission is considered a quality measure of hospital performance and a driver of healthcare costs. Studies have shown that patients with diabetes are more likely to have higher early readmissions (readmitted within 30 days of discharge), compared to those without diabetes (American Diabetes Association, 2018; McEwen & Herman, 2018). To reduce early readmission, one solution is to provide additional assistance to patients with high risk of readmission. For this purpose, the US Department of Health would like to know how to identify the patients with high risk of readmission using the collected clinical records of diabetes patients from 130 US hospitals between 1999 and 2008.

The *Diabetes* database contains information about readmission (*readmitted*) and the potential associated risk factors, including demographics (*age, gender, race*), primary diagnosis (*diag*), measurement of hemoglobin A1c at admission (*HA1C result*), admission source (*admission_source*), discharge disposition (*discharge_disposition*), medical specialty of the admitting physician (*medical_specialty*), and time spent in hospital (*time_in_hospital*). The aim of this case study is to consider all the available variables in order to assist the US Department of Health identifying the diabetes patients with high risk of readmission. The following tasks need to be addressed:

1. Based on *Sections 9.3.1* and *9.3.2*, identify the dependent variable (outcome variable), the independent variables (predictor variables) and their data types (continuous or categorical). For more details, the reader can also refer to *Chapter 8: Data Analytics and Data Visualisation*.
2. Choose an appropriate regression method based on the data type of the dependent variable and the number of independent variables. (See *Sections 9.3.3* and *9.3.4*).
3. Conduct the selected regression analysis (relevant Python code can be found in *Section 9.5*).
4. Identify the risk factors that significantly affect the risk of readmission using *p-values* (see *Section 9.2.2*)
5. Interpret the regression results and make suggestions to the US Department of Health (see *Section 9.5*).

Solution

The objective of this case study was to identify the diabetes patients with high risk of early readmission using the clinical records. The *Diabetes.csv* file was used for the analysis.

1. The identification of dependent and independent variables is based on the research question and the study objectives. In this case, since the purpose of the analysis was to identify patients with high risk of early admission, it was clear that the dependent variable (outcome variable) would be *readmitted*, and the independent variables (predictor variables) the potential risk factors: *Age, gender, race*, primary diagnosis (*diag*), measurement of hemoglobin A1c at admission (*HA1C result*), admission source (*admission_source*), discharge disposition (*discharge_disposition*), medical specialty of the admitting physician (*medical_specialty*), and time spent in hospital (*time_in_hospital*).

To identify the data type of each variable, one could firstly use the *read_csv()* function to import (read in) data from the CSV file into the data frame, and the *shape* and *columns* functions to check the number of records and the list of available variables in the dataset. Secondly, function *describe()* could be used to obtain brief summaries of all the variables.

```
1    import pandas as pd
2    dataset = pd.read_csv('Diabetes.csv')
3    print ("Number of records", dataset.shape[0])
4    print ("Number of columns", dataset.shape[1])
5    print ("List of columns", dataset.columns)
```

Output: Case Study Chapter 9.a

```
Number of records 101766
Number of columns 50
List of columns Index(['encounter_id', 'patient_nbr', 'race', 'gender', 'age', 'weight',
       'admission_type_id', 'discharge_disposition_id', 'admission_source_id',
       'time_in_hospital', 'payer_code', 'medical_specialty',
       'num_lab_procedures', 'num_procedures', 'num_medications',
       'number_outpatient', 'number_emergency', 'number_inpatient', 'diag_1',
       'diag_2', 'diag_3', 'number_diagnoses', 'max_glu_serum', 'HA1Cresult',
       'metformin', 'repaglinide', 'nateglinide', 'chlorpropamide',
       'glimepiride', 'acetohexamide', 'glipizide', 'glyburide', 'tolbutamide',
       'pioglitazone', 'rosiglitazone', 'acarbose', 'miglitol', 'troglitazone',
       'tolazamide', 'examide', 'citoglipton', 'insulin',
       'glyburide-metformin', 'glipizide-metformin',
       'glimepiride-pioglitazone', 'metformin-rosiglitazone',
       'metformin-pioglitazone', 'change', 'diabetesMed', 'readmitted'],
      dtype='object')
```

```
1     import pandas as pd
2     dataset = pd.read_csv('Diabetes.csv')
3     print(dataset.readmitted.describe(), "\n")
4     print(dataset.age.describe(), "\n")
5     print(dataset.gender.describe(), "\n")
6     print(dataset.race.describe(), "\n")
7     print(dataset.number_diagnoses.describe(), "\n")
8     print(dataset.HA1Cresult.describe(), "\n")
9     print(dataset.medical_specialty.describe(), "\n")
10    print(dataset.time_in_hospital.describe(), "\n")
```

Output: Case Study Chapter 9.b

```
count       101766
unique           3
top             NO
freq         54864
Name: readmitted, dtype: object

count       101766
unique          10
top        [70-80)
freq         26068
Name: age, dtype: object

count       101766
unique           3
top         Female
freq         54708
Name: gender, dtype: object

count       101766
unique           6
top      Caucasian
freq         76099
Name: race, dtype: object

count    101766.000000
mean          7.422607
std           1.933600
min           1.000000
25%           6.000000
50%           8.000000
75%           9.000000
max          16.000000
Name: number_diagnoses, dtype: float64
```

```
count       101766
unique           4
top           None
freq         84748
Name: HA1Cresult, dtype: object

count       101766
unique          73
top              ?
freq         49949
Name: medical_specialty, dtype: object

count    101766.000000
mean          4.395987
std           2.985108
min           1.000000
25%           2.000000
50%           4.000000
75%           6.000000
max          14.000000
Name: time_in_hospital, dtype: float64
```

As shown, the diabetes dataset contains 101,766 records and 50 columns. The dependent variable (**readmitted**) is a categorical variable, as it contains two levels (unique=2). "<30" signifies that the patient was readmitted within 30 days of discharge and "Other" covers both readmission after 30 days and no readmission at all. Regarding the independent variable, only **time_in_hospital** is continuous, while the rest of the variables are categorical.

2. Since the dependent variable (**readmitted**) is categorical and the number of potential risk factors is more than one, therefore, logistic regression is considered as an appropriate method for testing the relationship between the readmission and the potential risk factors.

3. To conduct the logistic regression in Python, the independent variable needs to be converted into a binary value (i.e., 0/1). Next, the *formula* function can be used to define the dependent and independent variables and the *logit()* function from the *statsmodels.formula.api* package to run the logistic regression. For easier interpretation, the *exp()* function from the *Numpy* package can be used to turn the coefficients into odd ratios. More details in relation to this coding can be found in Section 9.5.12.

```
1    import pandas as pd
2    import statsmodels.formula.api as smf
3    import statsmodels.api as sm
4    import numpy as np
5    dataset = pd.read_csv('Diabetes.csv')
6
7    # Create a new dependent variable that contains only 0 and 1
8    dataset['readmitted2'] = dataset.readmitted
9    dataset.readmitted2 = dataset.readmitted2.replace('Other', 0)
10   dataset.readmitted2 = dataset.readmitted2.replace('NO', 0)
11   dataset.readmitted2 = dataset.readmitted2.replace('>30', 0)
12   dataset.readmitted2 = dataset.readmitted2.replace('<30', 1)
13
14   # Logistic regression
15   formula = "readmitted2~C(age, Treatment('[60-70)')) + \
16               C(gender, Treatment('Male')) + \
17               C(race) + C(medical_specialty) + \
18               C(number_diagnoses, Treatment(8)) + \
19               C(HA1Cresult, Treatment('None')) + time_in_hospital"
20   Model = smf.logit(formula, data = dataset).fit()
21   print(Model.summary())
22
23   # Odds ratios
24   Odds = pd.DataFrame({
25        'coef': Model.params.values,
26        'odds ratio': np.exp(Model.params.values),
27        'name': Model.params.index })
28   print(Odds, "\n")
```

4. The output of the logistic regression is partially presented below. The results show that age, race, primary diagnosis, HbA1c measurement, medical speciality, and time in hospital were statistically significant ($p < 0.05$). Comparing to individuals with an age lower than 60 years, those aged over 60 were found to be 1.4 times more likely to have readmission (coefficient > 0, $p < 0.001$, odds ratio $= 1.4$). Caucasians were also 1.2 times more likely to have readmission compared to African-Americans (coefficient > 0, $p = 0.001$, odds ratio $= 1.2$). It was found that compared to primary diagnosis with diabetes, individuals with primary diagnosis of digestive, genitourinary, neoplasms and respiratory diseases were less likely to have readmission (coefficient < 0, $p < 0.05$). Patients spent longer time in hospital were also less likely to be readmitted (coefficient < 0, $p < 0.001$, odds ratio $= 0.97$).

Output: Case Study Chapter 9.c

```
Warning: Maximum number of iterations has been exceeded.
          Current function value: 0.345781
          Iterations: 35

C:\Users\xaran\anaconda3\lib\site-packages\statsmodels\base\model.py:566: ConvergenceWarning: Maximum Likelihood optimization failed to converg
e. Check mle_retvals
  warnings.warn("Maximum Likelihood optimization failed to "
                         Logit Regression Results
==============================================================================
Dep. Variable:            readmitted2   No. Observations:           101766
Model:                          Logit   Df Residuals:               101658
Method:                           MLE   Df Model:                      107
Date:                Mon, 13 Dec 2021   Pseudo R-squ.:             0.01162
Time:                        13:36:28   Log-Likelihood:             -35189.
converged:                      False   LL-Null:                    -35602.
Covariance Type:            nonrobust   LLR p-value:              1.026e-111
==============================================================================
```

	coef	std err	z	P>|z|	[0.025	0.975]
Intercept	-2.3974	0.087	-27.712	0.000	-2.567	-2.228
C(age, Treatment('[60-70)'))[T.[0-10)]	-0.9267	0.626	-1.481	0.139	-2.153	0.299
C(age, Treatment('[60-70)'))[T.[10-20)]	-0.1753	0.189	-0.929	0.353	-0.545	0.195
C(age, Treatment('[60-70)'))[T.[20-30)]	0.5145	0.076	6.759	0.000	0.365	0.664
C(age, Treatment('[60-70)'))[T.[30-40)]	0.1346	0.057	2.367	0.018	0.023	0.246
C(age, Treatment('[60-70)'))[T.[40-50)]	0.0057	0.040	0.144	0.885	-0.072	0.084
C(age, Treatment('[60-70)'))[T.[50-60)]	-0.1257	0.034	-3.742	0.000	-0.192	-0.060
C(age, Treatment('[60-70)'))[T.[70-80)]	0.0391	0.029	1.357	0.175	-0.017	0.096
C(age, Treatment('[60-70)'))[T.[80-90)]	0.0384	0.032	1.196	0.232	-0.025	0.101
C(age, Treatment('[60-70)'))[T.[90-100)]	-0.0633	0.064	-0.983	0.325	-0.190	0.063
C(gender, Treatment('Male'))[T.Female]	0.0009	0.020	0.043	0.966	-0.039	0.041
C(gender, Treatment('Male'))[T.Unknown/Invalid]	-18.3454	1.9e+04	-0.001	0.999	-3.72e+04	3.72e+04
C(race)[T.AfricanAmerican]	0.2826	0.080	3.522	0.000	0.125	0.440
C(race)[T.Asian]	0.2089	0.153	1.369	0.171	-0.090	0.508
C(race)[T.Caucasian]	0.2793	0.078	3.598	0.000	0.127	0.432
C(race)[T.Hispanic]	0.2394	0.106	2.256	0.024	0.031	0.447
C(race)[T.Other]	0.1282	0.117	1.096	0.273	-0.101	0.358
C(medical_specialty)[T.AllergyandImmunology]	1.8133	0.765	2.369	0.018	0.313	3.313
C(medical_specialty)[T.Anesthesiology]	-0.2007	1.047	-0.192	0.848	-2.252	1.851
C(medical_specialty)[T.Anesthesiology-Pediatric]	-0.1258	1.044	-0.120	0.904	-2.173	1.921
C(medical_specialty)[T.Cardiology]	-0.3392	0.053	-6.416	0.000	-0.443	-0.236
C(medical_specialty)[T.Cardiology-Pediatric]	1.0819	1.118	0.968	0.333	-1.109	3.273
C(medical_specialty)[T.DCPTEAM]	-46.0067	1.08e+10	-4.24e-09	1.000	-2.12e+10	2.12e+10
C(medical_specialty)[T.Dentistry]	-12.1713	515.851	-0.024	0.981	-1023.220	998.878
C(medical_specialty)[T.Dermatology]	-15.0808	5073.588	-0.003	0.998	-9959.335	9929.155

C(number_diagnoses, Treatment(8))[T.15]	0.0884	0.733	0.859	0.402	-0.590	2.121
C(number_diagnoses, Treatment(8))[T.16]	-0.3707	0.526	-0.704	0.481	-1.402	0.661
C(HA1Cresult, Treatment('None'))[T.>7]	-0.1765	0.055	-3.194	0.001	-0.285	-0.068
C(HA1Cresult, Treatment('None'))[T.>8]	-0.1402	0.039	-3.557	0.000	-0.217	-0.063
C(HA1Cresult, Treatment('None'))[T.Norm]	-0.2259	0.049	-4.571	0.000	-0.323	-0.129
time_in_hospital	0.0332	0.003	9.978	0.000	0.027	0.040

```
==============================================================================
        coef  odds ratio                                              name
0   -2.397399    0.090954                                         Intercept
1   -0.926666    0.395871          C(age, Treatment('[60-70)'))[T.[0-10)]
2   -0.175310    0.839197          C(age, Treatment('[60-70)'))[T.[10-20)]
3    0.514491    1.672787          C(age, Treatment('[60-70)'))[T.[20-30)]
4    0.134552    1.144024          C(age, Treatment('[60-70)'))[T.[30-40)]
..        ...         ...                                               ...
103 -0.370746    0.690220       C(number_diagnoses, Treatment(8))[T.16]
104 -0.176518    0.838184          C(HA1Cresult, Treatment('None'))[T.>7]
105 -0.140221    0.869166          C(HA1Cresult, Treatment('None'))[T.>8]
106 -0.225863    0.797827        C(HA1Cresult, Treatment('None'))[T.Norm]
107  0.033237    1.033795                                  time_in_hospital

[108 rows x 3 columns]
```

5. The results suggest that older diabetes patients with high HbA1c level, primary diagnosis of diabetes, and who stayed shorter periods in hospital were more likely to have early readmission. Therefore, greater attention should be paid to such patients in order to reduce readmissions, improve patient safety, and lower the costs of inpatient care.

Note that the dataset was originally derived from the UCI Machine Learning Repository (Frank & Asuncion, 2010; Strack et al., 2014). The data were further modified for purposes of this example.

CHAPTER 10 – MACHINE LEARNING WITH PYTHON

DESCRIPTION

Use dataset *dataset.csv* to write a Python script that predicts whether a patient will be readmitted or not within 30 days. The application should do the following:

1. Read the dataset and create a data frame with the following categories: *gender, race, age, admission type id, discharge disposition id, admission source id, max glu serum, A1Cresult, change, diabetesMed, readmitted* (categorical), *time in hospital, number of lab procedures, number of procedures, number of medications, number of outpatients, number of emergencies, number of inpatients, number of diagnoses* (numerical).
2. Apply the following ML algorithms and calculate their accuracy: *logistic regression, k-NN, SVM, Kernel SVM, Naïve Bayes, CART Decision Tree, Random Forest.*

SOLUTION

The *dataset.csv* contains data related to diabetes from 130 US hospitals for the years 1999–2008. The source can be found at: https://archive.ics.uci.edu/ml/datasets/diabetes+130-us+hospitals+for +years+1999-2008#. It includes over 50 features representing patient and hospital outcomes. The information extracted to satisfy the following criteria:

1. It is an inpatient encounter (a hospital admission).
2. It is a diabetic encounter during which any kind of diabetes was entered to the system as a diagnosis.
3. The length of stay was at least 1 day and at most 14 days.
4. Laboratory tests were performed during the encounter.
5. Medications were administered during the encounter.

The data contains such attributes as patient number, race, gender, age, admission type, time in hospital, medical specialty of admitting physician, number of lab test performed, HbA1c test result, diagnosis, number of medications, diabetic medications, number of outpatient, inpatient, and emergency visits in the year before the hospitalization, etc. The following tables lists the various categorical and continuous variables:

Categorical Variables

*race	6 levels	?, Caucasian, AfricanAmerican, Hispanic, Other, Asian
*gender	3 levels	Male, Female, Unknown/Invalid
*age	10 levels	0-, 10-, 20-, 30-, 40-, 50-, 60-, 70-, 80-, 90-
*weight	10 levels	drop the variable
*admission_type_id	8 levels	1, 2, 3, 4, 5, 6, 7, 8
*discharge_disposition_id	30 levels	1, 2, 3,…, 30
*admission_source_id	26 levels	1, 2, 3,…, 26
*payer_code	28 levels	?, MD,…, FR
*medical_specialty	72 levels	?, Allergy,…, Urology
*max_glu_serum	4 levels	None, Norm, >200, >300 (Glucose serum test result)
*A1Cresult	4 levels	None, Norm, >7, >8 (A1c test result)
*change	2 levels	No, Ch (Change of medications)
*diabetesMed	2 levels	Yes, No (any diabetic medication prescribed)

(Continued)

*readmitted	3 levels	>30, <30, No (Days to inpatient readmission)
*metformin	4 levels	No, Steady, Up, Down (medical drug 1)
*repaglinide	4 levels	No, Steady, Up, Down (medical drug 2)
*nateglinide	4 levels	No, Steady, Up, Down (medical drug 3)
*chlorpropamide	4 levels	No, Steady, Up, Down (medical drug 4)
*glimepiride	4 levels	No, Steady, Up, Down (medical drug 5)
*acetohexamide	4 levels	No, Steady, Up, Down (medical drug 6)
*glipizide	4 levels	No, Steady, Up, Down (medical drug 7)
*glyburide	4 levels	No, Steady, Up, Down (medical drug 8)
*tolbutamide	4 levels	No, Steady, Up, Down (medical drug 9)
*pioglitazone	4 levels	No, Steady, Up, Down (medical drug 10)
*rosiglitazone	4 levels	No, Steady, Up, Down (medical drug 11)
*acarbose	4 levels	No, Steady, Up, Down (medical drug 12)
*miglitol	4 levels	No, Steady, Up, Down (medical drug 13)
*troglitazone	4 levels	No, Steady, Up, Down (medical drug 14)
*tolazamide	4 levels	No, Steady, Up, Down (medical drug 15)
*examide	4 levels	No, Steady, Up, Down (medical drug 16)
*citoglipton	4 levels	No, Steady, Up, Down (medical drug 17)
*insulin	4 levels	No, Steady, Up, Down (medical drug 18)
*glyburide-metformin	4 levels	No, Steady, Up, Down (medical drug 19)
*glipizide-metformin	4 levels	No, Steady, Up, Down (medical drug 20)
*glimepiride-pioglitazone	4 levels	No, Steady, Up, Down (medical drug 21)
*metformin-rosiglitazone	4 levels	No, Steady, Up, Down (medical drug 22)
*metformin-pioglitazone	4 levels	No, Steady, Up, Down (medical drug 23)

Continuous Variables

*time_in_hospital	continuous – from 1 to 14
*num_lab_procedures	continuous – from 1 to 121
*num_procedures	continuous – from 0 to 6
*num_medications	continuous – from 0 to 85
*number_outpatient	continuous – from 0 to 36
*number_emergency	continuous – from 0 to 64
*number_inpatient	continuous – from 0 to 15

The first step in solving the suggested problem, as always, is to import the relevant libraries as follows:

```
1    # Import general libraries
2    import numpy as np
3    import matplotlib.pyplot as plt
4    import pandas as pd
5    import scipy
6    from scipy import stats
7
8    # Import the libraries for the encoding of the independent variables
9    from sklearn.compose import ColumnTransformer
10   from sklearn.preprocessing import OneHotEncoder
11   from sklearn.preprocessing import LabelEncoder
```

```
12
13   # Import the libraries for the Machine Learning processes
14   from sklearn.model_selection import train_test_split
15   from sklearn.preprocessing import StandardScaler
16   from sklearn.linear_model import LogisticRegression
17   from sklearn.neighbors import KNeighborsClassifier
18   from sklearn.svm import SVC
19   from sklearn.naive_bayes import GaussianNB
20   from sklearn.tree import DecisionTreeClassifier
21   from sklearn.ensemble import RandomForestClassifier
22
23   # Import the libraries for the confusion matrix and the accuracy score
24   from sklearn.metrics import confusion_matrix, accuracy_score
```

Apparently, the above yields no output.

The second step, as in most similar cases, is to read the data from the suggested file and clean it to improve it. The following script, the actual Step 1: Data Processing, reads the relevant "diabetic_data.csv" file, counts the number of initial observations, finds, and removes the duplicate records, and finishes with the cleansed version. The dataset includes several encounters of patients that have the same patient_nbr which might introduce bias.

```
1    # Step 1: Data Processing
2    # Read the dataset
3    ds = pd.read_csv('diabetic_data.csv')
4    # Current total number of observations (n = 101766)
5    print("Initial number of observations: ", len(ds.index))
6
7    # Remove duplicates; Sort data by patient_nbr and time_in_hospital
8    ds = ds.sort_values(["patient_nbr", "time_in_hospital"],
9                         ascending = (True, True))
10   Ds = ds.drop_duplicates('patient_nbr', keep = 'first')
11   # Updated number of observations (n = 71518)
12   print("Updated number of observations after removing the duplicates: ",
13         len(Ds.index))
```

Output: Case Study Chapter 10.a

```
Initial number of observations:  101766
Updated number of observations after removing the duplicates:  71518
```

In the next step, Step 2, it is necessary to make several conversion, replacements, and regroupings to bring the dataset in a form useful for further analysis. The script and its output follow:

```
1    # Step 2: Data Processing: Replace, clean, and regroup data
2    # Start with race: Replace? with NA; regroup race
3    print("\nThe races before cleaning and regrouping are:")
4    print(ds.race.value_counts())
5    ds.race = ds.race.replace('?', np.nan)
6    ds.race = ds.race.replace('Hispanic', 'Other')
7    ds.race = ds.race.replace('Asian', 'Other')
```

```
8    print("\nThe races after cleaning and regrouping are:")
9    print(ds.race.value_counts())
10
11   # Continue with gender: Replace Unknown/Invalid with NA
12   print("\nThe genders before cleaning and regrouping are:")
13   print(ds.gender.value_counts())
14   ds.gender = ds.gender.replace('Unknown/Invalid', np.nan)
15   print("\nThe genders after cleaning and regrouping are:")
16   print(ds.gender.value_counts())
17
18   # Continue with age: Regroup age ranges
19   print("\nThe age ranges before cleaning and regrouping are:")
20   print(ds.age.value_counts())
21   ds.age = ds.age.replace('[0-10)', '[50-]')
22   ds.age = ds.age.replace('[10-20)', '[50-]')
23   ds.age = ds.age.replace('[20-30)', '[50-]')
24   ds.age = ds.age.replace('[30-40)', '[50-]')
25   ds.age = ds.age.replace('[40-50)', '[50-]')
26   ds.age = ds.age.replace('[50-60)', '[50-60]')
27   ds.age = ds.age.replace('[60-70)', '[60-70]')
28   ds.age = ds.age.replace('[70-80)', '[70-80]')
29   ds.age = ds.age.replace('[80-90)', '[80+]')
30   ds.age = ds.age.replace('[90-100)', '[80+]')
31   print("\nThe age ranges after regrouping are:")
32   print(ds.age.value_counts())
33
34   # Continue with weight (in pounds):
35   print("\nThe weights before cleaning and regrouping are:")
36   print(ds.weight.value_counts())
37   ds.weight = ds.weight.replace('?', np.nan)
38   print("\nThe weights after cleaning and regrouping are:")
39   print(ds.weight.value_counts())
40
41   # Continue with admission type:
42   print("\nThe admission types before cleaning and regrouping are:")
43   print(ds.admission_type_id.value_counts())
44   ds.admission_type_id = ds.admission_type_id.replace([
45   1, 2, 7], 'Emergency')
46   ds.admission_type_id = ds.admission_type_id.replace([3, 4],
47   'Non-emergency')
48   ds.admission_type_id = ds.admission_type_id.replace([
49       5, 6, 8], np.nan)
50   print("\nThe admission types after cleaning and regrouping are:")
51   print(ds.admission_type_id.value_counts())
52
53   # Continue with discharge_disposition:
54   print("\nThe discarge disposition ids before cleaning and "
55         "regrouping are:")
56   print(ds.discharge_disposition_id.value_counts())
57   ds.discharge_disposition_id = ds.discharge_disposition_id. \
58       replace([1, 6, 8, 13], 'Home')
59   ds.discharge_disposition_id = ds.discharge_disposition_id. \
```

```
60       replace([2, 5, 29, 9, 10, 15, 10, 12, 16, 17], 'Hospital')
61   ds.discharge_disposition_id = ds.discharge_disposition_id. \
62       replace([3, 4, 22, 23, 24, 27, 28, 30, 13, 14], 'Care')
63   ds.discharge_disposition_id = ds.discharge_disposition_id. \
64       replace([11, 19, 20, 21], 'Death')
65   ds.discharge_disposition_id = ds.discharge_disposition_id. \
66       replace(7, 'AMA')
67   ds.discharge_disposition_id = ds.discharge_disposition_id. \
68       replace([18, 25, 26], np.nan)
69   print("\nThe discarge disposition ids after cleaning and "
70         "regrouping are:")
71   print(ds.discharge_disposition_id.value_counts())
72
73   # Continue with the admission source:
74   print("\nThe admission source ids before cleaning and "
75         "regrouping are:")
76   print(ds.admission_source_id.value_counts())
77   ds.admission_source_id = ds.admission_source_id.replace(
78   [7, 8], 'Emergency')
79   ds.admission_source_id = ds.admission_source_id.replace([1, 2,
80       3, 4, 5, 6, 10, 18, 19, 22, 25, 11, 13, 14, 23], 'Non-emergency')
81   ds.admission_source_id = ds.admission_source_id.replace([
82       9, 15, 17, 20, 21], np.nan)
83   print("\nThe admission source ids after cleaning and regrouping are:")
84   print(ds.admission_source_id.value_counts())
85
86   # Continue with player code:
87   print("\nThe payer codes before cleaning and regrouping are:")
88   print(ds.payer_code.value_counts())
89   ds.payer_code = ds.payer_code.replace('?', np.nan)
90   ds.payer_code = ds.payer_code.replace(['MC', 'HM', 'BC',
91       'SP', 'MD', 'CP', 'UN', 'CM', 'OG', 'PO', 'DM', 'WC', 'CH',
92       'OT', 'SI', 'MP', 'FR'], 'Insured')
93   print("\nThe payer codes after cleaning and regrouping are:")
94   print(ds.payer_code.value_counts())
95
96   # Continue with medical speciality:
97   print("\nThe medical specialties before cleaning and "
98         "regrouping are:")
99   print(ds.medical_specialty.value_counts())
100  # DM related
101  ds.medical_specialty = ds.medical_specialty.replace([
102      'InternalMedicine', 'Family/GeneralPractice', 'Cardiology',
103      'Orthopedics', 'Orthopedics-Reconstructive', 'Urology',
104      'Nephrology', 'Dentistry', 'DCPTEAM', 'Endocrinology-Metabolism',
105      'Neurology', 'Podiatry', 'Endocrinology', 'Ophthalmology'], 1)
106  # Pregancy related
107  ds.medical_specialty = ds.medical_specialty.replace(
108  ['ObstetricsandGynecology', 'Perinatology', 'Obstetrics'], 2)
109  # Pediatrics
110  ds.medical_specialty = ds.medical_specialty.replace([
111      'Pediatrics', 'Pediatrics-CriticalCare',
```

```
112        'Anesthesiology-Pediatric', 'Pediatrics-Pulmonology',
113        'Surgery-Pediatric', 'Pediatrics-Neurology',
114        'Cardiology-Pediatric', 'Pediatrics-Endocrinology',
115        'Pediatrics-Hematology-Oncology', 'Pediatrics-EmergencyMedicine',
116        'Pediatrics-AllergyandImmunology'], 3)
117 # Psychiatry
118 ds.medical_specialty = ds.medical_specialty.replace([
119        'Psychiatry', 'Psychiatry-Child/Adolescent',
120        'Psychiatry-Addictive', 'Psychology',
121        'Psychiatry-Child/Adolescent',
122        'PhysicalMedicineandRehabilitation'], 4)
123 # surgery
124 ds.medical_specialty = ds.medical_specialty.replace([
125        'Surgeon', 'Surgery-Cardiovascular',
126        'Surgery-Cardiovascular/Thoracic','Surgery-Colon&Rectal',
127        'Surgery-General', 'Surgery-Maxillofacial',
128        'Surgery-Plastic', 'Surgery-PlasticwithinHeadandNeck',
129        'Surgery-Neuro', 'Surgery-Vascular', 'SurgicalSpecialty',
130        'Surgery-Thoracic'], 5)
131 # Cancer
132 ds.medical_specialty = ds.medical_specialty.replace([
133        'Radiologist', 'Obsterics&Gynecology-GynecologicOnco',
134        'Osteopath', 'Hematology/Oncology','Oncology',
135        'Radiology'], 6)
136 # Other/ungrouped
137 ds.medical_specialty = ds.medical_specialty.replace([
138        'Emergency/Trauma', 'Pulmonology', 'Proctology', 'Dermatology',
139        'SportsMedicine', 'Speech', 'Neurophysiology', 'Resident',
140        'AllergyandImmunology', 'Anesthesiology', 'Pathology',
141        'OutreachServices', 'Rheumatology', 'Gastroenterology',
142        'Gynecology', 'Hematology', 'Hospitalist',
143        'InfectiousDiseases', 'Otolaryngology'], 7)
144 # missing
145 ds.medical_specialty = ds.medical_specialty.replace(['?',
146 'PhysicianNotFound'], np.nan)
147 print("\nThe medical specialties after cleaning and regrouping are:")
148 print(ds.medical_specialty.value_counts())
149
150 # Continue with diagnosis through max_glu_serum
151 print("\nThe diagnoses through max_glu_serum before cleaning and "
152        "regrouping are:")
153 print(ds.max_glu_serum.value_counts())
154 ds.max_glu_serum = ds.max_glu_serum.replace(['>200'], '200-')
155 ds.max_glu_serum = ds.max_glu_serum.replace(['>300'], '300-')
156 print("\nThe diagnoses through max_glu_serum after cleaning and "
157        "regrouping are:")
158 print(ds.max_glu_serum.value_counts())
159
160 # Continue with diagnosis through A1Cresults
161 print("\nThe diagnoses through A1Cresults before cleaning and "
162        "regrouping are:")
163 print(ds.A1Cresult.value_counts())
```

```
164  ds.A1Cresult = ds.A1Cresult.replace(['>8'], '8-')
165  ds.A1Cresult = ds.A1Cresult.replace(['>7'], '7-')
166  print("\nThe diagnoses through A1Cresults after cleaning and "
167        "regrouping are:")
168  print(ds.A1Cresult.value_counts())
169
170  # List the percentage of missing values per column
171  print("\nThe percentages of missing values per column are:")
172  print(ds.isna().sum()/len(dataset)*100, "%")
173
174  # Remove the variables that have over 40% of missing data
175  ds = ds.drop(columns = ['weight', 'payer_code',
176      'medical_specialty'])
177  # numeric columns
178  ds.fillna(ds.select_dtypes(
179      include = 'number').mean().iloc[0], inplace = True)
180  # categorical columns
181  ds.fillna(ds.select_dtypes(
182      include = 'object').mode().iloc[0], inplace = True)
183
184  # Remove unnecessary columns
185  ds = ds.drop(columns = ['encounter_id', "patient_nbr",
186      'diag_1', 'diag_2', 'diag_3', 'metformin', 'repaglinide',
187      'nateglinide', 'chlorpropamide', 'glimepiride', 'acetohexamide',
188      'glipizide', 'glyburide', 'tolbutamide', 'pioglitazone',
189      'rosiglitazone', 'acarbose', 'miglitol', 'troglitazone',
190      'tolazamide', 'examide', 'citoglipton', 'insulin',
191      'glyburide-metformin', 'glipizide-metformin',
192      'glimepiride-pioglitazone', 'metformin-rosiglitazone',
193      'metformin-pioglitazone'])
194  print("\nThe final version of the dataset is:")
195  print(ds)
```

Output: Case Study Chapter 10.b

```
The races before cleaning and regrouping are:
Caucasian          53517
AfricanAmerican    12898
?                   1913
Hispanic            1516
Other               1170
Asian                504
Name: race, dtype: int64
```

```
The races after cleaning and regrouping are:
Caucasian               53517
AfricanAmerican         12898
Other                    3190
Name: race, dtype: int64

The genders before cleaning and regrouping are:
Female                  38025
Male                    33490
Unknown/Invalid             3
Name: gender, dtype: int64

The genders after cleaning and regrouping are:
Female      38025
Male        33490
Name: gender, dtype: int64

The age ranges before cleaning and regrouping are:
[70-80)     18176
[60-70)     15949
[50-60)     12424
[80-90)     11706
[40-50)      6819
[30-40)      2676
[90-100)     1961
[20-30)      1121
[10-20)       533
[0-10)        153
Name: age, dtype: int64
```

```
The age ranges after regrouping are:
[70-80]     18176
[60-70]     15949
[80+]       13667
[50-60]     12424
[50-]       11302
Name: age, dtype: int64

The weights before cleaning and regrouping are:
?           68676
[75-100)     1188
[50-75)       785
[100-125)     563
[125-150)     131
[25-50)        89
[0-25)         42
[150-175)      33
[175-200)       8
>200            3
Name: weight, dtype: int64
```

```
The weights after cleaning and regrouping are:
[75-100)     1188
[50-75)       785
[100-125)     563
[125-150)     131
[25-50)        89
[0-25)         42
[150-175)      33
[175-200)       8
>200            3
Name: weight, dtype: int64
```

```
The admission types before cleaning and regrouping are:
1       36779
3       13940
2       12965
6        4441
5        3061
8         304
7          19
4           9
Name: admission_type_id, dtype: int64

The admission types after cleaning and regrouping are:
Emergency       49763
Non-emergency      13949
Name: admission_type_id, dtype: int64
```

```
The discarge disposition ids before cleaning and regrouping are:
1       44701
3        8526
6        8009
18       2473
2        1596
22       1305
11       1300
5         876
25        683
4         551
7         452
13        276
14        270
23        239
28         93
8          65
15         32
24         28
9          13
17          7
19          7
10          6
16          4
27          3
12          2
20          1
Name: discharge_disposition_id, dtype: int64
```

```
The discarge disposition ids after cleaning and regrouping are:
Home        53051
Care        11015
Hospital     2536
Death        1308
AMA           452
Name: discharge_disposition_id, dtype: int64

The admission source ids before cleaning and regrouping are:
7       38526
1       21812
17       4923
4        2539
6        1805
2         920
5         581
20        145
3         138
9         101
8          12
10          6
22          4
14          2
25          2
11          1
13          1
Name: admission_source_id, dtype: int64
```

```
The admission source ids after cleaning and regrouping are:
Emergency          38538
Non-emergency      27811
Name: admission_source_id, dtype: int64

The payer codes before cleaning and regrouping are:
?        30648
MC       20788
HM        4075
BC        3444
SP        3309
MD        2206
CP        1942
UN        1881
CM        1349
OG         664
PO         458
DM         381
WC         120
CH         119
OT          65
SI          36
MP          32
FR           1
Name: payer_code, dtype: int64

The payer codes after cleaning and regrouping are:
Insured     40870
Name: payer_code, dtype: int64
```

```
The medical specialties before cleaning and regrouping are:
?                                     34445
InternalMedicine                      10981
Family/GeneralPractice                 5103
Emergency/Trauma                       4468
Cardiology                             4295
                                ...
Speech                                    1
Neurophysiology                           1
Pediatrics-AllergyandImmunology           1
Dermatology                               1
Psychiatry-Addictive                      1
Name: medical_specialty, Length: 72, dtype: int64

The medical specialties after cleaning and regrouping are:
1.0     24294
7.0      5800
5.0      3797
6.0      1275
4.0       828
2.0       609
3.0       461
Name: medical_specialty, dtype: int64

The diagnoses through max_glu_serum before cleaning and regrouping are:
None      68083
Norm       1780
>200        972
>300        683
Name: max_glu_serum, dtype: int64
```

```
The diagnoses through max_glu_serum after cleaning and regrouping are:
None      68083
Norm       1780
200-        972
300-        683
Name: max_glu_serum, dtype: int64

The diagnoses through A1Cresults before cleaning and regrouping are:
None      59148
>8         5897
Norm       3732
>7         2741
Name: A1Cresult, dtype: int64

The diagnoses through A1Cresults after cleaning and regrouping are:
None      59148
8-         5897
Norm       3732
7-         2741
Name: A1Cresult, dtype: int64
```

```
The percentages of missing values per column are:
encounter_id              0.000000
patient_nbr               0.000000
race                      2.674851
gender                    0.004195
age                       0.000000
weight                   96.026175
admission_type_id        10.914735
discharge_disposition_id  4.412875
admission_source_id       7.227551
time_in_hospital          0.000000
payer_code               42.853547
medical_specialty        48.175285
num_lab_procedures        0.000000
num_procedures            0.000000
num_medications           0.000000
number_outpatient         0.000000
number_emergency          0.000000
number_inpatient          0.000000
diag_1                    0.000000
diag_2                    0.000000
diag_3                    0.000000
number_diagnoses          0.000000
max_glu_serum             0.000000
A1Cresult                 0.000000
metformin                 0.000000
repaglinide               0.000000
nateglinide               0.000000
chlorpropamide            0.000000
glimepiride               0.000000
```

```
glyburide                 0.000000
tolbutamide               0.000000
pioglitazone              0.000000
rosiglitazone             0.000000
acarbose                  0.000000
miglitol                  0.000000
troglitazone              0.000000
tolazamide                0.000000
examide                   0.000000
citoglipton               0.000000
insulin                   0.000000
glyburide-metformin       0.000000
glipizide-metformin       0.000000
glimepiride-pioglitazone  0.000000
metformin-rosiglitazone   0.000000
metformin-pioglitazone    0.000000
change                    0.000000
diabetesMed               0.000000
readmitted                0.000000
dtype: float64 %
```

```
The final version of the dataset is:
                  race  gender      age admission_type_id  \
4780         Caucasian  Female  [50-60]         Emergency
5827         Caucasian  Female  [50-60]     Non-emergency
67608        Caucasian  Female    [80+]         Emergency
17494        Caucasian  Female    [80+]         Emergency
2270    AfricanAmerican  Female   [50-]         Emergency
...              ...     ...      ...               ...
99863        Caucasian  Female    [80+]         Emergency
95282            Other    Male  [60-70]         Emergency
93651        Caucasian  Female    [80+]         Emergency
101748       Caucasian  Female    [50-]         Emergency
96147        Caucasian    Male    [50-]         Emergency

        discharge_disposition_id admission_source_id  time_in_hospital  \
4780                        Home           Emergency                 3
5827                        Home       Non-emergency                 2
67608                       Care           Emergency                 4
17494                       Home           Emergency                 3
2270                        Home           Emergency                 5
...                          ...                 ...               ...
99863                       Home           Emergency                 1
95282                       Home           Emergency                 3
93651                       Home           Emergency                 3
101748                      Care           Emergency                14
96147                       Home       Non-emergency                 5
```

```
         num_lab_procedures  num_procedures  num_medications  \
4780                     31               1               14
5827                     49               1               11
67608                    68               2               23
17494                    46               0               20
2270                     49               0                5
...                     ...             ...              ...
99863                    73               1               11
95282                    56               1                8
93651                    39               0               18
101748                   69               0               16
96147                    35               4               23

         number_outpatient  number_emergency  number_inpatient  \
4780                     0                 0                 1
5827                     0                 0                 0
67608                    0                 0                 0
17494                    0                 0                 0
2270                     0                 0                 0
...                    ...               ...               ...
99863                    0                 0                 0
95282                    0                 0                 0
93651                    0                 0                 0
101748                   0                 0                 0
96147                    0                 0                 0
```

```
        number_diagnoses max_glu_serum A1Cresult change diabetesMed readmitted
4780                   5          None      None     Ch         Yes        >30
5827                   3          None      None     No          No         NO
67608                  9          None        7-     No         Yes         NO
17494                  9          None        8-     Ch         Yes         NO
2270                   3          None      None     No         Yes         NO
...                  ...           ...       ...    ...         ...        ...
99863                  9          None      None     No          No         NO
95282                  7          None      None     No         Yes         NO
93651                  9          None      None     Ch         Yes         NO
101748                 5          None        7-     Ch         Yes        >30
96147                  8          None      None     Ch         Yes         NO

[71518 rows x 19 columns]
```

Step 3 lists the categorical and numerical variables and calculates their statistical significance. The script and its output follow:

```
1    # Step 3: Data Analysis
2    # List and significance of categorical feature
3    categorical_features = ['gender', 'age', 'race', 'admission_type_id',
4    'discharge_disposition_id', 'admission_source_id', 'max_glu_serum',
5    'A1Cresult', 'change', 'diabetesMed', 'readmitted']
6    print("\nThe following lists the cagegorical features and their "
7         "significance:")
8    for col in categorical_features :
9        data_crosstab = pd.crosstab(ds['readmitted'],
10           ds[col], margins = False)
11       stat, p, dof, expected = scipy.stats.chi2_contingency(data_crosstab)
12       if p < 0.4 :
13           print(p, col, 'is significant')
14       else:
15           print(p, col, 'is not significant')
16
17   # List and significance of numeric features
18   numeric_features = ['time_in_hospital', 'num_lab_procedures',
19       'num_procedures', 'num_medications', 'number_outpatient',
20       'number_emergency', 'number_inpatient', 'number_diagnoses']
```

```
21   print("\nThe folowing lists the numerical features and their "
22        "significance:")
23   for col in numeric_features :
24       rho, pval=scipy.stats.spearmanr(ds['readmitted'],ds[col])
25       if pval < 0.4 :
26           print(col, 'is significant')
27       else :
28           print(col, 'is not significant')
29           rejected_features.append(col)
```

Output: Case Study Chapter 10.c

```
The following lists the cagegorical features and their significance:
0.001871265376094707 gender is significant
9.575718385478755e-63 age is significant
3.4259476417207416e-34 race is significant
3.8967484632375537e-50 admission_type_id is significant
0.0 discharge_disposition_id is significant
1.7890554046642271e-66 admission_source_id is significant
0.10735254299651757 max_glu_serum is significant
1.1659473072886347e-12 A1Cresult is significant
9.341854944909139e-05 change is significant
8.144133994685768e-38 diabetesMed is significant
0.0 readmitted is significant

The following lists the numerical features and their significance:
time_in_hospital is significant
num_lab_procedures is significant
num_procedures is significant
num_medications is significant
number_outpatient is significant
number_emergency is significant
number_inpatient is significant
number_diagnoses is significant
```

Step 4 prepares the dataset for the ML model. The script and its output follow:

```
1    # Step 4: Prepare for Machine Learning
2    # Update >30 days as none
3    ds.readmitted = ds.readmitted.replace('>30', 'NO')
4    X = ds.iloc[:, :-1].values
5    y = ds.iloc[:, -1].values
6
7    ds.head()
8    ds.readmitted.value_counts()
9
10   X = pd.DataFrame(X, columns = ['race', 'gender', 'age',
11       'admission_type_id', 'discharge_disposition_id',
12       'admission_source_id', 'time_in_hospital ',
13       'num_lab_procedures', 'num_procedures', 'num_medications',
14       'number_outpatient', 'number_emergency', 'number_inpatient',
15       'number_diagnoses', 'max_glu_serum',
16       'A1Cresult', 'change', 'diabetesMed'])
17
18   X = pd.get_dummies(X, columns = ['gender', 'race', 'age',
19       'admission_type_id', 'discharge_disposition_id',
20       'admission_source_id', 'max_glu_serum',
21       'A1Cresult', 'change', 'diabetesMed'])
```

```
22   X = np.array(X)
23
24   # Encode for dependent variables
25   le = LabelEncoder()
26   y = le.fit_transform(y)
27   print(y)
```

Output: Case Study Chapter 10.d

```
[1 1 1 ... 1 1 1]
```

Following up from the previous Step 4, the dataset is split into train and test parts, the train part is transformed and trained to the relevant ML model, and, finally, the predictions are made based on the various ML models, the confusion matrices and the accuracy scores are calculated. (Note that only two ML models are active, the rest are deactivated, for practical reasons related to the time it takes to make the calculations for all the ML models). The script and its output follow:

```
1    # Step 4: Split the dataset to train and test parts
2    X_train, X_test, y_train, y_test = train_test_split(X, y,
3    test_size = 0.2, random_state = 1)
4
5    sc = StandardScaler()
6    X_train = sc.fit_transform(X_train)
7    X_test = sc.transform(X_test)
8
9    # Train the model for logistic regression
10   classifier1 = LogisticRegression(random_state = 0)
11   classifier1.fit(X_train, y_train)
12
13   # Train the model for the KNN
14   classifier2 = KNeighborsClassifier(n_neighbors = 5,
15       metric = 'minkowski', p = 2)
16   classifier2.fit(X_train, y_train)
17
18   """
19   # Train the model for the SVM
20   classifier3 = SVC(kernel = 'linear', random_state = 0)
21   classifier3.fit(X_train, y_train)
22
23   # Train the model for the Kernel SVM
24   classifier4 = SVC(kernel = 'rbf', random_state = 0)
25   classifier4.fit(X_train, y_train)
26
27   # Train the model for the Naive Bayes
28   classifier5 = GaussianNB()
29   classifier5.fit(X_train, y_train)
30
31   # Train the model for the Decision Tree
32   classifier6 = DecisionTreeClassifier(criterion = 'entropy',
33       random_state = 0)
34   classifier6.fit(X_train, y_train)
```

```
35
36   # Train the model for Random Forest
37   classifier7 = RandomForestClassifier(n_estimators = 10,
38       criterion = 'entropy', random_state = 0)
39   classifier7.fit(X_train, y_train)
40   """
41
42   # Predit the results
43   y_pred1 = classifier1.predict(X_test)
44   y_pred2 = classifier2.predict(X_test)
45   #y_pred3 = classifier3.predict(X_test)
46   #y_pred4 = classifier4.predict(X_test)
47   #y_pred5 = classifier5.predict(X_test)
48   #y_pred6 = classifier6.predict(X_test)
49   #y_pred7 = classifier7.predict(X_test)
50
51   # Report the confusion matrix
52   cm1 = confusion_matrix(y_test, y_pred1)
53   print("\nThe confusion matrix for Logistic Regression is:")
54   print(cm1)
55   print("\nThe accuracy score for the Logistic Regression is:",
56       accuracy_score(y_test, y_pred1))
57
58   cm2 = confusion_matrix(y_test, y_pred2)
59   print("\nThe confusion matrix for KNeighbors is:")
60   print(cm2)
61   print("\nThe accuracy score for the KNeighbors is:",
62       accuracy_score(y_test, y_pred2))
63
64   """
65   cm3 = confusion_matrix(y_test, y_pred3)
66   print("\nThe confusion matrix for SVM is:")
67   print(cm3)
68   print("\nThe accuracy score for the SVM is:",
69       accuracy_score(y_test, y_pred3))
70
71   cm4 = confusion_matrix(y_test, y_pred4)
72   print("\nThe confusion matrix for Kernel SVM is:")
73   print(cm4)
74   print("\nThe accuracy score for the Kernel SVM is:",
75       accuracy_score(y_test, y_pred4))
76
77   cm5 = confusion_matrix(y_test, y_pred5)
78   print("\nThe confusion matrix for Naive Bayes is:")
79   print(cm5)
80   print("\nThe accuracy score for the Naive Bayes is:",
81       accuracy_score(y_test, y_pred5))
82
83   cm6 = confusion_matrix(y_test, y_pred6)
84   print("\nThe confusion matrix for Decision Tree is:")
85   print(cm6)
86   print("\nThe accuracy score for the Decision Tree is:",
```

```
87         accuracy_score(y_test, y_pred6))
88
89  cm7 = confusion_matrix(y_test, y_pred7)
90  print("\nThe confusion matrix for Random Forest is:")
91  print(cm7)
92  print("\nThe accuracy score for the Random Forest is:",
93         accuracy_score(y_test, y_pred7))
94  """
```

Output: Case Study Chapter 10.e

```
The confusion matrix for Logistic Regression is:
[[    8  1131]
 [    9 13156]]

The accuracy score for the Logistic Regression is: 0.9203020134228188

The confusion matrix for KNeighbors is:
[[   22  1117]
 [   63 13102]]

The accuracy score for the KNeighbors is: 0.9175055928411633

'\ncm3 = confusion_matrix(y_test, y_pred3)\nprint("\nThe confusion matrix for SVM is:")\nprin
t(cm3)\nprint("\nThe accuracy score for the SVM is:", accuracy_score(y_test, y_pred3))\n\ncm4
= confusion_matrix(y_test, y_pred4)\nprint("\nThe confusion matrix for Kernel SVM is:")\nprin
t(cm4)\nprint("\nThe accuracy score for the Kernel SVM is:", accuracy_score(y_test, y_pred
4))\n\ncm5 = confusion_matrix(y_test, y_pred5)\nprint("\nThe confusion matrix for Naive Bayes
is:")\nprint(cm5)\nprint("\nThe accuracy score for the Naive Bayes is:", accuracy_score(y_tes
t, y_pred5))\n\ncm6 = confusion_matrix(y_test, y_pred6)\nprint("\nThe confusion matrix for De
cision Tree is:")\nprint(cm6)\nprint("\nThe accuracy score for the Decision Tree is:", accura
cy_score(y_test, y_pred6))\n\ncm7 = confusion_matrix(y_test, y_pred7)\nprint("\nThe confusion
matrix for Random Forest is:")\nprint(cm7)\nprint("\nThe accuracy score for the Random Forest
is:", accuracy_score(y_test, y_pred7))\n'
```

CHAPTER 11 – INTRODUCTION TO NEURAL NETWORKS AND DEEP LEARNING WITH PYTHON

DESCRIPTION

1. Phase 1: Build the necessary logical gates (i.e., AND, OR, and XOR) that will act as the building blocks for the perceptron modeling in the ANN. Follow the hints below:
   a. User real numbers instead of integers,
   b. Implement 1D vectors for weights (w) and inputs (x) as follows: $z = \vec{w} \cdot \vec{x}$,
   c. Feed the sum of the vectors to the sigmoid activation function.
2. Phase 2: Implement the following tasks:
   a. Task 1: (Perceptron Engineering):
      – Write a sigmoid function,
      – Write a method to send values to the weights.
   b. Task 2: (Validating it):
      – Develop a method/function to write values for the weights.
      – Develop a method/function to provide a sample to the network.
      – Then, test your network with the XOR gate-based weights.
   c. Task 3: Implement the multi-layer perceptron Class.
   d. Task 4: Develop a backpropagation algorithm by following the next steps:

      – Feed a sample to the network: $y = \begin{bmatrix} 0 \\ 1 \end{bmatrix}$,

      – Calculate the MSE: $MSE = \dfrac{1}{n} \sum\limits_{i=0}^{n-1} (y_i - Output_i)^2$

      – Calculating the error terms of each Neuron's output:

      $$\sigma_k = Output_k * (1 - Output_k) * (y_k - Output_k)$$

      $Output_k * (1 - Output_k)$, is derivative of sigmoid function

      – Repeatedly compute the errors terms in the hidden layers:

      $$\sigma_{Hid} = Output_{Hid} * (1 - Output_{Hid}) * \sum\limits_{k \in outputs} w_{kHid}\sigma_k$$

      – Applying the delta rule:

      $$\Delta w_{ij} = \gamma * \sigma_i * x_{ij}$$

      – Adjust the weights for the best model outcome:

      $$w_{ij} = w_{ij} + \Delta w_{ij}$$

   e. Task 5: Validate the class.

SOLUTION

1. Phase 1: Build the necessary logical gates (i.e., AND, OR, and XOR) that will act as the building blocks for the perceptron modeling in the ANN. Follow the hints below:
   a. User real numbers instead of integers,
   b. Implement 1D vectors for weights (w) and inputs (x) as follows: $z = \vec{w} \cdot \vec{x}$,
   c. Feed the sum of the vectors to the sigmoid activation function.

Recall the basic concepts of AND, OR, and XOR gates. The following table is a brief review:

A	B	A AND B	A OR B	A XOR B
0	0	0	0	0
0	1	0	1	1
1	0	0	1	1
1	1	1	1	0

```
1    import numpy as np
2
3    class PerceptronEngineering:
4
5        def sigmoidFunction(self, x):
6            return 1/(1+np.exp(-x))
7
8        def setTheWeights(self, list_Init):
9            #Use list_Init to set the weights using numpy array method
10           self.weights = np.array(list_Init)
11
12       def __init__(self, inputs, thebias = 1.0):
13           # Create Perceptron object with specified number of inputs
14           # 2 is a scaling factor and shift of -1
15           self.weights = (np.random.rand(inputs + 1) * 2) -1
16           self.thebias = thebias
17
18       def runThePerceptron(self, myList):
19           # Run the perceptron. myList includes the input values.
20           netSum = np.dot(np.append(myList, self.thebias), self.weights)
21           return self.sigmoidFunction(netSum)
22
23   # Test the class
24   myNeuron = PerceptronEngineering (inputs = 2)
25   myNeuron.setTheWeights([10, 10, -15])
26
27   print("Building AND gate as Perceptron")
28   print ("0 0 = {0:.10f}".format(myNeuron.runThePerceptron([0,0])))
29   print ("0 1 = {0:.10f}".format(myNeuron.runThePerceptron([0,1])))
30   print ("1 0 = {0:.10f}".format(myNeuron.runThePerceptron([1,0])))
31   print ("1 1 = {0:.10f}".format(myNeuron.runThePerceptron([1,1])))
```

Output: Case Study Chapter 11.a

```
Building AND gate as Perceptron
0 0 = 0.0000003059
0 1 = 0.0066928509
1 0 = 0.0066928509
1 1 = 0.9933071491
```

```
1    print("Building OR gate as Perceptron")
2    myNeuron = PerceptronEngineering (inputs = 2)
3    myNeuron.setTheWeights([15,15,-10])
4    print ("0 0 = {0:.10f}".format(myNeuron.runThePerceptron([0,0])))
5    print ("0 1 = {0:.10f}".format(myNeuron.runThePerceptron([0,1])))
6    print ("1 0 = {0:.10f}".format(myNeuron.runThePerceptron([1,0])))
7    print ("1 1 = {0:.10f}".format(myNeuron.runThePerceptron([1,1])))
```

Output: Case Study Chapter 11.b

```
Building OR gate as Perceptron
0 0 = 0.0000453979
0 1 = 0.9933071491
1 0 = 0.9933071491
1 1 = 0.9999999979
```

```
1    class MultiLayerPerceptronEngineering:
2        # innerlayers represents a python list with the number of elements
3        # per layer. thebias represents the bias term used for all neurons.
4        # LRate represents the learning rate.
5
6        def __init__(self, innerlayers, thebias = 1.0, LRate = 0.5):
7            self.innerlayers = np.array(innerlayers, dtype = object)
8            self.thebias = thebias
9            self.LRate = LRate
10
11       # Create an internal network for a list of lists of neurons
12            self.network = []
13       # Hold the list of lists of output values
14            self.values = []
15       # The list of lists of error terms
16            self.delta = []
17
18            for i in range(len(self.innerlayers)):
19                self.values.append([])
20                self.delta.append([])
21                self.network.append([])
22                self.values[i] = [0.0 for j in range(self.innerlayers[i])]
23                self.delta[i] = [0.0 for j in range(self.innerlayers[i])]
24
25                # No of neuros for input layer at network[0]
26                if i > 0:
27                    for j in range(self.innerlayers[i]):
28                        self.network[i].append(PerceptronEngineering(
29                            inputs = self.innerlayers[i-1],
30                            thebias = self.thebias))
31
32            self.network = np.array([np.array(x) for x in self.network],
33                dtype = object)
34            self.values = np.array([np.array(x) for x in self.values],
35                dtype = object)
36            self.delta = np.array([np.array(x) for x in self.delta],
37                dtype = object)
```

```
38
39       def setTheWeights(self, Linit):
40           # We will be Seting the weights. Linit is a list of
41           # lists with the weights for all excluding input layer.
42           for i in range(len(Linit)):
43               for j in range(len(Linit[i])):
44                   self.network[i+1][j].setTheWeights(Linit[i][j])
45
46       def printTheWeights(self):
47           print()
48           for i in range(1, len(self.network)):
49               for j in range(self.innerlayers[i]):
50                   print("Layer", i+1, "Neuron", j,
51                       self.network[i][j].weights)
52           print()
53
54       def runTheMLPerceptron(self, x):
55           # We will feed a sample x into the MultiLayer Perceptron.
56           x = np.array(x,dtype=object)
57           self.values[0] = x
58           for i in range(1, len(self.network)):
59               for j in range(self.innerlayers[i]):
60                   self.values[i][j] = self.network[i][
61                       j].runThePerceptron(self.values[i-1])
62           return self.values[-1]
63
64       # This method will do the magic of Training
65       def backPropogation(self, x, y):
66           # We will run a single (x,y) pair with the backpropagation
67           # algorithm.
68           x = np.array(x, dtype = object)
69           y = np.array(y, dtype = object)
70
71           # STEP 1: Feeding a sample to the network.
72           theOutputs = self.runTheMLPerceptron(x)
73
74           # STEP 2: Calculating the MSE
75           theError = (y - theOutputs)
76           MSE = sum( theError ** 2) / self.innerlayers[-1]
77
78           # STEP 3: Calculating error terms of each Neuron's output.
79           self.delta[-1] = theOutputs * (1 - theOutputs) * (theError)
80
81           # STEP 4: Repeatedly compute the errors terms in the
82           # hidden layers
83           for i in reversed(range(1, len(self.network) -1 )):
84               for h in range(len(self.network[i])):
85                   forwardError = 0.0
86                   for k in range(self.innerlayers[i+1]):
87                       forwardError += self.network[i+1][k].weights[h] * \
88                       self.delta[i+1][k]
89                   self.delta[i][h] = self.values[i][h] * \
90                   (1-self.values[i][h]) * forwardError
```

```
 91
 92            # STEPS 5 & 6: Calculating the deltas and finally updating
 93            # the weights
 94            for i in range(1, len(self.network)):
 95                for j in range(self.innerlayers[i]):
 96                    for k in range(self.innerlayers[i-1] + 1):
 97                        if k == self.innerlayers[i-1]:
 98                            Delta=self.LRate*self.delta[i][j]*self.thebias
 99                        else:
100                            Delta = self.LRate * self.delta[i][j] * \
101                            self.values[i-1][k]
102                        self.network[i][j].weights[k] += Delta
103            return MSE
104
105    # Testing
106
107    myMLP = MultiLayerPerceptronEngineering(innerlayers = [2, 2, 1])
108    myMLP.setTheWeights([[[-10, 10, 15],[15, 15, -10]],[[10, 10, -15]]])
109    myMLP.printTheWeights()
110
111    print("MLP: ")
112
113    print ("0 0 = {0:.10f}".format(myMLP.runTheMLPerceptron([0, 0])[0]))
114    print ("0 1 = {0:.10f}".format(myMLP.runTheMLPerceptron([0, 1])[0]))
115    print ("1 0 = {0:.10f}".format(myMLP.runTheMLPerceptron([1, 0])[0]))
116    print ("1 1 = {0:.10f}".format(myMLP.runTheMLPerceptron([1, 1])[0]))
```

Output: Case Study Chapter 11.c

```
Layer 2 Neuron 0 [-10  10   15]
Layer 2 Neuron 1 [ 15  15  -10]
Layer 3 Neuron 0 [ 10  10  -15]

MLP:
0 0 = 0.0066958493
0 1 = 0.9928471901
1 0 = 0.9923558642
1 1 = 0.9933071286
```

```
 1    #testing code
 2
 3    myMLP = MultiLayerPerceptronEngineering(innerlayers = [2, 2, 1])
 4
 5    print("NN Training as represented as XOR Gate")
 6
 7    for i in range(3000):
 8        MSE = 0.0
 9        MSE += myMLP.backPropogation([0, 0], [0])
10        MSE += myMLP.backPropogation([0, 1], [1])
11        MSE += myMLP.backPropogation([1, 0], [1])
12        MSE += myMLP.backPropogation([1, 1], [0])
13        MSE = MSE / 4
14
15        if (i % 100 == 0):
```

```
16          print(MSE)
17
18    myMLP.printTheWeights()
19
20    print("Now Executing MLP: ")
21
22    print("0 0 = {0:.10f}".format(myMLP.runTheMLPerceptron([0, 0])[0]))
23    print("0 1 = {0:.10f}".format(myMLP.runTheMLPerceptron([0, 1])[0]))
24    print("1 0 = {0:.10f}".format(myMLP.runTheMLPerceptron([1, 0])[0]))
25    print("1 1 = {0:.10f}".format(myMLP.runTheMLPerceptron([1, 1])[0]))
```

Output: Case Study Chapter 11.d

```
NN Training as represented as  XOR Gate
0.2720144753790936
0.26516388883931435
0.2642471818935426
0.26375243548870353
0.26338061345554764
0.2630292647266842
0.2626149187727961
0.2619857695745938
0.2607539070156049
0.2577238005584778
0.24804753678446606
0.20882354102924933
0.11055425595068445
0.048145493529100976
0.025853561319580914
0.016517733940049677
0.011764347063461073
0.008985111239726713
0.0071966309213706385
0.005963825170241792
0.005069450094831479
0.004394576337740649
0.003869265583541877
0.003449978534060723
0.003108326340736022
0.0028250764119500007
0.0025867736013498203
0.0023837436242575183
0.0022088649262205686
0.0020567866835392115

Layer 2 Neuron 0 [-5.31939108 -5.32394354  2.05825628]
Layer 2 Neuron 1 [-4.57679603 -4.57460582  6.8218647 ]
Layer 3 Neuron 0 [-8.30355219  7.79544761 -3.61037903]

Now Executing MLP:
0 0 = 0.0396641896
0 1 = 0.9582989195
1 0 = 0.9581861245
1 1 = 0.0511529754
```

CHAPTER 12 – VIRTUAL REALITY APPLICATION DEVELOPMENT WITH PYTHON

Description

You are given the following complete script from Chapter 12:

```
1    # Import libraries
2    import viz
3    import random
4    import vizact
5
6    # Create an empty window for the 3D environment
7    viz.setMultiSample(4)
8    viz.go(viz.FULLSCREEN)
9
10   # Add the 3D model of the plaza and enable collision detection
11   piazza = viz.add('piazza.osgb')
12   viz.MainView.collision(viz.ON)
13
14   # Create a plant object and the Jane avatar
15   plant = viz.add('plant.osgb')
16   jane = viz.add('vcc_female.cfg')
17
18   # Set the positions and orientation of the 3D objects and the camera
19   jane.setPosition([0, 0, 8])
20   jane.setEuler([180, 0, 0])
21   plant.setPosition([-3, 0, jane.getPosition()[2] + 3])
22   viz.MainView.setPosition(jane.getPosition()[0], 0,
23                            jane.getPosition()[2] - 11)
24
25   # Create and position multiple pigeon objects
26   pigeons_no = 30
27   pigeons = []
28
29   def create_pigeons(no_of_pigeons, pigeon_list):
30       for i in range(no_of_pigeons):
31           pigeon = viz.add('pigeon.cfg')
32           pigeon.setPosition([random.uniform(-5, 5), 0,
33                               random.uniform(2, 7)])
34           pigeon.state(random.choice([1, 3]))
35           pigeon_list.append(pigeon)
36
37   create_pigeons(pigeons_no, pigeons)
38
39   # Initialize & randomize the animation states & movement of the pigeons
40   def pigeon_state_move(rand_pigeon):
41       random_switch = random.choice([1, 2])
42       if random_switch == 1:
43           rand_pigeon.clearActions()
44           x = int(random.choice([1, 3]))
45           rand_pigeon.state(x)
```

```
46          else:
47              walk = vizact.walkTo([random.randint(-5, 5), 0,
48                                   random.randint(0, 8)])
49              rand_pigeon.addAction(walk)
50
51  pigeon_state_timer = vizact.ontimer(1, pigeon_state_move,
52                                      vizact.choice(pigeons))
53
54  # Randomize the animation states of Jane
55  def jane_state():
56  x = int(random.choice([1, 9]))
57  jane.state(x)
58
59  jane.state(1)
60  jane_state_timer = vizact.ontimer((random.randint(10, 20)), jane_state)
61
62  # Jane walking and behaviour
63  def jane_walk():
64      jane_state_timer.setEnabled(0)
65      random_switch = random.choice([1, 2])
66      if random_switch == 1:
67          walk_to_camera=vizact.walkTo([viz.MainView.getPosition()[0], 0,
68          viz.MainView.getPosition()[2] + 3])
69          jane.addAction(walk_to_camera)
70      else:
71          jane.clearActions()
72          jane_state_timer.setEnabled(1)
73
74  vizact.onkeydown('w', jane_walk)
75
76  # Jane feeding the pigeons
77  def jane_feed_pigeons(no_of_pigeons):
78      jane_state_timer.setEnabled(0)
79      pigeon_state_timer.setEnabled(0)
80      jane.clearActions()
81      jane.state(15)
82      for i in range(no_of_pigeons):
83          walk = vizact.walkTo([jane.getPosition()[0], 0,
84                               jane.getPosition()[2]])
85          pigeons[i].addAction(walk)
86
87  vizact.onkeydown('f', jane_feed_pigeons, pigeons_no)
88  vizact.onkeyup('f', jane_state)
```

Improve the existing script by adding the following features:

1. Allow Jane to feed the pigeons at random times without being instructed by the user. The feeding must stop automatically after a random amount of time has elapsed, and Jane should return to her normal *states* cycle.
2. Allow Jane to walk to random positions without the user's instruction. Make sure to restrict Jane's walking area so she does not get outside the 3D world boundaries or pass through 3D objects.

3. Make the pigeons spread around while eating instead of concentrating at the same single point as Jane.
4. Make the pigeons walk faster towards Jane when feeding is triggered.
5. Make the pigeons change back to their normal states at random times after they have walked towards Jane for feeding.
6. Add a general ambience background sound to the 3D world.
7. Make Jane whistle once when each round of the pigeon feed commences. This must emanate from the position Jane is at any given moment rather than being omnipresent. Note that for this sort of task the audio file needs to be in MONO rather than STEREO.
8. Add pigeon chirping sounds to the pigeons. The sounds must be allocated to random pigeons. As above, for this sort of task the audio file needs to be in MONO rather than STEREO.
9. Add some variety to the pigeons' chirping by adding two more chirping sounds and randomly switch between the three different sounds at run-time. Obviously, the three chirping sounds need to be distinguishably different to each other in order for this exercise to have practical value.
10. Add some randomness to the intervals at which the different pigeon chirps are triggered to improve realism.

SOLUTION

1. Simply trigger the existing `jane_feed_pigeons` and `jane_state` functions at random times:

```
random_feed_timer = vizact.ontimer(random.randint(30, 60),
jane_feed_pigeons, pigeons_no)
random_state_timer = vizact.ontimer(random.randint(20, 30), jane_state)
```

It may be a good idea to also add the following line at the beginning of the `jane_state` function, in order to make sure that previous actions are stopped before triggering new states:

```
jane.clearActions()
```

2. Add the following function:

```
def jane_random_walk():
   jane.clearActions()
   walk_to_random = vizact.walkTo([random.randint(-5, 5), 0,
   random.randint(0, 8)])
   jane.addAction(walk_to_random)
```

and a timer that triggers it randomly:

```
jane_random_walk = vizact.ontimer(random.randint(30, 60),
jane_random_walk)
```

3. Replace the `for` loop in the `jane_feed_pigeons` function with the following:

```
for i in range(no_of_pigeons):
   x1 = jane.getPosition()[0] -2
   x2 = jane.getPosition()[0] + 2
   y1 = jane.getPosition()[2] -2
   y2 = jane.getPosition()[2] + 2
   walk = vizact.walkTo([random.uniform(x1,x2), 0, random.uniform(y1,y2)])
   pigeons[i].addAction(walk)
```

This for loop creates a 4×4 units rectangular area around the position Jane is at the moment the feeding starts. Next, it randomizes the pigeon movements within this area using random.uniform() in the arguments of the walkTo() method.

4. Add a decimal number (0.4–0.5) as an argument at the end of the walkTo() method in the jane_feed_pigeons function. This argument dictates the movement speed in meters per second:

```
walk = vizact.walkTo([random.uniform(x1, x2), 0, random.uniform(y1, y2)], 0.5)
```

5. Add the following line at the end of the jane_feed_pigeons function. Make sure that the command is *outside* the for loop:

```
pigeon_state_timer.setEnabled(1)
```

6. Source a suitable *ambience* sound audio file (.wav), rename it to *Ambience.wav* and move it to the folder where the current script is saved. Add the following lines at the start of the script, *but after the 3D world has been instantiated*. For example, the code can be added after the plaza model is created and the viewpoint collision is turned on:

```
ambience = viz.addAudio('Ambience.wav')
ambience.loop()
ambience.play()
```

7. Source a suitable *whistling* audio file (.wav). As mentioned, the file needs to be in MONO. Rename the file to *Whistle.wav* and move it to the folder where the current script is saved. Add the following line to the jane_feed_pigeons function, just before the for loop and after the jane.state(15) command:

```
jane.playsound('Whistle.wav', viz.PLAY)
```

This is an alternative way to work with audio and it is used for 3D sounds (i.e., sounds that can be placed at specific points within the 3D world or attached to 3D objects). The command loads and attaches the sound directly to Jane. Flags can be passed as arguments to control the sound behavior.

8. Source a suitable pigeon *chirping* audio file (.wav). The file needs to be in MONO. Rename the file to *Chirp_1.wav* and move it to the folder where the current script is saved. Create the following function:

```
def random_pigeon_sound(rand_pigeon):
    rand_pigeon.playsound('Chirp_1.wav', viz.PLAY)
```

As usual, the function can be triggered by a timer that picks random pigeons:

```
random_pigeon_sound_timer = vizact.ontimer(5, random_pigeon_sound,
vizact.choice(pigeons))
```

9. Source two more *pigeon chirping* audio files (.wav). Rename the files to *Chirp_2.wav* and *Chirp_3.wav* respectively and move them to the folder where the current script is saved. Modify the random_pigeon_sound function to the following:

```
def random_pigeon_sound(rand_pigeon):
    random_switch = random.choice([1, 2, 3])
    if random_switch == 1:
        rand_pigeon.playsound('Chirp_1.wav', viz.PLAY)
    elif random_switch == 2:
        rand_pigeon.playsound('Chirp_2.wav', viz.PLAY)
    else:
        rand_pigeon.playsound('Chirp_3.wav', viz.PLAY)

random_pigeon_sound_timer = vizact.ontimer(5,
random_pigeon_sound, vizact.choic
```

10. This can be done in a number of ways, depending on the depth one wants to go to. A simple solution that provides a pseudo-randomized triggering is to use a number of different timers that trigger the chirping sounds of random pigeons at randomly chosen intervals:

```
vizact.ontimer(random.randint(7, 11), random_pigeon_sound, vizact.
choice(pigeons))
vizact.ontimer(random.randint(15, 20), random_pigeon_sound, vizact.
choice(pigeons))
vizact.ontimer(random.randint(25, 30), random_pigeon_sound, vizact.
choice(pigeons))
```

This choice provides a different random pattern of triggering between the timers every time the application is run. However, it has the disadvantage of repeating the same pattern while the application is running. Adding more timers will add more complexity to the pattern and, thus, improve realism, but at the same time it will increase computational power demands. If the reader is interested in the subject of controlling timed processes and flow in more detail, a good area to start is the *Tasks* and *Sequences* sections in the Vizard online reference manual (WorldViz, 2019).

REFERENCES

American Diabetes Association. (2018). Economic costs of diabetes in the US in 2017. *Diabetes Care*, *41*(5), 917–928. https://doi.org/https://doi.org/10.2337/dci18–0007.

Frank, A., & Asuncion, A. (2010). *UCI Maching Learning Repository*. University of California, School of Information and Computer Science. https://archive.ics.uci.edu/ml/datasets/Diabetes+130-US+hospitals+for+years+1999–2008.

McEwen, L. N., & Herman, W. H. (2018). Health care utilization and costs of diabetes. *Diabetes in America*, 3rd Edition. 40–1–40–78. https://www.ncbi.nlm.nih.gov/books/NBK567979/.

Strack, B., DeShazo, J. P., Gennings, C., Olmo, J. L., Ventura, S., Cios, K. J., & Clore, J. N. (2014). Impact of HbA1c measurement on hospital readmission rates: Analysis of 70,000 clinical database patient records. *BioMed Research International*, 2014, 781670. https://doi.org/10.1155/2014/781670.

WorldViz. (2019). *Vizard 6 (64-Bit)*. https://www.worldviz.com/vizard-virtual-reality-software.

Index

<Button-1>, <Button-3>, <Double-Button-1> 118
3D Cartesian coordinate system 496
3D game engine 487
__dict__ attribute 66
3D object lists 505
<key>, event.keycode 159
@property decorator 77
["values"] 137

absolute *versus* relative positioning 501
abstract class 91, 92, 94
access modifiers 72
activebackground, troughcolor, bg, fg, resolution 130
add_cascade() 181
add_command(),add_checkbutton(),add_radiobutton() 201
add_separator(), underline, hot keys, accelerator 174
add() and addChild() 491
after() 159
algorithm(s) 6, 10, 11
ALTER TABLE statement 289
animation *vs.* movement 514
ANOVA test 396
anti-aliasing 517
Apache mxnet 452
API 169
append(), delete(), clear() 145
apriori 438
askcolor(), asksaveasfile(), askopenfile(), askdirectory() 169
auto-encoder 483

the backslash special character ("\") 113
bar chart, plot.bar() 357
Basic Widgets 109
batch_size parameter 477
big data 320
binary search 208, 230, 233
binary search tree 208, 262, 263
binary tree 208, 261, 262
box and whisker plot, boxplot(), grid, figsize, labels 354
break 45, 46
bubble sort 44, 217, 218
button 119

Caffe 452
camel case 62
camera collision on/off 94
camera or viewpoint 487
CART 419
casting 31
categorical and continuous data 339, 350
categorical variables (nominal, ordinal) 379
checkbutton 126, 138, 144, 145
Chi-square Test 397
choice() 513
circular queue 250
class attribute 64
class keyword 62

clear() 137, 145
clearActions() 517
collision and collision detection 493
command 131, 145
comment(s) 13, 17
compound, left, right, center 118
condition 30, 31, 32, 33, 34, 35
confidence 439
confidence intervals 378
connecting to a database 279
constructor method 68, 69, 70
continue 45, 46
continuous variables (interval, ratio) 380
Convolutional Neural Network (CNN) 483
count() 350
create a new column in a dataset 333
create a new column using np.where() or np.select() 333
create a new CSV file 334
create a new excel file 334
create a table with a single primary key but no foreign key 283
create a table with combined primary key but no foreign key 283
create a table with one or more foreign keys 285
CREATE TABLE Statement 280
create tables with no primary or foreign key 280
current 137
curselection() 145

data analytics 320
data frame 320, 321, 322
data science 321
data structures 208
data type(s) 10, 14, 15, 16
data visualization 320, 321, 352
database 274, 276
database schema, database instance 274
DecisionTreeClassifier() 425
deep learning 450
DELETE statement 303
delete() 137, 145, 146
dependent and independent variables 380
DESC statement 296
describe() 350
descriptive statistics 339, 374
destroy(), exit() 144
destructor method 71
determinant 456
dictionary 16, 32, 34, 35, 42, 208, 215, 216
dot 452, 453
drop NaN or empty values 324
DROP TABLE statement 294

eigenvalue, eigenvector 459
encapsulation 72
Enter, Leave 178
entropy, Gini index 425
entry/text 120

epochs parameter 477
Euler Angles and the setEuler() method 499
evaluate() 481
event 108
event-driven (or visual) programming 108
exp() 456
expand, foreground, background, font, anchor 113
expression(s) 11, 18, 21, 22, 23

fill NaN or empty values 326
for loop 40, 42, 43, 44
foreign key 283
frame 108, 109, 110
from_=, to = 130
function 50, 51, 52, 53, 54

get() 137
getPosition() 501
getters/setters 72, 74
graph 208, 262, 267, 268
gravity() 494
grid() 111, 112, 113, 126
grid_remove() 202
groupby() 336

handling exception 98
Head Mounted Display (HMD) 490
head(n), tail(n) 331
histograms, plot.hist(), subplots, layout, grid, xlabelsize,
        ylabelsize, xrot, yrot, figsize, legend 352
horizontal 114, 130, 131, 152
hypothesis or statistical significance testing 377

identity matrix 456
idletasks() 155, 156
if statement 30, 31, 32, 35
import statement 95
indentation 30, 31
inferential statistics 374
infix, postfix, prefix 245, 246
inheritance 62, 78, 83
input 12, 29, 30
input and output datasets 419
insert records 296
insert()42, 137
insertion sort 208, 220
instance method 66
integer encoding 423
interactive *versus* linear systems 519
interface 76, 91, 94
IntVar() 137
inverse matrix 456

keywords 13
k-means clustering 435
k-NN 444
Kruskal-Wallis test 392
kurtosis 347

labelframe 131, 136
labels 111
lambda 118
len(), columns, shape 329
line chart, plot.line() 356

linear regression 400, 411
linked list 208, 242, 254
list 16, 208, 209
listbox, combobox 126, 131
loc[], iloc[] 331
logistic regression 402, 414
loop 218, 220, 222

machine learning process 409
magic/dunder methods 88
mainloop() 110
Mann-Whitney U test 391
map or 3D world 492
mean (arithmetic) 340
measures of central tendency 340
measures of spread 343
median 340
Menu() 171
merge sort 208, 230, 238
messagebox 138, 146, 162
messagebox, showinfo(), showerror(), showwarning() 138,
        146, 162, 164
messagebox with Options, askokcancel(), askretrycancel(),
        askyesno(), askquestion() 164
messagebox with User Input, askstring(), askinteger(),
        askfloat(), simpledialog 166
method 108
method loss parameters 475
method overloading 70, 85, 86
methods rand(), randn(), mean(), var(), std() 461
min_length 439
min_lift 439
min(), max() 343
mode 340
module, 94, 95, 96, 97
movement within the map 491

Naïve Bayes classifier 431
nested loops 42, 45
neuron 466
Notebook() 181, 185
null hypothesis 377

object data (attributes) 63
object instantiation 63
onkeydown() and onkeyup() 519
ontimer() and ontimer2() 511
onvalue, offvalue 144
operands 21, 23, 24, 87, 88, 245
operator overloading 87, 89
operators 11, 21, 22, 24, 25, 26, 28, 32, 35, 87, 88
option 108
orient 130, 151
output 6, 11, 12, 17, 29
overfit, underfit 482
overloading built-in methods 90

package 96
Paired t-Test 393
Pandas library 322
parameter maximum_depth 426
parameter min_samples_leaf 426
pass keyword 62, 78
passing 3D objects as arguments 507

passing values to arguments 52, 53
Pearson's Chi-square test 399
Pearson's correlation 398
perceptron 467
pie chart, pie() 363
polymorphism 85
population, sample 376
pop-up 175, 178, 181
prefabricated animations 510
prefabricated objects (prefabs) 491
primary key 276, 282
process_time() 137
processing 12, 31
progressbar, determinate, interminate 152
property method 76
Python GUI modules 109
Pytorch 452

quartiles 343
queue 208, 242, 248
quicksort 208, 230, 235

radiobutton 126, 138, 144, 145
raising exceptions 101
randint() 131
random forest 443
randomization 506
random.uniform() 506
read_excel(), read_csv(), read_html() 322
read-only attributes 75
rectifier linear unit (ReLU) function 471
Recurrent Neural Network (RNN) 483
recursion 208, 230, 231
relief, borderwidth 118
relx, rely 130, 156
rename() 327
Resize(), ANTIALIAS 118

sample characteristics 376
sample() 473
scale 126, 130, 131
scatter plot, plot.scatter() 364
scrollbar, xview, yview, xscrollcommand,
        yscrollcommand 137
SELECT statement 306
selecting the appropriate test 385
selection 32, 37, 45
selection sort 208, 220
selection_set() 202
Self-Organizing Map (SOM) 483
sequential approach 474
set 16, 208, 209, 214
setEnabled() 516
set_index(), reset_index() 329
setPosition() 496

shaker sort 208, 227
shell sort 208, 225
show 126
SHOW TABLES statement 279
sigmoid function 470
skewness 347
Software Development Kit (SDK) 523
solve() 461
sort_values() 336
spinbox 126, 130, 131, 185
splash screen 168
stack 208, 240, 242, 243
standard deviation (SD) 343
start(), stop() 189
state 126, 145
state, NORMAL, DISABLED 145
state() 510
statement(s) 11, 12, 13, 21
statistics 374
stochastic gradient descent (SGD) 475
StringIO, Graphviz 420
Student t-test 395
supervised learning 410
support 438

TensorFlow 452
textvariable 121, 137
Theano 452
Thread 162, 186
timer 511
tk_popup(event.x_root, event.y_root) 181
toolbar 162, 171, 175
tooltip 175, 178
tuple 16, 208, 209, 214
types of errors 98
types of scripting in relational databases 274
types of statistical analysis 381

unique() 332
unsupervised learning 411
UPDATE statement 301

validating data 74
variable(s) 10, 13, 14, 15, 379
variance() 343
vertical 114, 130, 134, 138, 152
viz library 491
vizconnect library 524
viz.go() method 491
VR hardware 523

walkTo(), turn() and addAction() 515
while loop 11, 36, 37
widget 108
Wilcoxon Signed-Rank test 391

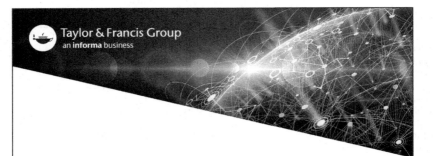